Power Systems

For further volumes:
http://www.springer.com/series/4622

Konstantin O. Papailiou
Frank Schmuck

Silicone Composite Insulators

Materials, Design, Applications

Springer

Konstantin O. Papailiou
Malters
Switzerland

Frank Schmuck
Eschenbach
Switzerland

ISSN 1612-1287 ISSN 1860-4676 (electronic)
ISBN 978-3-642-15319-8 ISBN 978-3-642-15320-4 (eBook)
DOI 10.1007/978-3-642-15320-4
Springer Heidelberg New York Dordrecht London

Library of Congress Control Number: 2012942718

Printed on acid-free paper

Springer is part of Springer Science+Business Media (www.springer.com)

Foreword

This book summarises the experience of the authors in high-voltage overhead transmission lines, built up over 35 years, the majority of which were spent researching the innovative technology of silicone composite insulators. At their time of introduction, composite insulators were rather specialised products, being rather expensive owing to the low numbers in which they were produced and were thus limited to special applications, for example in areas with extreme pollution. There were hardly any standards or long-term operational experience, and design errors sometimes resulted in corresponding failures. Since then, the composite insulator has been continuously developed as a quality alternative to porcelain and glass insulators, and in the meantime it has acquired a positive reputation owing to good operational experience in global use and accordingly high production numbers. Important pioneering work was carried out by a few international manufacturers and interested users, who recognised the potential of the composite insulator. This work included contributions to the understanding of the three different material components (metal fitting, housing material and core material) as well as of the characteristic interfaces of a composite insulator. Research institutions were involved in the identification and derivation of models which describe the dynamic interactions between the housing material and the surrounding environment and which, together with correct design, have virtually eradicated the need for maintenance. This increased level of understanding alongside growing production numbers has also led to optimisation and automation of production processes so that high quality can be provided at an attractive cost (economies of scale). In hindsight, the authors have been lucky enough to gain their practical engineering experience at one of the above-named companies, PFISTERER, which has always supported the composite insulator as an innovative product. Due to their managerial positions, the authors were able to make some lasting decisions in favour of composite insulator development. In addition, they were given the opportunity since many years to actively take part in the thematically relevant working groups of CIGRE and the IEC and they have been involved, right from the start, both in the development of basic scientific principles (CIGRE) and in the drafting of standards (IEC). The desire to collate and pass on their collective experience in a specialist book stemmed from this, particularly since there is currently no comparative documentation at this level of detail.

For the authors, the fascination with the composite insulator, which consists "simply" of the three aforementioned components, lies in the fact that interdisciplinary aspects of mechanics, electrical engineering, chemistry, physics, process engineering and even biology are combined in a single product. This book has also been structured accordingly:

After a short introduction, in which an overview is given regarding the global use of composite insulators, the mechanical behaviour and applications of the individual types of composite insulators are described in particular. More specifically, these are composite longrod insulators, composite post insulators, insulated crossarms for compact lines, interphase spacers and composite hollow core insulators.

In the case of composite longrod insulators, attention is paid specifically to their long-term behaviour under tensile load. This is closely linked to appropriate design of the metal fittings. In the case of composite post insulators, the authors have concentrated on the physical explanation and determination of the damage limit load, since this insulator-specific parameter is crucial for operational safety of the transmission networks where the insulators are subject to a wide range of loads. Both analytical and simulation methods are presented (and compared) with regard to both insulator types, these methods enabling safe dimensioning of the insulators.

The two subsequent chapters deal with two important specific features of modern networks to which the authors have made significant contributions: insulators for compact lines and for interphase spacers. The former are "insulated crossarms", which need not only a reduced right of way but at the same time help limit the values of the electric and magnetic field, an increasingly important feature for line route planning. Compact lines with insulated crossarms are complex mechanical systems, and comprehensive experience was required to enable reliable design. In addition to a suitable selection of composite insulators for the crossarm itself, the problem of the stability of compact lines is also discussed in detail.

Interphase spacers are used where problems with the interphase spacing between the conductors are to be expected. Such problems are caused by ice shedding from iced conductors, conductor approach caused by short-circuit loads, and galloping. They may also be necessary if tower head geometry is designed for line compaction and the normal span lengths have to be retained. Interphase spacers are particularly slender elements and require special calculation methods, which will also be presented.

As housings for electrical apparatus in, such as bushings, cable terminations, surge arresters, etc., composite hollow core insulators are indispensable. This is due, on the one hand, to their uncritical mechanical behaviour in the case of overpressure or earthquake, and to the effect of hydrophobicity in the case of strong pollution layer stress on the other hand. The relevant chapter in the book focuses on their mechanics.

Material selection and production methods are closely interlinked in the case of composite insulators and may have a strong influence on their operational behaviour and reliability. Due to the acknowledged operational acceptance of silicone rubber and its widespread use up to the highest of voltage levels, this material is dealt with in detail in one of the further chapters.

Corona and power arc protection is still an important topic since the wide range of composite insulator designs makes it difficult to develop unified standardisation, and the market now offers individual solutions where possible. Cost pressure, a lack of product knowledge compared to glass and porcelain technology, and, in the case of large tenders, the tendency of inexperienced suppliers to offer complete insulator strings as a commodity product without appropriate testing can lead and has led to problems during operation. These problems are not to be attributed to composite insulator technology, which in the meantime has been fully developed. For this reason, the current state of knowledge on corona and power arc protection is presented in detail and is summarised over two chapters.

The topic of the chapter on the evaluation of "old" composite insulators from the network, is mainly based on PFISTERER products with more than 30 years' field experience.

The final chapter of the book deals with the standardisation and testing of composite insulators. Standardisation is presented comparatively in the case of the IEC Standards of conventional insulators and of composite insulators. Presented are also methods for evaluating further properties of polymer housing materials mainly from the point of view of key properties which are relevant to operation, such as hydrophobicity and hydrophobicity transfer as well as a current CIGRE topic, the material fingerprint.

Generally, much material is used from the authors' own work, whether this is taken from their own publications or from publications of their CIGRE and IEC working groups where they have actively contributed. A further main focus is provided by their participation in two industry projects carried out with the collaboration of EPFL—ÉcolePolytechniqueFédérale de Lausanne (Federal Institute of Technology, Lausanne) as well as from a wide range of test reports compiled over many years of product development and product support at PFISTERER SEFAG. The work of many colleagues in this field accumulated over many years of experience was also used to supplement the own contributions and, hopefully, has been cited correctly. Figures and diagrams which do not originate from the their works and contributions have been provided with corresponding references.

It is hardly possible to thank everybody who, over the years, has helped the authors in one way or another to develop a deep knowledge in this field. As a key figure, the authors would like to name Dr. Claude de Tourreil, long-time Convenor of the CIGRE working group "Insulators", who unfortunately passed away in March 2006 after a short period of illness and who, with his product knowledge and ethical understanding, was a valued colleague and friend.

Thanks also go to Prof. Klaus Fröhlich, who, as Chairman of the Technical Committee of CIGRE, agreed to review this work critically and to write the foreword.

Knowledge is not much use if it cannot be passed on in written form of a stimulating nature and made available to a wider public. Both of these tasks were easily achieved with the help of the renowned Springer publishing house, on whose Senior Engineering Editor, Dr. C. Baumann, the authors were able to rely during all phases of the book. The authors would also like to extend their thanks to Professor Steffen Großmann, as well as to Professors J. Pilling, R. Bärsch, V. Hinrichsen, L. Gaul and P. Ermanni for their meticulous study of the manuscript and for their critical comments.

The authors would like to thank in advance any readers wishing to draw our attention to any errors in or wishing to provide any suggestions for further amendments and updates.

Last but not least, the authors would like to thank our families, who have always supported our professional activities and who have motivated us to bring the work to a successful end.

Malters, August 2012

Konstantin O. Papailiou
Frank Schmuck

Preface

For more than a hundred years, energy transmission by means of electricity has by far provided the most environmentally sound solution compared to other energy carriers. It is therefore of no surprise that a relatively high level of attention has been paid to the development and installation of transmission lines within the electrical energy sector. As early as 1891, the first high-voltage line of 10 kV was constructed in Germany over a length of 176 km. At that time, of course, it was unimaginable that 1,100 kV AC overhead lines would span distances of 1,000 km and more or that several thousand kilometres would be able to be bridged by 800 kV of DC lines at 6,000 MW. Starting with 110 kV in 1912, an operating voltage of 765 kV had already been achieved by the 1970s in what is a remarkable history of events. Apart from a few unsuccessful attempts with a considerably higher voltage, development stalled at this level until recently; this can be partly explained by the stagnation of electrical energy requirements over the last 30 years. The flawless introduction of the 1,100 kV level in China in 2009 refreshed optimism however, and transmission of even higher voltages is now being contemplated and quite a few innovative projects are presently concerned with an increase in voltage level.

A good example is the introduction of an "eight bundle" conductor, and other major improvements in line technology. With regard to the construction and design of towers, considerable progress has been also made. Simulation and testing methods have been developed in order to understand and master the mechanical and electrical phenomena which occur upon interaction between the electrical network and the surrounding environment. Since the beginning, the traditional material for insulators has been porcelain for longrod and cap and pin insulators. For cost reasons, and partly due to technical considerations, glass was later also used for cap and pin insulators.

Among the various technical improvements and advances in development which have been made over the last hundred years of overhead transmission line history, one component stands out and, for sure, constitutes one of the most significant innovations for the high-voltage electrical insulating system. This element, known as the composite insulator, heralds a change from natural insulating materials, such as porcelain and glass, to composites, which is associated with a wide range of advantages. Composite materials are easier to work with and demonstrate improved tolerances compared to conventional components. Improvement in the mechanical properties, pollution behaviour and comparatively low weight lead to small dimensions. For example, the latter has enabled the development of "compact lines", which are slimmer and also more cost effective in cases of restricted right of way. Composite insulators are more resistant to the bullet shots of vandalising gunfire, which increases the availability of substations and overhead lines in critical regions affected by this problem.

As is the case with all technical developments, the path of success of the composite insulator has been full of twists and turns. In the past, a series of substantial obstacles had to be overcome. Key examples include brittle fracture, the design of interfacial zones, the initial sensitivity to corona or solar radiation, the occasional loss of hydrophobicity after heavy rainfall, power arc resistance and, last but not least, damage caused by birds. Engineers have also long been preoccupied with the question of how such new insulators are to be tested. These challenges have now been overcome and the technology is now fully developed and polished. As an important, key element of a high-voltage or extremely high voltage overhead transmission lines, the composite insulator enjoys increasing popularity among line designers and constructors. The manufacturers of electrical apparatus also appreciate the use of composite hollow core insulators in many fields. A good proportion of composite hollow insulators are thus now used as supporting and insulating housings for bushings, instrument transformers, surge arresters and also circuit breakers.

The present book investigates the many aspects of composite insulators in a detailed and comprehensive manner. The properties of the materials used and the corresponding production processes are presented clearly. Advantages of composite insulators and the initial phases of the learning process are also described. Topics such as the mechanics of the metal fittings and of the glass fibre rod, corona resistance and power arc protection are dealt with in detail. Examples of design and application of overhead transmission lines are also given, even for specialised cases such as compact lines and interphase spacers. Correct testing of composite insulators is considered. A successful mixture of the theory behind an extremely wide range of technological aspects and suitable practical application is thus provided. Newcomers will gain knowledge in many fields of overhead transmission line construction, in particular with regard to composite insulators. Many important details are collated and summarised and would only otherwise be accessible to an experienced high-voltage engineer as a result of in-depth study of the vast technical literature. Personally, I have no knowledge of any other document which discloses such a well-rounded and comprehensive overview of the state of the art technology of composite insulators and their application.

Prof. Klaus Fröhlich
CIGRE President

Contents

1 Introduction 1
1.1 Advantages and Development of Composite Insulators 1
1.2 Experience with Composite Insulators 2
1.3 Use of Composite Insulators in High-Voltage Overhead Transmission Lines 5
1.4 Use of Composite Insulators in Electrical Apparatus and Outdoor Substations 5
1.5 Current Status of Standardisation 7
References 8

2 Composite Long Rod Insulators 9
2.1 Applications of Composite Long Rod Insulators 10
2.2 Behaviour of Composite Long Rod Insulators Under Mechanical Load 11
2.2.1 Long-Term Behaviour of Composite Long Rod Insulators 14
2.3 Behaviour of Composite Long Rod Insulators Under Dynamic Load 24
2.4 Design and Assembly of End Fittings for Composite Long Rods 28
2.4.1 Development and State of the Art Technology of Metal Fittings 28
2.4.2 Basic Considerations Regarding the Design of Crimped Fittings 29
2.4.3 Assembly of Crimped Fittings 32
2.4.4 Calculations 32
2.4.5 Simple Analytical Method 33
2.4.6 Complex Analytical Method 34
2.4.7 Numerical Simulation Methods 42
References 51

3 Composite Post Insulators 53
3.1 Key Properties 55
3.2 Applications of Composite Post Insulators 56
3.3 Behaviour of Composite Post Insulators with Bending 60
3.3.1 General Considerations 60
3.3.2 Adoption of a Damage Limit 60
3.4 Manufacturer Specifications with Regard to the SCL/MDCL Concept 68
3.5 Safe Failure Mode of Composite Post Insulators 70
3.6 Combined Loads 73
3.6.1 Load Diagrams 73
3.6.2 Examples 75
3.6.3 Computer Simulation 76
3.6.4 Tests 77
3.7 Dynamic Loads 77
3.7.1 Test Specimens 77
3.7.2 Test Method 78
3.7.3 Test Results 79
3.8 Constructional Requirements of the End Fittings 79
3.9 Analytical Calculation Methods 81
3.9.1 Simple Analytical Method 81
3.9.2 Complex Analytical Method 82
3.10 Numerical Simulation 86
3.10.1 Finite Element Model 87
3.10.2 Testing Arrangement 88
3.11 Behaviour of Composite Post Insulators in the Event of Failure 89
3.12 Sensitivity Analysis 93
References 94

4 Insulated Cross-Arms for Compact Lines 97
4.1 Introduction 98
4.2 Principles of Compaction 99
4.2.1 Influence of Conductor Suspension on the Tower 99
4.2.2 Options for Line Compaction 99
4.3 Mechanical Design of Insulated Cross-Arms 102
4.3.1 Rigid Insulated Cross-Arms 105
4.3.2 Pivoted Insulated Cross-Arms (Horizontal Vs) 107
4.3.3 Dynamic Loads 111
4.3.4 Stability Tests 113
4.4 Innovative Applications of Compact Lines 120
4.4.1 400 kV Line with Hollow Core Insulators in Switzerland 121

4.4.2 Emergency Restoration Systems with Composite Insulators 122
4.4.3 420 kV Double Circuit Line with Solid Core Composite Cross-Arms 124
References 125

5 Interphase Spacers 127
5.1 Introduction 129
5.2 CIGRE Survey 129
5.2.1 Assessment of Survey Responses 129
5.2.2 Operational Experience 133
5.3 Attachment Techniques 133
5.4 Mechanical Design of Interphase Spacers 135
5.4.1 Galloping 136
5.4.2 Shedding of Ice Loads 142
5.4.3 Electrodynamic Short-Circuit Loads 144
5.4.4 Buckling Behaviour of Interphase Spacers 149
5.5 Electrical Design of Interphase Spacers 153
5.5.1 Minimum Length 153
5.5.2 Corona Effects 155
5.5.3 Pollution 156
5.6 Applications 157
5.6.1 Compact Line for Medium Voltage 157
5.6.2 Interphase Spacers in the Event of Galloping Caused by Ice Shedding 160
5.6.3 The Tennis Racket Tower 161
References 162

6 Composite Hollow Core Insulators 165
6.1 Main Properties of Composite Hollow Core Insulators 166
6.2 Composite Insulators in Outdoor Substations 167
6.2.1 Bushings 167
6.2.2 Surge Arresters 168
6.2.3 Outdoor Terminations for Cables with Extruded Insulation 168
6.2.4 Current and Voltage Transformers 169
6.2.5 Outdoor Circuit Breakers 170
6.3 Service Experience 170
6.4 Main Components of Composite Hollow Core Insulators 173
6.4.1 FRP Tube 173
6.4.2 End Fittings 173

6.5 Mechanical Behaviour of Composite Hollow Core Insulators Under Bending . . . 175
6.5.1 Failure Mechanisms of Composite Hollow Core Insulators and Diagnostic Methods . . . 175
6.5.2 The Concept of the "Damage Limit Load" in Composite Hollow Core Insulators . . . 176
6.6 Testing of Composite Hollow Core Insulators . . . 176
6.6.1 Definitions of the Loads . . . 176
6.6.2 Bending Loads . . . 179
6.6.3 Pressure Loads . . . 179
6.6.4 Examples of Practical Tests According to IEC 61462 . . . 182
6.7 Mechanical Design of Composite Hollow Core Insulators . . . 185
6.7.1 Simple Analytical Method . . . 187
6.7.2 Numerical Simulation Methods . . . 188
References . . . 194

7 Material Selection and Manufacturing Processes for Composite Insulators with Silicone Rubber Housing . . . 197
7.1 Introduction . . . 198
7.2 Components/Semi-Finished Parts for Composite Insulators . . . 200
7.2.1 (End) Fittings . . . 200
7.2.2 Internal Insulation . . . 205
7.2.3 Outer Insulation–Silicone Rubber Grades for the Insulator Housing . . . 224
7.3 Processes . . . 247
7.3.1 Fitting Assembly . . . 247
7.3.2 Applying the Insulator Housing and Seal . . . 255
7.4 Summary . . . 275
References . . . 276

8 Composite Insulator Design from the Perspective of Corona Protection . . . 285
8.1 Introduction . . . 286
8.2 Corona as a Design Problem: Recent Examples . . . 288
8.2.1 An Example for 525 kV-Double Tension Set on a Lattice Tower . . . 288
8.2.2 An Example for 245 kV-Double Tension String at a Station Entry Point . . . 290
8.2.3 Examples for 115/138/145 kV-Various Insulator Sets . . . 292
8.3 Analysis of the Electrical Field of Composite Insulator Sets Being Installed in Service . . . 295
8.4 Current Standardisation . . . 295

8.5 Water Droplet Corona on Hydrophobic Housing Materials . . . 298
8.5.1 Formation of Water Droplet Corona 299
8.5.2 Effect of Corona on Polymeric Surface, Especially Silicone Rubber 304
8.5.3 Corona: Potential for Damage 314
8.6 Additional Requirements for Composite Insulators or Composite Insulator Sets 315
8.6.1 Material and Geometry Differences 315
8.6.2 Calculating the Electrical Field Stress 317
8.7 Empirical Threshold Values for Corona Prevention 324
8.8 420 kV Composite Insulator Sets After 10 Years in Service 325
8.8.1 Overmoulded Design 326
8.8.2 Modular Design 330
8.9 Summary 334
References 335

9 Power Arc Protective Fittings for Composite Long Rod Insulators 339
9.1 Introduction 340
9.2 Power Arcs as a Physical Phenomenon 340
9.2.1 Causes of Flashover on a Composite Insulator String/Set 340
9.2.2 The Power Arc 341
9.2.3 Principles of Power Arc Protection 343
9.3 The Power Arc Effects 348
9.3.1 Porcelain Long Rod and Porcelain as well as Glass Cap and Pin Insulators 348
9.3.2 Composite Insulators 349
9.3.3 Summary Comparison of Effects 353
9.3.4 Peripheral Protection 353
9.4 Designing Power Arc Protective Fittings 355
9.4.1 Material Selection 355
9.4.2 Density of Fault Current in the Elements of an Insulator String 358
9.5 Tests related to Power Arc Protection 359
9.5.1 Materials Testing of Housing Materials (Design Test) 359
9.5.2 String/Set Test from a Current Density Perspective: Short Circuit Test 362
9.5.3 String/Set Test from a Power Arc Effect Perspective: Power Arc Test 363

9.6 A Selection of Project Examples 366
9.6.1 Misuse of Power Arc Protective Fittings for Cap and Pin Insulator Strings in Composite Insulator Strings 366
9.6.2 Coordination Between a Corona Ring and Power Arc Protective Fitting 369
9.6.3 Direct Mounting of Power Arc Protective Fittings onto Composite Insulator End Fittings 370
9.6.4 Effect of the String/Set Design on the Tower. 372
9.7 Summary. 375
References 376

10 In-Lab Evaluation of Composite Insulators following their Withdrawal from the Network 379
10.1 Introduction 380
10.1.1 Is it Necessary to Evaluate Composite Insulators? . . . 381
10.1.2 Interface Areas as a Key Difference for Composite Insulators 382
10.2 Reasons for Evaluating Composite Insulators Following Their Withdrawal from the Network. 382
10.2.1 Evaluation of an Insulator Installation in Terms of its (Ageing) State 382
10.2.2 Evaluation of an Insulator Failure or Insulator Type with a High Risk of Failure. 383
10.2.3 Evaluation of Composite Insulators for Research Purposes. 385
10.3 Composite Insulator Failure Rate and Failures 385
10.3.1 USA/EPRI. 385
10.3.2 CIGRE Survey, Published in 2000 386
10.3.3 Service Experiences in China. 388
10.4 Commonly Used Test Strategies. 389
10.4.1 Identification of Test Specimens. 389
10.4.2 Test Sequence 390
10.4.3 Summary of Test Methods and Failure Criteria 409
10.5 Examples of Testing Programmes and Their Results. 409
10.5.1 Example 1—a 420 kV Composite Insulator (Stepped Injection Moulding Variant) After 10 Years in Service 410
10.5.2 Example 2—a 420 kV Composite Insulator (Modular Variant) After 10 Years in Service 420
10.5.3 Example 3—15 kV Composite Insulators After 15 Years in Service in a Railway Tunnel System 429

10.5.4 Example 4—Analysis of Zinc Layer Thicknesses and the Hydrophobic Effect After 30 Years in a 15 kV Service in a Railway Tunnel 434
10.5.5 Example 5—Evaluation of 123 kV Insulators for the Purpose of Product Qualification 439
10.5.6 Example for Deriving Reliability Figures 444
10.6 Summary 447
References 447

11 Overview of Standards and Tests Concerning Composite Insulators 451
11.1 Introduction 452
11.2 Current IEC Standardisation for Composite Insulators and Similarities in Comparison to Conventional Insulators and Insulator Strings/Sets 454
11.3 Special Flame Resistance Tests 456
11.4 Test Methods for Evaluating Certain Properties of Polymeric Housing Materials 456
11.5 Inclined Plane Test 459
11.5.1 Erosion and Tracking Resistance 459
11.5.2 Inclined Plane Test Principle for Evaluating the Resistance of Hydrophobicity 473
11.5.3 The Inclined Plane Test Principle for Evaluating the Hydrophobicity Transfer 474
11.6 Finger-Print Analysis of Polymeric Housing Materials 476
11.6.1 Background 476
11.6.2 Overview of Processes 477
11.6.3 Processes and Examples: A Closer Look 478
11.7 Summary 482
References 484

Correction to: Composite Post Insulators C1

Index 489

About the Authors

Dr. Konstantin O. Papailiou was CEO of the PFISTERER Group in Winterbach (Germany) until his retirement at the end of 2011. He studied electrical engineering at Braunschweig University of Technology and civil engineering at the University of Stuttgart. He received his doctorate degree from the Swiss Federal Institute of Technology Zurich and his post doctoral qualification as lecturer (Dr.-Ing. habil.) from the Technical University of Dresden.

From 1975 to 1985 he was Head of Technology and International Sales at Gesellschaft für elektrische Anlagen (GEA), in Fellbach, and was CEO of PFISTERER SEFAG AG in Malters (Switzerland) from 1986 to 2002.

He is Chairman of the CIGRE Study Committee for Overhead Transmission Lines and has published numerous papers in professional journals, such as CIGRE ELECTRA, IEEE Transactions, Elektrizitätswirtschaft and SEV Bulletin. He is also co-author of the "EPRI Transmission Line Reference Book—Wind Induced Conductor Motion".

Dr. Frank Schmuck is a member of the Management Board of PFISTERER SEFAG AG in Malters (Switzerland). He studied for his degree at the University of Applied Sciences in Zittau, specialising in high-voltage insulating technology, before going on to earn his doctorate degree. He received his post doctoral qualification as lecturer (Dr.-Ing. habil.) from the technical University of Dresden.

From 1992 to 1994 he was a scientific assistant at the Technology Centre North at Rheinmetall AG (Germany), before becoming Product Manager and Head of Development for epoxy resin-reinforced electrical components and composite hollow core insulators from 1994 to 1998 at Cellpack (Switzerland), Field of Advanced Composites. Since 1998 he has been employed by PFISTERER SEFAG, since 2009 in the capacity of Head of Development and Technology of the Centre of Competence for Overhead Transmission Lines and for OEM Products.

Since 1994 he has been involved in international CIGRE and IEC working groups, and since 2006 he has been the Convenor of the CIGRE working group for insulators. Since the beginning of 2012 he is also Chairman of the Swiss TC 36, Insulators. He has more than 70 publications in professional journals of CIGRE and the IEEE, and since 2007 he has been a columnist for the international insulator magazine "Insulator News and Market Report".

Chapter 1
Introduction

High-voltage insulators are of key importance for the operational safety and operating efficiency of transmission systems of electrical power, and it is therefore of no surprise that these components must meet particularly high demands in terms of reliability. In addition to both long rod insulators of conventional design made of porcelain and cap and pin insulators made of glass or porcelain, as were previously used as standard, composite insulators have gained great importance in recent years in the field of insulation technology for medium and high-voltage overhead transmission lines and substations.

1.1 Advantages and Development of Composite Insulators

The concept of the composite insulator was actually developed in the USA in the 1950s. However, this technology was not taken seriously until the 1970s, when major progress had been made in terms of the fibre reinforced materials, which form the basic core of the composite insulators, and the polymer, which form the housing of these insulators. In particular, the above-average insulation behaviour of silicone when subject to pollution [1] and the lack of sensitivity of the composite insulators to breakage when subject to impact loads during operation, contributed to their wide acceptance and these important properties soon won over many users who were initially sceptical. Composite insulators have convincing qualities when designed appropriately, both in terms of construction and in terms of material selection. Whilst it is not possible to optimise mechanical and electrical performance separately in conventional insulators (porcelain or glass), these properties can be considered individually when selecting materials for composite insulators. The resultant structure can be divided into three parts [2] (Fig. 1.1):

- Rod of high mechanical strength for taking up the external loads.
- Polymeric, shed-like housing surrounding the rod, designed to provide creepage distance and electrical insulation.
- Metal fittings at either end of the rod for transmission of the mechanical loads.

K. O. Papailiou and F. Schmuck, *Silicone Composite Insulators*, Power Systems,
DOI: 10.1007/978-3-642-15320-4_1,

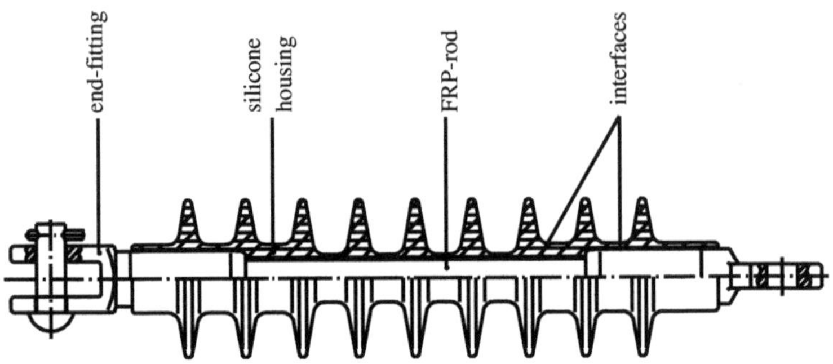

Fig. 1.1 Design and components of composite insulators

It is particularly important to perfect the interfaces provided between the different components of a composite insulator, since this is where electrical and mechanical problems may often occur [3].

A further advantage of the composite insulator is its low weight, which facilitates transport and assembly and allows construction of lighter towers. In the case of high-voltage lines, the weight of conventional insulators may account for almost 20 % of the entire vertical load on a tower. By contrast, composite insulators allow a weight saving of up to more than 90 % compared to conventional insulators. The sensitivity of the composite structure to impact loads caused by power arc, dynamic mechanical forces is much lower than that of a conventional insulator. Owing to its slim design, the composite insulator also forms a target which is much more difficult to hit when fired at (vandalism), which is of considerable importance in countries outside Europe. The composite insulator is also well suited for use in catenary operation for railways, since in this case the low weight makes it possible to avoid uncontrolled vibrations of the catenary system to a large extent, especially in view of the increasingly faster travelling speeds.

1.2 Experience with Composite Insulators

The first overhead transmission line insulators made of organic materials to be used as standard on a relatively large scale were those used more than 50 years ago in the USA. These were cast resin insulators; many of them had to be replaced after a short period of use due to various problems. Building on this experience, the first "real" composite insulator with porcelain sheds supported by an FRP rod (FRP = fibre reinforced plastic—resin in general) then followed; such insulators were mostly used in medium-voltage lines. A few years later, manufacturers from Germany, the US, France, England, Italy, and later also from Switzerland, began to manufacture composite insulators in the form we know today. Due to the above-mentioned advantages, this first generation of composite insulators received a great level of interest worldwide among line operators. Many power utilities soon began to experiment with composite insulators and started using them over short line sections to start with, especially in locations where these insulators were suited owing to their advantageous properties. Unfortunately, there various problems were encountered with some first-generation insulator designs, such as:

- Prolonged tracking and surface erosion in the shed material
- A drop in resistance to power arc, in particular an age-related loss of the original particularly good pollution layer properties of the shed surface
- A loss of the original mechanical strength, sometimes with the rod slipping out from the metal end fitting
- Adhesion problems and therefore breakdown at the interfaces between the different materials of the insulator
- Penetration of moisture, which, together with insufficient adhesion and inappropriate electrical dimensioning of the metal fittings, led to brittle fractures in the supporting glass fibre rod.

Around 30 years ago, these problems led to another re-think concerning the use of composite insulators. The clear advantages of such insulators motivated manufacturers across the globe to tackle and solve these problems and to introduce to the market a vastly improved, second generation of composite insulators. Since then, the better understanding of possible failure mechanisms achieved as a result of consolidated research and long-term testing, the optimisation of production processes, and a greater awareness of quality assurance have led to "third-generation" composite insulators, which are highly reliable since they are mass-produced industrially and have an expected service life of more than 50 years. It is therefore of no surprise that composite insulators have enjoyed increased use worldwide in recent years and, in the meantime, have been accepted as an equivalent alternative to porcelain and glass. This is backed up by survey results and statistics published by CIGRE—Conseil International des Grands Réseaux Électriques (International Council on Large Electric Systems) and the IEEE (Institute of Electrical and Electronics Engineers), which will be summarised below [4].

In particular, CIGRE working group 22.03, with which the authors collaborated for many years, carried out valuable work in gathering operational experience throughout the world and has conducted the first survey of electric utilities, the results of which were published in 1990 [5]. A further survey was started in 1997, the results of which were published in 2000 [6]. The following results provide a summary of this survey, which was limited to transmission voltages greater than 100 kV.

In particular, the reasons given by operators for use of composite insulators are interesting (Fig. 1.2).

The number of installed units (Fig. 1.3) clearly reflects the trend from conventional to composite insulators. Unfortunately, the survey does not include the largest market for composite insulators—the USA—and therefore the total number of almost 700,000 resulting from the current assessment is actually much greater in reality (compared to just under 50,000 in 1990).

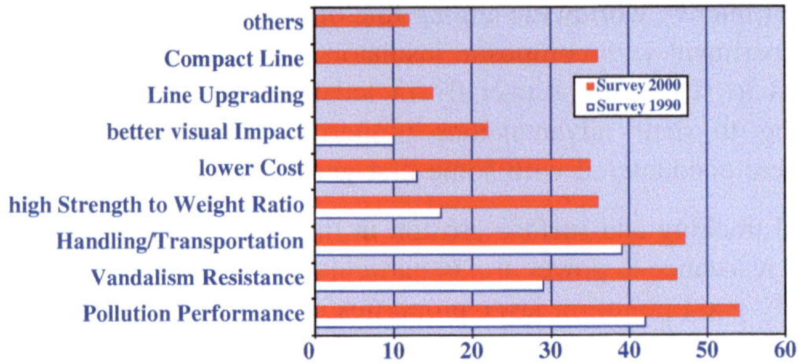

Fig. 1.2 Reasons for use of composite insulators

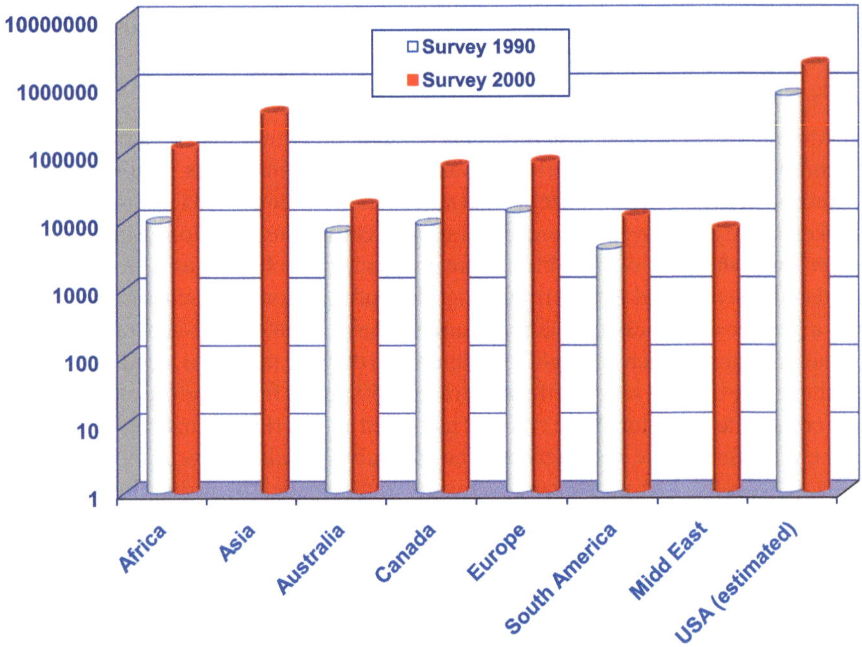

Fig. 1.3 Installed composite insulators according to the survey

Worldwide estimations stand at an estimated total number of 5 million composite insulators, and 20 million if the voltage level up to 50 kV is included. High growth is evident on all continents, whilst the disproportionately high increase in Asia can be attributed to many large electrification projects in China.

Approximately 5 % of the composite insulators included in the survey are post insulators. This, seemingly subordinate, insulator type plays a key role in line compaction and in voltage upgrade The philosophy of compaction is followed strongly in the USA, and the data provided by the assessment thus showed that 50 % of all composite post insulators could be attributed to American energy suppliers. On the other hand the world's first 400 kV compact line with insulated cross-arms with composite insulators was actually designed, manufactured and erected in Switzerland in 1996 and is described in detail in [7] and in Sect. 4.4.1.

1.3 Use of Composite Insulators in High-Voltage Overhead Transmission Lines

The wide range of applications of composite insulators in overhead transmission lines will be described in detail in the following chapters. It is to be added that, in addition to the above-mentioned advantages of composite insulators, environmental demands in terms of the quality and quantity of high-voltage overhead transmission lines have continued to grow in recent years, particularly in industrialised countries. For example, it is now of crucial importance that the appearance and environmental compatibility of overhead transmission lines and towers are considered as early as the planning stages. A large electricity company in West Switzerland already implemented these principles a few years back in its development of new 400 kV lines [8]. The widespread use of silicone composite insulators proved to be favourable; in that application with a coupling length of 3.0 m, the composite insulator can be manufactured in one piece and is almost 1.5 times shorter than the previously used porcelain strings, each having three porcelain long rod insulators type LG 85/22/1470. The way in which the shorter length of suspension insulators has a positive effect on the tower structure is shown in Fig. 1.4. Shorter insulators enable the use of shorter cross-arms, without any risk of contact between the conductors and the tower in the event of insulator swing. This means that the torsional load in the cross-arms and in the entire tower is reduced, thus allowing a lightweight structure which looks more attractive and costs less.

1.4 Use of Composite Insulators in Electrical Apparatus and Outdoor Substations

Following the successful use of composite insulators in overhead transmission lines up to the highest voltages, composite insulators have also been used increasingly in electrical apparatus and outdoor substations for about the last 20 years. It is therefore now possible to design practically an entire outdoor substation on the basis of composite technology (Fig. 1.5).

Due to the greater demands placed on operational safety and also for damage risk minimisation (to people and property), the majority of bushings are therefore now also manufactured using composite technology, particularly at higher voltage levels (Fig. 1.6).

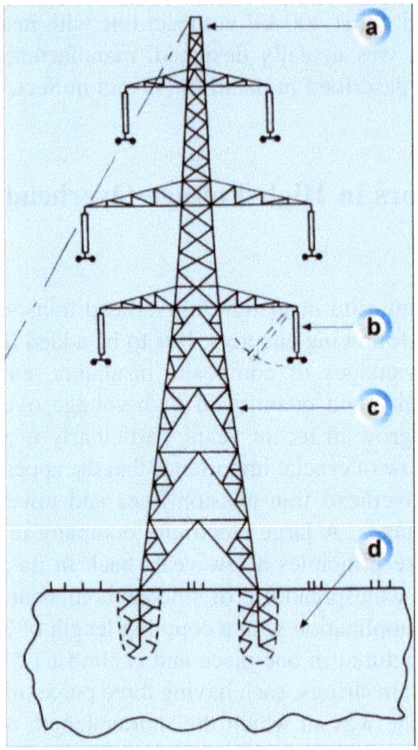

Fig. 1.4 Influence of insulator length on the tower structure. **a** Shorter earth wire peak (because of b). **b** Shorter cross-arms, hence reduced torsional load. **c** Reduced load, hence less steel and more attractive appearance. **d** Reduced bending and torsional moments, hence smaller foundations

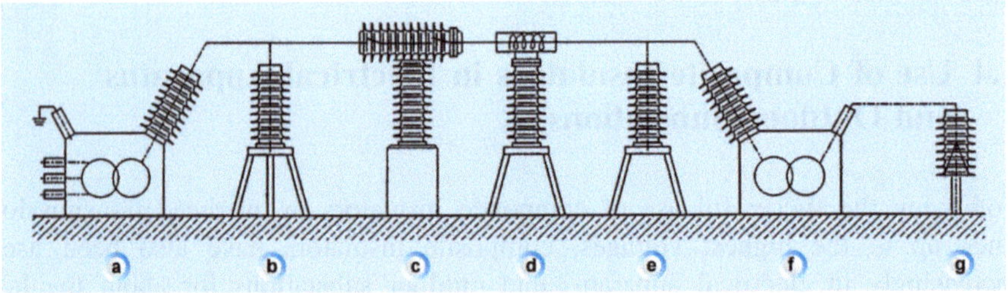

Fig. 1.5 Possible points of use for composite insulators in outdoor substations. **a** Bushing for machine transformer. **b** Surge arrester. **c** Circuit breaker. **d** Current transformer. **e** Voltage transformer. **f** Bushing for distributor transformer. **g** Cable termination

The same applies to cable terminations of XLPE cables. The many users and advocates of high-voltage cables with extruded plastic insulation were calling for terminations which are suitably adapted to the features and possibilities of such

Fig. 1.6 800 kV HVDC wall bushing

cables. Silicone terminations which can be slipped on were thus developed as early as the 1970s and have been used successfully worldwide (Fig. 1.7).

Composite insulators for electrical apparatus are primarily hollow core insulators, consisting of an FRP (fibre reinforced plastic) tube and a silicone housing. The "mechanics" thereof, which are far from trivial, including corresponding FEM (finite element method) simulations, were examined in depth in an industrial project carried out in collaboration with EPFL—École Polytechnique Fédérale de Lausanne (Swiss Federal Institute of Technology, Lausanne), with which the authors were involved [9] and which will be presented and discussed in Chap. 6.

1.5 Current Status of Standardisation

The high global demand for composite insulators has led to the need for international standards to ensure product quality and, consequently, operational reliability [4]. A wide range of IEC (International Electrotechnical Commission) standards are now available and concern the design, relevant testing, and

Fig. 1.7 170 kV Ixosil cable termination

acceptance criteria of composite insulators. Details can be found in the respective sections and in the overview provided in Chap. 11.

References

1. Papailiou KO (1990) Silikon in der Hochspannungsisoliertechnik. SEV-Bulletin, Jg. 81, Heft 15
2. Papailiou KO, Fluri W, Wermelinger A (1990) Erfahrung mit Einsatz von Verbundisolatoren. SEV-Bulletin, Jg. 81, Heft 19
3. Papailiou KO (1999) Grenzflächen bei Silikon-Verbundisolatoren. SEV-Bulletin, Jg. 99, Heft 21
4. Fluri W, Schmuck F, Papailiou KO (2002) Zum Einsatz von Verbundisolatoren und resultierenden Aktivitäten in IEC und CIGRE. ETG Fachtagung
5. CIGRE (1990) Worldwide experience with HV composite insulators. Electra 130, Paris
6. CIGRE (2000) Worldwide service experience with HV composite insulators. Electra 191, Paris
7. Ammann M, Papailiou KO, Dalleves P, Leva M, Villa S (1998) A new 400 kV line with compact towers and composite insulated cross-arms. paper 22/33/36-06, CIGRE, Paris
8. Papailiou KO (1997) Verbundisolatoren auf dem Vormarsch—Anwendungen von Silikonverbundisolatoren in der Schweiz. SEV-Bulletin, Jg. 88, Heft 7
9. Bonhôte P, Gmür T, Botsis J, Papailiou KO (2003) Stress and damage analysis of composite-aluminium joints used in electrical insulators subject to traction and bending. Compos Struct 64(2004):359–367

Chapter 2
Composite Long Rod Insulators

Symbols and Abbreviations

3D	Three-dimensional
CIGRE	Conseil International des Grands Réseaux Électriques (International Council on Large Electric Systems)
DLL	Damage limit load
FE mesh	Finite element mesh
FEM	Finite element method
FE model	Finite element model
FRP	Fibre reinforced plastic
IEC	International Electrotechnical Commission
IEEE	Institute of Electrical and Electronics Engineers
IREQ	Institut de Recherche d'Hydro-Québec (Québec-Hydro's Research Institute)
LVDT	Linear variable differential transformer
RML	Routine mechanical load
SML	Specified mechanical load
UHV	Ultra-high voltage
δ	Half length of the linear region for the shear stress
ΔR	Radius reduction of the FRP rod during crimping
ϑ	Coordinate in circumferential direction of the FRP rod
k	Danger factor
μ	Coefficient of friction between metal and FRP
v_{LT}	Poisson number "axial-transverse" to the FRP rod
v_{TT}	Poisson number "transverse–transverse" to the FRP rod
$\sigma_{\vartheta\vartheta}$	Circumferential stress in the FRP rod
σ_M	Standard deviation
$\sigma_{shear}^{\max}$	Shear strength of the FRP rod
σ_{rr}	Radial stress in the FRP rod
σ_z	Tensile stress in the cross-section of the end fitting
σ_{zul}	Tensile strength of the end fitting

K. O. Papailiou and F. Schmuck, *Silicone Composite Insulators*, Power Systems,
DOI: 10.1007/978-3-642-15320-4_2,

σ_{zz}	Axial stress in the FRP rod
τ	Critical shear stress in the end fitting
τ_f	Shear stress in the FRP rod
τ_{max}	Shear strength in the FRP rod
$\tau_{rz}\ (z, R)$	Shear stress at the surface ($r = R$) of the FRP rod
τ_{zul}	Max. permissible shear stress of the end fitting
b	Wall thickness of the end fitting
d	Diameter of the FRP rod
dz	Length of the differential beam element
D_o	Outer diameter of the end fitting
Di	Inner diameter of the end fitting
D_k	Ball size (ball diameter)
DS	Stiffness of the fibre
e	Location of critical shear of the end fitting
E	Modulus of elasticity (Young's modulus) of the fibre
E_L	Modulus of elasticity (Young's modulus) in axial direction of the FRP rod
E_T	Modulus of elasticity (Young's modulus) in transverse direction of the FRP rod
F	Cross-section of fibre
F_{max}	Failing load in tension of the insulator
F_N	Pressure load between metal and FRP
F_z	Tensile load in the FRP rod
G_{TT}	Shear modulus "transverse–transverse" of the FRP rod
G_{LT}	Shear modulus "axial-transverse" of the FRP rod
L, L_{cr}, L_p	Crimp length of the end fitting
M_{96}	Average 96 h failing load
M_{av}	Average failing load of the assembled core
p	Contact pressure on the FRP rod
r	Radial coordinate for the FRP rod
R	Radius of the FRP rod before crimping
S	Contact (interface) surface between FRP rod and end fitting
x, y, z	Geometry parameters of the end fitting
x_{cr}	Critical crimp length
x_{max}	Limit crimp length
z	Axial coordinate of the FRP rod

2.1 Applications of Composite Long Rod Insulators

The term "long rod" is actually used for a specific design of porcelain insulator which was first introduced in Germany in the 1920s as an alternative to cap-and-pin insulators (and also for critical ambient conditions, for example high pollution). The construction of a composite long rod is shown in Fig. 2.1. The fibre reinforced core, the metal fittings and the silicone housing can be seen.

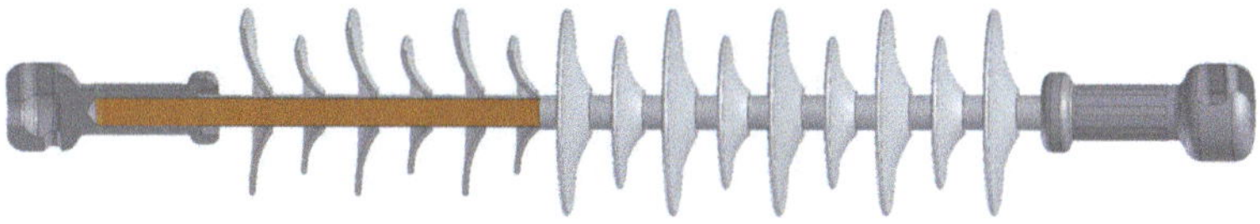

Fig. 2.1 Construction of a composite long rod insulator

Composite long rod insulators benefit from the manufacturing possibilities of the FRP (fibre reinforced plastic) rod, and in particular from the fact that FRP rods can be produced practically endlessly in one piece in lengths of up to 20 m which is not possible with conventional insulator materials such as porcelain and glass. Composite long rod insulators are currently produced in lengths of 10–100 cm for low-voltage and medium-voltage network systems, and in lengths up to 10 m and more for today's UHV network systems (Figs. 2.2 and 2.3).

Composite long rod insulators are primarily used in suspension strings in straight-line supports and as tension strings in anchor towers and dead-end towers (Figs. 2.4 and 2.5). They are also used in the jumpers or portals of outdoor substations. In some cases, composite long rods are used in the guys of wooden poles, and more rarely in the guys of steel towers.

Since, in contrast to porcelain and glass insulators, composite long rod insulators can be formed in one piece up to the highest of voltages, they only weigh a fraction of conventional insulators (at 400 kV they weigh approximately 10 % of a comparable porcelain long rod).

Their use is therefore also favoured in special tower designs such as guyed cross-rope suspension (chainette) towers (Canada, South Africa, Argentina—see Fig. 2.5), or in floating dead-ends to reduce tower height (Fig. 2.6).

With the increasing use of compact lines, even for voltages of 400 kV and above, these insulators can be found as support elements in insulated cross-arms (Fig. 2.7); see also Chap. 4.

Millions of composite long rods have been used for many years at medium-voltage level in a wide range of tower types (wooden, steel, concrete) and for a large number of applications (see Fig. 2.8—straight-line supports, angle towers, dead-end towers, pole mounted transformer stations).

Composite insulators, and in particular their advantages under extreme pollution, have been "discovered" in the meantime by railway operators and are thus also being used increasingly in railway catenary systems.

2.2 Behaviour of Composite Long Rod Insulators Under Mechanical Load

As is evident from the above-mentioned applications of composite long rods, they are subject primarily to tensile load. However, these insulators may also sometimes be subject to torsional load during the line construction and erection process.

Fig. 2.2 Production of a 6 m long composite long rod

Fig. 2.3 Suspension string

For this reason, this type of load will also be discussed briefly, especially since FRP rods react rather sensitively to torsion. Long-term behaviour, damage mechanisms, and the associated long-term tests will also be discussed. The

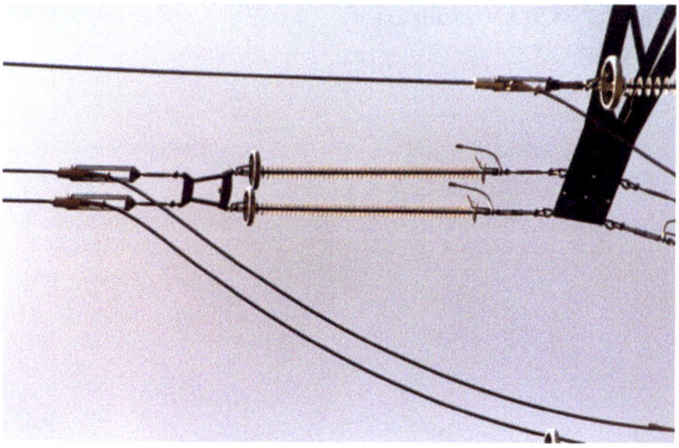

Fig. 2.4 Tension string

Fig. 2.5 Cross-rope suspension (chainette)

Fig. 2.6 Floating dead-end

conventional analytical methods used today will then be presented, from the simple formulas for "day-to-day use" to computer-based simulations. The mechanical behaviour of composite long rods is often determined by the quality and application processes of the respective fittings, which will be described in detail. Reference will lastly be made to the necessary mechanical tests, as stipulated in the international standards.

The material for the core rods of the composite insulators concerned in this instance is a fibre reinforced plastic (FRP). Such rods are normally produced by a pultrusion method (see Chap. 7), that is to say the glass fibres of these rods are oriented uniaxially and in the direction of the rod axis. The rod therefore has different

Fig. 2.7 420 kV insulated cross-arm

Fig. 2.8 20 kV insulators on a wooden tower

Table 2.1 Elasticity constants of the FRP rod

Modulus of elasticity		Shear modulus		Poisson number	
E_L (GPa)	E_T (GPA)	G_{TT} (GPA)	G_{LT} (GPA)	v_{TT} (–)	v_{LT} (–)
44	10.3	4.3	5.1	0.5	0.32

material properties in the rod direction and transverse thereto; the rod is said to be "orthotropic" (*orthos* = Greek for perpendicular, *tropos* = Greek for property).

The elasticity constants and the strengths of the rods used in this instance were established by complex tests [1] and are summarised in Tables 2.1 and 2.2.

2.2.1 Long-Term Behaviour of Composite Long Rod Insulators

Despite the many advantages of composite insulators compared to porcelain and glass, it took many years of persuasion, mainly by the manufacturers of composite insulators, before they were widely accepted. In the early years of this technology—i.e. during the 1970s—it was accepted both by the manufacturers and by users

Table 2.2 Strengths of the FRP rod

Strength value	(MPa)
Longitudinal (tension)	1,038
Longitudinal (compression)	−794
Transverse (tension)	32
Transverse (compression)	−141
Longitudinal-transverse (shear)	54

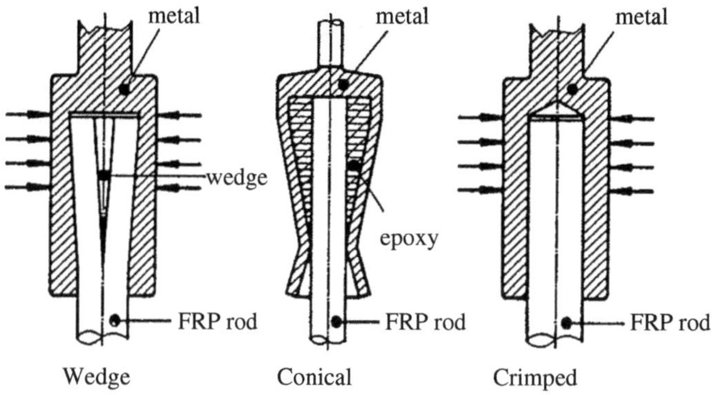

Fig. 2.9 Different types of fittings of composite long rods

(primarily network operators) that, ultimately, suitable qualification tests were necessary in order to demonstrate both the resistance of the insulation material under adverse ambient conditions (corresponding accelerated tests were developed for this purpose) and the long-term behaviour of the composite insulators under mechanical load. The pioneering work in this field was carried out by Claude de Tourreil and his CIGRE working group, and will be presented hereinafter.

De Tourreil and his colleagues at IREQ–Institut de Recherche d'Hydro-Québec (Hydro-Québec's Research Institute)–tested insulators provided by three different manufacturers A, B and C under rather complex long-term test conditions (some individual tests lasted up to three years) [2], these insulators differing substantially in terms of the technology of the end fittings: Insulator A uses a conical (potted) fitting, insulator B uses a crimped fitting, and insulator C uses a wedge fitting (Fig. 2.9).

The aim of the first series of tests was to determine the failing load of the insulators as a function of ambient temperature within a range of −25–100 °C. The results are shown in Fig. 2.10.

As can be seen, the ultimate tensile stress in all three insulator types increases with decreasing temperature, which is beneficial in particular with their use in cold environments, since the insulators then also have to withstand greater tensile loads as a result of an increase in line tension. It can also be seen that the force-temperature behaviour differs substantially between the three insulator types. Insulators having crimped fittings react less sensitively to changes in temperature, even at low temperatures, at which a negative influence on failing load would be expected as a result of the different coefficients of thermal expansion of the FRP rod and steel.

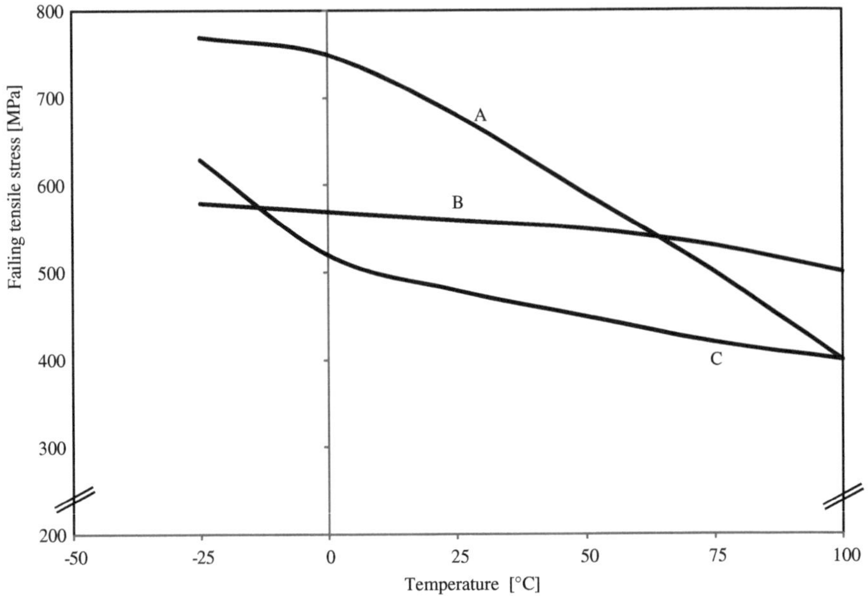

Fig. 2.10 Failing tensile stress of three composite insulators having different fittings as a function of temperature [2]

As more recent CIGRE tests have shown (see Sect. 9.6.3), line temperatures of 200 °C and more which occur in modern high-temperature conductors only have a small effect on the tensile strength of composite insulators.

The aim of the second series of tests was to determine the so-called load-time curve of the insulators. The time until failure under constant tensile load was established in each case. The results of this test are shown in a graph in Fig. 2.11. The insulator used in this instance had crimped fittings, these being the most common fittings used currently. This type of fitting also demonstrated the best long-term behaviour, wherein the relatively broad scattering of the measurement results for a specific load is noticeable and may extend up to three orders of magnitude over the time scale.

It is also interesting that the rate of decline of the load-time curve increases slightly with temperature, as is shown in Fig. 2.12 for insulator B (having crimped fittings), which also behaves better in this regard than insulators A (conical) and C (wedge).

Operational use was simulated in an outdoor test station, where the insulators were subjected to a constant load with simultaneous application of an alternating voltage and climatic influences, the alternating voltage (60 Hz) being approximately 10 % higher than the nominal voltage of the insulators, and the ambient conditions ranging from winter temperatures down to −30 °C, to summer temperatures up to +30 °C, through sunshine, rain, and snow and ice. Although the load-time curve retained its (negative) slope under these conditions (Fig. 2.13), it shifted in parallel "downwards" until at the level of the 50 °C curve (see also Fig. 2.12).

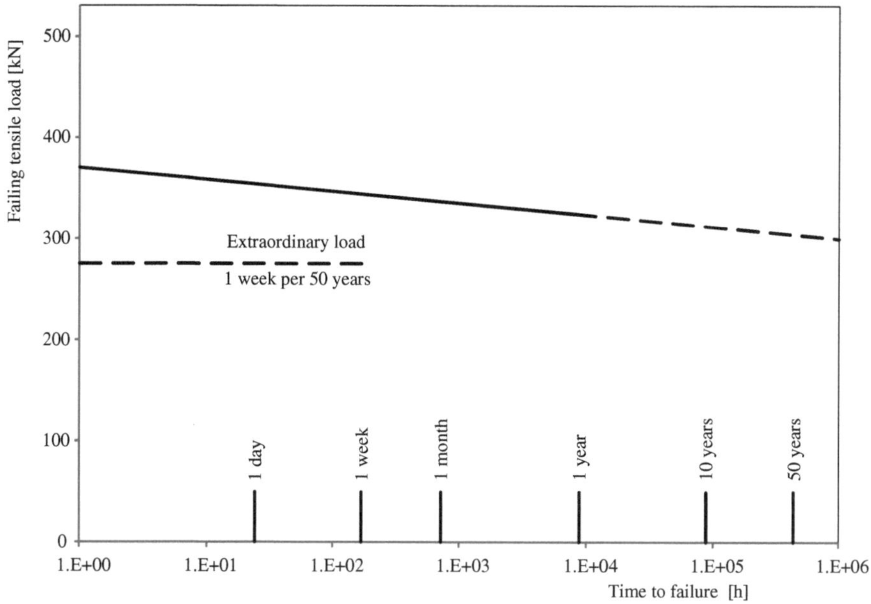

Fig. 2.11 Load-time diagram of an insulator having a crimped fitting at 23 °C [2]

De Tourreil later carried out similar tests [3], but this time with smaller rod diameters (up to 20 mm), the results of which are summarised in Table 2.3. By contrast, the measured load-time curves of all these test specimens are shown in Fig. 2.14, normalised to the respective failing load (100 % value). In these tests, too, the best results were achieved by the insulators having crimped fittings, and in particular the insulator having a swaged fitting.

These ground-breaking tests clearly demonstrated that a composite insulator subjected to constant tensile load (which exceeds a specific threshold value—the damage limit) will fail after a certain period of time, the duration of which depends on the magnitude of the applied load. The failure occurs, however, at a load level which is somewhat lower than the static failing load of the insulator.

This finding is of fundamental importance to the dimensioning of such insulators, since not only the load, but also the duration thereof, should now be considered. This has understandably led to a certain amount of uncertainty among experts in this field, in particular among utility engineers, and further testing was introduced with the aim of better understanding the physical relationships involved. A new test was also introduced into IEC standard 61109/92 (new at that time) on the standardisation of tests for composite insulators, the objective of this test being to ensure that composite insulators would not fail suddenly, even after long periods of load. The first test proposal was unfortunately unrealistic, and has only been replaced in recent years by a more "sensible" test. This development must be borne in mind below [4].

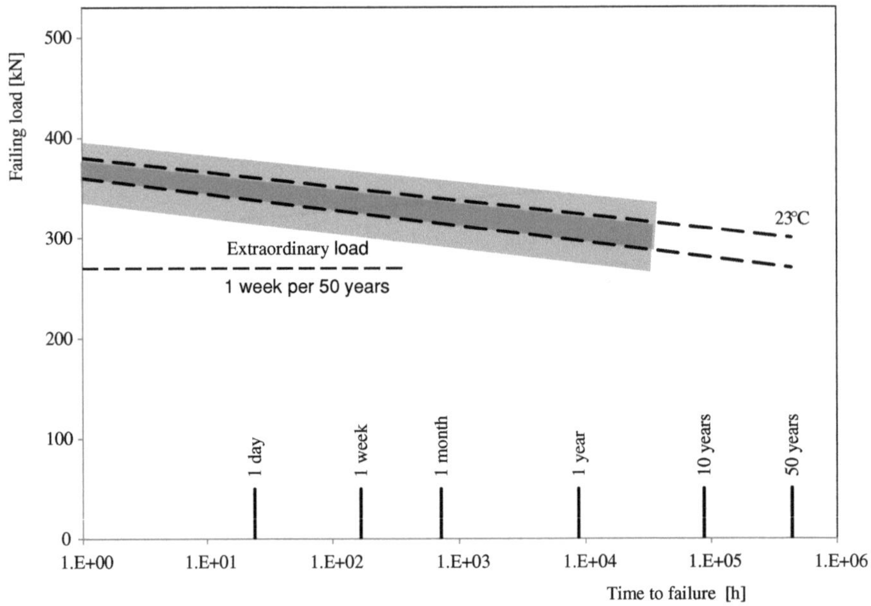

Fig. 2.12 Comparison of the load-time curve for an insulator having a crimped fitting at 23 and 50 °C [2]

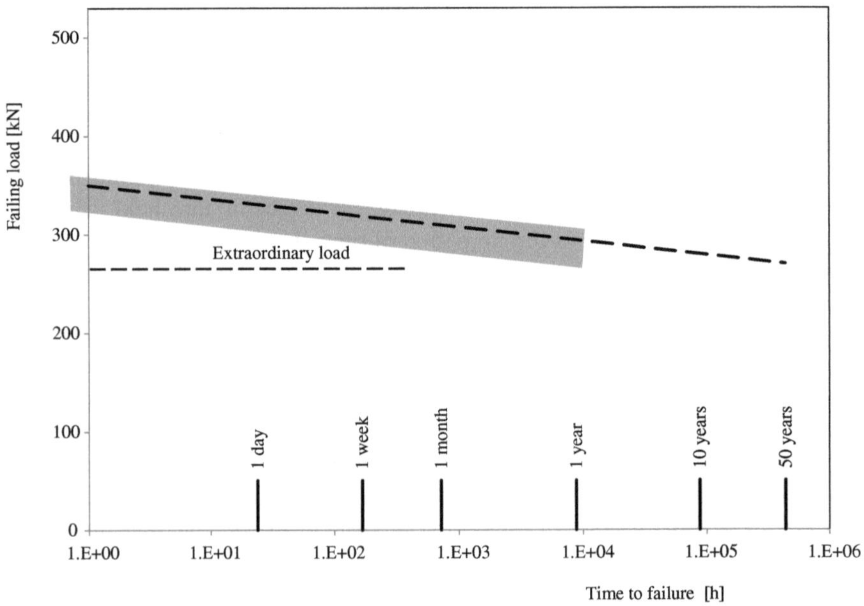

Fig. 2.13 Load-time curve of an insulator having a crimped fitting under mechanical, electrical and climatic load [2]

Table 2.3 Comparison of the failing loads and of the failing stresses of composite long rod insulators having different fittings

Fitting type	Average failing load [kN]	FRP failing stress [MPa]
Conical fitting	176 ± 6	823
Polygonal crimped fitting	219 ± 1	796
Coaxial fitting	338 ± 4	1,076

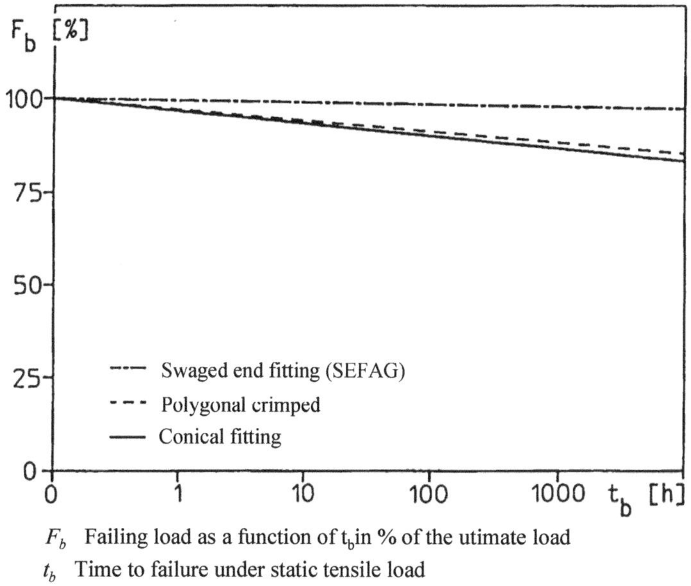

Fig. 2.14 Average load-time curves of composite long rods having different fittings, normalised to the respective failing load (see also Table 2.3)

2.2.1.1 Load-Time Curve of Composite Insulators According to IEC 61109/92

In principle, three tests are provided in the "old" IEC 61109/1992 to ascertain the mechanical strength of composite insulators: Annex A details the long-term behaviour of composite insulators within the scope of a *design test*, and describes the associated test procedures. It is also noted that the tensile strength of composite insulators decreases over time; this decrease can be assumed to be linear with the duration of load application.

This is shown in a graph by the straight line *a* in Fig. 2.15. The value F_{Br} represents the average *failing load* of three test specimens established in the tensile test (1 min test). The standard also requires the slope of the straight line *a* to be 8 % at most per decade. This is to be checked by a 96 h test of a further three test specimens at 60 % of the previously determined average failing load F_{Br}, as indicated in Fig. 2.15.

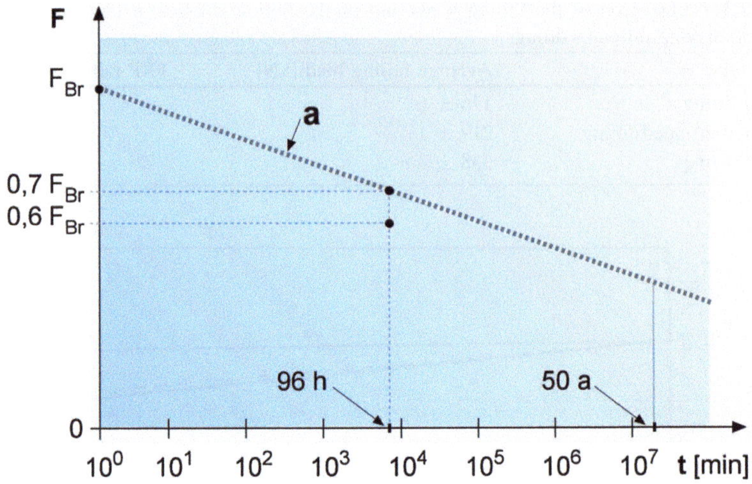

Fig. 2.15 Determining of the load-time curve "a" of a composite insulator according to the "old" IEC 61109/92

The "specified mechanical load" (*SML*) is also introduced in the standard. It is defined as the load which, after a 96 h test at 70 % of the specified mechanical load of the insulator (specified by the manufacturer), is retained in a subsequent 1 min tensile test, and is to be determined in the *type test* (Fig. 2.16). Lastly, a *routine test* at *routine mechanical load* (*RML*) is provided, which is 50 % of the specified mechanical load and has to be demonstrated in all insulators during the tensile test prior to delivery.

As important as it was to demonstrate this specific feature of the mechanical long-term behaviour of composite insulators when this standard was created, its depiction irritated users and often led to redundant over dimensioning. There are two main reasons for this:

Firstly, the linear fall of the withstand load curve in the standard suggests that this failing load of a composite insulator decreases continuously over time. In theory it is often ignored that this curve only indicates the service life until failure of a composite insulator when the insulator is loaded *continuously* by a specific tensile load. Secondly, the undefined linear fall of the curve implies that the tensile strength of the composite insulator would fall practically to zero after a finite, even if rather long, operational period, which leads to the absurd conclusion that a composite insulator which has been stored away and "forgotten" will break into two parts at some point in the future.

The fact that this cannot be true is confirmed not only by sound engineering expertise, but has also been proven by the experimental findings and physical considerations, as detailed hereinafter.

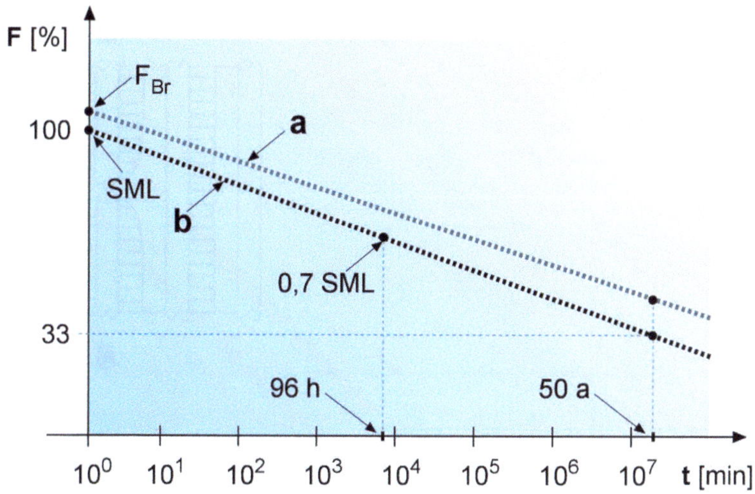

Fig. 2.16 Determining the specified mechanical load (*SML*) of a composite long rod insulator

2.2.1.2 Mechanical Model for the Long-Term Behaviour of Composite Insulators

The mechanical behaviour, and in particular the time-dependent failure mechanism, of an FRP rod under continued tensile load can be explained qualitatively as follows: Since the resin matrix of the rod can stretch considerably, but in contrast the glass fibres are hardly able to undergo plastic deformation (brittle material), the load is taken over practically completely by the glass fibres within a very short time after application of the external tensile load. It must be taken into account that the properties of the glass fibres vary widely in a random manner. The approximately 800,000 individual glass fibres in the cross-section of an FRP rod of 16 mm diameter not only have different tensile strengths. The fibre cross-section F, and to a certain extent the modulus of elasticity E of the fibres also vary from fibre to fibre, the overall tensile load acting on the FRP rod thus being distributed over the individual fibres in accordance with the tensile stiffnesses DS of said fibres ($DS = E \cdot F$)—that is to say the greater the stiffness of a fibre, the more load it must bear, and since, as mentioned, the tensile strength of the fibre is subject to a certain level of random scattering, some weaker glass fibres will break.

The time-dependent failure mechanism of the FRP rod, which is particularly interesting in this instance, is based on the load transfer from the broken fibres to adjacent fibres (Fig. 2.17). This occurs in such a way that when, for whatever reason, a glass fibre breaks, the resin matrix enveloping the glass fibre transfers the axial load to the cylindrical surface of the broken fibre via the distribution of shear stress, this break therefore going "unnoticed" at said cylindrical surface from a certain distance from the area of the break. Secondly, the matrix transfers the

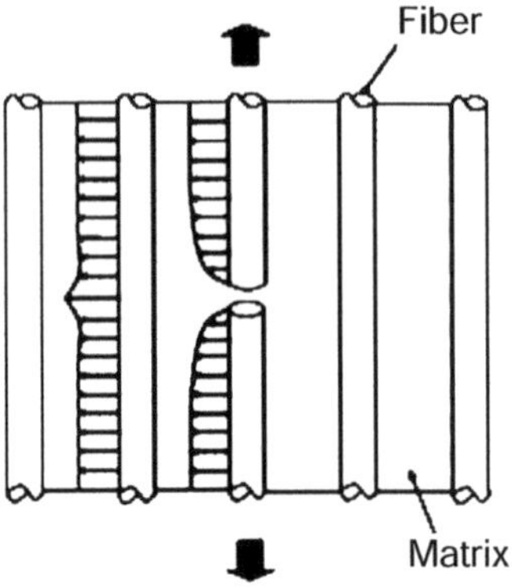

Fig. 2.17 Load transfer processes in the fibres of an FRP rod under axial load

original load from the broken fibre to the healthy adjacent fibres, which are thus loaded increasingly, which in turn results in these adjacent fibres being stressed to a greater extent at the point of break of the broken fibre. It has been attempted to illustrate this physical process schematically in Fig. 2.17.

However, since the adhesion between glass fibres and resin matrix is subject to relaxation, this leads to a decrease in the "load transfer capability" of the fibre matrix composite over time, especially as the interface between fibre and resin matrix is loaded. The load which could still be borne by the broken fibres is increasingly transferred completely to the healthy fibres, which will also fail by the same mechanism after a certain period of time until all fibres, and therefore the FRP rod, have broken. It makes sense that the higher the load, the more fibres will break over a rather short period of time and the quicker the described relaxation processes will occur. However, if the overall initial load lies below a certain threshold value, for example the limit value for the first fibre breakages (*damage limit load* or *DLL*), then hardly any fibres will break on the basis of this model, even if the load is applied to the insulator for an infinite period of time. Conservative estimations indicate that this damage limit load can be set at least at 50 % of the specified mechanical load. These considerations are only true for failure of the rod, without consideration of the fitting.

This model was confirmed by a number of independent tests which showed that the permanent load curve of suitably designed composite insulators is not linear [5], contrary to the illustration in Fig. 2.16. It can be seen that the measured curves fall relatively steeply at the start, but asymptotically approach a lower threshold value of the damage limit load after a relatively long period of time. This means that a composite insulator retains a significant proportion of its original failing load F_{Br} for an "infinite" period of time.

For example, the results of a number of tests are plotted together in Fig. 2.18, these tests having been carried out by IREQ in the 1980s. In this case, F is the failing load normalised to the 1 min value, and t is the time until failure in minutes (plotted logarithmically); the dashed straight line indicates the DLL (damage limit load) at 60 %.

These assumptions have also been confirmed by practical experience. De Tourreil [6] reports that he measured the 1 min average failing load (M_{av}) on

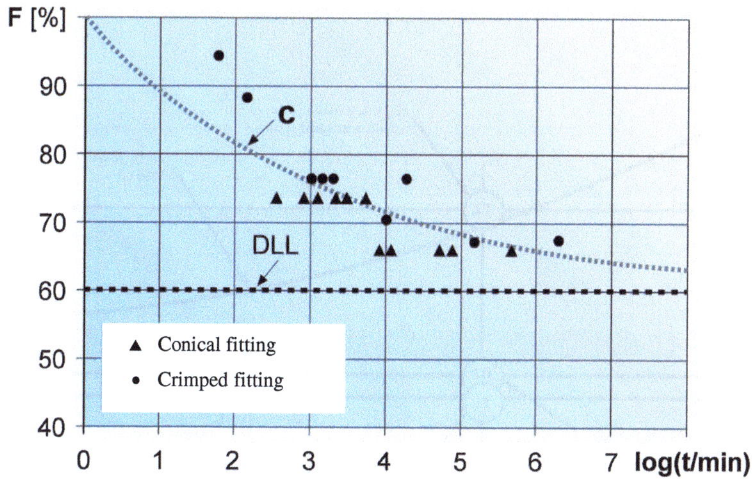

Fig. 2.18 IREQ test results of static long-term behaviour of insulators from various manufacturers

composite insulators which had been in operation for 12 years. No significant difference from the failing load of the same insulator family when new could be established. Later results [7] of mechanical tests on 132 and 400 kV insulators after more than 20 years in operation also confirmed these findings.

The improved understanding of the physical processes during tensile loading of composite insulators as well as the extensive tests and positive operational experience were taken into account during the standardisation and resulted in the design and type tests in IEC 61109, Ed 02 being adapted accordingly. The new edition of IEC 61109 (2008) requires (Fig. 2.19):

- a 96 h test as a design test to determine the "position" of the load-time curve of the insulator (value pair D1 and D2).
- a limit load test as a type test to establish the limit load once the insulator has been loaded for 96 h by a constant load of 0.7 SML (value pair T1 and T2).

The design test establishes M_{av} (av = *average failing load of the assembled core*) and thus the starting point of the actual load-time curve of the insulator, and also the minimal limit load below which, according to the statements above, no mechanical damage to the insulator should occur, and does so as a result of a long-time test in which a load of 0.60 M_{av} has to be maintained over a period of 96 h, the choice of this test parameter being clarified below.

Taking into account the practical capabilities of testing laboratories, the test period of 96 h was selected because it lies in the middle of the logarithmic time scale of 1 min to 50 years. On the other hand, the load was defined by 60 % M_{av} because, in the case of three test specimens which passed this 96 h test at 0.60 M_{av}, there is a 90 % probability that the average failing load at 96 h will be at least 0.70 M_{av}. This results from the assumed Gaussian distribution for three test specimens:

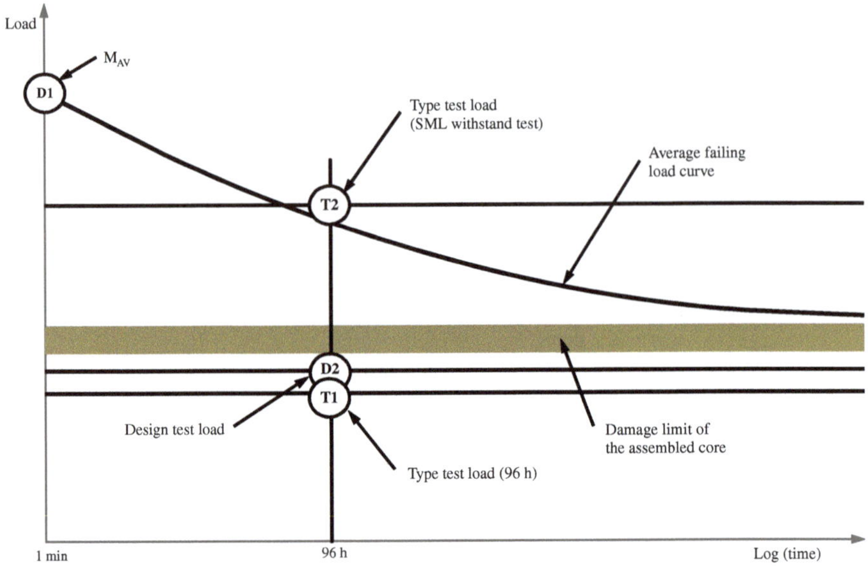

Fig. 2.19 Test loads according to the "new" IEC 61109, Ed 02 (2008)

$$M_{96} - 1.820\sigma_M$$

where M_{96} is the average 96 h failing load and σ_M is the standard deviation.

If the rather conservative value for the standard deviation of $\sigma_M = 0.08$ is used, it follows that

$$0.7 \cdot M_{av}(1 - 1.820 \cdot 0.08) = 0.60 \cdot M_{av}$$

thus justifying the assumption of 0.60 *M*av.

However, the standard cited above [IEC 61109, Ed. 2, 2008] allows the use of the "real" standard deviation, as established from tensile tests, in order to establish the 96 h withstand load if more than ten such tests have been carried out.

It should be noted in this regard that, when tested, modern composite insulators provided by well-known manufacturers should have standard deviations for their failing load of no more than 5 %.

2.3 Behaviour of Composite Long Rod Insulators Under Dynamic Load

Outdoor composite insulators are subject to the same loads, in particular the same dynamic loads, as the other components of an overhead transmission line. Loads caused by wind-induced vibrations, such as Aeolian vibrations, sub-span oscillations and galloping are the main dynamic loads. These types of vibration are described in detail in [8], for example. The behaviour of these insulators under dynamic load will now be described below in greater detail.

In early experiments carried out by IREQ the insulator load was simulated for galloping [3]. According to estimations at that time, a conductor which "gallops" could load the insulator with a pulsed tension of ±15 kN at a frequency of 0.5 Hz. These parameters were also selected for testing the insulators with pulsed tension loads. It is important to note that the static tensile load (average tension) was selected in such a way that the time until failure of an insulator was within the time window available for the tests (100 h at most). Unrealistically high static tensile loads (compared to the tensile loads experienced by an insulator during operation) were thus created, and these tests are thus better suited to providing a benchmark between different insulators and different fitting technologies.

Figure 2.20 shows the four tested insulators. Insulators A and D1 use a conical fitting, which was conventional at the time (see also Fig. 2.9), whereas insulators B and D2 use a crimped fitting, which is more common nowadays.

The FRP rods failed in all cases, the breakages starting at the transition to the metal fitting. The breakages typically started on the surface of the rod in insulators A and D1, and propagated to the centre of the rod at an angle of 45° to the rod axis with subsequent delamination along the glass fibres. Insulators B and D2 failed either as a result of the rod slipping out of the metal fitting (pull-out) or as a result of extensive delamination.

Rod breakages at an angle of 45° to the rod axis (as in A and D1) were observed less frequently. These different breakage patterns have since been attributed to and explained by the type of load applied to the rod by the metal fitting (see Sect. 2.4.7.4 below, Fig. 2.47).

The test results are illustrated in Fig. 2.21. It can be seen that the design and method of crimping are very important for the behaviour of such insulators under pulsed tensile load. Whereas insulators A and D1, both with a conical fitting, exhibit a similar decline in their failing load after 100 h (compared to their static failing load) of approximately 18–20 %, insulators B and D2, both with crimped fittings but using different crimping techniques, demonstrate considerable differences in this regard. Also insulator B shows a prominent fall of 43 %, insulator D2 "gets by" with "only" 25 %. It is worth mentioning that the same insulator D2 also achieved the best results in the above-described load-time tests (see also Fig. 2.14 and Table 2.3).

The behaviour of composite long rods under pulsed load, as may occur as a result of Aeolian vibrations, has also been tested [9]. The test rig can be seen in Fig. 2.22 during the testing of a 220 kV insulator.

The pulsed tension test started at a maximum pulsed load approximately equal to the SML, which had been determined previously on insulators of similar design. The pulsed load was reduced until failure no longer occurred after 2 million load cycles (assumed to be realistic for service conditions). Figure 2.23 summarises the test results of insulators provided by four different manufacturers (referred to here anonymously by A, B, C and D), the number of load cycles until failure being plotted on the X-axis and, for improved comparison, the ratio of F_{max}/SML (in %) being plotted on the Y-axis.

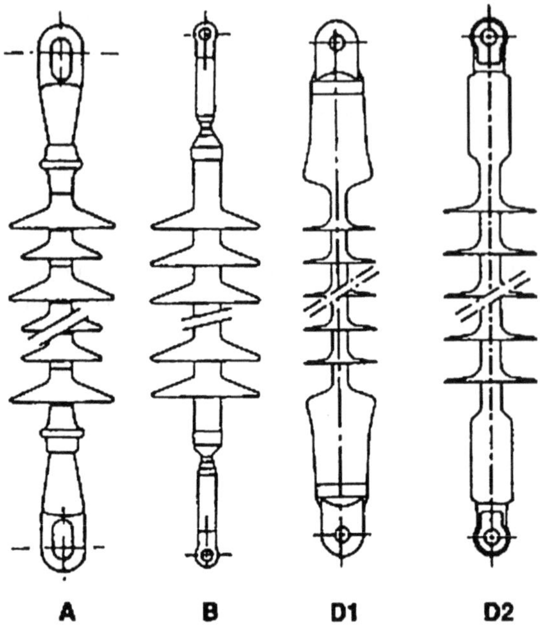

Fig. 2.20 Insulators for the dynamic tests [3]

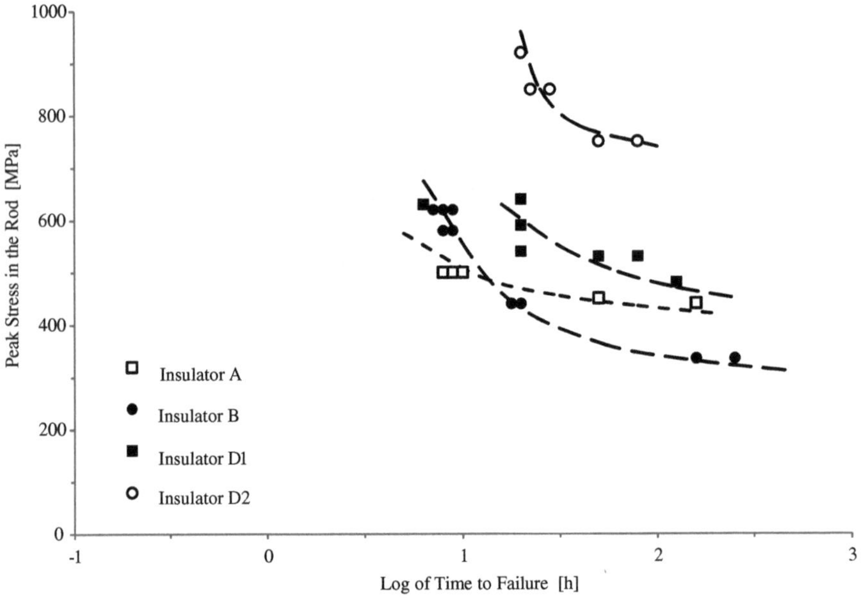

Fig. 2.21 Ultimate pulsed tension stress over time to failure during the dynamic tests [3]

Fig. 2.22 Test station during testing of a 220 kV composite insulator [9]

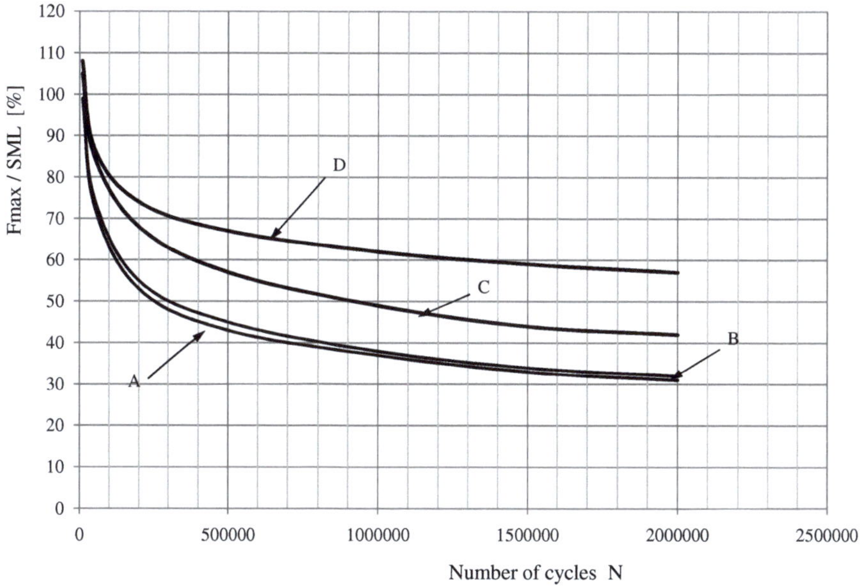

Fig. 2.23 S-N curve of composite long rods [9]

This illustration is known as the S–N curve.

Cracks were observed over the entire surface of the rod, and not just in the areas bordering the metal fitting (Fig. 2.24). The damage to the 25 mm FRP rod of a 400 kV insulator in the vicinity of the upper fitting can be seen (SML 210 kN, test frequency 7 Hz, test load approximately 60 % × *SML*, number cycles until failure 1.5 million).

It should be noted however that, in accordance with estimations based on [10], the pulsed loads applied during these tests are probably higher than the pulsed loads which would be experienced by the corresponding conductor before fatigue failure. This means that, under such extreme loads, the conductor would fail first. Because of that, conductor vibrations causing similar loads are usually dampened sufficiently in practice [11].

Tests carried out on 120 kN composite long rods having ball-and-socket fittings lead to similar results [12]. In these tests, insulators of similar type but with reinforced end fittings were additionally tested to establish the fatigue properties of the FRP rod itself. The fatigue values of the FRP rod are greater than those of the ball fittings, which failed earlier as a result of the notch effect. In these tests, too, a flattening of the S–N curve of the insulator was observed from approximately 2 million cycles (Fig. 2.25).

This means that the fatigue behaviour of the metal fittings may be crucial under pulsed load, as could occur in the event of wind-induced vibrations.

2.4 Design and Assembly of End Fittings for Composite Long Rods

2.4.1 Development and State of the Art Technology of Metal Fittings

Composite insulators are loaded by rather high tensile loads of up to 500 kN, and more in special cases. These forces have to be transferred both to the tower and to the conductor. This is achieved with the aid of end fittings, which have to be attached to the FRP rod of the insulator in a suitable manner.

Since this type of insulator has to withstand high mechanical loads, its fittings are made of steel (generally hot-dip galvanised) or ductile cast iron. Aluminium is occasionally used if there is a need to reduce weight (railway applications), albeit for rather smaller loads (up to approximately 100 kN). Bronze is used in some special cases where the ambient conditions are extremely corrosive. Typical fitting types for composite long rod insulators are as in Fig. 2.26 from left to right: socket, tongue, clevis, ball, eye and Y-clevis.

The international standards IEC 61466: "Standard Strength Classes and End Fittings" define the dimension and other parameters for these fittings.

The insulator manufacturers have developed different assembly methods over the years for assembly of the end fittings on the FRP rod. The fittings were originally potted to the FRP rod, then in following generations of composite insulators the fittings were wedged in place, and in recent years the fittings have almost exclusively been crimped (see also Fig. 2.9). This type of fitting will therefore be described in greater detail hereinafter.

Fig. 2.24 Damage to the FRP rod [9]

2.4.2 Basic Considerations Regarding the Design of Crimped Fittings

As already mentioned, practically all composite long rod insulators now use crimped fittings. Whether cast or forged, they are characterised by low manufacturing costs, in particular if they can be manufactured in bulk. They are applied to the FRP rod using commercially available crimping equipment. Fittings and crimping methods have been constantly optimised over the years by the insulator manufacturers. The main parameters of the fittings will be defined and explained before discussing the mathematical considerations when designing the fittings.

In Fig. 2.27 the FRP rod and the end fitting are illustrated schematically together with all the dimensions relevant for the crimp [13]. Since practically all dimensions and material properties influence the quality of the crimp to a greater or lesser extent, a large number of tests have to be carried out in order to establish the relationships between said dimensions and properties and the quality of the crimp.

The main information obtained from these tests is summarised as follows:

1. Influence of the crimp length *L*: It has been found that the failing load of a composite insulator increases approximately proportionally with the crimp length *L* of the metal fitting, that is to say the longer the fitting, the greater the tensile load which can be withstood by the insulator. In any case, the crimp length should not exceed a certain length, because the electrical values of the insulator might then not be met and, as is known from experience, extremely long crimps do not necessarily result in an proportional increase of the insulator failing load.
2. Influence of the geometry parameters *x*, *y* and *z*: These parameters have a considerable influence on the crimp. For example, a well designed length x reduces the concentration of stress at the opening in the fitting, which in turn has a favourable effect on the failing load M_{av} of the insulator, as can be seen in

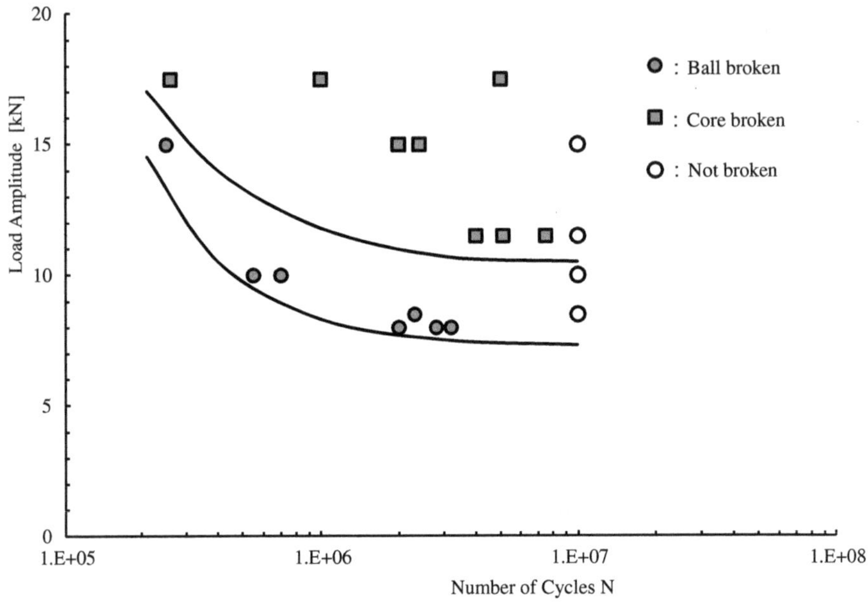

Fig. 2.25 Fatigue properties of 120 kN composite long rods [12]

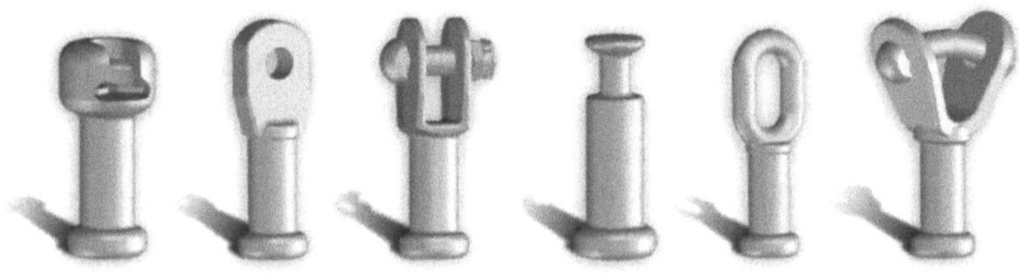

Fig. 2.26 Typical end fittings for composite long rod insulators

Fig. 2.28. It can be also seen that the failing load M_{av} is considerably reduced below a critical length x_{cr}. On the other hand there is a certain limit x_{max}, above which there is no significant increase in M_{av}.

3. Influence of the wall thickness of the end fitting b: This is strongly related to the material properties of the end fitting. In the case of a high plastic limit of the fitting, the duration of the crimping process, i.e. the application of the pressure load, increases with the wall thickness b of the end fitting.
4. Influence of the roughness of the inner surface of the fitting: Many tests have demonstrated that a smooth inner surface reduces the failing load of the insulator, since in this case the coefficient of friction between the metal fitting and the FRP rod, which is decisive for the damage limit load of the insulator, is also reduced considerably. If, by contrast, the inner surface of the fitting is very rough, the asperities of this surface will damage the surface of the FRP rod,

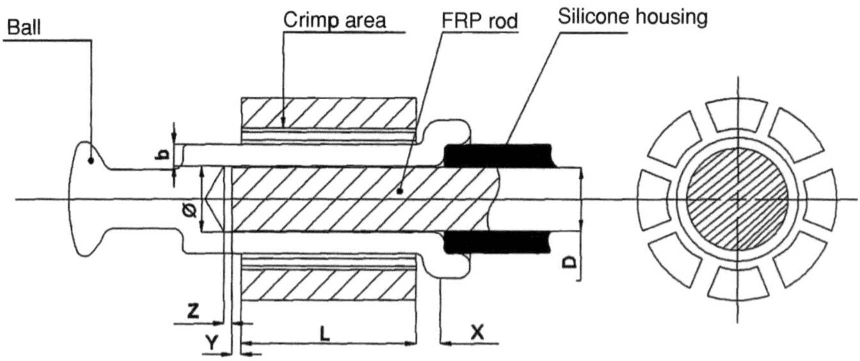

Fig. 2.27 Critical geometrical parameters of the end fitting [13]

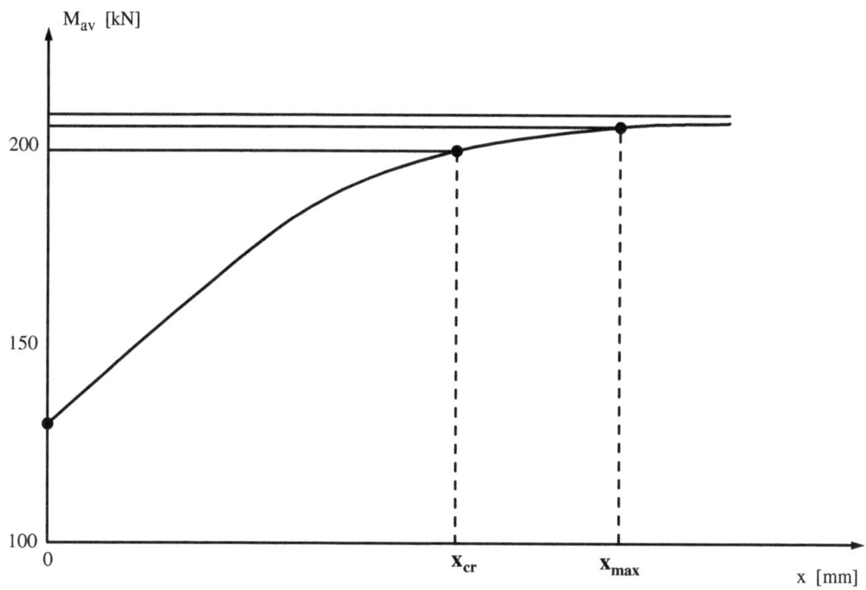

Fig. 2.28 Dependence of M_{av} over x (schematic view) [13]

which will in turn have a negative influence on the damage limit load of the insulator.

5. Influence of the tolerance between the metal fitting and the FRP rod: Sufficient crimping (and thus sufficient damage limit load of the insulator) can be generated by the suitable selection of this tolerance, it being necessary to adjust the crimping parameters accordingly (for example peak value and time dependence of the applied contact pressure).
6. Influence of the material properties of the fitting: As already mentioned above (point 3), the stress–strain curve of the fitting material, and in particular the plastic limit of the fitting material, play a key role during the crimping process.

7. Influence of the material properties of the FRP rod: The modulus of elasticity transverse to the rod axis also plays an important role for production of an effective crimp (Table 2.1). The ultimate strength of the FRP rod should also be taken into account during the crimping process, since a low ultimate strain transverse to the rod axis limits the possible deformation of the metal fitting during the crimping process (Table 2.1).

2.4.3 Assembly of Crimped Fittings

As mentioned at the outset, the crimping technique is now the preferred method for fixing the end fittings to the FRP rod. With this technique, the FRP rod is inserted into the end fitting, which is then crimped onto the rod by applying a radial external pressure, normally using a set of eight crimping jaws.

The crimping jaws are driven either by individual hydraulic cylinders, or by a central cylinder and a sliding system of deflecting cams. These crimping jaws have socket fittings for different crimping tools, which can be interchanged quickly and easily so that fittings of different sizes can be crimped using the same press. The press assembly is shown in Fig. 2.29.

This method leads to a relatively homogeneous distribution of stress and deformation in the fitting and in the FRP rod, since the radial pressure applied to the fitting by the hydraulic tool can also be assumed to be homogeneous over the circumference of the fitting. Although this method calls for relatively narrow tolerances for the dimensions of the metal fitting and of the FRP rod as well as for the roughness of the inner surface of the fitting, it is possible to compensate for small deviations since contact pressure and its time path during the crimping process are monitored as abort criteria. For long fittings, which require high contact pressures, two short crimping areas are provided instead of one long crimping area, since the maximum permissible forces per crimping jaw are fixed by the dimensions of the crimping tools.

2.4.4 Calculations

The calculation of the mechanical stresses in the FRP metal-composite joint of a composite insulator is certainly not simple, since the different basic material behaviours of metal and FRP have to be taken into account and the interface between metal and FRP has to be modelled as realistically as possible.

As is so often the case in practice, the calculation methods develop over time with increasing experience. For the purposes of solving the problem addressed in this instance, a distinction is made between simple and complex analytical calculation methods and numerical simulation methods. The analytical methods are intuitive and can be implemented relatively easily. They are based on practical experience and are regularly checked against the test results. The numerical

Fig. 2.29 Details of the hydraulic press for crimping of the fitting

methods use the finite element method (FEM) exclusively. They were developed to obtain a better understanding of the stress distribution and failure mechanisms of the FRP rod in particular. They are also used if special applications, such as those involving extremely high tensile loads, are to be designed.

2.4.5 Simple Analytical Method

In the case of the composite long rods, the simple analytical method is primarily used:

1. To establish the FRP rod diameter necessary to achieve the required specified mechanical load (SML) and to pass the relevant tests.
2. To ensure that the mechanical fittings withstand the limit load.

The first objective is achieved by setting a maximum sustainable shear stress of the FRP rod of approximately $\sigma_{shear}^{\max} = 40\,\mathrm{N/mm^2}$. This shear stress is assumed to be constant over the crimping area, that is to say over the interface between the FRP rod and the metal fitting. If d is the rod diameter and L_{cr} is the crimp length, the specified mechanical load (SML) is given as follows:

$$SML < \pi \cdot d \cdot L_{cr} \cdot \sigma_{shear}^{\max}$$

The second objective ensures that the metal fitting will withstand the necessary specified mechanical load (SML). There are two critical zones in a crimped end fitting which have to be considered:

(a) The cross-sectional area of the fitting (hollow cylinder), Fig 2.30.
(b) The transition zone from the crimped part to the connection part of the fitting.

With regard to (a):

The tensile stress in the cross-sectional area of the fitting is:

$$\sigma_{\mathrm{Z}} = \frac{\mathrm{SML}}{\pi\left(\mathrm{D}_{\mathrm{a}}^{2} - \mathrm{D}_{\mathrm{i}}^{2}\right)/4} < \sigma_{\mathrm{zul}}$$

where D_a is the outer diameter of the fitting, D_i is the inner diameter of the fitting and σ_{zul} is the tensile strength of the fitting.

With regard to (b):

The dimensions of the transition zones can be seen in the drawing of the end fitting (Fig. 2.30). The hatched areas are subject to shear load and should also withstand the specified mechanical load (SML), with τ_{zul} being the max. Permissible shear stress of the fitting, that is to say:

$$\tau = \frac{SML}{\pi \cdot D_i \cdot \ell} < \tau_{zul}$$

It is not necessary to check the connection element of the fitting (ball, clevis, tongue, etc.) if the standard dimensions specified in IEC 61466 have been used for these parts.

2.4.6 Complex Analytical Method

Such calculation methods are useful since they can be incorporated into the normal software tools used, including table calculation programs, and since, due to their analytical formalism they make it possible to better identify the relationships between the individual geometrical and material values and their effects on the state of stress of the insulator. The methods presented below were carried out within the scope of a research project [1], in which the authors took part in their capacity as representatives of the industry partner.

In the case of crimped fittings, there are three different phases which have to be examined:

1. Crimping: In this phase a pressure is applied to the interface between the FRP rod and the metal fitting by a predefined pressure at the surface of the metal fitting (Fig. 2.31).

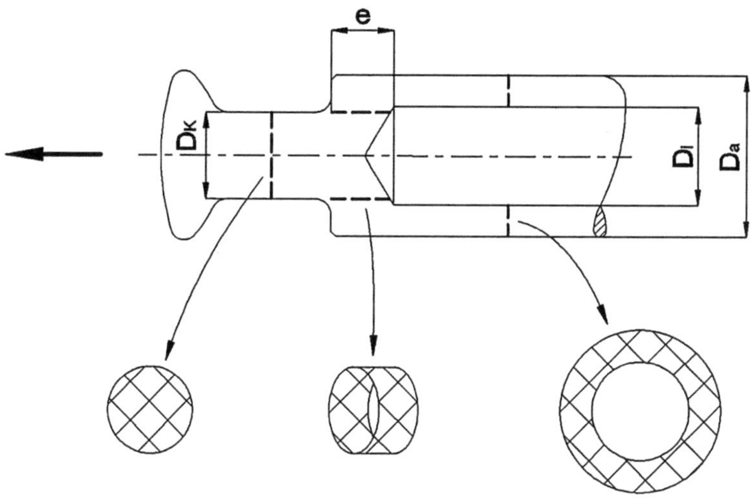

Fig. 2.30 Critical cross-sections when dimensioning an end fitting

2. Relaxation: In this phase the external pressure is removed and the metal fitting is partly relieved of its elastic deformation; the remaining plastic deformation in the metal fitting still exerts a pressure on the FRP rod, which ultimately ensures that the joint between the FRP rod and the metal fitting withstands the external tensile loads (Fig. 2.32).
3. Tensile load: This is the phase in which an external load is applied to the insulator and is increased until the rod slides out from the insulator (pull-out).

2.4.6.1 Crimping

As mentioned above, a radial force is applied to the metal-composite joint during the crimping phase, and therefore the pressure is distributed as uniformly as possible over the outer circumference of this joint.

The external radial pressure p acts on the FRP rod similarly to a hydrostatic pressure. It causes a reduction ΔR from the radius of the FRP rod. This results in a radial σ_{rr} and circumferential stress $\sigma_{\theta\theta}$ of equal magnitude in the FRP rod, and also an extension of the rod in the axial direction owing to the Poisson effect, which in turn leads to a shear stress τ at the metal/FRP interface (Fig. 2.33).

Hooke's law in cylindrical coordinates is as follows:

$$\frac{\Delta R(z)}{R} E_T = \sigma_{rr}(z) - \nu_{TT}\sigma_{\vartheta\vartheta}(z)$$

which, taking into account that $\sigma_{rr}(z) = \sigma_{\vartheta\vartheta}(z)$ can be rewritten as follows:

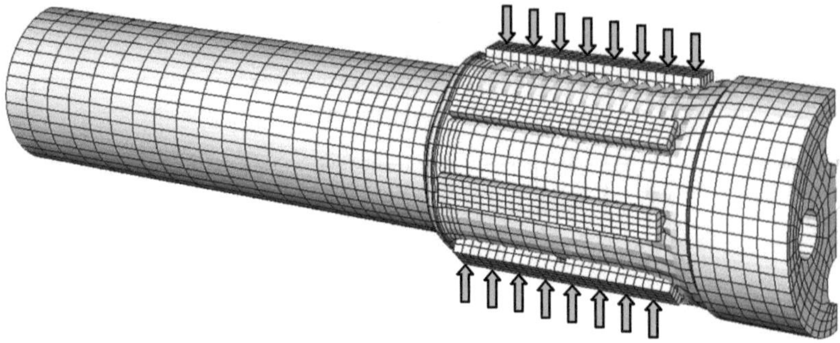

Fig. 2.31 Crimping (schematic view)

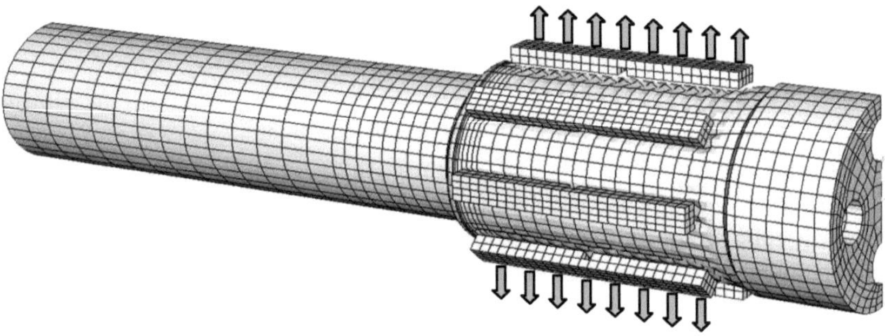

Fig. 2.32 Relaxation (schematic view)

$$\sigma_{rr}(z) = \sigma_{\theta\theta}(z) = \frac{\Delta R(z)}{R} \frac{E_T}{1 - \nu_{TT}}$$

where E_T is the transverse modulus of elasticity and ν_{TT} is the "transverse–transverse" Poisson number of the FRP rod (see Table 2.1).

As a result of the Poisson effect an axial-plastic deformation, which transfers an elastic-longitudinal deformation to the FRP rod via the frictional forces at the metal/FRP interface, also occurs in the metal fitting during the crimping process in addition to the radial deformation.

Owing to the symmetry of the arrangement, this deformation starts at the "centre" of the metal fitting and creates shear stresses τ_f which behave linearly at an interval $(-\delta, +\delta)$ until they reach the threshold value $\tau_{rz,\max}$ set by the FRP material (Fig. 2.34). To simplify matters, $\delta = 0$ is assumed in the analytical calculation, the progression of shear stress over the length of the fitting appearing to be stepped.

This distribution of shear stress over the surface of the rod causes an axial stress in the rod. This can be calculated as follows from the equilibrium of forces in a differential rod element of length dz (Fig. 2.35):

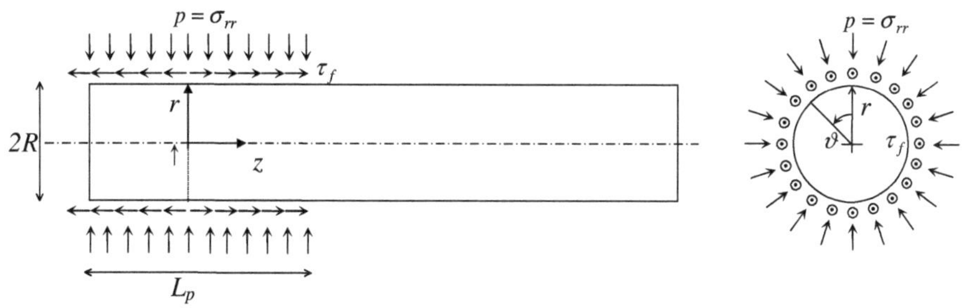

Fig. 2.33 Geometry and loading of the FRP rod during the crimping process [1]

$$\sum F: \; F_z(z) + \left(\frac{\mathrm{d}F_z}{\mathrm{d}z}\right)\mathrm{d}z - \tau_f(z) \cdot 2\pi R \cdot \mathrm{d}z - F_z(z) = 0$$

from which it follows that:

$$\tau_f(z) = \frac{\mathrm{d}F_z}{\mathrm{d}z} \cdot \frac{1}{2\pi R}$$

The tensile load in the rod $F_z(z)$, under consideration of the boundary conditions $F_z(0) = 0$ and $F_z(L_p) = 0$ and the progression of shear stress $\tau_f(z)$, according to Fig. 2.34 to give:

$$F_z(z) = F_z(z) - F_z(0) = \int_{F_z(0)}^{F_z(z)} \mathrm{d}F_z = 2\pi R \int_0^z \tau_f(z) \cdot \mathrm{d}z = 2\pi R \cdot \int_0^z (-\tau_{\max}) \cdot \mathrm{d}z$$
$$= -2\pi R \cdot \tau_{\max} \cdot z$$

for the range $0 \le z \le L_p/2$, and:

$$F_z(z) = F_z(L_p/2) - F_z(z) = \int_{F_z(z)}^{F_z(L_p/2)} \mathrm{d}F_z = 2\pi R \int_z^{L_p/2} \tau_f(z) \cdot \mathrm{d}z$$
$$= 2\pi R \cdot \int_z^{L_p/2} \tau_{\max} \cdot \mathrm{d}z = 2\pi R \cdot \tau_{\max} \cdot \left(\frac{L_p}{2} - z\right)$$

for the range $L_p/2 \le z \le L_p$.

On the other hand, the tensile load $F_z(z)$ can also be established from the integral of the axial stress σ_{zz} over this cross-section of the rod. A parabolic Ansatz is adopted for the distribution of σ_{zz} over this cross-section (Fig. 2.36), the maximum stress occurring at the point $r = R$, that is to say at the outer fibres of the rod:

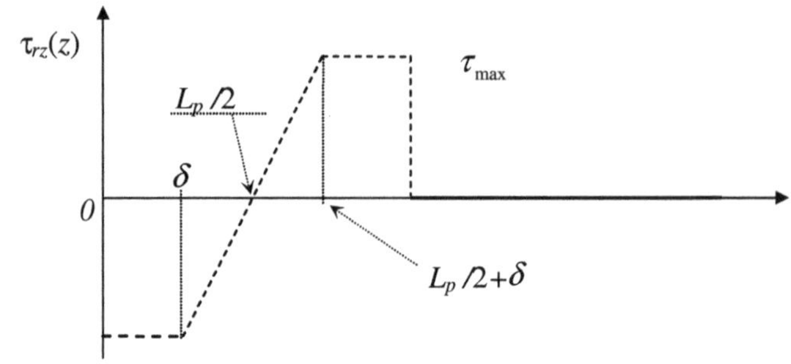

Fig. 2.34 Distribution of shear stress over the FRP surface in the area of the fitting during the crimping process [1]

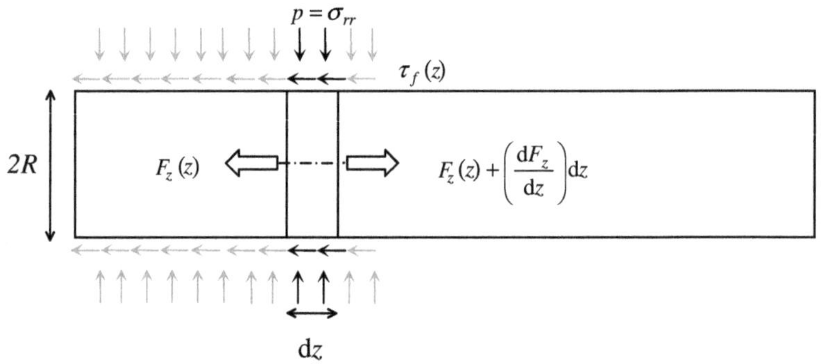

Fig. 2.35 Equilibrium at the differential beam element during crimping [1]

$$\sigma_{zz}(r,z) = \frac{\sigma_{zz}(R,z)}{R^2} r^2$$

It is noted, incidentally, that this "intuitive" approach is based on knowledge of the numerical simulation and, as will also be shown, delivers rather "sensible" results.

From the integration of $\sigma_{zz}(r,z)$over the cross-section, it follows that:

$$F_z(z) = \int_0^{2\pi}\int_0^R r\cdot\sigma_{zz}(r,z)\cdot dr\cdot d\theta = \frac{\pi}{2}\cdot R^2\cdot\sigma_{zz}(R,z)$$

If this result is equated with the previously obtained expression for $F_z(z)$,

$$F_z(z) = -2\pi R\cdot\tau_{\max}\cdot z \quad \text{or} \quad F_z(z) = 2\pi R\cdot\tau_{\max}\cdot\left(\frac{L_p}{2} - z\right)$$

$\sigma_{zz}(R,z)$ is ultimately given as follows:

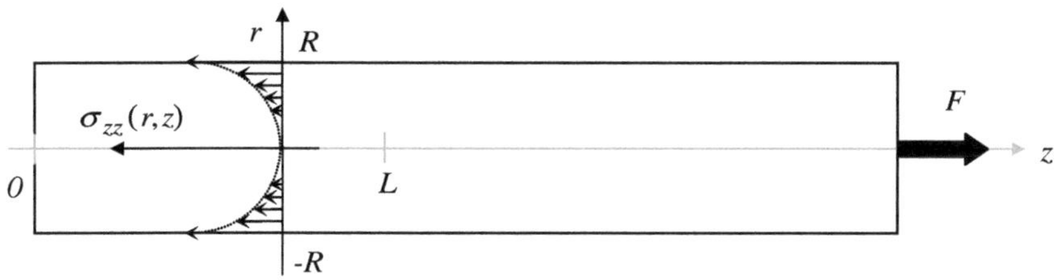

Fig. 2.36 Approach for the distribution of axial stress over the cross-section of the rod during crimping and under tensile load [1]

$$\sigma_{zz}(R,z) = -\frac{4 \cdot z}{R}\tau_{\max} \text{ or } \sigma_{zz}(R,z) = \frac{4 \cdot \left(\frac{L_p}{2} - z\right)}{R}\tau_{\max}$$

The portions of axial stress which originate from the "longitudinal-transverse" Poisson effect therefore have to be included, that is to say those portions which originate from the previously established radial and circumferential stresses in the rod:

$$\sigma_{zz}(z) = -\frac{4 \cdot z}{R}\tau_{\max} + \nu_{LT}[\sigma_{rr}(z) + \sigma_{\vartheta\vartheta}(z)]$$

and:

$$\sigma_{zz}(z) = \frac{4 \cdot \left(\frac{L_p}{2} - z\right)}{R}\tau_{\max} + \nu_{LT}[\sigma_{rr}(z) + \sigma_{\vartheta\vartheta}(z)]$$

And, as already illustrated above:

$$\sigma_{rr}(z) = \sigma_{\vartheta\vartheta}(z)$$

from which it follows that:

$$\sigma_{zz}(z) = -\frac{4 \cdot z}{R}\tau_{\max} + 2\ \nu_{LT}\ \sigma_{rr}(z) \text{ for } 0 \le z \le L_p/2$$

$$\sigma_{zz}(z) = \frac{4 \cdot \left(\frac{L_p}{2} - z\right)}{R}\tau_{\max} + 2\ \nu_{LT}\ \sigma_{rr}(z) \text{ for } L_p/2 \le z \le L_p$$

where ν_{LT} is the "longitudinal-transverse" Poisson number of the FRP rod (see Table 2.1).

Figure 2.37 shows the distributions of, $\sigma_{rr}(z,R)$, $\sigma_{\vartheta\vartheta}(z,R)$, $\sigma_{zz}(z,R)$ and $\tau_{rz}(z,R)$, that is to say the various stresses at the outer fibres of the FRP rod, over the crimp length of the fitting L_p for a typical 18.57 mm rod. The results of the numerical simulation (see Sect. 2.4.7) for the same variables are also plotted in the same figure [1]. The good correlation between the analytical and numerical simulations can be seen.

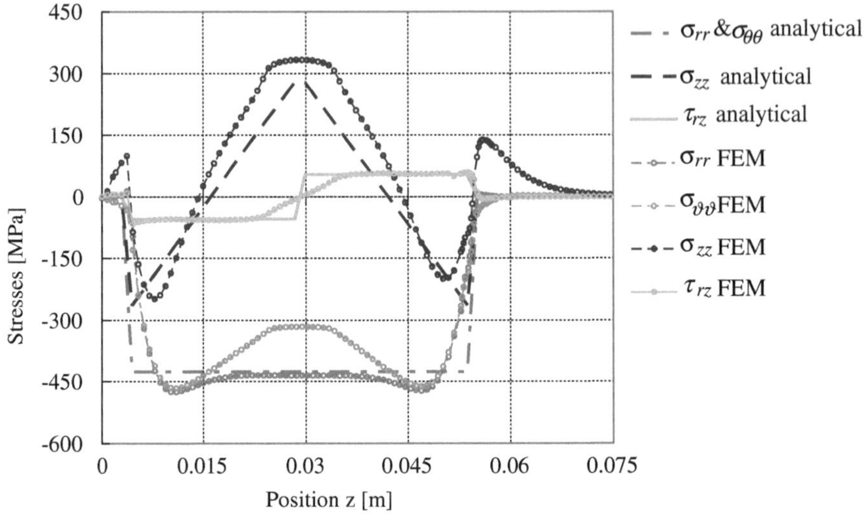

Fig. 2.37 Stress comparison (analytical/FEM) of σ_{zz}, σ_{rr}, $\sigma_{\vartheta\vartheta}$ and τ_{rz} at the outer fibres during crimping [1]

2.4.6.2 Relaxation

Once the maximum crimping pressure has been reached, it is set back to zero, as described above. The metal fitting is partly relieved of elastic deformation. The remaining radial plastic deformation continues though to apply a pressure, likewise radially, to the circumference of the FRP rod, but this pressure is not as high as the pressure produced during the crimping process. The stresses in the FRP rod are calculated accurately using the same formulas as above, wherein all stresses are approximately 30 % lower than during crimping.

2.4.6.3 Tensile Load/Pull-Out

In this phase the insulator is subjected to tensile load until failure. Failure ideally occurs in such a way that the FRP rod slips out from the metal fitting (pull-out). Before discussing the calculation of stress, we will first consider the maximum tensile load which can be sustained.

This tensile strength depends on the magnitude of the residual contact pressure after relaxation and on the coefficient of friction between metal and FRP (this was established by tests to be $\mu \cong 0.35$). It is:

$$F_{R.} = \mu \cdot F_N = \mu \cdot \sigma_{rr} \cdot S$$

where S is the contact (interface) surface between FRP rod and metal fitting (R: rod diameter, L_p: crimp length) as follows:

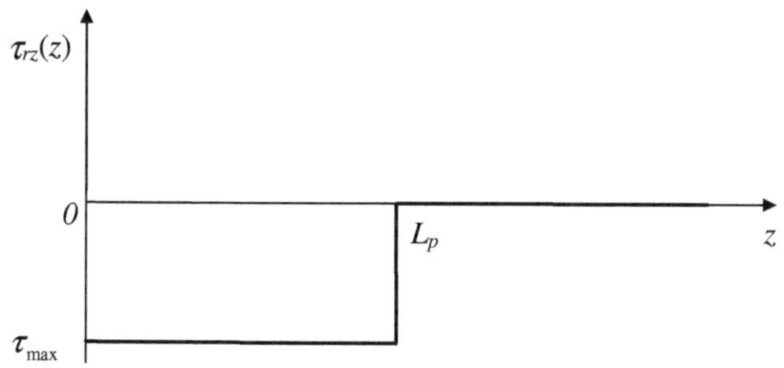

Fig. 2.38 Distribution of the shear stress over the FRP surface in the region of the fitting at the maximum tensile load possible (pull-out) [1]

$$S = 2\pi \cdot R \cdot L_p$$

However, this load is limited by the shear strength τ_{max} of the FRP rod:

$$F_{max} = \tau_{max} \cdot S$$

A maximum tensile load F_{max} of 158 kN results where $R = 9.285$ mm, $L_p = 50$ mm and $\tau_{max} = 51 \mathrm{N/mm}^2$. This compares well with the measured value of 148 kN (see Sect. 2.4.7.4).

The stress is calculated accurately by the same pattern as for the crimping and relaxation phases. However, the progression of shear stress $\tau_{rz}(z)$, as occurs under tensile load, has to be taken into account when establishing the axial stress $\sigma_{zz}(z)$ in the rod (Fig. 2.38).

The stress distributions for loading by the maximum tensile load are plotted in Fig. 2.39 together with the results from the numerical simulation. In this case, too, a relatively good correlation between the analytical and numerical simulations can be seen.

To summarise:

(a) For a composite long rod insulator which is subjected purely to tensile load, it is possible to establish, for all three phases of crimping, relaxation and tensile load, both the tensile strength of the FRP rod and the stress distribution in the FRP rod (this being the critical mechanical component of the insulator) using a simple analytical model.
(b) The most unfavourable stress distribution, and therefore the greatest potential for damage to the FRP rod occurs during crimping, which is why particular attention is to be paid to this process.

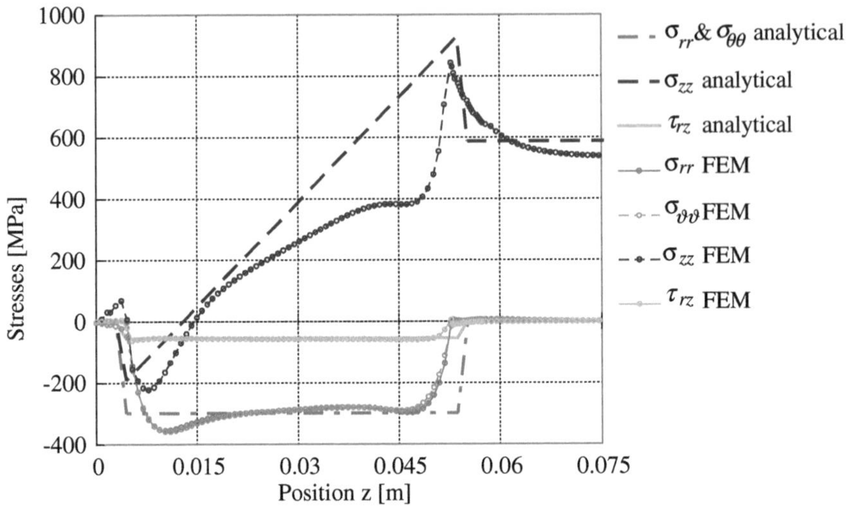

Fig. 2.39 Stress comparison (analytical/FEM) of σ_{zz}, σ_{rr}, $\sigma_{\vartheta\vartheta}$ and τ_{rz} at the outer fibres under maximum tensile load [1]

2.4.7 Numerical Simulation Methods

Numerical simulation methods using the finite element method (FEM) are the current state of the art, when non-isotropic materials, plastic deformation, contact problems and non-linearities have to be considered, as is the case here. The basic principles and approach for applying the FEM in the case of composite long rod insulators under tensile load, as described in [14], will be presented hereinafter.

2.4.7.1 Finite Element Modelling

A complete 3D beam FE model of the entire metal-composite joint was created for numerical simulation using commercial software. Non-linear simulations were carried out using this model for the three phases already described above, namely: (a) crimping of the metal fitting on the FRP rod, (b) relaxation of the crimping pressure, and (c) tensile load of the joint.

For a rod of 18.6 mm, which is typical for composite long rods, the FE model consisted of a dense mesh of hexahedral, square 20-node beam elements. Owing to the rotational symmetry of the arrangement, only one sixteenth of the structure was modelled. The resultant 3D FE mesh consisted of 180 elements for the crimping jaws, 2,800 elements for the metal fitting and 1,407 elements for the FRP rod (Fig. 2.40).

Coulomb's friction was applied both for the contact between the crimping jaws and the metal fitting, and for the contact between the metal fitting and the FRP rod,

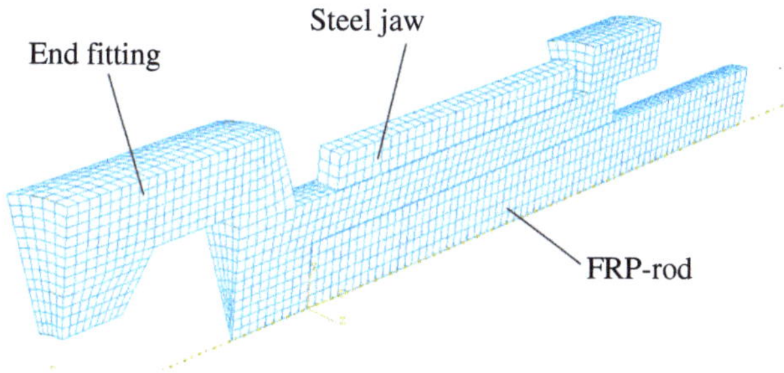

Fig. 2.40 FE model of an insulator segment in the region of the fitting

with a coefficient of friction in both cases of 0.25 being established by suitable tests; this value correlates well with the values from the literature for instances of contact with similar mounting. In addition, a maximum permissible shear stress of the FRP rod of 50 MPa, as resulted from the corresponding tests on the FRP material, was applied at the interface between metal and FRP.

2.4.7.2 Results of the Simulation

The non-linear numerical simulation of the crimping process was carried out using the above-described 3D model for the entire insulator with use of the elastoplastic material law for the metal fitting, as established by way of experiment; linear-elastic behaviour was taken as a basis for the FRP rod.

Figure 2.41 shows the stress distribution in the FRP rod at the interface between FRP and metal. It can clearly be seen that the radial stress σ_{rr} and the circumferential stress $\sigma_{\vartheta\vartheta}$ are not evenly distributed; both experience a double hump at approximately 470 MPa. In the case of average values over the entire interface, this double hump occurs between 350 and 400 MPa. There is a sound theoretical explanation for this double hump: It is caused by the load situation for a cylinder under a band of pressure [15]. The longitudinal stress is distributed more or less parabolically, near the centre of the compression area, with a maximum stress of 320 MPa. When compared to the above axial stresses, the shear stresses are negligible with the exception of the radial-axial shear stress τ_{rz}, which, due to the Poisson effect, has a quasi-sinusoidal distribution with a peak value of $\pm$50 MPa.

In the second phase (known as relaxation), the external pressure is relieved. The stress distributions are similar to those in the crimping phase, but generally with smaller peak values.

In the final phase of numerical simulation, a tensile load is applied to the FRP rod (while simultaneously keeping the end of the metal fitting rigidly fixed) and is increased until the FRP rod, once it has overcome the frictional forces, begins to slide out of the metal fitting. The associated stress distribution at the interface is shown in Fig. 2.42. While distribution of the radial stress σ_{rr} and of the circumferential stress $\sigma_{\vartheta\vartheta}$ is similar to that in the compression phase (Fig. 2.41), the

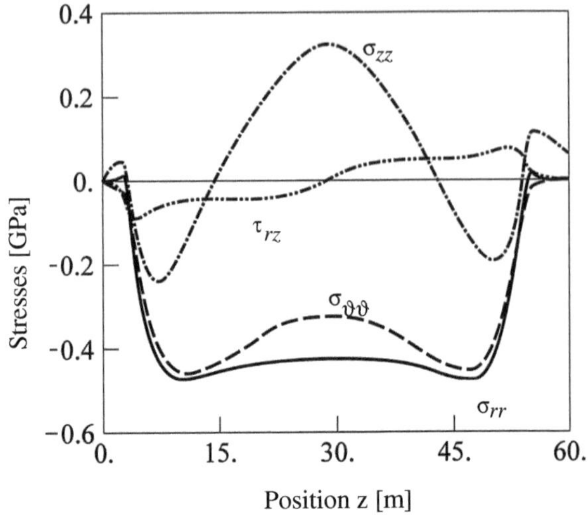

Fig. 2.41 Stresses σ_{zz}, σ_{rr}, $\sigma_{\vartheta\vartheta}$ and τ_{rz} on the outer fibres during crimping, numerical simulation

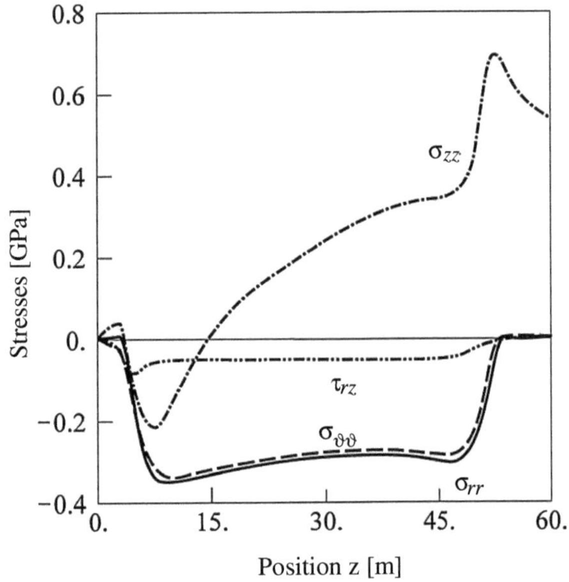

Fig. 2.42 Stresses σ_{zz}, σ_{rr}, $\sigma_{\vartheta\vartheta}$ and τ_{rz} on the outer fibre under maximum tensile load, numerical simulation

maximum values are now approximately 25 % lower. The axial stress σ_{zz} increases up to a value of 690 MPa before falling to 450 MPa, which corresponds to the axial stress in the "free" part of the FRP rod. Lastly, it should be noted that the shear stress τ_{rz} is distributed reasonably evenly, except at the fitting/rod interface where end effects manifest themselves and values of up to 50 MPa are attained, which correspond to the rod's shear failure limit.

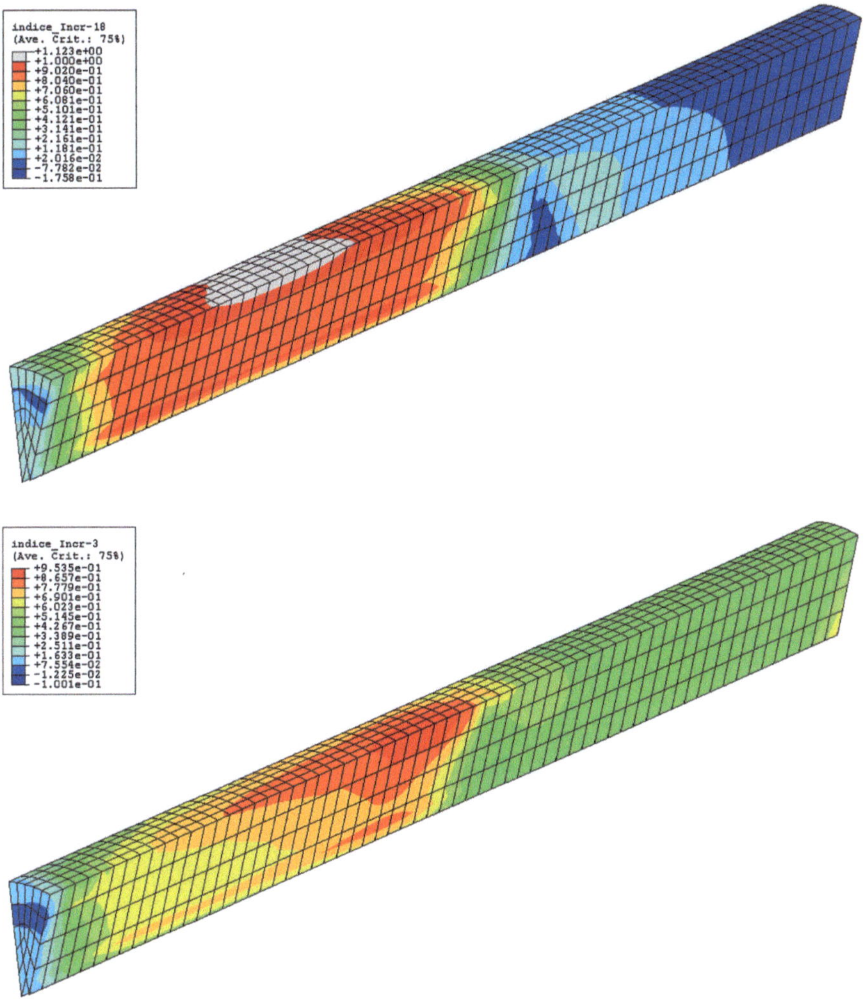

Fig. 2.43 Distribution of the danger factor κ during crimping (*top*) and under maximum tensile load (*bottom*)

2.4.7.3 Numerical Simulation Results in the Event of Failure

It is evident from the above that the critical state of inner stress, which may cause the FRP rod to split in two, occurs during crimping. To confirm this quantitatively, the failure hypothesis developed in [1] is applied to the FRP rod. The material limit values determined in suitable tests are used here for tension (1,038 MPa) and compression (−794 MPa) in the fibre direction of the rod and for tension (32 MPa) and compression (−140 MPa) transverse to fibre direction. Such tests have estimated the limit stress under biaxial load to be −500 MPa. This value is slightly

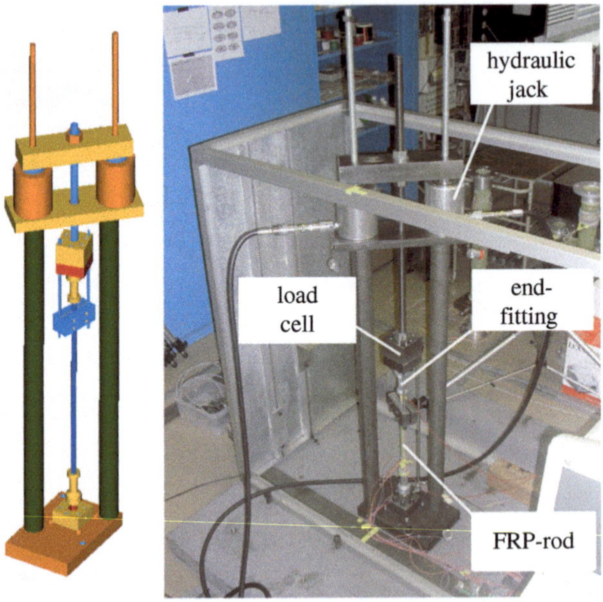

Fig. 2.44 Testing arrangement for the tensile tests

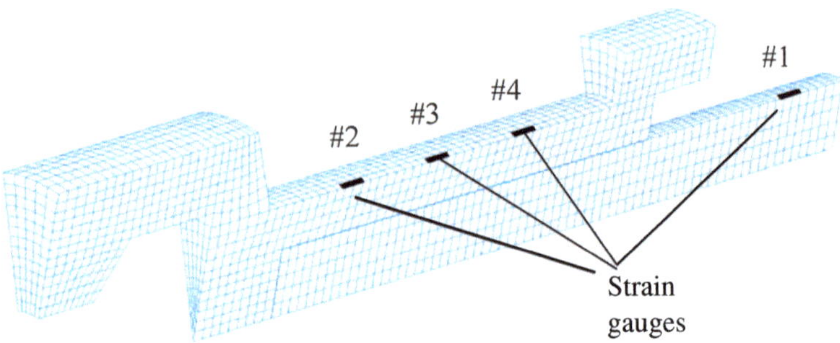

Fig. 2.45 Arrangement of the DMS on the fitting and FRP-rod

higher (absolutely) than the numerically determined value of −470 MPa for radial stress, which should be relatively close to the limit stress at which the first cracks occur in the FRP rod.

The state of stress in the FRP rod is determined using a danger factor κ, the inverse $1/\kappa$ of which represents the maximum possible value by which the load on the insulator (i.e. the crimping pressure during the crimping process or the tensile load during pull-out) can be multiplied until the FRP rod fails. This danger factor is shown in Fig. 2.43. It is important to note that a danger factor of 1.12 occurs close to the metal fitting/FRP rod interface during crimping (see Fig. 2.4, top). However, this transgression (which occurs if the danger factor is greater than one)

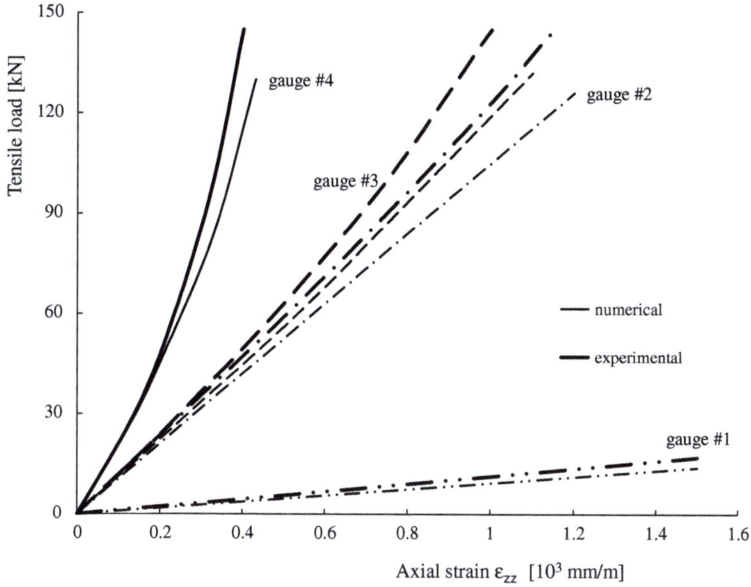

Fig. 2.46 Comparison of measurement/numerical simulation

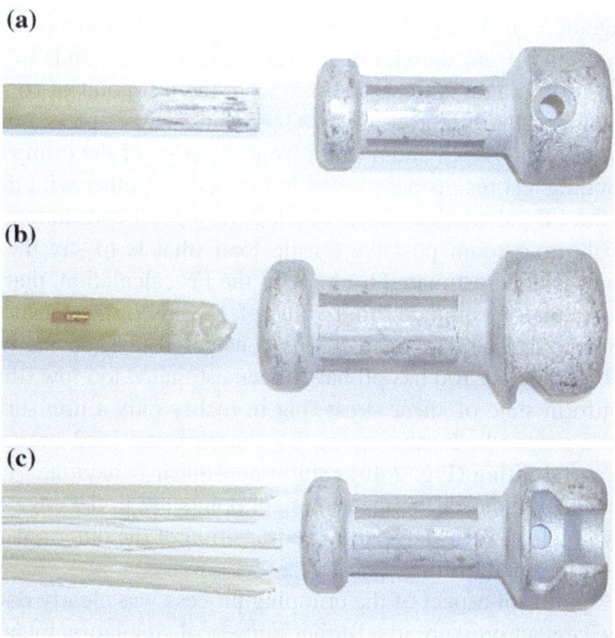

Fig. 2.47 **a** "Pull-out" (preferred), **b** Separation failure, **c** Delamination

is highly localised, thus making it improbable that a latent crack will propagate. The progression of the danger factor during tensile loading and, in particular, when the rod begins to slide out of the metal fitting (which occurs when the maximum bearable tensile load is applied) is shown in Fig. 2.43 (bottom); the values shown here are lower than those during crimping. In this case, the danger factor in the outer glass fibres of the FRP rod reaches relatively high values of 0.95, but therefore still always remains below the threshold value of 1 for the onset of material failure, which also explains the fact that the FRP rod slides out "cleanly" from the fitting during the tensile test without experiencing any delamination (see also Fig. 2.47a).

2.4.7.4 Test Results

The above-described calculation models were validated over a series of tests. Two variables which could be measured relatively accurately, namely some of the measurements of accessible stresses (via the strains) and the maximum pull-out force, which was established during the tensile test, were used for this purpose.

The test rig (Fig. 2.44) was formed of two servo hydraulic cylinders capable to deliver a total of 200 kN, a linear variable differential transformer (LVDT), a load cell and a number of strain gauges (DMS), which were applied to different points of the FRP rod and to the surface of the metal fitting.

The arrangement of the DMS can be seen in Fig. 2.45: DMS#1 was applied to the FRP rod, 88 mm from the end of the rod, which corresponds to 23 mm from the edge of the crimp, that is to say 23 mm from the last point of contact between the metal fitting and the FRP rod. DMS #2, #3 and #4 were placed on the surface of the metal fitting, at 12, 28 and 44 mm from the edge of the crimp respectively. The corresponding test results are plotted in Fig. 2.46 together with the load-strain curves from the FE calculation, from which a very good correlation can be seen, even though the maximum possible tensile load (that is to say the tensile load during pull-out) is underestimated by 11 % in the FE calculation, that is a value of 132 kN was calculated compared to a value of 148 kN in the tensile test.

This difference can be explained if it is considered that, on the one hand, the shear strength of the FRP rod has probably been estimated too low (the calculation assumes a uniform state of shear stress, but in reality only a thin surface layer of the FRP rod is stripped). Furthermore, the experimental load-strain curves with regard to the metal fitting (Fig. 2.46) exhibit non-linear behaviour which becomes more pronounced, the closer the measurement points are to the edge of the crimp, which clearly points towards a progressive transition of the outer tensile load from the FRP rod to the metal fitting.

The most significant aspect of the crimping process was clearly documented by increasing the contact pressure in a further numerical simulation by approximately 25 %. The danger factor reached values of approximately 1.27, which inevitably results in irreversible material damage to the rod; the tensile strength thus falls sharply to 56 kN. The associated damage pattern is shown in Fig. 2.47b; there is a

Table 2.4 Value ranges of the parameters for the sensitivity analysis

Parameter	Coefficient of friction (%)	Tolerances (%)	Crimp length (%)	Wall thickness (%)	Reduction in diameter (%)
Lowest value	−20	−60	−15	−15	−25
Highest value	+60	+60	+15	+15	+25

clean separation failure of the FRP rod. If, by contrast, the crimp pressure is increased to a slightly lesser extent, for example by approximately 15 %, delamination of the FRP material is observed (Fig. 2.47c). If correctly crimped, the rod should always pull out "cleanly" from the metal fitting (Fig. 2.47a).

2.4.7.5 Sensitivity Analysis

A sensitivity analysis of the primary variables for a crimped fitting makes it possible to validate the values selected originally, and also makes it possible to optimise the load bearing capacity of the fitting/insulator assembly. In [14], the effect of the following dimensions was examined: Coefficient of friction between FRP rod and fitting, tolerance between FRP rod and fitting, length of the crimp zone, wall thickness of the fitting, and reduction in diameter of the FRP rod during the crimping process. The variation range of these parameters are listed in Table 2.4, each one being divided into five intervals and the corresponding calculations were made using the 3D FE Model presented above.

The results are summarised in Fig. 2.48 with the von Mises stress as a reference for the load of the FRP rod. It can be seen that the parameters to which this assembly reacts most sensitively are the tolerance between rod and fitting, the crimp length, and the reduction in diameter of the FRP rod during the crimping process. For example, if the tolerance is increased by 60 %, the von Mises stress reduces by 16 or 12 % (crimping and max. tensile load respectively), but at the same time the maximum pull-out force reduces by 11 % from 132 to 118 kN. On the other hand, if the crimp zone is extended by 15 %, the pull-out force increases by 20 % from 132 to 158 kN, albeit at the cost of a von Mises stress which is approximately 13 % higher in the event of rod pull-out. The rod is damaged internally to such an extent that there is forced rupture upon pull-out (Fig. 2.47b). As expected, the sharp reduction in rod diameter during crimping by 25 % results in a massive increase in the von Mises stress by 28 %, but only in a marginal increase in the maximum tensile load by 5 % from 132 to 139 kN. A reduction in crimping with regard to rod diameter by 25 % (compared to the initial state of 100 %) results in a dramatic decrease in the failing load of the insulator by a whole 28 %, that is to say from 132 to 95 kN. By contrast, neither the coefficient of friction between FRP rod and metal fitting, nor the wall thickness of the fitting

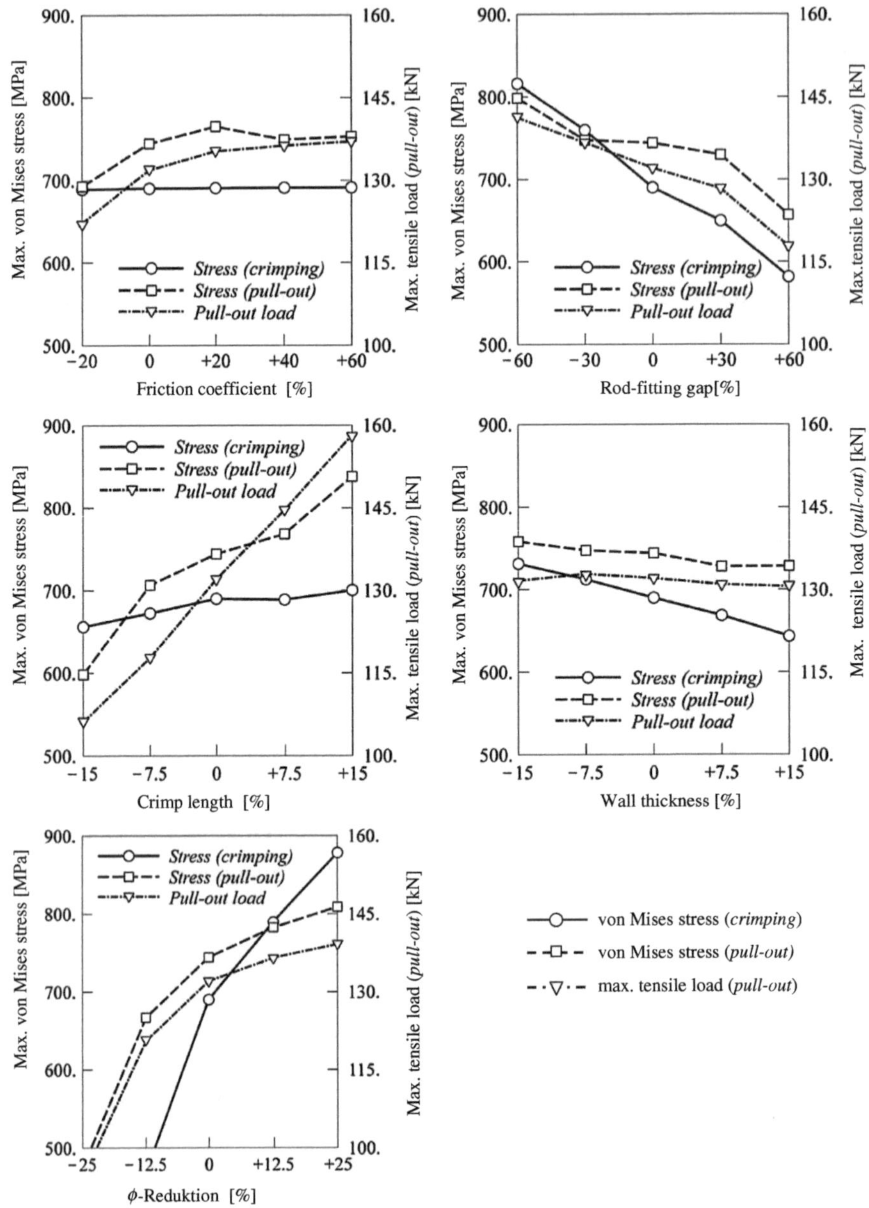

Fig. 2.48 Results of the sensitivity analysis

have a substantial effect on the stress regime of the rod or on the failing load of the insulator, although a moderate decrease in the von Mises stress (during crimping) is to be observed with an increase in the wall thickness of the fitting.

References

1. Prenleloup A (2008) Analyse de l'état de contrainte et de l'endommagement d'assemblages sertis en matériau mixte métal-composite sollicités en traction ou en flexion. PhD Thesis No. 4005. Lausanne Ecole polytechnique fédérale de Lausanne
2. De Tourreil C, Roberge R, Bourdon P (1985) Long-term mechanical properties of high voltage composite insulators. IEEE Trans Power Apparatus Syst PAS- 104(10):2918–2921
3. De Tourreil C (1990) Response of composite insulators to dynamic mechanical loads. IEEE Trans Power Delivery 5(1):379–383
4. Lumb C, Papailiou KO (1997) Unterschätzte Dauerkraft der Verbundisolatoren, SEV/VSE Bulletin 7/97
5. Paris L, Pargamin L, Dumora D, Parraud R (1994) Rating of composite suspension insulators related to the long-term mechanical strength of rods. IEEE Trans Power Delivery 9(4):2055–2063
6. De Tourreil C (1994) Discussion to [5]. IEEE Trans Power Delivery 9(4):2062
7. Ammann M, Papailiou KO, Ansorge S, Schmuck F (2008) Zur Bewertung von Verbundisolatoren und deren relevanten Interfacebereichen nach Freilufteinsatz. ETG Fachtagung Grenzflächen in elektrischen Isoliersystemen (ETG FB-112), Sept
8. CIGRE Technical Brochure 251 (2004) The mechanical behaviour of conductors and fittings (CD + Brochure). Paris
9. Wankowicz J, Bielecki J, Struzewska E (2006) Damage limit of composite long rod insulators subjected to cyclic loads, paper B2-308. CIGRE Session, Paris
10. Papailiou KO (1997) On the bending stiffness of transmission line conductors. IEEE Trans Power Delivery 12(4):1576–1588
11. De Weck P, Ammann M, Papailiou KO (1990) Two years vibration measurements and their evaluation for an optical ground wire (OPGW) installed on a 400 kV transmission line, paper 22-203. CIGRE Session, Paris
12. Yoshida S (2006) Fatigue characteristics on composite long rod and line post insulators. Discussion contribution SC B2, CIGRE Session, Paris
13. Fan J, Liang X, Yin Y, Wang C, Chen L (2000) Application of acoustic emission technology on structure design and quality control of composite insulators. In: Proceedings of the IEEE 6th international conference on properties and applications of dielectric materials, Xi'an, China
14. Preneleloup A, Gmür T, Papailiou KO, Botis J (2006) Acoustic emission study and strength analysis of crimped steel-composite joints under traction. Compos Struct 74(2006):370–378
15. Timoshenko S, Goodier JN (1970) Theory of elasticity. McGraw Hill, New York

Chapter 3
Composite Post Insulators

Symbols and Abbreviations

AC	Alternating current
CFL	Cantilever failing load
CIGRE	Conseil International des Grands Réseaux Électriques (International Council on Large Electric Systems)
DC	Direct current
DLL	Damage limit load
FEM	Finite element method
FE model	Finite element model
FE mesh	Finite element mesh
FRP	Fibre reinforced plastic
Horizontal V	Insulated cross-arm
IEC	International Electrotechnical Commission
ISO	International Organisation for Standardisation
LVDT	Linear variable differential transformer
LPC	Last point of contact
MDCL	Maximum design cantilever load
ROW	Right of way
SBB	Schweizerische Bundesbahnen (Swiss Federal Railways)
SCL	Specified cantilever load
β	Horizontal angle of the post insulator
Δf	Change of the deflection over time
ν_{LT}	Poisson number "axial-transverse" to the FRP rod
σ	Max. allowable stress (tensile or compressive)
$\sigma_{\vartheta\vartheta}$	Circumferential stress in the FRP rod
$\hat{\sigma}_{L,Druck}$	Axial compressive strength of the FRP material
σ_{max}	Ultimate bending stress in the FRP rod
σ_{rr}	Radial stress in the FRP rod
$\sigma_{rr,max}$	Ultimate radial bending stress in the FRP rod
σ_{zz}	Axial stress in the FRP rod

The original version of the chapter was revised: Incorrect function in equation has been corrected. The correction to the chapter is available at https://doi.org/10.1007/978-3-642-15320-4_12

K. O. Papailiou and F. Schmuck, *Silicone Composite Insulators*, Power Systems, DOI: 10.1007/978-3-642-15320-4_3, © Springer-Verlag Berlin Heidelberg 2013

τ_{max}	Shear strength of the FRP material
$\tau_{max}(z, y)$	Shear stress in the FRP rod during bending
$\tau_{max}(z, y = 0)$	Maximal shear stress (at the neutral plane of the FRP rod)
φ	Maximal allowable angle of rotation during torsional loading of the insulator
d	Half crimp length
d	Rod diameter
D	Rod diameter
D_{min}	Minimal required diameter of the post
E	Modulus of elasticity (Young's modulus) of the FRP rod
V, V'	Vertical load
f	Deflection of the post
f_{max}	Maximal deflection of the post
F	Bending load
F_b	External load
F_{max}	Ultimate bending load of a composite post insulator
G	Shear modulus
H, H'	Compressive insulator load (perpendicular to the tower axis)
I	Moment of inertia of the FRP rod
I_p	Polar moment of inertia of the FRP rod
l	Length of the post (insulator)
L, l	Bending length
L_o	Position of the strain gauge on the FRP rod
L_p	Crimp length of the end fitting
M, M_Z, M_H	Crimp length of the end fitting
M_o	Reaction moment at the fixed end
M_t	Torsional moment
p_1, p_2	Peak values of the line loads
R	Radius of the FRP rod
Ro	Reaction force at the fixed end
s	Lever arm
t	Time
T, T'	Transverse load
W	Section modulus of the FRP rod
W_p	Polar section modulus of the FRP rod
y	Radial coordinate of the rod cross-section
z	Axial coordinate of the FRP rod
Z	Tensile insulator load

Fig. 3.1 230 kV composite post insulator

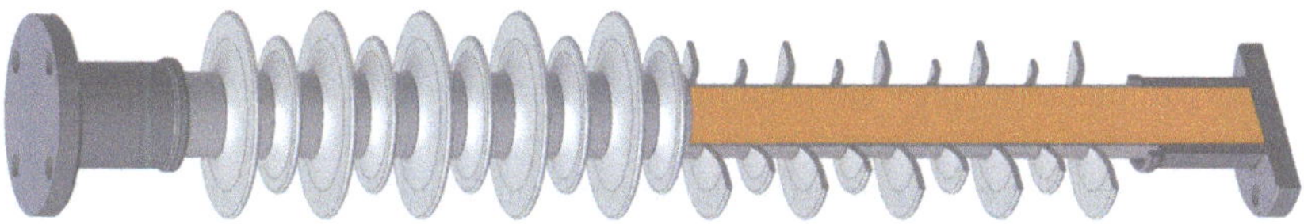

Fig. 3.2 Construction of a composite post insulator

3.1 Key Properties

In contrast to composite long rods, composite post insulators are subjected mainly to loads which act perpendicular to the longitudinal axis of the insulator and thus produce bending stresses over the cross-section of the insulator. These bending stresses reach their maximum value at the point at which the FRP rod is fixed in the metal fitting (circled in red in Fig. 3.1), the latter being located where the insulator is attached to the support structure, whether this be the cross-arm of a steel lattice tower for an overhead transmission line, the coupling plate of a tubular steel or concrete pole, or the platform in a pole-mounted substation. Calculation of these bending stresses is discussed in detail in Sect. 3.9.

As a result of this bending load, composite post insulators use FRP rods of substantially greater diameter compared to composite long rods. Whereas a diameter of 32 mm (1.5 inches) is almost always sufficient for composite long rods, composite post insulators require FRP rods having a diameter of up to 150 mm (6 inches) in some applications (for example 400 kV post insulators as

Fig. 3.3 Medium-voltage post insulator

busbar supports in outdoor substations or as struts of insulated cross-arms). Figure 3.2 shows the construction of a composite post insulator.

3.2 Applications of Composite Post Insulators

Historically, composite post insulators were first used in distribution networks up to 35 kV. In that case, they are used in large quantities as conductor supports fixed on wooden or steel cross-arms often in a horizontal position (Fig. 3.3).

Line post insulators in transmission networks normally have a nominal voltage above 69 kV. Similarly to post insulators used in distribution networks, they are installed vertically or horizontally. Since they have to bear considerably greater loads than post insulators used in distribution networks, they use FRP rods typically measuring 63 mm (2.5 inches), 76 mm (3 inches) and (more rarely) 88 mm (3.5 inches) in diameter.

Line post insulators contribute to compaction and to a more attractive appearance of the line and also to a considerable reduction in the necessary line corridor (right of way)—advantages which are not to be underestimated at a time when authorisation procedures for new lines are rather protracted. Figure 3.1 shows an example of horizontal installation in the case of a 230 kV overhead transmission line.

By contrast, a vertical installation is shown in Fig. 3.4 in the case of the "trident" arrangement (*trident* refers to the three-pronged spear of the God of the sea, Poseidon, in Greek mythology).

At voltages greater than 220 kV (and in most cases even greater than 132 kV), the mechanical loads are so strong that they cannot be withstood by a single horizontal post insulator.

The post insulator is then anchored using a tension insulator. This design is called an insulated cross-arm (also known as a braced line post or horizontal V) (Fig. 3.5). In this case, the post insulator is primary loaded by buckling. This

Fig. 3.4 132 kV line with trident cross-arms in scottish highlands

Fig. 3.5 Horizontal post insulator in an insulated cross-arm

design is typical for compact lines, which due to their increasing importance will be discussed in greater detail in a separate chapter.

A further specific application in which composite insulators have to withstand compression loads is interphase spacers, which are used both for line compaction and to control conductor galloping. Chapter 5 is dedicated to interphase spacers and the respective specific loads to which they are subject (forces from galloping, short circuit or ice shedding).

Fig. 3.6 150 kV bus bar post insulator

Whilst the above-mentioned applications of composite post insulators in overhead transmission lines at all voltage levels has been the "state of the art" for many years, their use in outdoor substations as station posts is still relatively new. In spite of this, they are encountered as bus bar insulators (Fig. 3.6), apparatus insulators (isolators, resonance coils) and also as support insulators for cable terminations, both for high voltage as well as in relatively large quantities for some time already, for medium voltage.

A similar application is their use in jumper locations in high-voltage overhead transmission lines. The first application for 800 kV DC [1] was also observed recently, wherein the high loads required a special triple-leg construction measuring over 10 m tall and having a specified mechanical load (bending) of 10 kN (Fig. 3.7).

Composite insulators were used quite early on in railway engineering, specifically as post insulators for pantographs or conductor rails, but above all as an integral component of the arms of the catenary supports, both in tunnels (Fig. 3.8) and over open track (Fig. 3.9).

Fig. 3.7 800 kV triple-leg construction

Fig. 3.8 15 kV railway insulator in the Lötschberg tunnel

Fig. 3.9 25 kV railway insulator over a high-speed track

3.3 Behaviour of Composite Post Insulators with Bending

3.3.1 General Considerations

As explained in the section above, composite post insulators are subject primarily to bending. The mechanical behaviour of composite post insulators under such bending loads will be discussed in greater detail hereinafter, paying particular attention to the failure mode and development over the years of test procedures for such insulators. Although insulators are generally primarily designed for static loads and are tested in this respect, post insulators, which may be increasingly subject to vibratory loads (for example horizontal post insulators with vibrating conductors), have been tested extensively with regard to such dynamic loads. An analytical tool will also be presented in this chapter, the aim of which is to make it possible to design such insulators as realistically as possible on the basis of relative simple formulas, and to facilitate the understanding of the complex mechanical processes during bending of the insulator.

The chapter will conclude with a presentation of an FE simulation concept and a brief explanation of current test procedures.

3.3.2 Adoption of a Damage Limit

In recent years, composite insulators and composite post insulators in particular have increasingly been able to replace conventional insulators made of porcelain

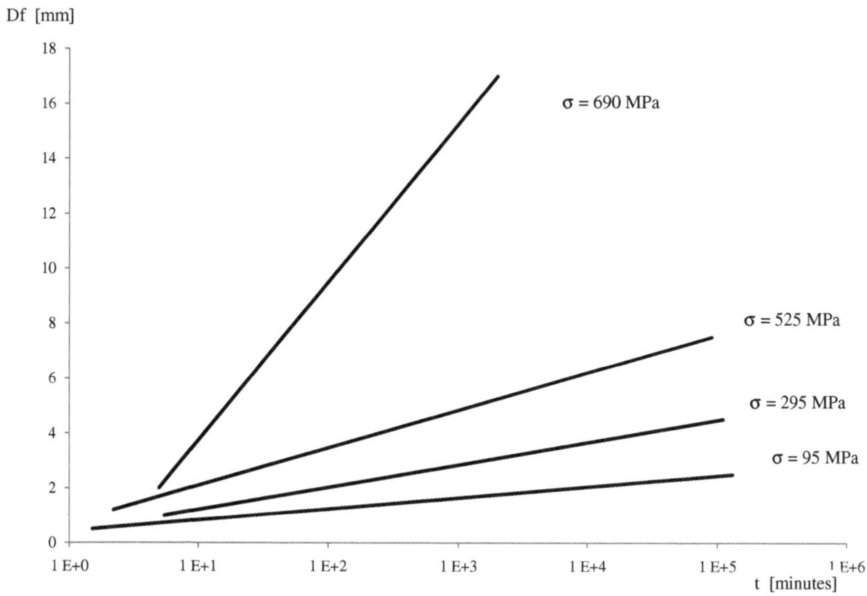

Fig. 3.10 Change in deflection over time with bending stress as a parameter [2]

and glass. A specific feature of composite post insulators is that their load-bearing component, that is to say the FRP rod is ductile, which is why there is no physical separation of the insulating body in the event of mechanical failure of the insulator (for example as the case with a porcelain insulator), and why there is thus no risk of a catastrophic line dropping. This also means that the "failure load" cannot be defined so clearly. In particular, the FRP rod of the insulator may be damaged before there is a noticeable change to the bending behaviour of the insulator. This sort of damage (for example microcracks in the FRP material) might not influence the short-term behaviour of the insulator, but may have a negative impact on its service life, since partial discharges may be triggered in these microcracks and have a negative impact on the electrical (and ultimately mechanical) strength of the material. It was therefore important to develop a test method in order to establish the mechanical loads at which these microcracks start to form in the FRP rod.

This method is based on the damage limit concept, which was presented for the first time in [2]. This concept builds on the observation that FRP exhibits considerable creep as soon as it has been subjected to a constant load. In the case of composite post insulators subjected to bending load, this leads over time to a noticeable increase in their maximum deflection.

To better understand this phenomenon, 1 m long FRP rods measuring 63 mm (2.5 inches) (in this case a composite post insulators before application of the silicone housing) were subject purely to bending load for a period lasting a few weeks. The change Δf of the deflection was measured over time t. If the

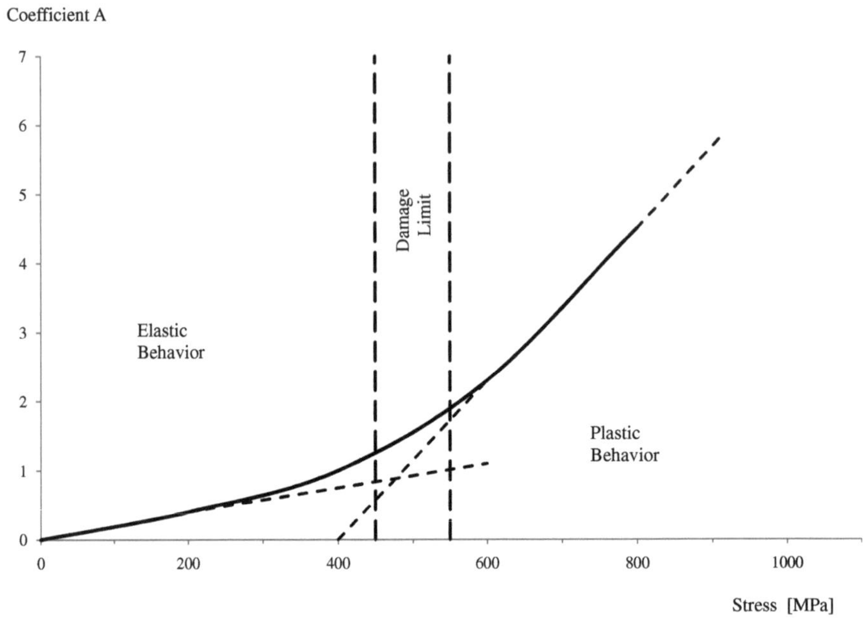

Fig. 3.11 Creep coefficient over bending stress [2]

measurement results are plotted semi-logarithmically (Fig. 3.10), there are a family of curves which obey the following formula: $\Delta f = A \log t$.

The values of the coefficient A thus established are plotted in Fig. 3.11. The inclination of the respective curve changes with (nominal) bending stresses of approximately 500 MPa, calculated at the point of fixing. In the lower stress range, the FRP rod returns to its initial position after a certain relaxation time. For bending stresses above 500 MPa, the creep coefficient A exhibits non-linear behaviour. This means that greater loads (but in any case still below the failing load of approximately 800 MPa) lead to failure of the FRP rod within a few days. If these insulators are relieved after a few hours, their deflection indeed noticeably subsides, but they retain a permanent deformation, which is dependent on the magnitude and duration of the load applied.

To summarise, under just bending load (and at ambient temperature), FRP rods, and therefore composite post insulators, exhibit similar behaviour to a ductile, metal material, that is to say there is purely elastic deformation at lower loads and quasi-plastic deformation at higher loads. The bending stress which separates these two ranges is called the damage limit stress (Fig. 3.11).

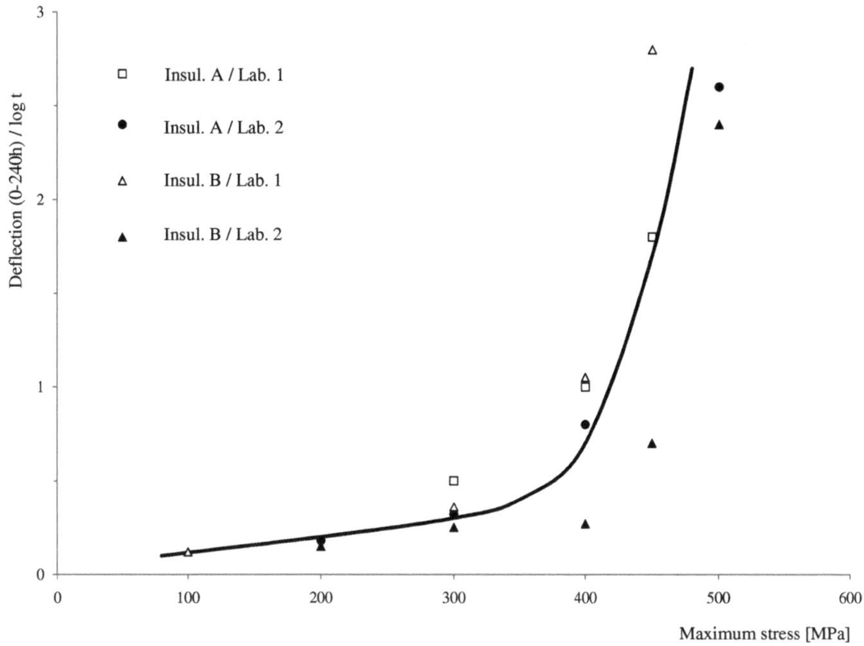

Fig. 3.12 Deflection over the maximum bending stress for "short" insulators

3.3.2.1 Tests for Determining the Damage Limit

This concept was adopted within CIGRE working group 22.3 (insulators), with the objective of developing a workable test method [3]. A number of tests were carried out on different types of insulators for this purpose.

In order to better compare the different types of insulators tested, the results were presented according to the nominal bending stress σ, which can be illustrated using the known formula of a beam fixed at one end:

$$\sigma = \frac{F_b \cdot l}{\pi \cdot d^3/32}$$

wherein:

σ maximum tensile or compressive stress
F_b external load (bending)
l bending length
d rod diameter

It must be emphasised here that the bending stress thus calculated should only be used for comparison. As documented in Sect. 3.10, the actual state of stress in the insulator is much too complex to be illustrated by the simple formula above.

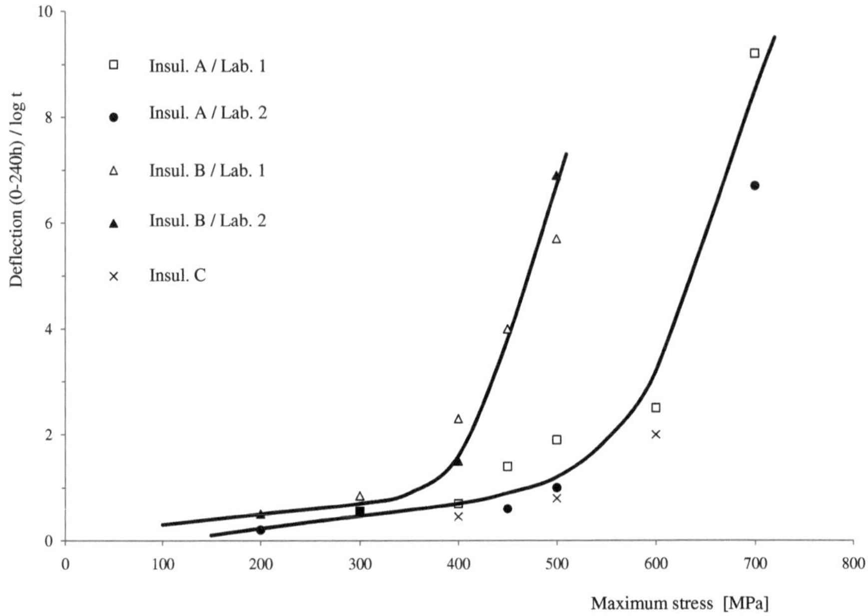

Fig. 3.13 Deflection over maximum bending stress for "long" insulators

In the first series of tests, the damage limit was established both by measuring the change in deflection over time, and by visual inspection of the surface of the rod for cracks, which is why the test insulators were manufactured without a silicone housing. In the interests of better comparability, all insulator manufacturers involved in the test provided insulators of equal dimensions, that is to say 45 mm rod diameter and 0.3 m (bending) length ("short insulators"), and 63 mm rod diameter and 1.2 m bending length ("long insulators").

These tests were carried out using the methods described in Sect. 3.2. Over the course of these (non-destructive) methods, the parameter $\Delta f/\log t$ (where t lasts from 0 to 240 h) is established as a result of 10 days of continuous application of a constant bending load, and the damage limit is determined therefrom.

Since the insulators were not provided with a housing, it was possible to check the surface of the FRP rods for visible damage once the load had been relieved.

The damage limits thus established are presented in Fig. 3.12 for "short" insulators and in Fig. 3.13 for "long" insulators. It can be seen that, with values between 325 and 425 MPa, the damage limits for short insulators are lower than for most long insulators, which have values between 475 and 600 MPa.

After each test, the insulators were also examined visually for localisation of surface defects. Cracks on the surface of the rod were observed at bending stresses above 400 MPa and up to 450 MPa, indicating delamination of the composite material. In most cases there was good correlation with the measurements

Fig. 3.14 Crack on the compressed side of the rod

established for change in deflection, that is to say no damage to the surface of the rod was visible below the damage limits established by this measurement.

In the second series of tests, commercially available insulators which had already been provided with a silicone housing were examined. There were two rod diameters of 45 and 63 mm, and the bending length was between 0.41 and 0.71 m. The insulators were installed and subjected to realistic bending loads. After 4.5 days, the load was relieved and the FRP rod, including metal fitting, was carefully cut open along the plane defined by the rod axis and the direction of load. The interface was polished and examined for the presence of cracks.

The test results for the insulators having a silicone housing clearly showed that both types of rod (having a diameter measuring 45 or 63 mm) were damaged below the previously established damage limit. The first cracks are almost perpendicular to the longitudinal axis of the FRP rod and are located on the compressed side of the rod, almost exactly at the point where the metal fitting ceases to contact the rod (Fig. 3.14). It is therefore easy to see why these cracks remained hidden during the deflection measurements and the subsequent inspection of the surface of the rod. The physical background of these cracks is explained in Sect. 3.11.

It is interesting to note that whilst an insulator without a silicone housing exhibited such cracks on the compressed side at 400 MPa, no cracks were formed in two similar insulators, albeit shedded, up to 500 MPa. This indicates the positive effect of a close-fitting silicone shed in the vicinity of the critical point formed by the FRP rod/end fitting/shed.

The findings detailed above can be summarised as follows:

- The maximum bending load of a post insulator during operation is defined as the bending load which can be applied to the insulator without any damage to the

FRP rod of the insulator. Extensive tests and inspections have to be carried out by the manufacturer of the insulator in order to establish this load.
- If composite post insulators experience bending loads above the maximum bending load (see above), cracks will start to form in the FRP rod, more specifically in the vicinity of the base end fitting and before it is possible to detect any external signs in the mechanical behaviour of the insulator. These cracks can only be found as a result of special "internal" inspection, as described above.
- With the exception of "long" post insulators, the failure begins on the compressed side, just within the base end fitting. In long post insulators however, longitudinal delamination of the FRP rod may occur beforehand (see also Sect. 3.11).

These findings have been implemented in international regulations as design tests. In practice, composite post insulators are designed and tested in accordance with IEC Standard 61952 "Composite Line Post Insulators". These following test loads are defined in this standard:

- The SCL (**Specified Cantilever Load**) is the bending load during testing which an insulator can withstand at the conductor side. This value is specified by the manufacturer. It is to be noted that the SCL is to be considered a withstand value. This means that the insulator can withstand this load during testing.
- The CFL (**Cantilever Failing Load**) is the maximum load which is achieved during testing of the insulator; it should be higher than the specified cantilever load.
- The MDCL (**Maximum Design Cantilever Load**) is the load above which damage to the core is initiated and which is the limit load for the loads occurring during operation. The MDCL is specified by the manufacturer on the basis of the above-mentioned IEC Standard. The associated test procedure will be described in the following section. The MDCL is normally approximately 25 % below the actual DLL (**Damage Limit Load**).

3.3.2.2 Test for Determining the Maximum Design Cantilever Load

Three insulators provided with the standard end fittings are to be tested by the manufacturer. The total length of the insulators must correspond to 15–18 times the core diameter unless the manufacturer does not possess equipment suitable for the manufacture of insulators of this length. In this case, the insulator length must be as close as possible to the range stipulated above.

The base end fitting must be fixed rigidly during this test. The insulator is gradually loaded up to 1.1 times the MDCL at a temperature of 20 ± 10 °C. This load is maintained for 96 h. The load shall be applied to the insulators at the conductor position, perpendicular to the direction of the conductor, and perpendicular to the core of the insulators.

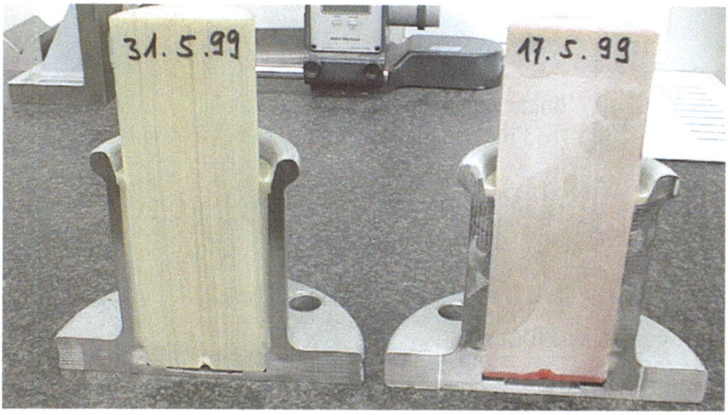

Fig. 3.15 No cracks, "test passed": *left* without dye, *right* with dye [4–12]

It is recommend that the deflection of the insulator at the point of the load application be recorded at 24, 48, 72 and 96 h in order to provide additional, useful information.

The following steps are necessary once the load has been removed:

- Visual inspection of the base end fitting for cracks or permanent deformation and
- Measurement of residual deflection

Each insulator is cut in two at a distance of approximately 50 mm from the base end fitting at an angle of 90° to the axis of the core, then the bottom fitting is cut into two halves in the longitudinal direction in the plane in which the bending load was previously applied. The interfaces have to be smoothed using an abrasive cloth (particle size 180). The interface halves are then inspected visually for cracks and delamination. Dye penetration testing is then carried out in accordance with ISO 3452 in order to detect any cracks at the interfaces. Failure while the load is applied, cracks, permanent deformation or defective threads in the end fitting, or the presence of cracks or delamination in the rod of a test specimen constitute failure of the test.

The dye penetration test is extremely suitable for "discovering" fine cracks and for providing visual confirmation of the maximum design cantilever load (MDCL). The test specimen in Fig. 3.15 has passed the test, whereas the test specimen in Fig. 3.14 has clearly failed.

3.3.2.3 Determining the MDCL and CFL in the Same Test

The test procedure just described for determining the MDCL is rather cumbersome (duration: four days) and is thus unfortunately often "overlooked". This is undesirable because the point of the test is to ensure the quality of the metal fitting,

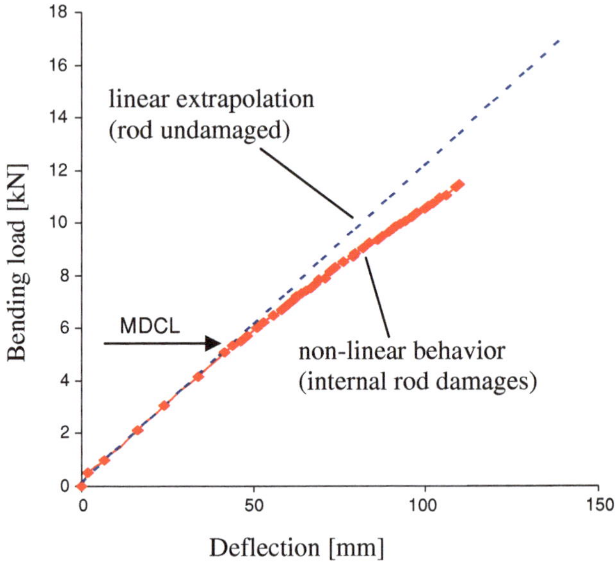

Fig. 3.16 Load vs. deflection diagram for determining the MDCL and CFL in the case of a 38 mm post insulator

FRP rod and also of the crimping process provided by the manufacturer of the insulator. For this reason and since the MDCL plays a rather key role for faultless operation of the composite post insulator, a simple short-time test for determining MDCL was sought. This was achieved following extensive testing (discussed in Sect. 3.10) which showed that the load–deflection curve is suitable for establishing the MDCL [4]. In practice, it is thus sufficient to determine the MDCL from the slope increase in the load–deflection curve, that is to say the point at which the non-linearity of the curve starts (Fig. 3.16). The cantilever failing load (CFL) can also be established by continuing this relatively simple test until failure of the insulator.

3.4 Manufacturer Specifications with Regard to the SCL/MDCL Concept

The catalogues of a number of leading insulator manufacturers were inspected and their approaches for assessment of composite post insulators were compared. It was found out that there is a certain level of correlation between the (numerical) ratios of the test loads described above, as follows:

- The cantilever failing load (CFL) is normally 15–20 % higher than the specified cantilever load (SCL)

Fig. 3.17 15 kV railway insulators (l-r: silicone, porcelain, cast resin)

Table 3.1 Maximum allowable (nominal) stresses (in MPa) at maximum design cantilever load for various manufacturers

Rod diameter	44 mm	63 mm	76 mm	88 mm
Manufacturer A	345–320	345–375		330
Manufacturer B	390	345–375	380–365	
Manufacturer C	350–330	330–270		

- The damage limit load (DLL) is approximately 20–40 % smaller than the specified cantilever load (SCL)
- The maximum design cantilever load (MDCL) is approximately half the specified cantilever load (SCL)
- The damage limit load (DLL) is approximately 25 % higher than the maximum design cantilever load (MDCL)

The maximum allowable (nominal) stresses (in MPa) at maximum design cantilever load given below was calculated from the MDCL values and from other specifications in the manufacturers' catalogues on the basis of the formula for bending stress σ as provided in Sect. 3.6.2.1 (Table 3.1):

With the ratios specified above, stresses at the specified cantilever load are in the range between 540 and 780 MPa.

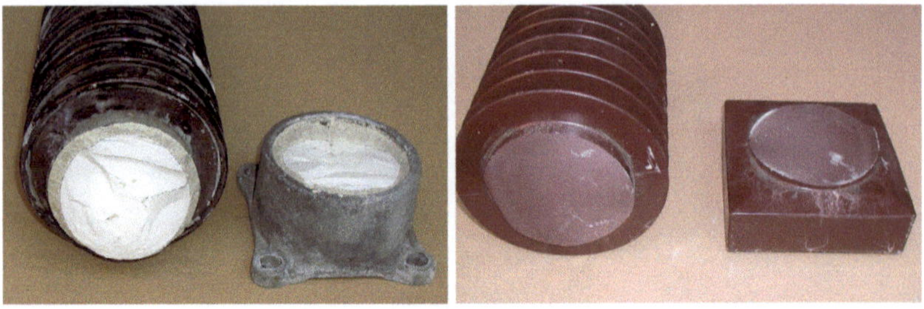

Fig. 3.18 Failure patterns of the porcelain post insulator (*left*) and of the cast resin post insulator

3.5 Safe Failure Mode of Composite Post Insulators

A key difference between composite line post insulators and those made of porcelain was discussed in [5] and is called the "safe failure mode". This concept was developed within the scope of a test programme which included three different types of insulator as used by the SBB—Schweizerische Bundesbahnen (Swiss Federal Railways) in their catenary support arms (phase-to-earth voltage: 15 kV AC, single phase). More specifically, these were a silicone composite insulator, a porcelain insulator (13 years of operation) and a cast resin insulator (2 years of operation), Fig. 3.17.

All three insulators were subjected, under the same conditions, to bending load until failure (this does not mean necessarily breakage). The cantilever failing load (CFL) was verified by the type test in IEC Standard 61952 as cited above. The test procedure has similarities to the corresponding test for establishing the cantilever failing load of porcelain post insulators. The bending load is increased rapidly, but uniformly from zero to approximately 75 % of the specified cantilever load (SCL). The test load is then increased within a time window between 30 and 300 s until failure of the insulator core (FRP rod) or of the base end fitting. This is a short-time test compared to the 96 h test described earlier for checking the maximum design cantilever load (MDCL),. It must be ensured that, where possible, the direction of load is perpendicular to the axis of the (unloaded) insulator during the entire test, which is not always easy due to the relatively large deflection caused by the test conditions.

The test results can be summarised as follows:

- The porcelain post insulator failed at 20 kN. The failure pattern shows a physical separation of the insulator into two parts, as expected, Fig. 3.18, to the left. The failing stress was 182 MPa.
- The cast resin post insulator demonstrated similar behaviour. A bending load of 20 kN was reached before failure, when the insulator broke into two parts, Fig. 3.18, to the right. The failing stress was calculated to be 133 MPa in this case.

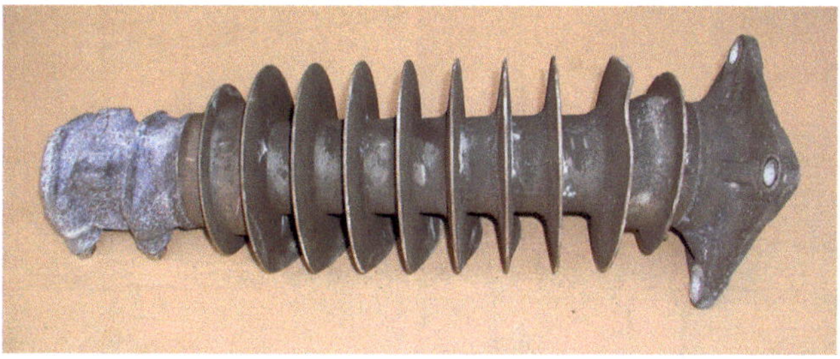

Fig. 3.19 Failure of the silicone post insulator: there is no physical separation

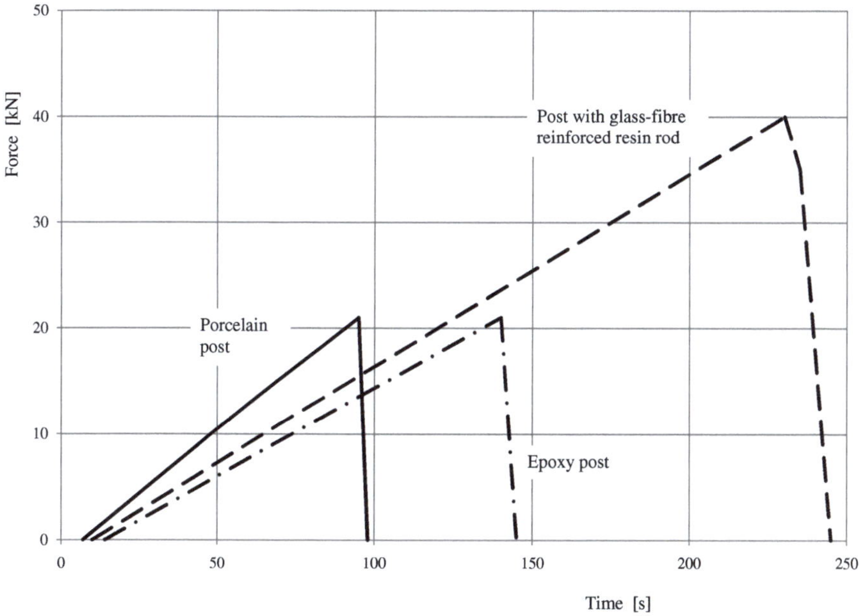

Fig. 3.20 Comparison of the loading diagrams of the three railway post insulators

The silicone post insulator (Fig. 3.17, to the left) had an FRP rod with mono-axially oriented glass fibres and a content of glass of approximately 65 % by weight, and had been installed for almost 15 years in an SBB catenary support arm in a tunnel (Fig. 3.8), that is to say in unfavourable ambient conditions (pollution, moisture, no cleaning by rain, etc.). The behaviour of the silicone post insulator upon failure was considerably different compared to that of the other insulators formed of conventional materials. Under identical test conditions it achieved a failing load of 40 kN (this corresponds to a calculated nominal failing stress of 540 MPa), without physical separation of the body of the insulator (Fig. 3.19).

The different behaviours of the insulators containing brittle materials (porcelain, cast resin) are compared to that of the composite post insulator (containing

Fig. 3.21 **a** Test with SCL = 10 kN (prior to damage), **b** subsequent 96 h test with MDCL = 5 kN

the "flexible" FRP rod) in Fig. 3.20. The insulator containing the FRP rod clearly demonstrates a higher safety margin.

After this short-time test, the same insulator was loaded by its operational load in a 96 h test with the aim of examining its behaviour after incurring the prior damage. The 96 h test period was selected since this is generally accepted to be a sensible compromise for a long-time mechanical test of composite insulators. It is interesting that neither during nor after this test were there were any visible signs of further damage to the insulator, for example caused by an increase in deflection, which is why the authors "christened" this particular behaviour of composite post insulators having an FRP core as "safe failure mode" in [5].

This, to a certain extent, unique and useful behaviour of composite post insulators was further confirmed within the scope of a self-developed test [6]. In this test, after sustaining prior damage as a result of the SCL test (Fig. 3.21a), an insulator used typically for voltages from 36 to 52 kV and having an SCL of 10 kN, but with no silicone housing (for improved visual inspection) was subjected to a 96 h test at 5 kN in accordance with the MDCL test philosophy of IEC 61952 (Fig. 3.21b) and passed this test successfully without any visible external damage to the FRP rod, let alone a physical separation thereof.

The many years of operational experience of composite post insulators and the execution of en extremely high number of bending tests including the detailed inspection of failure behaviour of the insulators, during which the above-described behaviour has been observed time and time again, can be summarised as follows in terms of the operational safety of the line:

- In the case of overload, a correctly dimensioned and manufactured composite post insulator will demonstrate safe behaviour in the event of failure (safe failure mode). There is no actual breakage resulting in a physical separation of the insulator. The overloaded insulator can be easily recognised during the course of routine inspection for maintenance purposes, especially from a distance, owing to its excessive deflection.

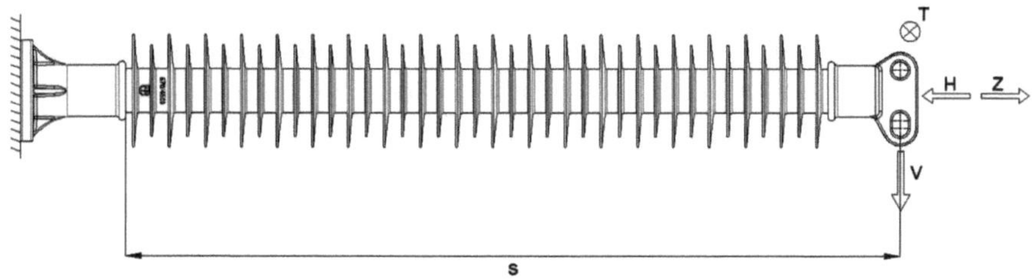

Fig. 3.22 Geometry and loading of a horizontal post insulator

- In spite of the strong deflection and owing to the relatively high residual strength of the insulator, the line conductor usually more or less retains its original position until the damage has been repaired.

3.6 Combined Loads

Until now, insulators subjected to a single bending load were considered. However, line post insulators are often subjected both to bending in the vertical and horizontal directions, and sometimes also to compression in the axial direction. The vertical bending load is primarily caused by conductor weight, whilst the longitudinal load is primarily caused by conductor tension. Axial compression loads, and also tensile loads, are produced by the wind perpendicular to line direction, and in the case of angle towers by the angular force caused by the change in line direction. Ultimately, all of these forces have to be taken into consideration when selecting a suitable insulator. In practice, this occurs by the use of load diagrams (application curves).

As is demonstrated in Sect. 3.6.2.1, there is a simple analytical formula for line post insulators, and this can be used with satisfactory results up to a maximum deflection of approximately 10 % of the length of the insulator. Computer simulations have to be used for greater deflections. Both calculation methods will be explained hereinafter, and the results obtained thereby will be compared to corresponding laboratory tests.

3.6.1 Load Diagrams

The correct choice of a composite post insulator, that is to say of an insulator which ensures a high level of operational safety, also depends on good cooperation between the user and the manufacturer of the insulator. The user must consider the line data and operating conditions to determine the loads at the tip of the insulator in the vertical, longitudinal and axial directions, including any factors of safety which take into account uncertainties when determining these loads. By contrast, the manufacturer, who will have previously established the maximum mechanical stress to which the FRP rod of the insulator may be subjected, must specify which combinations of these loads are permissible, that is to say with which combinations the max. Allowable stress is not exceeded. These permissible combinations are then presented in the form of load diagrams (application curves). These load diagrams are actually based primarily on the mechanical properties of the FRP rod. However, it may also be that the end fittings, in particular the base end fitting, are the limiting factor for a specific application. In the examples below, the hatched areas in the diagrams indicate that the strength of the fittings has been exceeded.

The following calculations [3] are based on the assumptions below (Fig. 3.22):

- Both metal fittings of the insulator are assumed to be completely rigid.
  The same assumption also applies to the tower to which the insulator is fixed. The loads are applied at a point at the tip of the insulator, arranged in line with the axis of the insulator.

- V is the vertical load; it is applied in the direction of gravity but is not necessarily perpendicular to the axis of the insulator.
- T is the longitudinal load; it is applied perpendicular to both the vertical load and the axis of the insulator.
- H and Z are transverse loads. H is a compression load which acts perpendicular to the axis of the tower, whereas Z is a tensile load (for the insulator) directed away from the tower.
- σ is the maximum allowable tensile or compressive stress (assuming there is no prior damage to the FRP rod) at the neck of the base end footing produced by any combination of all above loads and calculated by the formula $\sigma = M/W$.

As described above, its value is largely dependent on the material properties of the FRP rod, but also on the crimping method and on rod diameter. In the examples below, a value of $\sigma = 400$ MPa was selected for a rod diameter of D = 63 mm.

- E is the modulus of elasticity of the FRP rod. A value of $E = 37$ GPa was used in this case.
- s is the lever arm, that is to say the distance between the tip of the line-side fitting of the insulator, that is to say from the point of load application, to the point at which the FRP rod enters the base end fitting. This lever arm $s = 1{,}000$ mm in all the examples below.

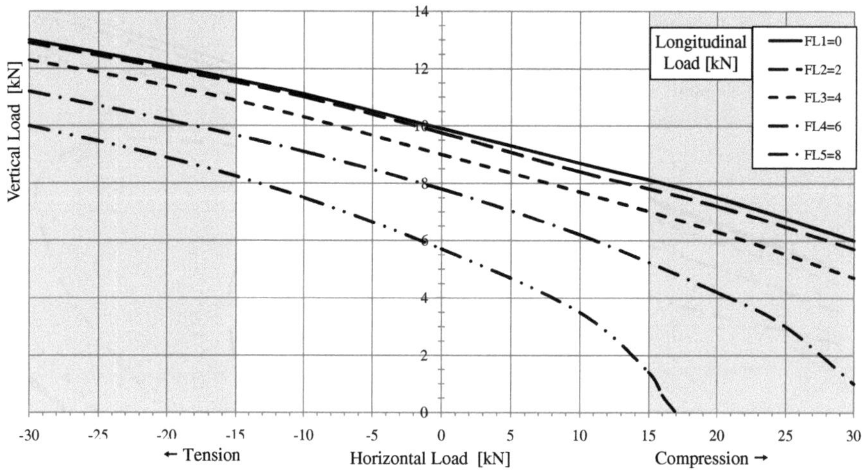

Fig. 3.23 Load diagram for a horizontal post insulator, created analytically ($s = 1{,}000$ mm, $D = 63$ mm, $\sigma = 400$ MPa)

3.6.2 Examples

The following load diagrams were created both analytically and by computer calculations, the results being practically identical. Two situations were examined: In the first case the insulator was arranged horizontally, but in the second case it was arranged at an angle of 15° to the horizontal so that the tip of the insulator was arranged above the base end fitting.

3.6.2.1 Horizontal Position of the Insulator

The bending moment at the base end fitting was established first of all using the formulas below. It should be noted that these formulas are only valid for small deformations (that is to say deflections of 10 % of the length of the insulator at most). In the case of compression load:

$$M_H = \sqrt{\frac{(V^2 + T^2)EI}{H}} \cdot \tan\left(s \cdot \sqrt{\frac{H}{EI}}\right)$$

and in the case of tensile load:

$$M_Z = \sqrt{\frac{(V^2 + T^2)EI}{Z}} \cdot \tanh\left(s \cdot \sqrt{\frac{Z}{EI}}\right)$$

where $I = D^4/64$ is the moment of inertia of the FRP rod

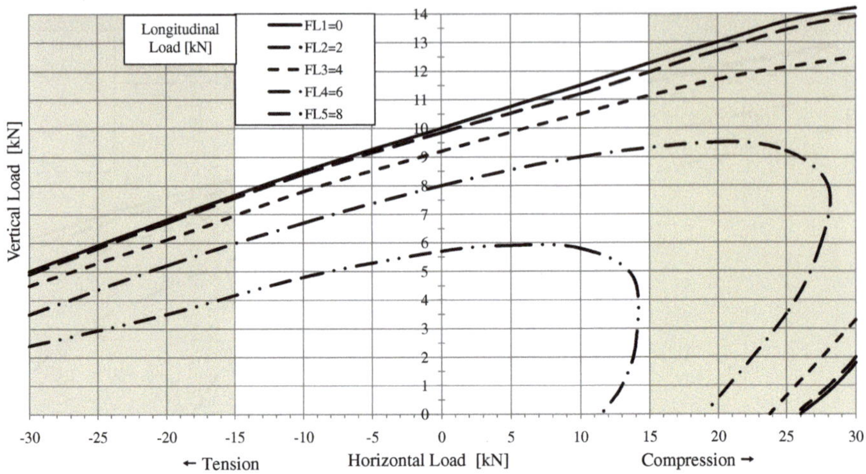

Fig. 3.24 Load diagram for a post insulator inclined at 15°, created analytically ($s = 1{,}000$ mm, $D = 63$ mm, $\sigma = 400$ MPa)

The stress on the FRP rod can therefore be calculated as follows: $\sigma = \frac{M_{H/Z}}{W}$ where $W = D^3/32$ of the section modulus of the rod. The load diagrams thus obtained are shown in Fig. 3.23.

3.6.2.2 Insulator at an Angle of $\beta = 15°$ to the Horizontal

In this case, the above formulas were adapted as follows (dashed variables are the loads in the global coordinate system, whilst un-dashed variables are the loads based on the coordinate system of the insulator)

$$V = H' \cdot \sin(\beta) + V' \cdot \cos(\beta)$$

$$H = H' \cdot \cos(\beta) + V' \cdot \sin(\beta)$$

$$Z = Z' \cdot \cos(\beta) + V' \cdot \sin(\beta)$$

Figure 3.24 shows the load diagrams thus created.

3.6.3 Computer Simulation

A commercially available software was used for the computer simulation. This finite element program can take into account significant deformations. It calculates iteratively, wherein the external load on the deformed structure (in this case the insulator) is assessed at each step of iteration. An advantage is that the program not only calculates the stresses along the insulator, but also the deflection of the insulator. As already mentioned, since there is practically no difference between the load diagrams created using analytical formulas and those created using computer simulation, the computer simulation is not implemented.

3.6.4 Tests

The calculations above were confirmed by laboratory tests. Since it is difficult to measure the maximum tensile or compressive stresses on the FRP rod (as is known, these stresses occur at the entrance of the rod to the base end fitting, where it is not at all easy to apply strain gauges for the measurement of such stresses), the maximum deflection of the insulator calculated by computer simulation was used as a comparison parameter.

Four different combined loads from *V*, *H* or *Z* were applied for each of four axial loads *L*, a total of 16 load situations being tested overall. The loads were selected in such a way that their vector sum produced a maximum stress of $\sigma = 320$ MPa, which is slightly lower than the damage limit established in Sect. 3.3.2.

The difference between measured and calculated deflections was 4.1 % on average, with a maximum value of 15 % [3].

3.7 Dynamic Loads

Just like other types of isolator, composite post insulators are also subjected to dynamic loads, such as wind-induced vibrations and ice shedding, in addition to static loads (primarily caused by the conductor weight and by wind). Such dynamic loads may lead to material fatigue, even if the stresses generated thereby remain below the (static) damage limit detailed above, which would in turn call into question the purely static test for determining the damage limit (Sect. 3.3.2). For this reason, CIGRE working group WG 22-03 decided to also examine the behaviour of FRP rods under dynamic loads [7]. The results will be discussed hereinafter.

3.7.1 Test Specimens

The tests were carried out on two different types of insulator which were similar to the insulators used for the static tests (Sect. 3.3.2.1). One type had no housing so as to facilitate a visual inspection of the surface of the rod during the dynamic tests.

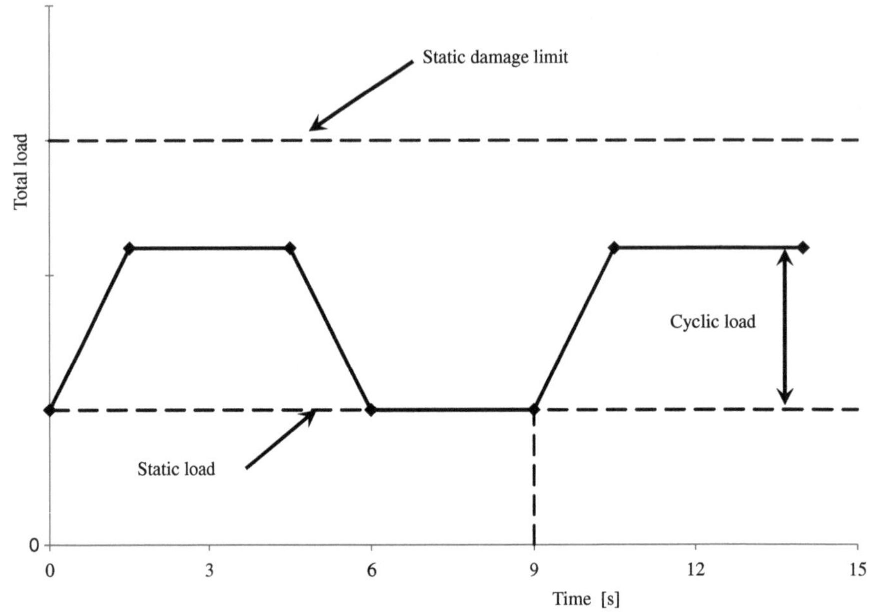

Fig. 3.25 Load cycle for the dynamic tests

These insulators has a rod diameter of 63 mm, crimped base end fittings and a bending length of 1.2 m. By contrast, the second type of insulators were complete with their housing. They also had a 63 mm FRP rod with crimped fittings, but were slightly longer at 1.4 m.

3.7.2 Test Method

The test specimens were mounted and loaded as they supposedly are during operation, i.e. a static load was first applied and this was then supplemented by a cyclic load, as shown in Fig. 3.23. The peak load (static + cyclic) was selected in such a way that the total stress thus calculated remained below the static damage limit of the rod (this damage limit was calculated to be approximately 400 MPa in Sect. 3.3.2.1) (Fig. 3.25).

One load cycle lasted approximately 9 s, whereby the 10^6 load cycles deemed to be necessary lasted approximately 4 months.

The load amplitudes were selected in such a way that they corresponded to the actual loads during operation for typical 45–138 kV lines in which such post insulators are primarily used. More specifically, the static loads were between 25 and 80 kN, and the dynamic loads were between 6 and 60 kN (peak–peak). The ambient temperature was set between 18 and 28 °C during the test. In most cases, two insulators were tested for each load level. At the end of the tests the insulators were examined for the presence of any damage (delamination, that is to say separation of the glass fibres from the resin matrix, cracks, or both), as described in Sect. 3.3.2.1.

3.7.3 Test Results

The test results can be summarised as follows:

- No damage could be determined with peak stresses in the rod up to 270 MPa (corresponding to a bending load of 37 kN (static) and 18.5 kN (dynamic).
- Damage was only determined at peak stresses in the rod above 390 MPa [corresponding to a bending load of 80 kN (static) and 18.5 kN (dynamic)].
- Additional tests may be useful for determining the behaviour of the insulators in the range between 300 and 400 MPa.

To summarise, it is found that, except for quite specific cases in which the dynamic load exceeds 30 % of the static load, it is not necessary to establish a new test to examine, specifically, the dynamic load behaviour of composite post insulators. The test described in Sect. 3.3.2.2, based on the MDCL concept, should be sufficient to cover both static and dynamic load situations.

3.8 Constructional Requirements of the End Fittings

The constructional design of end fittings for composite post insulators so that these insulators can demonstrate their maximum bending "ability" is a complicated process since all factors, that is to say end fitting type, assembly process, materials involved, dimensioning of the insulator and fixing design, have to "work together" so that the prescribed tests are passed, but also so that the insulator can enjoy faultless operation over many years, as is desired.

In the case of the composite post insulators considered in this instance, the constructional design of the end fittings, in particular of the base end fitting, is rather important since these fittings account for a considerable proportion of the overall cost of these insulators because they are larger and therefore require more material compared to the end fittings for long rod insulators. In addition, the preparations for assembly of the FRP rod and of the end fittings are significant for optimal design of the insulator in terms of cost. The corresponding experience, which has been obtained in almost all cases by costly and time-consuming "trial and error" methods, is therefore considered unsurprisingly by the manufacturers of insulators to be a major trade secret and is therefore not made public. In spite of this, we have attempted to summarise below some of the basic principles of the constructional design of these end fittings and in particular of the base end fittings,

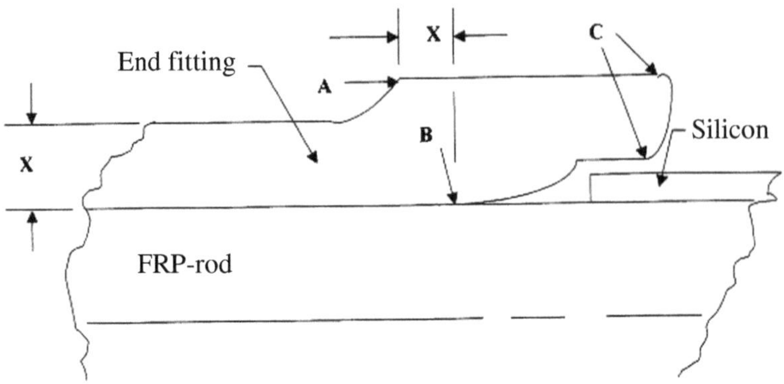

Fig. 3.26 Critical areas of the post insulator fitting

that is to say the fittings via which the post insulator is fixed to its supporting structure.

The critical points to which particular attention has to be paid in the constructive design of a fitting of this type so that the insulator can ultimately withstand the necessary maximum operational load and the necessary failing load according to Sect. 3.3.2.3 are as follows:

- Engagement length of the FRP rod into the end fitting
- Wall thickness of the fitting
- Machining requirements of the inner wall
- Preparation of the FRP rod
- Attachment method of the fitting on the FRP rod
- Profile of the mouth of the fitting
- Stability of the attachment hardware

One prerequisite is that the material properties of the FRP rod and of the metal fitting must satisfy the specific process and design requirements.

Generally, due to cost considerations the length of the FRP into the end fitting is kept to a minimum. Typical values for the ratio of engagement length to rod diameter are between 1.0 and 2.0, this ratio being largely dependent on the type and quality of the method for assembling the fitting on the FRP rod.

Nowadays, the end fittings are almost always crimped onto the FRP rod, the radial crimp usually being applied by a range of crimping tools or more rarely by swaging. Gluing is no longer used, as is also the case with long rod insulators.

In principle, thicker FRP rods, for example from 63 mm in diameter, do not require any surface preparation before assembly, in contrast to thinner rods, as are used in medium-voltage post insulators for example, which often have to be roughened.

With regard to the metal fitting, there are basically three critical points which need to be taken into consideration during the design of the fitting. These are the wall thickness of the fitting b, the crimp length (this in turn depends on the method for machining the surface of the FRP rod), and the geometry (A, B, C and D) of the mouth of the fitting, which has to be designed so that stress peaks are avoided, more specifically transverse to the fibre direction, which is particularly sensitive. A good design of a fitting should take into account the points shown in Fig. 3.26.

3.9 Analytical Calculation Methods

It has already been demonstrated a number of times that the mechanical design of composite post insulators is rather complex. There are two main reasons for this: Firstly, in contrast to composite long rods, the load of composite post insulators interferes with the rotational symmetry of the arrangement, and secondly the "collaboration" of two materials which differ greatly in terms of their material properties [for example the metal end fittings (isotropic) and the FRP rod (orthotropic)] is anything but simple and intuitive.

As in the case of long rods, the calculation methods for composite post insulators have also been developed over time and with increasing experience. For this type of insulator too, a distinction is made between analytical methods (of varying complexity) and numerical simulation methods, wherein the latter are often only implemented in specific, unusual circumstances, for example in the case of rods of very large diameter (>100 mm) or extremely high bending loads.

3.9.1 *Simple Analytical Method*

The simple analytical method of composite post insulators is based on conventional beam bending theory Essentially, it can be used to calculate the nominal bending stresses in the FRP rod. This method can also be used to calculate the maximum deflection of the insulator at the point of the load application, since in some applications of composite post insulators, such as in railway catenary supporting structures, it is this deflection, and not the allowable bending stress, which forms the basis of the assessment. For example, the allowable deflection of the railway insulators is often limited to 1 % of the length of the insulator.

For the maximum deflection f of a post insulator of length l, rod diameter D, modulus of elasticity E, and subject to the load F_b, the following can be applied to small deformations with good approximation:

$$f = \frac{F_b \cdot l^3}{3 \cdot E \cdot I}$$

With the moment of inertia I of the rod cross-section being:

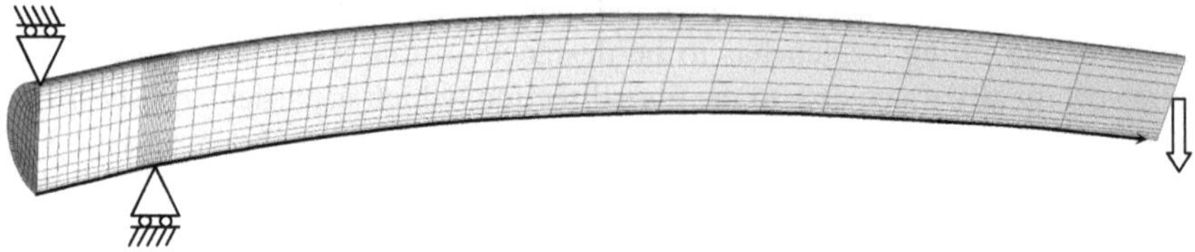

Fig. 3.27 Three-point bending of a composite insulator

$$I = \frac{\pi \cdot D^4}{64}$$

The necessary minimum diameter D_{min} with a predefined maximum deflection f_{max} thus follows:

$$D_{min} = \sqrt[4]{\frac{64 \cdot F_b \cdot l^3}{3 \cdot \pi \cdot E \cdot f_{max}}}$$

Composite materials are rather weak when subjected to torsional loading (shear strength $\tau_{max} = 55$ N/mm^2, shear modulus $G = 5{,}000$ N/mm^2). For this reason, it is sometimes necessary to determine the torsional strength and maximum allowable angle of rotation ϕ of composite insulators. The latter is important, for example for insulators which are used as supports for rotary disconnector switches.

This calculation is made as follows:

$$W_p = \frac{\pi}{16} \cdot D^3$$

$$I_p = \frac{W_p \cdot D}{2}$$

$$\tau_t = \frac{M_t}{W_p}$$

$$\varphi = \frac{M_t \cdot l}{G \cdot I_p} \text{ in rad and } \varphi = \frac{M_t \cdot l}{G \cdot I_p} \cdot \frac{180}{\pi} \text{ in Grad}$$

M_t is the torsional moment, l is the length of the post insulator, and I_p and W_p are the polar moment of inertia and the polar section modulus of the post insulator cross-section respectively (diameter D).

3.9.2 Complex Analytical Method

A similar method has already been presented in Chap. 2 for composite long rods purely under tensile stress. Such a method for composite post insulators under

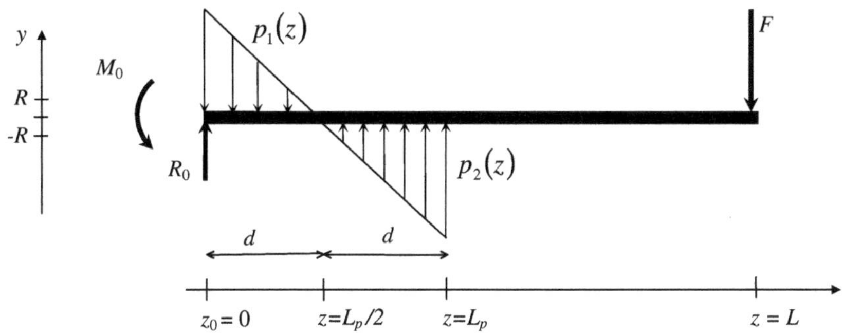

Fig. 3.28 Equivalent static system and loads [8]

bending load will now be described hereinafter [8]. As in the case of long rods, it was again possible to obtain and use some important basic information for the analytical model from knowledge provided by the numerical simulation (see Sect. 3.10).

The bending load at the tip of the insulator influences the contact at the interface between metal base fitting and FRP rod, since it results in inhomogeneous contact pressure at this interface. For this reason, the analytical model of the insulator presented here will be considered as a beam under "three-point bending". Two "points" are the supports in the metal fitting, and the third "point" is the tip of the insulator, where the load is applied (Fig. 3.27).

It is assumed that in a first approximation the pressure distribution at the interface between FRP rod and metal fitting is linear over the length of the crimp, that is to say where the rod is located inside the fitting (Fig. 3.28).

From static equilibrium conditions, that is to say when the sum of the forces and the sum of the moments equals zero, the following peak values p_1 and p_2 of the line loads can be calculated:

$$p_1 = F \cdot \frac{3 \cdot L - 5 \cdot d}{2 \cdot d^2} \text{ and } p_2 = F \cdot \frac{6 \cdot L - 2 \cdot d}{4 \cdot d^2}$$

The following is therefore given for the line loads:

$$p_1(z) = -\frac{p_1}{d} z + p_1 \text{ and } p_2(z) = \frac{p_2}{d} z - p_2$$

with which the bending moment and shear force of the FRP rod can be calculated not only in the fitting, that is to say in the range from $z = 0$ to $z = Lp$, but also in the "free" length of the FRP rod, which is though not of particular interest, since the rod material load is not stressed critically there.

3.9.2.1 Radial Compressive Stress at the Mouth of the Metal Fitting

The line load $p_2(z)$ leads to a radial compressive stress $\sigma_{rr}(z)$, the maximum value of which lies on the surface of the rod at the last point of contact (LPC) between the rod and the end fitting. To simplify this, a uniform pressure distribution over the circumference of the rod is assumed, which leads to the following formula for the radial compressive stress ($2R$ is the projected contact area per unit of length):

$$\sigma_{rr}(z) = \frac{p_2(z)}{2 \cdot R}$$

The peak value lies at $z = L_p$ and is:

$$\sigma_{rr,\max} = \frac{F \cdot (6 \cdot L - L_p)}{2 \cdot L_p^2 \cdot R}$$

3.9.2.2 Axial Stress in the FRP Rod Inside the Metal Fitting

This stress is produced by the bending moment $M(z)$ and is calculated in accordance with the conventional Bernoulli–Euler beam theory:

$$\sigma_{zz}(z, y) = \frac{M(z) \cdot y}{I}$$

The value of this varies not only along the rod (z-coordinate), but also over the cross-section of the rod (y-coordinate). It reaches its peak value at the surface of the rod (that is to say at the interface with the fitting at the last point of contact therewith), wherein the axial stresses at the compressed part of the cross-section ("concave") are the most dangerous for the FRP rod, since its compression strength in the axial direction is somewhat less than its corresponding tensile strength. This peak value is:

$$\sigma_{zz,\max} = \sigma_{zz}(z = L_p, y = -R) = \frac{4 \cdot F}{\pi \cdot R^3}(L - L_p)$$

To this stress should be added the axial stresses produced from the radial and circumferential stresses by the Poisson effect:

$$\sigma_{\max}(z) = \frac{4 \cdot F}{\pi \cdot R^3}(L - L_p) - \nu_{LT}[\sigma_{rr}(z) + \sigma_{\vartheta\vartheta}(z)]$$

Since, in the rod, $\sigma_{rr}(z) = \sigma_{\vartheta\vartheta}(z)$ (see also Sect. 2.4.6), the following is obtained by introducing the ultimate bending stress in the FRP rod σ_{max}:

$$\sigma_{\max} = \frac{4 \cdot F}{\pi \cdot R^3}(L - L_p) - 2 \cdot \nu_{LT} \frac{F \cdot (6 \cdot L - L_p)}{2 \cdot L_p^2 \cdot R}$$

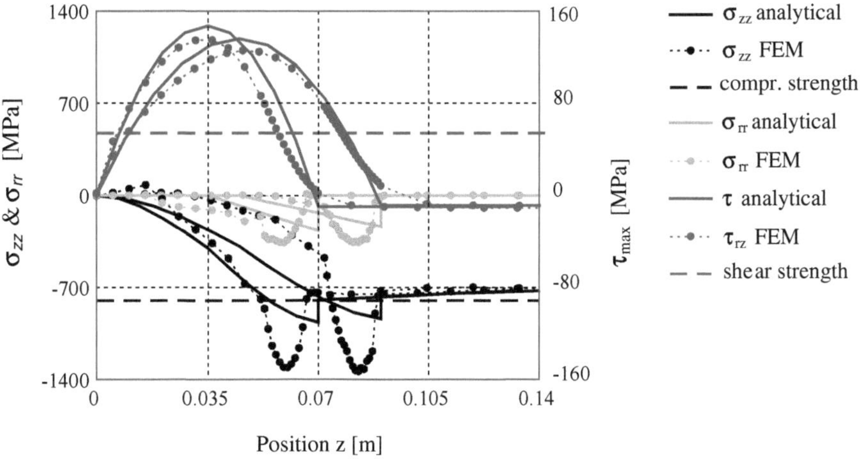

Fig. 3.29 Axial stresses at the outermost compressed fibres of the FRP rod and shear stress in the rod axis for rods 51 and 63 mm thick: *FEM* analytical comparison with external loading of 15 and 20 kN [8]

3.9.2.3 Shear Stress in the Neutral Plane of the Rod

This key stress is produced by the shear force in the rod and is:

$$\tau_{\max}(z, y) = \frac{4 \cdot Q(z)}{3 \cdot \pi \cdot R^2} \cdot \left[1 - \left(\frac{y}{R}\right)^2\right]$$

This shear stress acts not only in the cross-sectional plane of the rod, but also, as a result of the known symmetry of the shear stress in two directions perpendicular to one another, in the planes perpendicular to the rod cross-section, which include the neutral plane of the rod. As can be seen from the above equation, it's maximum value does not occur at the interface between the rod and fitting ($y = R$, where it disappears), but in the neutral plane of the rod ($y = 0$):

$$\tau_{\max}(z, y = 0) = \frac{4 \cdot Q(z)}{3 \cdot \pi \cdot R^2}$$

This expression varies in the z-direction and reaches its peak value precisely at the centre of the length of the rod clamped in the fitting, that is to say at $z = L_p/2$:

$$\tau_{\max}(z = L_p/2, y = 0) = \frac{F}{3 \cdot \pi \cdot R^2} \cdot \left[\frac{6 \cdot L}{L_p} - 5\right]$$

The stresses $\sigma_{rr,\max}$, $\sigma_{\max}$ and $\tau_{\max}$, which are decisive for the loading of the FRP rod, can be calculated analytically using the above formulas, and compared with the values established from the numerical simulation (Sect. 3.10).

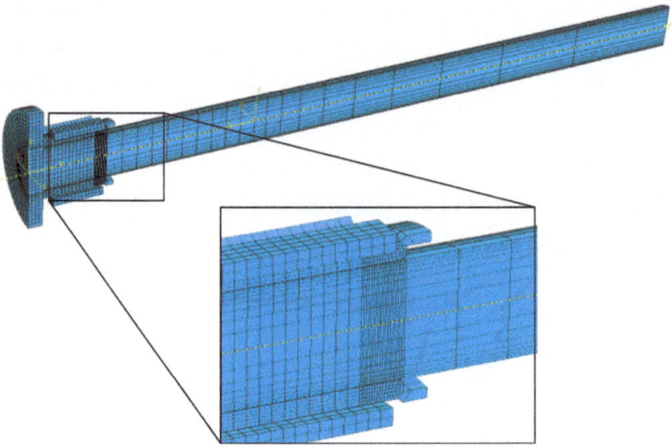

Fig. 3.30 Finite element model for the 63 mm insulator with detailed view

This comparison is shown in Fig. 3.29 for two typical rod diameters (51 and 63 mm) loaded by 15 kN and by 20 kN.

If, in the equation for $\sigma_{\max}$, this is replaced by $\hat{\sigma}_{L,Druck}$, that is to say by the axial compressive strength in the FRP material and is then solved for F, the following is obtained for the ultimate bending load $F_{\max}$ of a composite post insulator (with no consideration of the Poisson effect):

$$F_{\max} = \frac{\pi \cdot R^3 \left| \hat{\sigma}_{L,Druck} \right|}{4 \cdot \left(L - L_p \right)}$$

If $L = 760$ or 1030 mm, $Lp = 70$ or 90 mm, $R = 25.5$ or 31.7 mm, and $\hat{\sigma}_{L,Druck} = -794\text{MPa}$ (Table 2.2), $F_{\max} = 15$ or 20.7 kN respectively for the two insulators examined in this instance measuring 51 and 63 mm in diameter. These values are consistent with the values obtained by testing (Sect. 3.10.2) measuring 15.6 and 21.7 kN.

3.10 Numerical Simulation

In contrast to the long rod insulators examined, which have a small cross-section (16–31 mm) compared to length (up to 6 m and more), the dimensional ratios in composite post insulators are the exact reverse: These have relative large cross-sections (sometimes up to 150 mm) with lengths which are not excessively long (max. 2–3 m). This means that, in the case of composite post insulators, relatively small pressures for crimping are applied which are guaranteed not to overload the rod, the crimping process therefore not constituting a situation of critical load for the insulator, again in contrast to long rods. This is why the numerical situation is

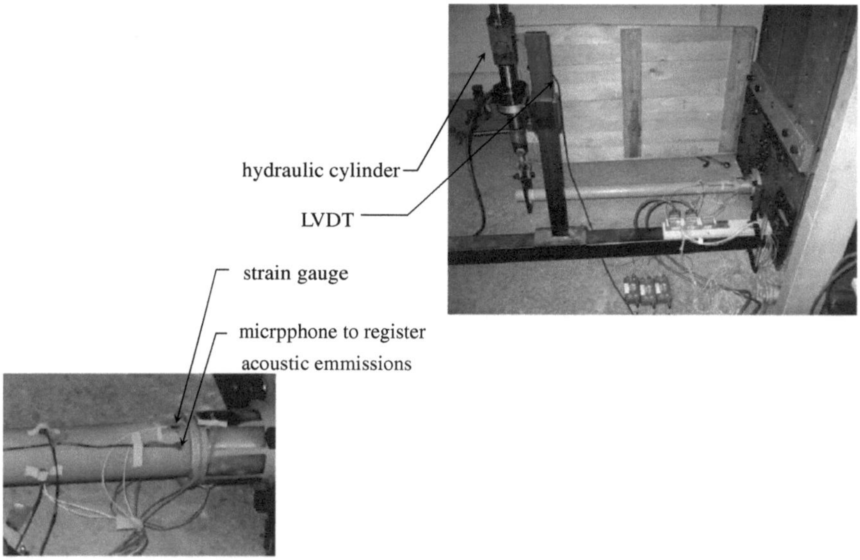

Fig. 3.31 Testing arrangement

focussed hereinafter to the examination of the internal state of stress in the FRP rod of the insulator during the bending phase [9].

Numerical models for mechanical testing of composite insulators are preferably based on the finite element method (FEM), the principles of which are sufficiently known and will not be repeated here. Only particular features of the FEM will be detailed hereinafter with regard to its use in the case of composite insulators subjected to bending load.

3.10.1 Finite Element Model

The FE model used in this case with commercial software, is formed of 3D hexahedral elements each having 20 nodes. Owing to the symmetry, only half the structure was modelled. The FE mesh consists of 4,915 elements for the metal fitting and of 6,912 elements for the (half) FRP rod. As is shown in Fig. 3.30, the mesh in the rod is refined in the end zone of the fitting, since stress concentrations are to be expected in this zone.

As is the case with long rods (Chap. 2), the quality of the FE model has to be checked using suitable tests by comparing the calculated stresses (strains) with those obtained by way of testing, by comparing the deflection of the insulator, and by comparing the behaviour of the structure in the event of failure. In the latter case it is advantageous to describe in detail the failure mechanism of a composite insulator under bending load from a phenomenological point of view (Sect. 3.11).

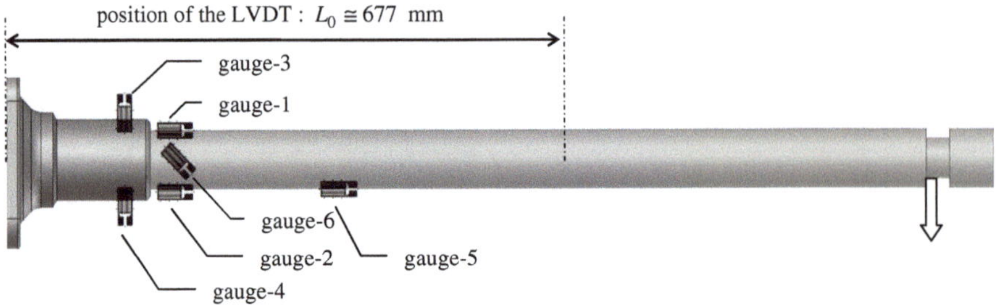

Fig. 3.32 Positioning of the DMS and of the linear variable differential transformer during the bending tests

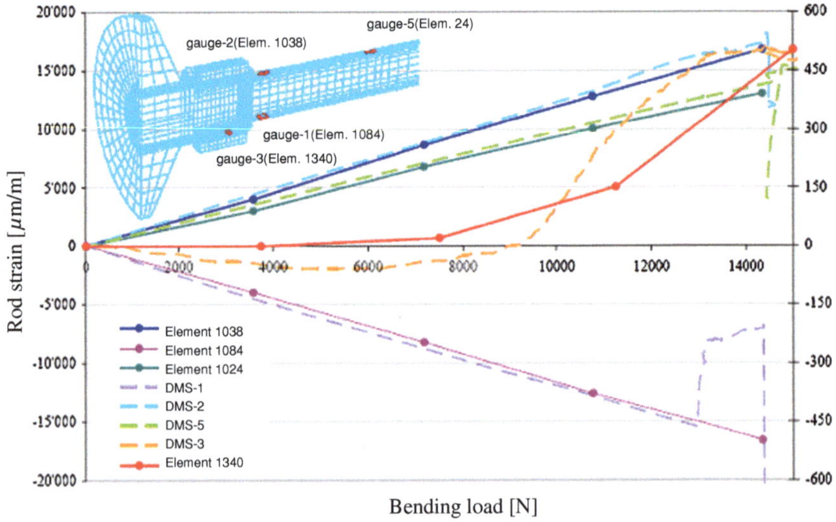

Fig. 3.33 Comparison between measured and calculated (*FEM*) strains

3.10.2 Testing Arrangement

In these tests the insulator is fitted in a test rig developed specifically for this purpose. The test specimen is manufactured without the silicone housing for the purposes of improved access and for unrestricted monitoring of the surface of the rod during the bending process. The test rig has a hydraulic cylinder equipped with a load cell. Additional instruments consist of a linear variable differential transformer (LVDT), six strain gauges (DMS) which are applied to the rod and to the base end fitting, and a six-channel data logger for measuring acoustic emissions which is to be used to gain a better understanding of the damage analysis with increasing bending load (Fig. 3.31).

The positioning of the DMS on the 51 mm rod can be seen in Fig. 3.32.

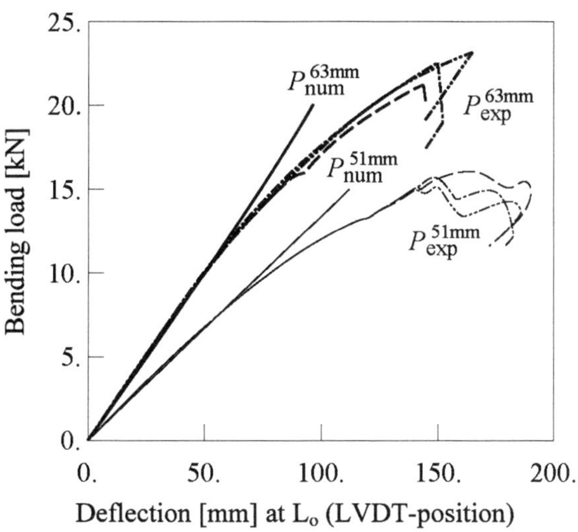

Fig. 3.34 Comparison between measured and calculated (*FEM*) deflection

L_0 Š · 677 :

The comparison between the calculated and measured strains is shown in Fig. 3.33:

Very good correlation was established between the calculation and test for the DMS on the free part of rod, and satisfactory correlation was established between the calculation and test for the DMS on the metal fitting (DMS no. 3).

The second approach for testing the quality of the FE simulation is the comparison between the measurement (*test*) and calculation (*num. sim.*) of the load–deflection curve. A comparison is shown in Fig. 3.34 for each of a 51 mm insulator and a 63 mm insulator.

There is very good correlation up until the point marked (a), where the non-linearity of the curve starts. This is precisely also where the failure of the rod begins, which will be discussed in detail in the next section.

3.11 Behaviour of Composite Post Insulators in the Event of Failure

The behaviour of the FRP material in the insulator core in the event of failure was discussed in detail in [8]. In this section, however, the behaviour of an entire post insulator in the event of failure will be examined. Not only will the different failure mechanisms which the insulator experiences when subjected to bending load be described, but these will also be quantified in a numerical simulation and compared to the knowledge gained from testing.

When examining the behaviour of an insulator in the event of failure, the insulator is loaded until failure and the metal fitting is then cut in half so as to

Fig. 3.35 Damage pattern of a 51 mm insulator

Fig. 3.36 Damage zones of a 51 mm insulator [8]

better observe not only the surface of the rod, but also the internal damage pattern (Fig. 3.35):

The close-up view in Fig. 3.36 shows the different damage zones of the 51 mm insulator. This damage pattern is typical for insulators having small to medium diameters.

The first damage zone (zone 1 in Fig. 3.36) is not of particular interest, since it only occurs after the actual failure of the insulator. The upper fibres of the FRP rod are stretched during the bending process and slip out of the metal fitting. Once the insulator has failed and is relieved of its load, these fibres cannot return to their original position owing to the friction occurring at this point, and may buckle.

The second damage zone (zone 2 in Fig. 3.36) is slightly less pronounced and consists of a few axial cracks which extend as far as the cut rod end in the base of the metal fitting. These cracks are produced transverse to the fibres as a result of axial stresses and lead to failure of the resin matrix. This zone lies at a point which plays no role in the functional efficiency of the insulator.

The third damage zone (zone 3 in Fig. 3.36) is of particular importance, since it is precisely here that the progression of damage to the insulator begins. Microscopic examinations (Fig. 3.37) show a strong crack in the rod shortly before it exits the metal fitting, the crack having cut through glass fibres and the resin matrix, as well as massive micro buckling of the glass fibres in the FRP material.

Figure 3.38 shows the development of the damage in zone 2 with increasing bending load:

It can be seen that irreversible damage which is not externally visible occurs in the FRP rod, even at loads below the failing load of 15.6 kN in this case. This

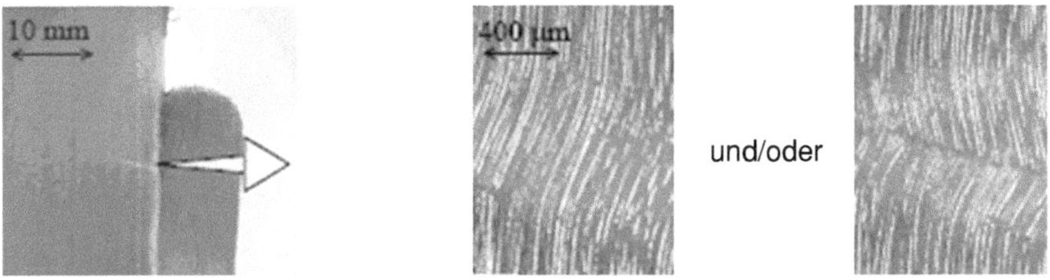

Fig. 3.37 Close-up views of the micro buckling of the fibres in the FRP rod of a 51 mm insulator (Zone 3) [8]

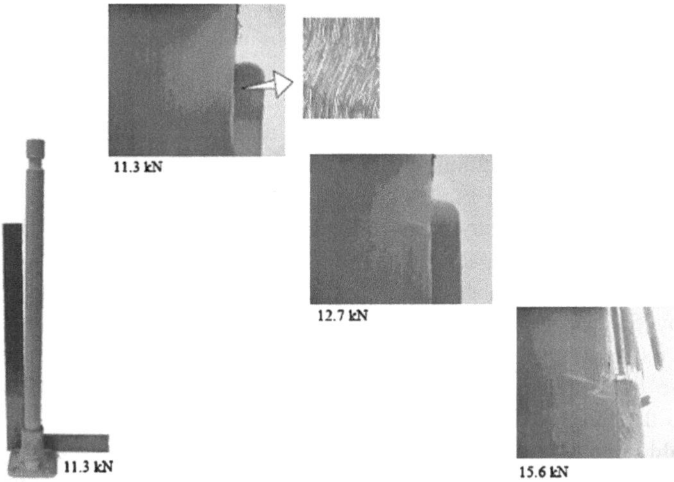

Fig. 3.38 Development of damage in a 51 mm insulator with increasing bending load [8]

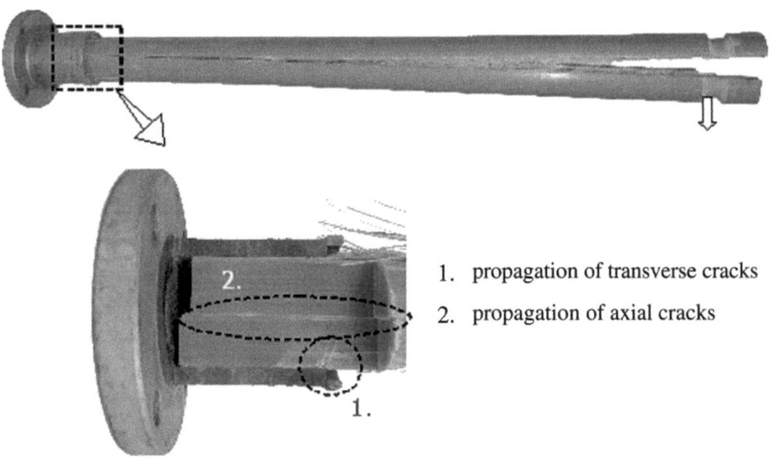

Fig. 3.39 Damage pattern of a 63 mm insulator [8]

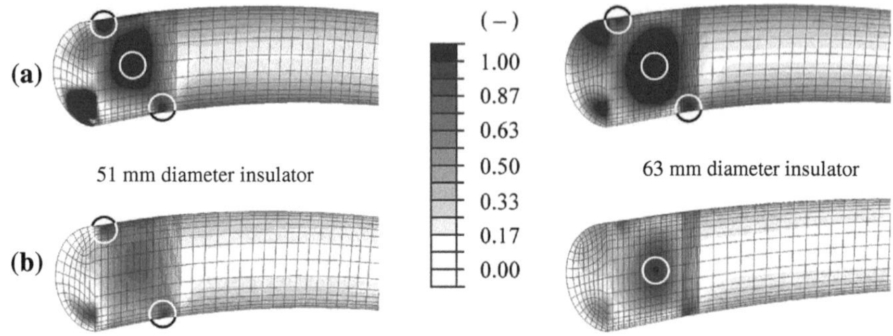

Fig. 3.40 Distribution of the danger factor for 51 and 63 mm insulators **a** maximum bending load (failure), **b** 60 % (51 mm) and 50 % (63 mm) of the max. load—the circles mark the highest values of the danger factor in each case

means that, even at a load of 11.3 kN (70 % of 15.6 kN), the insulator does not return completely to its original pre-load position once the load has been relieved, (to the left of the image in Fig. 3.38).

For insulators of greater cross-section (typically from 63 mm diameter), a further important damage pattern is observed (Fig. 3.39).

This damage pattern (labelled 2 in Fig. 3.39) is produced in the neutral axis of the rod by the shear stress already noted in the section above as a result of analytical calculations. Even if this damage is introduced earlier, it is only visible just before the total failure of the insulator when the delamination propagates quickly to the free rod end (top of Fig. 3.39).

It should be noted that this last damage pattern is only observed in the case of large rod diameters. This may therefore be linked to the fact that, comparatively, insulators of large rod diameter are not crimped so strongly. This means that the frictional contact between rod and fitting is also weak and sometimes allows movement in the "upper" part of the FRP rod. This in turn leads to high axial deformation and to associated high shear stresses in the neutral fibres of the rod. It can even be seen clearly in Fig. 3.39 that this partial movement has led to a "step" in the rod cross-section at the fixed end of the rod.

With the aid of the FE calculations presented in the previous section, the damage patterns observed in the tests also be reproduced mathematically with the aid of a danger factor as detailed in [8]. As explained there, this factor provides the numerical value by which the external load has to be multiplied so that the damage limit of the FRP material, in which the danger factor is assumed to be 1, is achieved. The results calculated for the danger factor of an insulator 51 mm thick and of an insulator 63 mm thick under bending load until failure are shown in Fig. 3.40a. There, it can be seen that the danger factor in the centre of the FRP rod largely exceeds the value 1, which is indeed the damage limit of the FRP material, under the maximal possible load. As explained above, the shear stresses are dominant at this point. In addition, the damage limit is exceeded in the vicinity of the two contact points between the rod and the metal fitting, since this is where

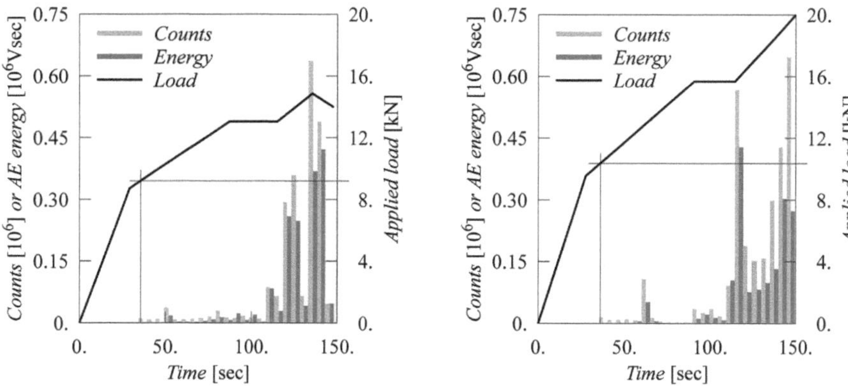

Fig. 3.41 Acoustic emissions in the event of bending of **a** 51 mm, and **b** 63 mm insulators

high radial stresses occur in the FRP rod. If the load is reduced to 60 % (for the 51 mm rod) or to 50 % (for the 63 mm rod), the danger factor also decreases sharply and the value of 1 is only exceeded very locally, if at all. Internal damage starts to occur in the FRP material at these points, even under relatively low load. This process can also be seen from the load–deflection diagram, since the non-linear region of the load–deflection curve starts at the corresponding bending loads (Fig. 3.34).

It is considerably easier to follow the initiation of damage in the FRP rod using the acoustic emission method [10]. A corresponding device is connected to the test rig, as mentioned above (Fig. 3.31). A number of insulators 51, 63 and 88 mm thick were loaded until failure, and the activity of the individual damage events was recorded using the acoustic emission equipment. The frequency and intensity of the acoustic pulses recorded and averaged over five seconds are shown in Fig. 3.41 for an insulator 51 mm thick (a) and for an insulator 63 mm thick (b) as a function of the external bending load. It can be seen that the first signs of acoustic emissions for both insulator types coincide with the onset of non-linearity in the respective load–deflection diagrams (see Fig. 3.34). This activity increases massively when the load is increased further and reaches its highest value just as the cantilever failing load (CFL) of the respective insulator is reached.

3.12 Sensitivity Analysis

In this case, too, a sensitivity analysis was carried out, as with the long rods (Sect. 2.4.7.5), but only for certain parameters, since it has been found that, in contrast to long rods, the other parameters tested in that instance (such as coefficient of friction between road and fitting, wall thickness of the fitting, and play between rod and fitting), do not have such a substantial influence on the

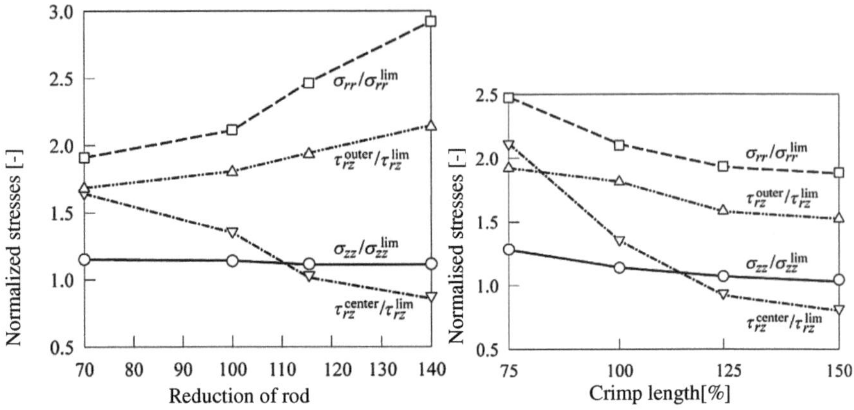

Fig. 3.42 Change to the critical stresses (normalised on the initial configuration) with variation of the radial compression (*left*) and crimp length (*right*) of a 51 mm rod

bending strength of post insulators, which mainly depends on axial stress σ_{zz} and shear stress τ_{rz} of the rod.

The 51 mm insulator was tested at a bending load of 9 kN, which still lies beneath its damage limit (Fig. 3.34). The results of the sensitivity analysis for the effect of radial compression are shown to the left in Fig. 3.42, and those for the effect of crimp length are shown to the right.

It can be seen that, when the compression, that is to say the reduction in diameter of the rod, varies between 70 and 140 %, the axial stress σ_{zz} hardly changes, and by contrast the shear stress τ_{rz} in the centre of the rod is reduced massively by up to 40 %. Unfortunately, this reduction is achieved at the cost of the radial stress at the outer fibres and the shear stress in the vicinity of the outer fibres, which then increase by the same amount, that is to say 40 %.

The variation of crimp length between 75 and 150 % of its nominal value also leads to predictable results. A considerable reduction in all key stresses is established with an increase in crimp length. In particular, the critical shear stress τ_{rz} in the rod axis falls by 40 with a 50 % extension of the crimp zone. The longer the end fitting, the later the damage process of the FRP rod is initiated.

References

1. Papailiou KO, Thaddey S, Schmuck F, Armschat C, Stankewitz JC (2010) Development of an 800 kV HVDC station post design based on the long-term experience with composite line post insulators. CIGRE Session, Paris
2. Dumora D, Feldman D, Gaudry M (1990) Mechanical behaviour of flexurally stressed composite insulators. IEEE Trans Power Deliv 5(2):1066–1072
3. CIGRE (1996) Cantilever load performance of composite line post insulators. Electra 169, Paris

4. Schmuck F, Aitken S, Papailiou KO (2010) A proposal for intensified inspection and acceptance tests of composite insulators as an addition to the guidelines of IEC 61109 Ed. 2 2008 and IEC 61952 Ed. 2 2008. IEEE Trans Dielectr Electr Insul 17(2):394–401
5. Schmuck F, Papailiou KO (2000) On the raising application of polymeric post insulators. CIGRE SC 33 conference, Prague, Czech Republic, 7–8 Sept
6. Cavelti G, Papailiou KO, Schmuck F (2001) Verbundisolatoren liegen im Trend. SEV/VSE Bulletin 15/01
7. CIGRE (2002) Guide for the evaluation of composite line post insulators subjected to combined mechanical loads. Electra 203, Paris
8. Prenleloup A (2008) Analyse de l'état de contrainte et de l'endommagement d'assemblages sertis en matériau mixte métal-composite sollicités en traction ou en flexion. PhD thesis no. 4005, Lausanne Ecole polytechnique fédérale de Lausanne
9. Prenleloup A, Gmür T, Botsis J, Papailiou KO, Obrist K (2009) Stress and failure analysis of crimped metal–composite joints used, in electrical insulators subjected to bending. Compos Part A 40:644–652
10. Prenleloup A, Gmür Th, Botsis J, Papailiou KO, Obrist K, Bonhôte Ph (2007) Acoustic emission inspection and analysis of crimped metal-composite joints subjected to bending. In: Proc 4th Int Conf NDT, pp 11–14, Chania (Crete), Greece

Chapter 4
Insulated Cross-Arms for Compact Lines

Symbols and Abbreviations

CIGRE	Conseil International des Grands Réseaux Électriques (International Council on Large Electric Systems)
FRP	Fibre reinforced plastic
FEM	Finite element method
Horizontal V	Insulated cross-arm
IEC	International electrotechnical commission
ROW	Line corridor, line width
V-string	Suspension string with two insulators at an angle to one another (>0°) at the point of conductor suspension
ΔT	Horizontal deflection due to the wind
ΔV	Vertical deflection due to the wind
α	Angle between brace and post
β	Horizontal angle of the post insulator
γ	Angle between axis of rotation of the insulated cross-arm and post
ϑ	Inclination angle to the axis of rotation of the insulated cross-arm
ϑ	Angle between axis of rotation of the insulated cross-arm and brace
ϕ	Angle of rotation of the cross-arm tip in direction of the line
B	Brace load
c	Deflection of the cross-arm in line direction
d	Lateral distance between the two conductor loads L
E	Modulus of elasticity (young's modulus) of the FRP rod
g	Acceleration of gravity
H	Insulator compression load (perpendicular to the tower axis)
I	Moment of inertia of the FRP rod
L	Buckling length
L	Conductor tension
M	Aligning torque
p	Radius of rotation of the tip of the cross-arm
P	Compressive load on the post

K. O. Papailiou and F. Schmuck, *Silicone Composite Insulators*, Power Systems,
DOI: 10.1007/978-3-642-15320-4_4, © Springer-Verlag Berlin Heidelberg 2013

P_{Euler}	Euler load
m_L	Weight mass for dynamic loads
s	Length of the post
T, T_1, T_2	Longitudinal load
T_{HW}	Load responsible for the displacement of the tip of the cross-arm in line direction
V	Vertical load

4.1 Introduction

Compact lines are an interesting alternative to conventional overhead transmission lines where wayleave and also the visual impact on the surrounding landscape are of importance. In addition, compact lines offer further advantages, such as higher power transfer and lower EMC load, and all at a cost which is not necessarily greater than that for conventional lines. Compact lines were first developed in the 1970s [1, 2], but only started to become more popular with the widespread introduction of composite insulators, which were really introduced in large numbers in the 1990s. The main reason for this is that the insulated cross-arms, which are indispensable for the installation of compact lines, are loaded primarily by compression and are subject to relatively large deformations, which can be better sustained by composite materials than by conventional insulating materials (porcelain and glass).

More specifically, the following properties of composite insulators are advantageous for use in insulated cross-arms:

- High bending strength
- Elastic limit in the region of ultimate strength
- High ultimate strain
- Not brittle

There are now many examples of compact lines which can even operate without difficulty at 420 kV [3–5].

The mechanical dimensioning of compact lines and, in particular, the mechanical dimensioning of insulated cross-arms with composite insulators will be discussed in detail in this chapter. Firstly, the principles of line compaction and the different options for this will be explained [6].

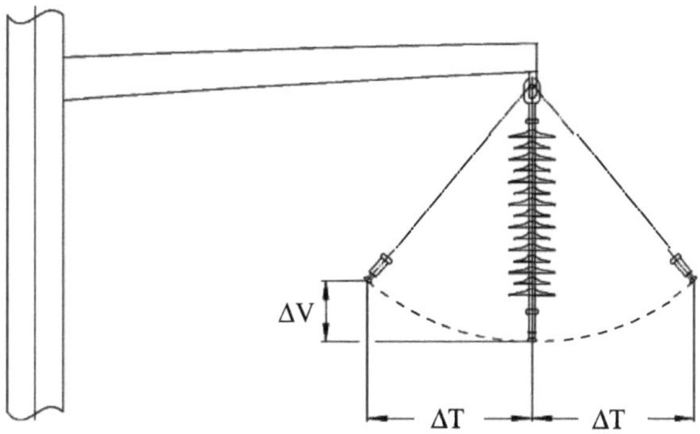

Fig. 4.1 Limits of line compaction [6]

4.2 Principles of Compaction

4.2.1 Influence of Conductor Suspension on the Tower

The type of conductor suspension has a key influence on tower geometry and therefore on tower dimensions. If the freedom of movement of the conductor is restricted by a suitable insulator arrangement, the line can be made more compact. Figure 4.1 shows a conventional suspension string which can move freely. The point at which the conductor is suspended will move accordingly, under the influence of the wind by a distance ΔT in the vertical direction and by a distance ΔV in the horizontal direction. These two paths have to be taken into account when determining the necessary electrical clearances to the mast and the right of way (ROW) along the line route.

4.2.2 Options for Line Compaction

If, by contrast, the conductor is "fixed" in its deflected position, the tower can be shorter owing to the reduction in the vertical direction ΔV. Additionally, a reduction transverse to the line direction ΔT minimises the necessary line corridor (at a specific voltage of the line) and also allows improved utilisation of an existing route enabling an increase in line voltage or by addition of a second line or circuit within the same route.

Over time, four different insulator arrangements have been used for line compaction: the V-string, horizontal post insulators, the combination of suspension strings with post insulators, and insulated cross-arms (horizontal V). These four arrangements will be presented briefly hereinafter.

Fig. 4.2 V-string with composite insulators

4.2.2.1 V-Strings

Insulator strings were used very early on in V-strings, generally transverse to line direction, to control movement of the conductors in the direction of the tower, which would have reduced the necessary electrical clearances (Fig. 4.2). V-strings are now also used successfully with composite insulators up to the highest voltage levels.

4.2.2.2 Line Post Insulators (Horizontal)

Line post insulators made of porcelain were introduced in the late 1950s and have proven to be advantageous when installed in existing lines of higher voltage. They also improve the aesthetics of the line and have thus helped the lines in question to gain wider acceptance. The use of line posts has increased massively in recent years owing to the development of accordingly designed composite insulators (in particular with core rods of large diameter) up to a voltage level of 220 kV (Fig. 4.3). These insulators, which are subject to bending load, have already been discussed in detail in Chap. 3.

Another important advantage of this design is shown in Fig. 4.4. In a "normal" suspension string, the insulator is deflected accordingly under the effect of the wind (or angular pull) on the conductor. The respective force acts on the tower at point "A". If the tip of the cross-arm is fixed by a line post insulator, the same load acts on the tower at point "B". Tower height can thus be reduced by the length of the path AB in the vertical direction, which in turn reduces the bending moments

Fig. 4.3 Line post insulator

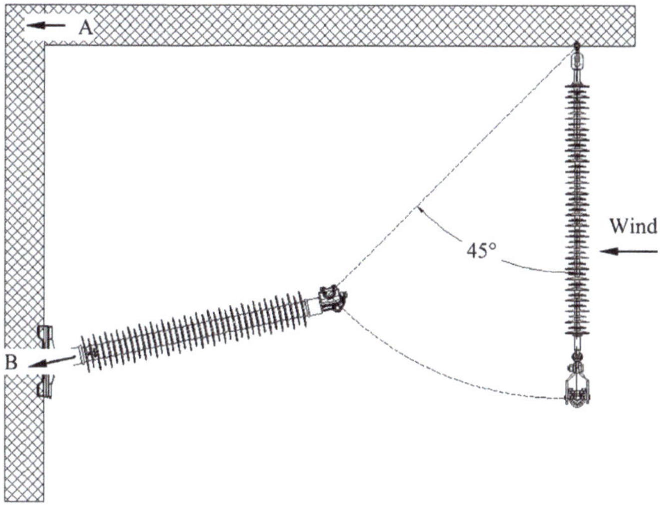

Fig. 4.4 Advantages achieved by the fixing the tip of the cross-arm [6]

on the tower and foundation and has a positive effect not only on the dimensions, but also on the cost of the supporting structure.

4.2.2.3 Suspended Line Post Insulators

So that the line post insulator is not additionally subjected to vertical loads (ice, conductor weight), these loads were taken up by a long rod insulator, which led to construction of suspended line post insulators (Fig. 4.5).

Fig. 4.5 420 kV suspended line post insulator

4.2.2.4 Insulated Cross-Arms

The concept of insulated cross-arms (horizontal Vs) was created on this basis in the 1960s. The innovative concept was to fix the suspension string (also called the brace) directly to the tower and to thus eliminate the necessity of a metallic cross-arm structure (Fig. 4.6). When loaded, the post works under compression (and very rarely under tension, although this type of load is not critical), whereas the brace always works under tension. The insulated cross-arm is often fixed to the tower in a pivoted manner so as to be able to rotate about an inclined axis, thus creating a restoring force should it be deflected out of its idle position in line direction for whatever reason (see Sect. 4.3.4).

4.3 Mechanical Design of Insulated Cross-Arms

This section deals with the mechanical properties and the design of insulated cross-arms [7].

When it comes to insulated cross-arms, a distinction is made between a rigid and a pivoted connection, depending on how the post insulator is fixed to the tower (Fig. 4.7). The main difference lies in the fact that, in the case of a rigid connection, the post insulator also has to withstand forces in line direction, since its rigid fixing, does not allow any rotation.

Fig. 4.6 420 kV insulated cross-arm

The loads which act on an insulated cross-arm (Fig. 4.8), are vertical loads *V* from the conductor, and, if present, from ice, horizontal loads *H* from the wind and, again if present, from angular pull, and longitudinal loads *T*, possibly from non-uniform conductor tension in the adjacent spans or from a conductor break, an extremely rare exceptional load.

In both cases, that is to say rigid or pivoted, the vertical loads are largely taken up by the brace, irrespective of the angle α between the brace and post.

By contrast, horizontal loads, when acting in compression, load the post in buckling. Parallel posts, posts of large rod diameter and even hollow core insulators are used for very high compression loads [8].

The insulator forces, that is to say the force *P* on the post and the force *B* on the brace, are calculated relatively easily using the following formulas (assuming $T = 0$):

$$P = H \cdot \frac{\cos(\varphi + \vartheta)}{\sin(\alpha)} + V \cdot \frac{\sin(\varphi - \vartheta)}{\sin(\alpha)} \tag{4.1}$$

$$\begin{aligned} B = H \cdot & \left[\frac{\cos(\alpha)}{\sin(\alpha)} \cdot \cos(\varphi + \vartheta) - \sin(\varphi + \vartheta) \right] \\ + V \cdot & \left[\frac{\cos(\alpha)}{\sin(\alpha)} \cdot \sin(\varphi - \vartheta) + \cos(\varphi - \vartheta) \right] \end{aligned} \tag{4.2}$$

The load in line direction *T* only needs to be taken into consideration in the case of a rigid connection, but may be the decisive factor in the dimensioning of the

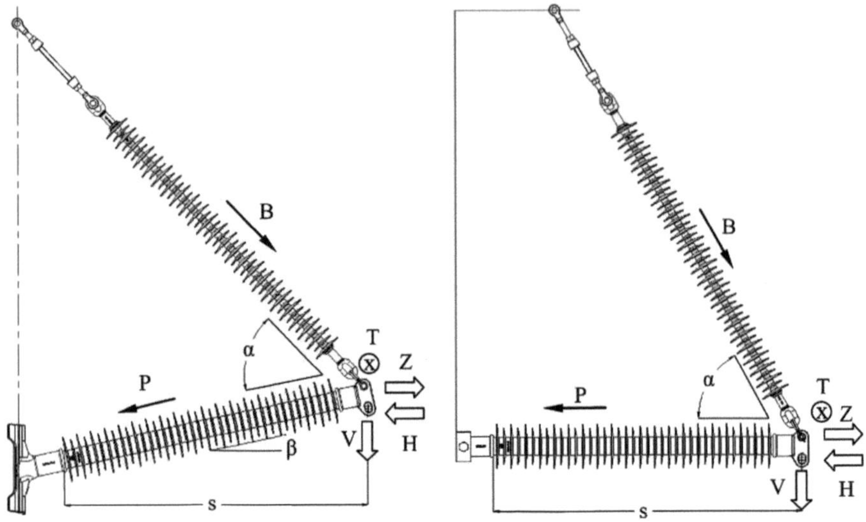

Fig. 4.7 Rigid (*left*) and pivoted (*right*) insulated cross-arm (*schematic view*)

Fig. 4.8 Geometry and forces in the case of an insulated cross-arm [7]

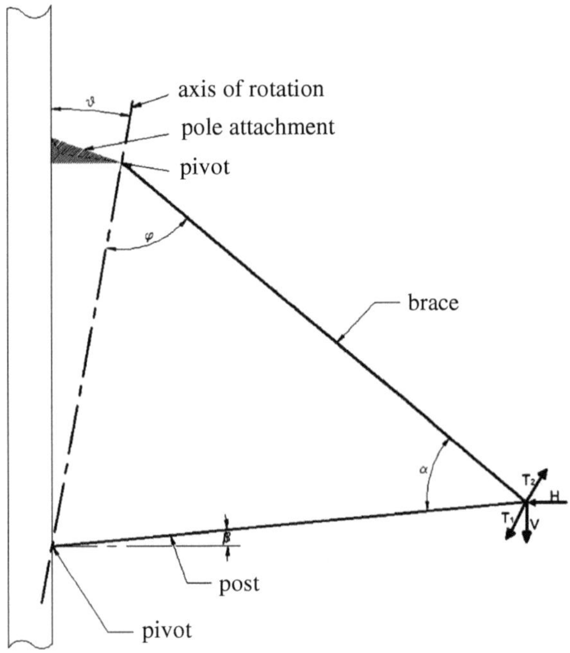

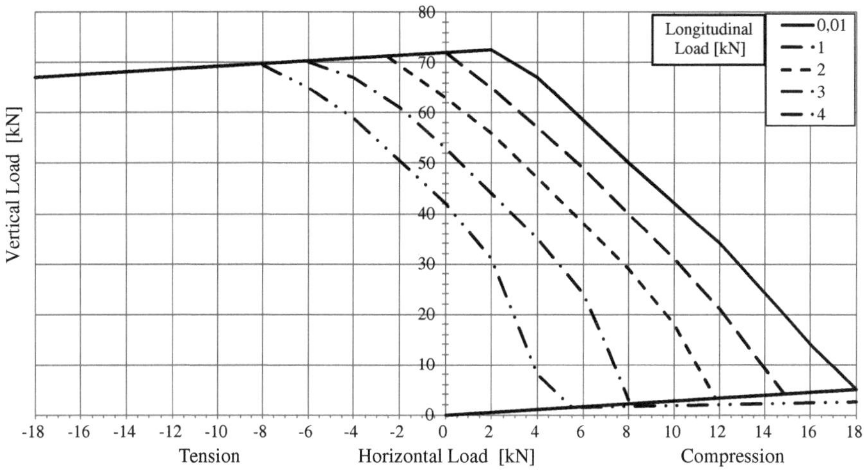

Fig. 4.9 Loading diagram for a rigid insulated cross-arm

cross-arm. In addition to the use of high-strength posts, the use of a bottom fitting having a fail-safe base and suspension of the line using a load release clamp may also be of use. Should the post of an insulated cross-arm fail nevertheless, there should follow no cascade, since the failing load of the remaining, healthy brace (composite long rod) is normally sufficient to prevent the risk of line release. If in doubt, a double tension string with composite long rods as the string is used (see also Fig. 4.19).

4.3.1 Rigid Insulated Cross-Arms

Since in insulated cross-arms having long posts and for lines of higher voltage significant deformations often occur, it is recommended to carry out a numerical simulation for the dimensioning of rigid insulated cross-arms. For example, the CIGRE working group for insulators used a commercial FEM calculation program to generate loading diagrams (application curves) as presented in the case of line post insulators in Chap. 3 [9]. This will be discussed briefly hereinafter.

For these calculations, it was assumed that all three forces ideally act at one point at the tip of the cross-arm. The inclination angle of the 63 mm post was 15° and its length 2,000 mm. The coupling angle of the 16 mm brace to the tower was 45°, said brace being assumed to be pivoted at either end, in contrast to the post, which was fixed in a rigid manner to the tower.

The corresponding loading diagrams are illustrated in Fig. 4.9. The load of the brace should not be negative (compression) so as to prevent a buckling of the brace leading to contact between the metal fittings of the two insulators; the latter is inadmissible and could give rise to RIV disturbances. With the horizontal angle of

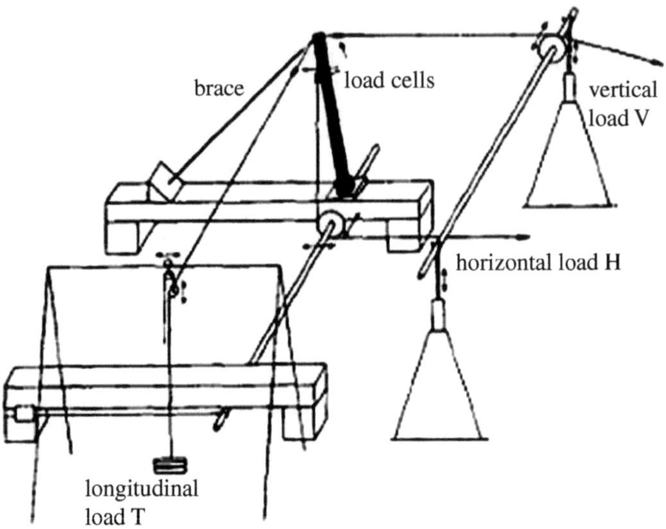

Fig. 4.10 Diagram of a test arrangement for static tests on insulated cross-arms [10]

the post insulator of 15° as used in this case, this condition leads to the inequality: $V > H \tan 15°$. In this diagram, the lower straight line corresponds to the equality $V = H \tan 15°$, that is to say the brace is not loaded along this line, or in other words the insulated cross-arm should not "work" below this line. The upper straight line in the diagram extends parallel to the lower straight line and corresponds to the maximum allowable tensile load of the brace.

The results of the calculation were checked by measuring the deflection of the tips of the cross-arm for nine different load situations. The average value of the deviation between the measured and calculated deflection was 8.1 %, with a maximum value of 15 % and a standard deviation of 4 %, which can be considered to be a relatively good correlation.

Similar tests were carried out in [10] using the test arrangement shown in Fig. 4.10.

In these tests, too, a good correlation with the calculations was established. In addition, the failure behaviour of such insulated cross-arms was able to be examined. It was observed that relatively short insulators, that is to say up to a length of approximately 1.5 m as are used up to 150 kV, fail, similarly to single horizontal posts (see Sect. 3.11). This means that the glass fibres on the compressed side of the post at the mouth of the bottom fitting fail first, followed next by the glass fibres on the tensile side, and followed lastly by delamination of the core owing to the high shear stress in the centre of the rod.

By contrast, in long, braced posts with lengths of over 2.0 m, as used at a voltage level of 220 kV and above, a completely different failure mechanism is observed. In this case, the FRP rods practically always fail as a result of fibre

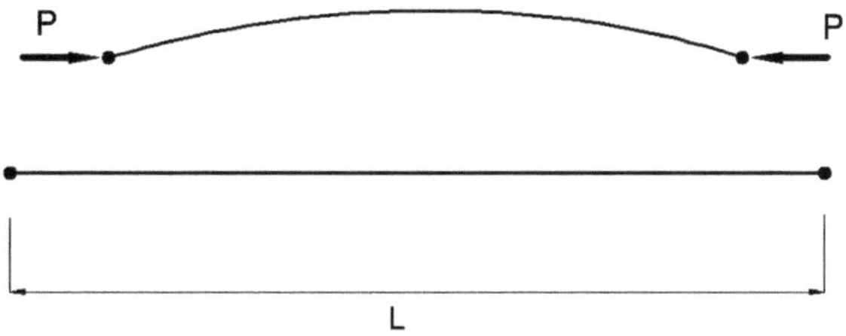

Fig. 4.11 Static system and loading of the cross-arm post

breakage on the convex side of the deflected insulator, at a distance of approximately 1/3 the length of the insulator from the bottom fitting. This is typical for the failure of relatively slender compressed rods with high resilient ultimate strain, as is the case with FRP, once the so-called Euler load has been exceeded, and is explained in detail in chapter 5 with reference to interphase spacers.

4.3.2 Pivoted Insulated Cross-Arms (Horizontal Vs)

This is the most commonly used arrangement, in particular for higher voltages, since it has a high level of mechanical strength and is also fault-tolerant.

The static calculation of the pivoted insulated cross-arm is similar to the calculation just described for rigid cross-arms. The external loads which act on the tip of the cross-arm in the horizontal and vertical direction (Fig. 4.8) are first decomposed into their respective components in the direction of the brace and of the post. With the assumption of an equilibrium of forces in the longitudinal (line) direction ($T_1 = T_2$) and with the same notation as in Fig. 4.8, the loads B for the brace and P for the post are given from the two equations above 4.1 and 4.2.

The composite long rod in the brace is loaded purely by tension and can be dimensioned relatively easily by the procedure detailed in Chap. 2. By contrast, the post is loaded by compression and, since it is articulated at both ends in this case (Fig. 4.11), can be calculated as a Euler buckling beam with the maximum tolerable compression load—the Euler load:

$$P_{Euler} = \frac{\pi^2 \cdot EI}{L^2}$$

However, when calculating buckling it must be noted, that by the practically unavoidable eccentric application of the compression load, (Fig. 4.12), the post insulator is additionally subjected to bending. The negative influence of this eccentricity can be seen in Fig. 4.13, where the results of buckling tests carried out on 63 mm rods are plotted in comparison to the maximum buckling load given as described above from the conventional buckling formula [7]. A considerable reduction of the measured failing load compared to the theoretical buckling load can be seen.

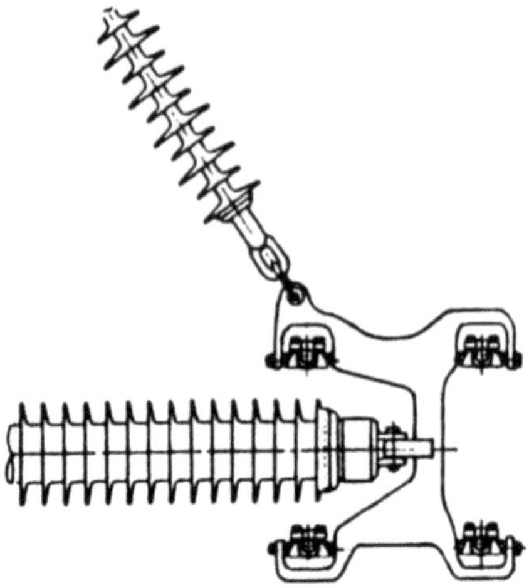

Fig. 4.12 Eccentric application of compression load via the conductor fitting in a quad bundle [11]

One possibility for reducing this harmful additional bending load is the use of special fittings, for example the so-called "boomerang" [11] (Fig. 4.14).

In practice, the unavoidable friction in the many joints of the cross-arm is also to be taken into consideration, since these may have a considerable effect on the mechanical behaviour of the cross-arm. Such effects can only be recorded sensibly by numerical simulation; this is now achieved using the finite element method (FEM). The results of such a calculation for a 420 kV insulated cross-arm are shown in Fig. 4.15 [4].

Due to all these uncertainties, it is recommended that such insulated cross-arms also be subjected to mechanical testing as a whole in addition to the type tests as stipulated in the various IEC Standards, some of which are unfortunately based only on the individual components (insulators) of the cross-arm. In such a mechanical test, the insulated cross-arm should withstand the maximum allowable loads without permanent deformation or other damage. This should be verified by the procedure described in Sect. 6.3.1 of IEC Standard 61952, that is to say by applying for 96 h load the maximum allowable load of the structure increased by a safety margin of 10 %. Once this test is complete, the composite post insulator should be cut and examined in accordance with the method described in IEC 61952 and here in Sect. 3.3.2.2 in order to check for any internal damage, both in the vicinity of its two fittings and in the centre of the rod, more specifically at the point at which the maximum deflection was observed during the test (see Sect. 3.3.2). This test can be carried out with insulators from the production preferably without silicone housing.

Figure 4.16 shows the test arrangement for a 420 kV cross-arm, whilst the test results are plotted in Fig. 4.17.

With the maximum load of $F\text{res} = 64$ kN as required in this case, a maximum deflection at the point of measurement y on the insulator of 23 mm was recorded. Up to this point (and beyond) the buckling of the composite post insulator extends in a very gentle manner, that is to say the insulator remains fully in the elastic

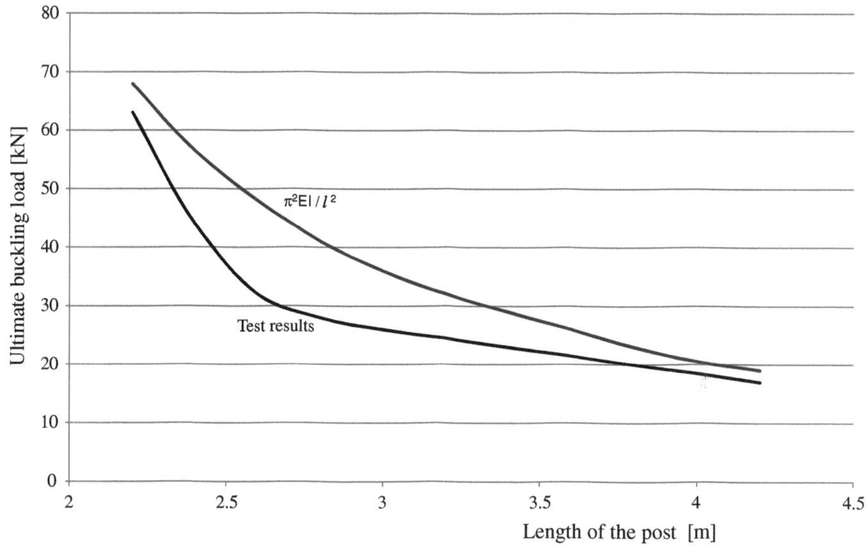

Fig. 4.13 Test results of buckling tests on insulated cross-arms having 63 mm posts of different length compared to the theoretical Euler load [7]

Fig. 4.14 Boomerang fitting ensures a central application of load [11]

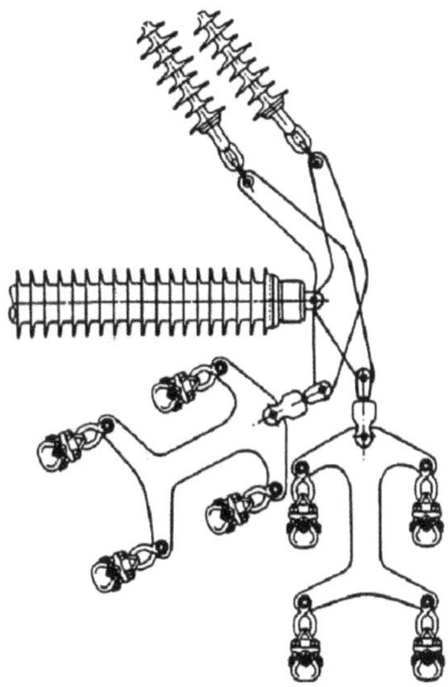

range and returns to its original position when the load is relieved, without any notable permanent deformation. The extension of 20 mm observed at the point of measurement x at maximum load of 64 kN is to be assigned to the -elastic- extension of the composite long rod of the brace.

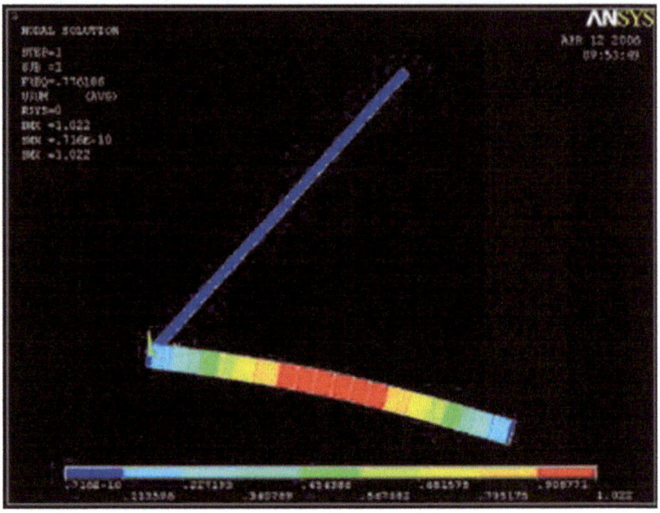

Fig. 4.15 Finite element calculation of an insulated cross-arm

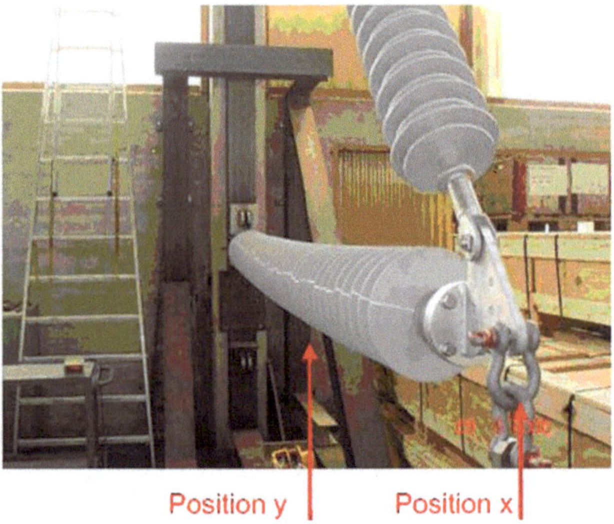

Fig. 4.16 Testing of a 420 kV insulated cross-arm

Application diagrams can be created for pivoted insulated cross-arms, similarly as for rigid cross-arms. Such an application diagram (Fig. 4.18) from [7] shows the permissible load combinations for a typical insulated cross-arm for 230 kV, wherein in this case, and in contrast to rigid cross-arms, no longitudinal loads have to be taken into consideration. The upper (solid) line is determined by the buckling

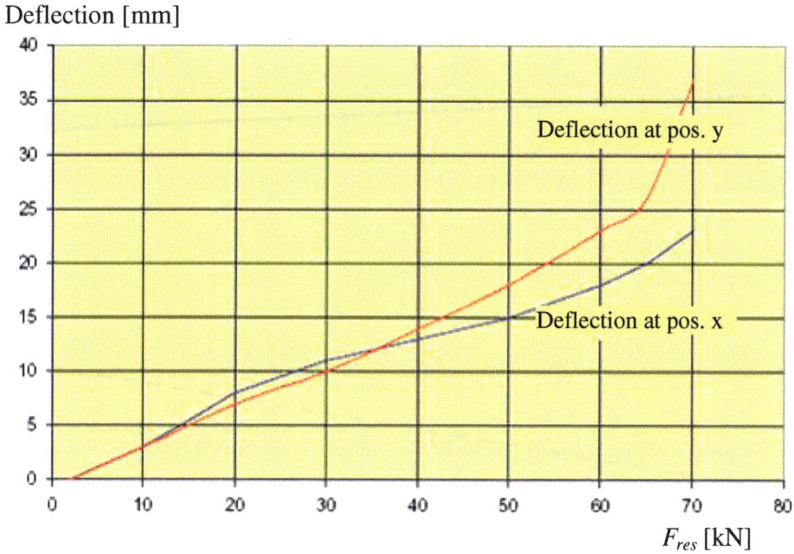

Fig. 4.17 Deformation of the insulated cross-arm with increasing load F_{res}

strength of the post. It is interesting to note that this strength is marginally dependent on the external vertical load. By contrast, the lower (dashed) line restricts the possible load combinations: Below this line, the brace would be subject to compression, which would have a negative effect on the strength of the entire cross-arm. This limit load can be determined from the Eq. (4.2) above (simplified for $\vartheta = 0°$), for the critical case that, specifically, the horizontal load H in the tower direction demonstrates:

$$B = V \cdot \frac{\cos(\beta)}{\sin(\alpha)} - H \cdot \frac{\sin(\beta)}{\sin(\alpha)} > 0$$

This leads to: $V/H > \tan\beta$ or $V > H \tan\beta$ for the minimum load V, which has to be effective so that the brace is not subject to compression.

4.3.3 Dynamic Loads

A key advantage of pivoted insulated cross-arms is their tolerance to longitudinal loads, as may occur as a result of unequal tensile forces in the adjacent spans of the cross-arm, since this then rotates in the direction of the greater tensile load. The mechanics of this load situation will be dealt with in the next section. We will first address how the high dynamic loads to which an insulated cross-arm is subjected can be established on the basis of experimental tests, more specifically in the rather

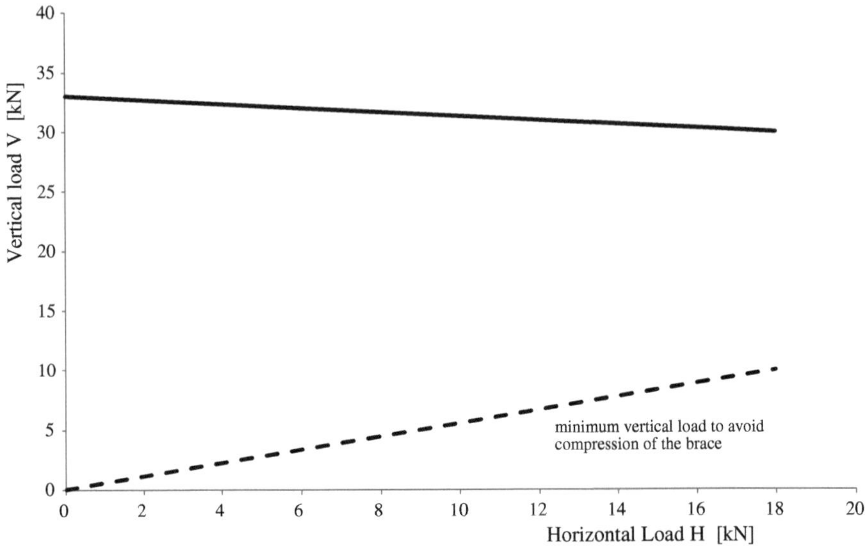

Fig. 4.18 Application diagram for a typical 230 kV insulated cross-arm (*horizontal V*) where = 65°, $\theta = 15°$ and $\alpha = 50°$, see Fig. 5.8 [7]

unusual case of a one-sided conductor failure. A sudden peak load is then produced, namely as a result of the acceleration caused by the rotation (according to Newton's laws), which has to be taken up by the cross-arm and in particular by the post insulator. The first documented tests and numerical simulations in this regard were carried out in the 1990s in [8] on the insulated cross-arm shown in Fig. 4.19, which, at that time, was also the first ever 420 kV insulated cross-arm to be put into operation.

In this case, a hollow core composite insulator was used for the post, since at this time, it was not possible to produce a solid core post insulator for the high compression loads required. The Euler load of a 76 mm (3 inch) post with a buckling length of 3.25 m (this length is necessary due to electrical considerations) is thus "only" 55 kN. By contrast, the compression load for the load situation which is critical in this case (wind at 0 °C, no ice, line angle of 1.8°) was 110 kN (including the required safety factor of 3.1). It is noted at this juncture that, in the meantime, it has become possible to produce FRP rods having diameters up to 150 mm, which makes it possible that such loads –and even higher- can mastered [4].

For these tests, the longitudinal load corresponding to conductor tension was generated by a weight, applied to the tip of the cross-arm by idler pulleys, and then suddenly dropped by means of an electromechanical device. The test arrangement is illustrated schematically in Fig. 4.20. The measured maximum value for transverse acceleration at the tip of the cross-arm was approximately 90 g. The

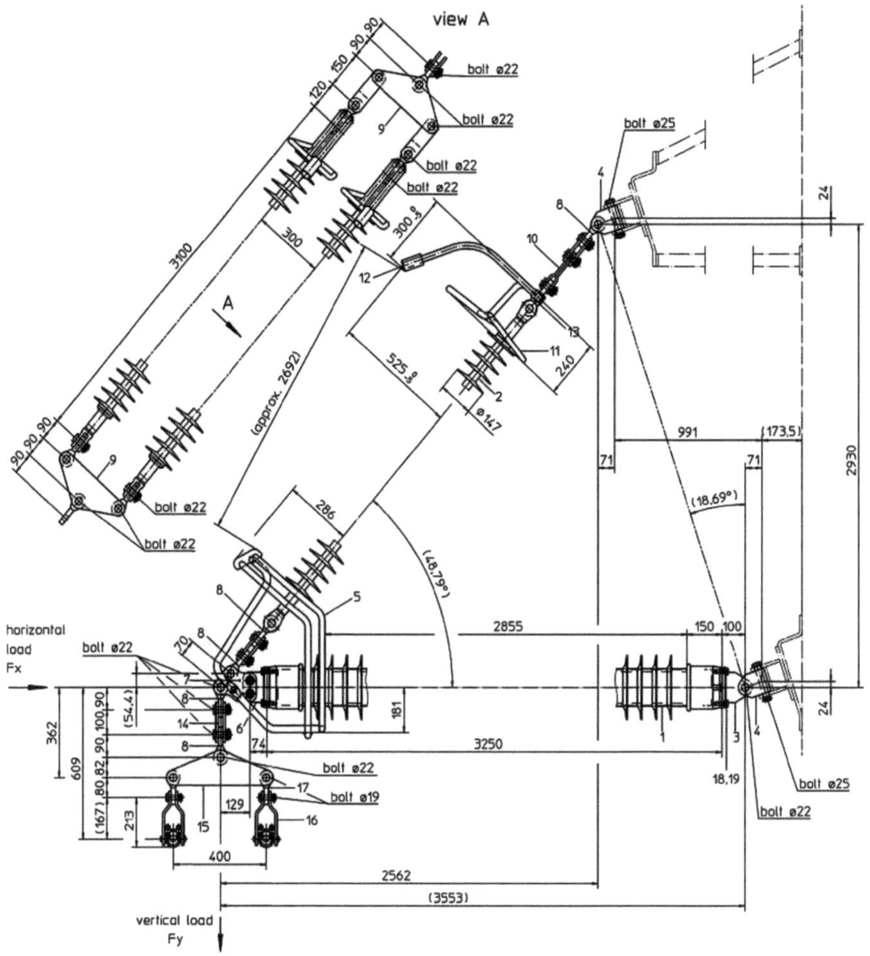

Fig. 4.19 The world's first 420 kV insulated cross-arm for EOS Lausanne (all dimensions in mm)

transverse acceleration over elapsed time is illustrated in Fig. 4.21 for two different weight masses mL of 2,700 and 3,600 kg.

4.3.4 Stability Tests

A particular advantage of pivoted insulated cross-arms is their ability to stabilise in the event of sudden conductor movements. Such movements take place when a temporary differential line tension occurs at the tip of a cross-arm within a line portion consisting of a number of spans, these movements being caused by gusts of

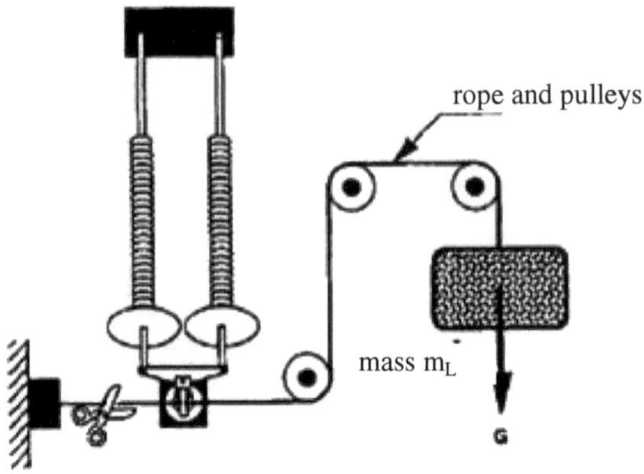

Fig. 4.20 Test arrangement for the dynamic test (*schematic view*)

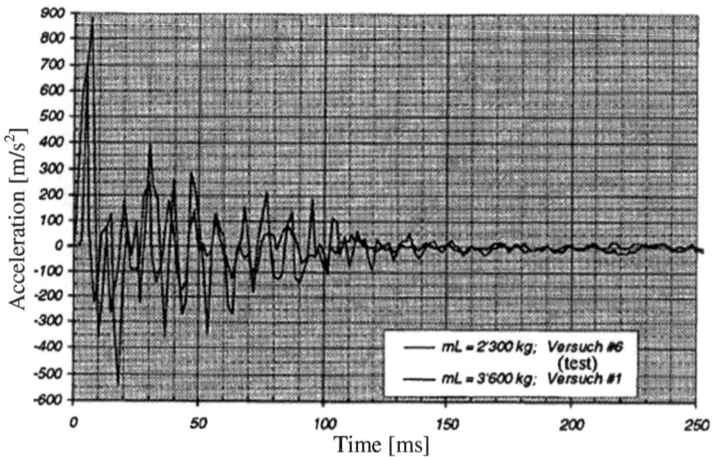

Fig. 4.21 Transverse acceleration over elapsed time at the tip of the cross-arm after load shedding

wind, irregular icing, spans of considerably different length, for example in mountainous terrain, and short circuit forces. It may be that, in particular in the case of long cross-arms for higher voltages, that the cross-arms are destabilised, which leads to considerable deflections with an associated reduction in the safety distances of the live conductor to the tower; in extreme cases the cross-arm could fail mechanically.

If the cross-arm/line system becomes destabilised, theoretically the cross-arm rotates until it reaches a second equilibrium position (Fig. 4.22). This is called "snap-through".

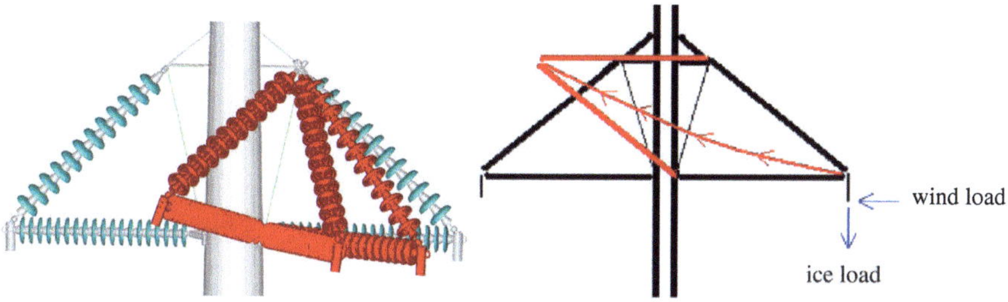

Fig. 4.22 Snap-through of an insulated cross-arm (*left*: model, *right*: *schematic view*)

Fig. 4.23 Snap-through of an insulated cross-arm in the event of 1:1 testing [12]

In most cases, the cross-arm regains stability however, if further rotation is prevented by blocking [12] (Fig. 4.23).

In a first step, the behaviour of a cross-arm in the event of differential line tension is to be examined [13, 14]. If there are differences in horizontal line tension in two adjacent spans, the tip of the cross-arm moves towards the higher tensile load. If, in this case, the angle of rotation of the insulated cross-arm is inclined to the vertical, the tip of the cross-arm is physically raised and the vertical line loads generate a restoring torque, which causes the equilibrium of forces in line direction to be re-established.

The insulated cross-arm is sketched in Fig. 4.24 in its equilibrium position, that is to say equal line tension prevails on both sides of the cross-arm. The angle ϑ denotes the inclination of the axis of rotation of the insulated cross-arm, and s is the length of the post with β the inclination thereof and V the weight of the conductor and any ice present plus the weight of the cross-arm. The weight force V can be split into two components; one is arranged perpendicular to the radius of rotation of the tip of the cross-arm p, and the other $V \sin \theta$ is arranged in the radial direction (in the direction of p) and thus perpendicular to the line.

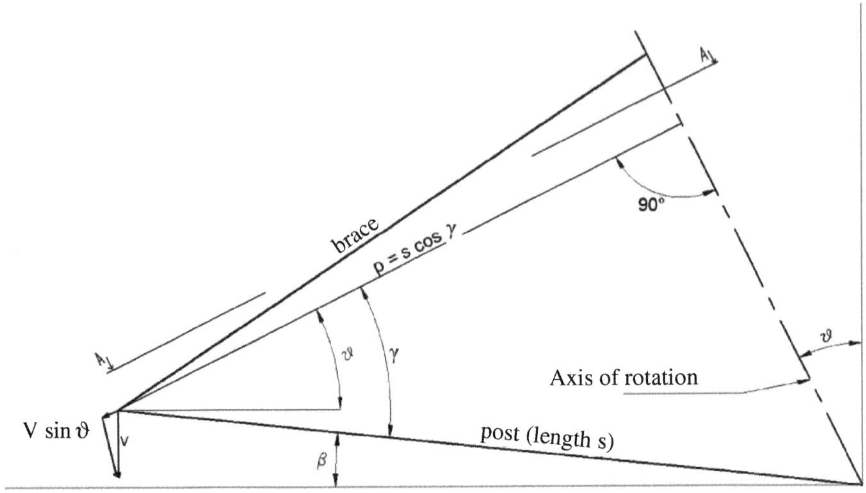

Fig. 4.24 Schematic diagram of the cross-arm in equilibrium position, *side view* [14]

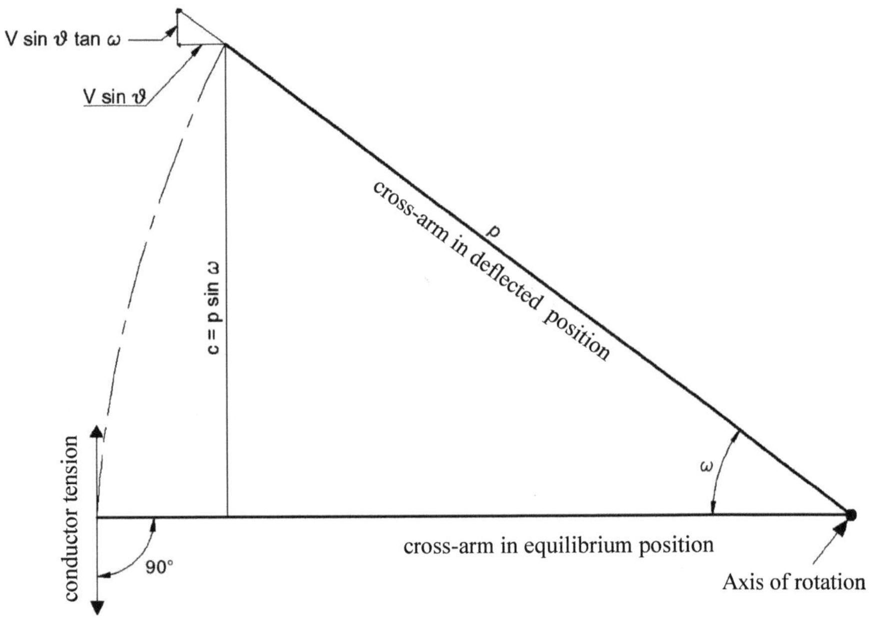

Fig. 4.25 Cross-arm in the deflected position, *plan view* (line of section A–A in Fig. 4.24) [14]

If the cross-arm is now deflected by an angle ω (Fig. 4.25), the force $V \sin \vartheta$ can then in turn be split into two components; one in the radial direction, that is to say in the direction of p, and the other, T, in line direction. The latter is also the force which causes the deflection c of the cross-arm in line direction. It follows that:

$$T = V \cdot \sin(\vartheta) \cdot \tan(\omega)$$
$$c = p \cdot \sin(\omega)$$

For a given insulated cross-arm, the variables *V*, ϑ and *p* are constant, and *T* and *c* are variable, since they are dependent on *V* and ω. For stability issues *H* has to be additionally considered, that is the force which is produced by wind transverse to line direction and/or by angular pull (if present), and in particular its component $H \cos \vartheta$ in direction of the radius of rotation p.

$$T_{HW} = (H \cdot \cos(\theta) + V \cdot \sin(\theta)) \cdot \tan(\omega)$$

If the force H cos θ is oriented in the same direction as the force V sin θ (Fig. 4.26, top), the stability of the cross-arm is increased. If the opposite is true (Fig. 4.26, bottom), this stability is reduced, that is to say the cross-arm becomes removed increasingly deflected from its equilibrium position.

The unavoidable friction effects in the pivot joints of the cross-arm were not initially taken into consideration in the discussion above. They have a stabilising effect: when the cross-arm is in the equilibrium position, the friction hinders deflection of the cross-arm and in case if the cross-arm is deflected, the frictional forces of the cross-arm counteract further deflection.

The section of an overhead transmission line which consists of a number of spans having a corresponding number of insulated cross-arms will be discussed in detail hereinafter. If a cross-arm which is free to rotate is deflected out of its equilibrium position due to the influence of wind and/or irregular line tension, as described above, the equilibrium of the other cross-arms in the line section will also be disturbed. Similarly, all cross-arms within a single section will be deflected if the line forces change from span to span due to irregularly distributed ice loads. However, this effect only occurs in long spans from 350 m, which are rather unusual for the compact lines dealt with in this instance, since the compression loads occurring in this case in the strut would be difficult to overcome.

As a result of the pivoting arrangement, via which the cross-arm post is fixed to the tower (Fig. 4.7, right), the entire cross-arm can deflect in the direction of the higher line tension. The line sag thus increases on one side and reduces on the other side of the cross-arm. This in turn leads to a greater line tension in the span in which line sag has been reduced, thus counteracting rotation. The resultant conductor load exerts an aligning force on the cross-arm and acts on the entire line portion in question in a self-stabilising manner.

Based on the work carried out by [15], a calculation program was developed which has made it possible to carry out such stability calculations for compact lines with insulated cross-arms. Different installation situations were thus examined in [14]. These will be presented briefly hereinafter.

A wind load on some of the spans was applied as critical load. Furthermore, a small wind component is also assumed in line direction, which corresponds to the actual situation encountered sometimes in reality. The mathematical algorithm is thus "triggered" and the critical wind speed at which the system becomes destabilised is established. For example, the critical wind speed for stability of a line

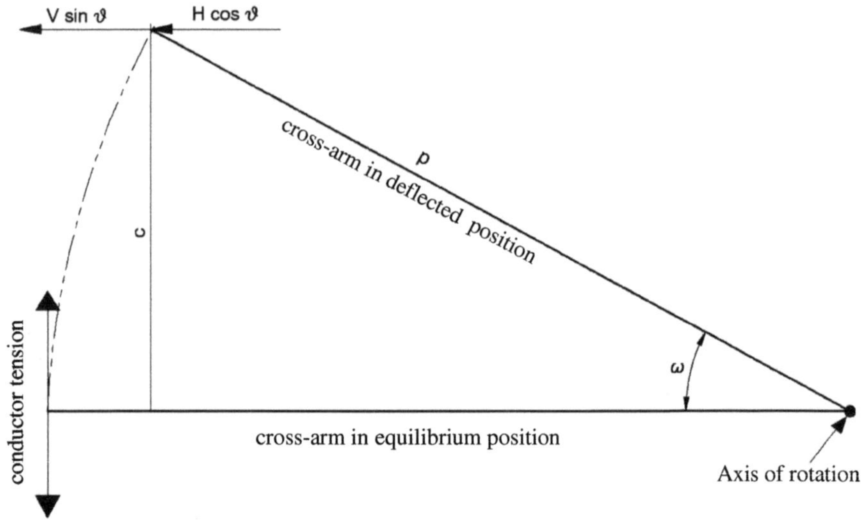

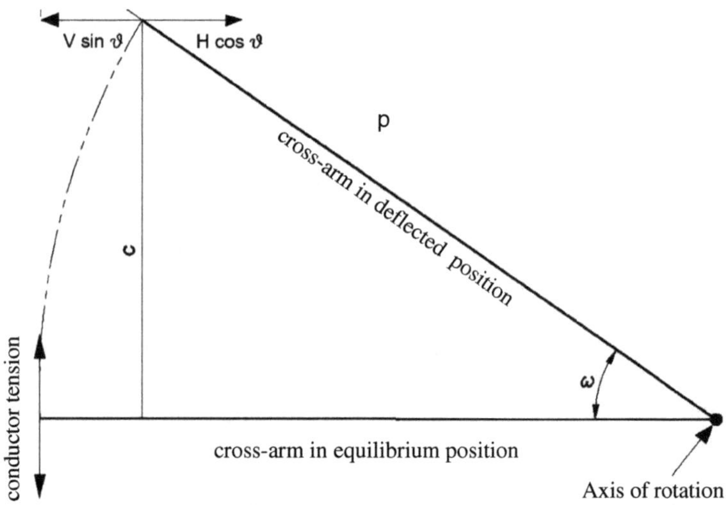

Fig. 4.26 Cross-arm deflected with the respective forces (plan view), *top* stabilised, *bottom* critical [14]

section of a 345 kV compact line was thus calculated as a function of the number of spans, and is illustrated in Fig. 4.27 with the inclination angle θ of the cross-arm as parameter. As can be seen, stability improves (that is to say the critical wind speed increases) with a decreasing number of spans and with an increase in the inclination angle of the cross-arm.

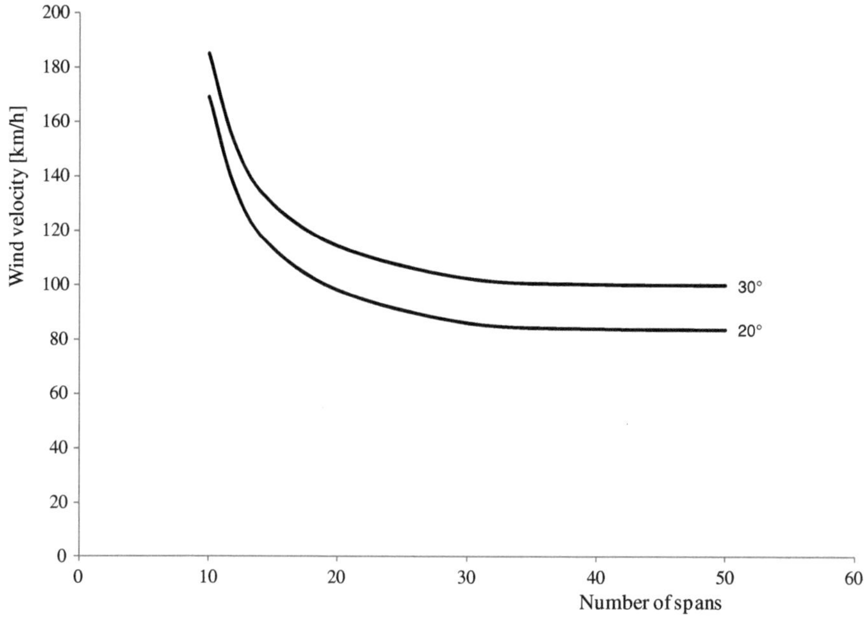

Fig. 4.27 Critical wind speed as a function of the number of spans with the inclination angle θ as a parameter [14]

As expected, wind direction in relation to line direction also has a considerable effect on stability. Figure 4.28 shows the critical wind speed for a line section formed of 20 spans and an inclination angle of the insulated cross-arms used of $\theta = 20°$. If the wind blows predominantly in line direction, the stability of the span section increases accordingly. The most unfavourable condition is when high winds blow perpendicular to line direction and also have a notable component parallel to line direction.

The wind stability of compact lines with insulated cross-arms can be improved by the following measures [7]:

- Increasing the inclination angle θ or the angle α between brace and post (Fig. 4.8)
- Increasing the vertical loads on the cross-arm, for example by additional weights
- Increasing conductor tension
- Reducing the individual span lengths and/or the number of spans within a line section
- Reducing the line angle in angle towers
- Using "stabilising cross-arms" in long line sections, the pivoted post of the normal cross-arm being replaced by two posts mounted rigidly at an acute angle to one another (Fig. 4.29).

Another solution for improving stability was proposed by [16]. With this solution, a somewhat differently designed stabilising cross-arm is used which

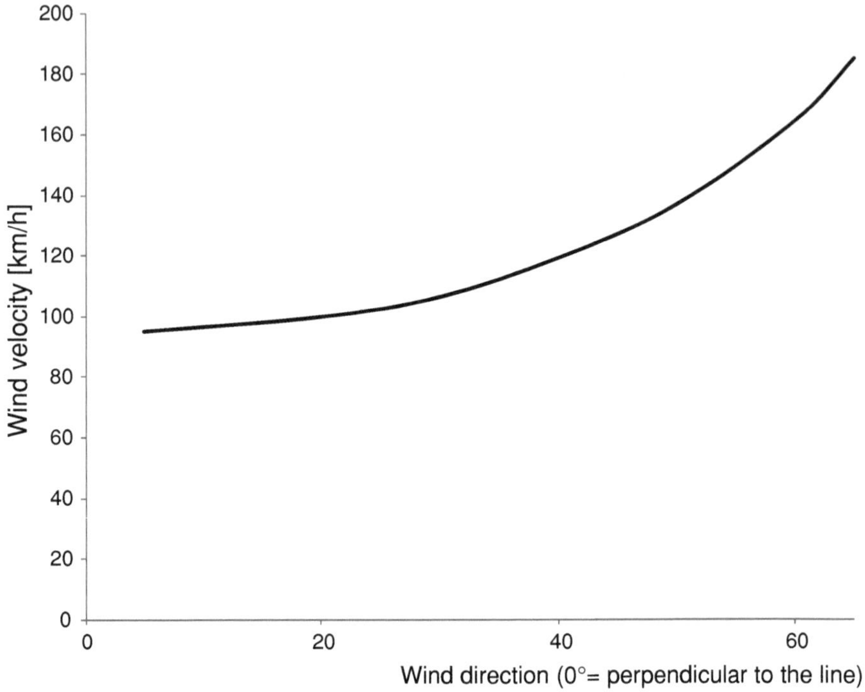

Fig. 4.28 Critical wind speed over the angle of attack of the wind on the line (0°: Wind perpendicular to the line) with a cross-arm with $\theta = 20°$ [14]

consists of two normal cross-arms arranged at a horizontal opening angle to one another (Fig. 4.30).

Depending on the results of the stability calculation, this cross-arm will replace one or more normal cross-arms in the examined line section. Its stabilising effect takes hold when it rotates, creating an aligning torque, given by $M = L \times d$, where L is conductor tension and d is the lateral distance between the two conductor loads L (Fig. 4.30, right).

4.4 Innovative Applications of Compact Lines

Some examples from the authors' personal practical experience of compact lines which were considered to be innovative at their time of implementation will be presented below.

Fig. 4.29 Special cross-arm with composite insulators (three-leg) to increase stability of a 345 kV compact line [7]

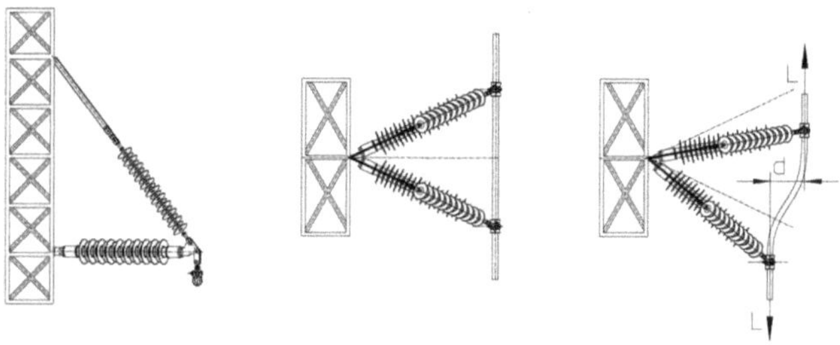

Fig. 4.30 Schematic diagram of a pendular twin cross-arm for stabilising rotation of the cross-arm [16]

4.4.1 400 kV Line with Hollow Core Insulators in Switzerland

At the end of the 1990s, a new 400 kV line had to be constructed in West Switzerland between Geneva and Lausanne over the same route as an existing 125 kV double line. In addition, two single-phase 132 kV circuits owned by SBB had to be suspended from the same towers. Topography and housing along the line route resulted in the fact that only one compact line in a line section would satisfy the necessary clearances and the prescribed EMF values [17]. In addition, the requests of local residents for an improved appearance of the new tower were met, as can be seen from the comparison shown in Fig. 4.31.

The high compression loads on the posts of the insulated cross-arms meant that a solution using hollow core insulators had to be developed, since at that time solid core post insulators were only available up to a maximum of 83 mm rod diameter.

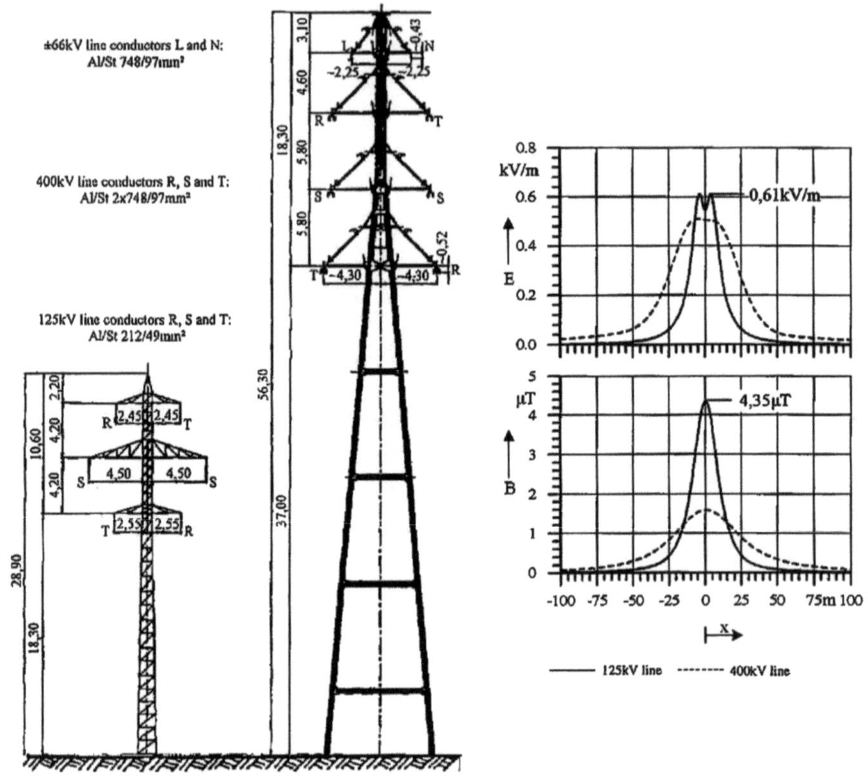

Fig. 4.31 Comparison between the existing 125 kV line and the new 400 kV compact line. *Left*: comparison of size and row *right*: comparison of the **E** and **B** fields

The solution developed in close collaboration with the utility and the manufacturer of the towers (Fig. 4.32) was tested extensively and has been in operation for almost 15 years without any problems being encountered.

Reference can be made to [8] and [18] as well as to Sect. 4.3.3 for further details.

4.4.2 Emergency Restoration Systems with Composite Insulators

Even high-voltage overhead transmission lines have not been spared by the forces of nature in recent years, which has led to the development and use of "emergency restoration systems". In addition, network operators are continuously encountering situations, for example in the case of re-routing or intersections, in which emergency restoration systems which can be installed quickly and reliably are required. Evidently these must be compact, which has led to the development of a family of

Fig. 4.32 Pictures of the world's first ever 400 kV *compact lines* with composite isolators

emergency restoration systems having insulated cross-arms with composite insulators [3]. Composite insulators are suitable for such emergency restoration systems because they are lightweight, and thus easy to handle in emergency situations, and are break-resistant. It is therefore no longer necessary to cater for spare replacement insulators, as was previously the case with porcelain and glass.

This is of particular importance for container-optimised transport of emergency restoration systems in difficult terrain.

Figure 4.33 shows a 420 kV emergency restoration system, as was used in the year 2000 in the re-routing of a 420 kV line in Holland. This re-routing was necessary since construction of a new steel lattice tower was required by the intersection with a stretch of high-speed railway line, and the resultant increase in tower height by approximately 10 m could not be implemented quickly enough within the time allowed by the necessary outages and interruption to rail operation. The conversion took place in a record-breaking 5 h.

Fig. 4.33 Emergency restoration systems with insulated cross-arms with re-arrangement of the quad bundle

4.4.3 *420 kV Double Circuit Line with Solid Core Composite Cross-Arms*

When planning a new connection between two substations in Dubai, the course of the new route over a 4.5 km stretch was restricted to 11 m by an existing 420 kV line and also by the city motorway. In addition, the power to be transferred measuring 2,600 MVA required a relatively strong quad bundle (Aldrey conductors with 28.4 mm˘), the high pollution encountered in this area required a high specific creepage distance of 55 mm/kV for the insulators, and the high short circuit current of 63 kA/s required insensitive insulating material and adapted short-circuit protection fittings.

All of these conditions were able to be met by the selection of insulated cross-arms having special protection fittings mounted on tubular steel poles. After initial considerations as to whether to use a double beam for the post owing to the high compression loads, this idea was dropped in favour of a single post (Fig. 4.6) thanks to the improved understanding of the mechanical behaviour of composite insulators under compression loads, the FE simulations developed specifically for this, and the associated tests (see Sect. 4.3.2), and also as a result of the technological progress made with regard to the manufacture of "thick" FRP rods (now

Fig. 4.34 The new 420 kV *compact line* (under conductor tension) next to the existing conventional 420 kV line, *right of the image* the city motorway, *in the background* the skyline of Dubai

possible up to 150 mm). This line has been in operation successfully since 2004 [4] (Fig. 4.34).

The current state of application of compact lines is discussed in [19].

References

1. Voyatzakis Y (1988) 150 kV transmission lines on tapered steel poles supporting insulator cross-arms in Greece. International conference on overhead transmission line design and construction, London Nov 1988
2. Havard DG, Nashid MS, Meale JR, Foty SM (1991) Transmission line compaction in Ontario Hydro, CIGRE symposium leningrad. Paper 200-05 Session 33–91
3. Schmuck F, Papailiou KO, Kolmeijer PJ, van Wolven JF (2004) Further developments of compact lines for 420 kV with silicone insulators and their advantages for applications in emergency restoration systems. CIGRE Paris, B2-304
4. George J, Iqbal M, Papailiou KO, Huiber W, Schmuck F (2007) Dewa's new 420 kV compact line as an example of maximizing capacity of overhead transmission lines in an urban environment. GCC Power 07 Dubai
5. Papailiou KO, Schmuck F (2011) The past experience and future trends with compact lines to solve right-of-way issues. CIGRE international symposium—the electric power system of the future, Bologna, Sept 2011

6. Pohlman JC, Pinkham TC, Baker AC (1991) A look at compacting transmission lines using armless construction under today's constrains. Paper 700-04, CIGRE Symposium, Leningrad
7. Baker AC, Bernstorf RA, del Bello E, Farzaneh M, Hill RJ, King B, Philips AJ, Powell DG, Shaffner D, Steward GA, Grisham T (2008) IEEE guide for braced insulator assemblies for over- head transmission lines 60 kV and greater. IEEE Trans Power Delivery 23(2):785–791
8. Ammann M, Papailiou KO, Dalleves P, Leva M, Villa S (1998) A new 400 kV line with compact towers and composite insulated cross-arms. Paper 22/33/36-06, CIGRE, Paris
9. CIGRE (2002) Guide for the evaluation of composite line post insulators subjected to combined mechanical loads Electra 203, CIGRE, Paris
10. Dumora D, Feldman D, Gaudry M (1990) Mechanical behaviour of flexurally stressed composite insulators. IEEE Trans Power Delivery 5(2):1066–1072
11. Paris L, Pargamin L Parraud R (1991) Application of composite insulators for overhead compact lines, CIGRE symposium Leningrad. Paper 400-04 Session 33–91
12. Bernstorf RA (2008) Braced line post ratings, panel discussion-braced line posts, IEEE Working Group 15.09.09, IEEE
13. Lummis J, Pinkham TA (1970) Mechanical behavior of horizontal-V arrays under unbalanced conductor forces. IEEE summer power meeting and EHV conference, Los Angeles, July 12–17 1970
14. Baker AC, Murray PE, Mozer JD (1982) Computer aided analysis of wind loads on horizontal Vee type transmission line systems. IEEE Trans PAS, PAS-101(8):2415–2419, Aug 1982
15. Campbell DB (1970) Unbalanced tensions in transmission lines. J Struct Division, Proceedings of the American society of civil engineers, 96(ST10), Oct 1970
16. Paris L, Comellini E (1969) Bi-dimensional structures for transmission lines, IEEE summer power meeting, Dallas, 2–27 June 1969
17. Filippopoulos G, Tsanakas D, Kouvarakis G, Voyatzakis J, Ammann M, Papailiou KO (2002) Optimum conductor arrangement of compact lines for electric and magnetic field minimization—calculations and measurements, Med Power, Athens, 4–6 Nov 2002
18. Papailiou KO (1998) Contribution to question 1.3, Session Proceedings Joint Session 22/33/36, CIGRE, Paris 1998
19. Papailiou KO, Schmuck F (2012) Past experience and future trends with compact lines to solve the right-of-way issue: CIGRE Bologna 2011, The electric power system of the future—integrating supergrids and microgrids. Paper 242

Chapter 5
Interphase Spacers

Symbols and Abbreviations

AGS	Armour Grip Suspension
BKW	Bernische Kraftwerke (Power Utility of Bern, Switzerland)
CIGRE	Conseil International des Grands Réseaux Électriques (International Council on Large Electric Systems)
DB	Deutschen Bahn (German Railways)
EDF	Électricité de France (Electricity of France)
ESCC	CIGRE Task Force on Effects of Short-Circuit Currents
EVU	Electric power utility
FEM	Finite element method
FRP	Fibre reinforced plastic
IEC	International Electrotechnical Commission
IPS	Interphase spacer
VEW	Vereinigte Elektrizitätswerke Westfalen (Electric power utility of Westfalia, Germany)
α	Maximal Euler angle at load application point ($s = 0$)
ϑ	Maximal tangential angle of the vibration curve at the location of the interphase spacer ($t = 0$, $x = 0$)
ϑ	Euler angle
$\kappa(x)$	Curvature of the IPS
v	Terminal velocity of the deflected conductor
$\sigma_{b,max}$	Maximal bending stress in the post-buckling regime
$\sigma_{b,zul}$	Ultimate bending stress of the FRP rod
σ_{max}	Maximal bending stress in the centre ($x/2$) of the IPS
τ	Time constant of the short-circuit path
φ	Angle of the short-circuit impedance
ϕ	Conductor (or sub-conductor) diameter
A	Conductor cross-section
A	Maximal amplitude of galloping in a sub-span
$A_{s\text{-}s}$	Galloping amplitude peak-to-peak

K. O. Papailiou and F. Schmuck, *Silicone Composite Insulators*, Power Systems,
DOI: 10.1007/978-3-642-15320-4_5, © Springer-Verlag Berlin Heidelberg 2013

b	Conductor acceleration
C_1, C_3	Constants (depend on line voltage)
d	Distance between two parallel conductors
d_f	Maximum sag in the sub-span between two IPS or between an IPS and the next tower
d_{sp}	Rod diameter of the IPS
E	Modulus of Elasticity (Young's modulus) of the conductor
E_b	Strain energy of the conductor
E_{sp}	Modulus of Elasticity (Young's modulus) of the IPS
$E(p)$	Complete elliptic integral of the second kind
f	Conductor sag
$F(t)$	Short-circuit force per unit length
g	Acceleration of gravity
$i(t)$	Short-circuit current
I	Moment of inertia of the FRP rod
I_{sp}	Moment of inertia of the IPS
I_{eff}	Initial short-circuit alternating current *Ik2* Permanent short-circuit current
I_{k2}	Permanent short-circuit current
k	Parameter
k	Correction factor depending on conductor type, conductor diameter and wind pressure
$K(p)$	Complete elliptic integral of the first kind
ℓ	Half-length of rod
L	Length of conductor, length of rod
L_{sp}	Length of IPS
M	Conductor mass per unit length
$M_{(X)}$	Bending moment along an IPS
M_{max}	Maximal bending moment in the centre ($x/2$) of an IPS
n_{sp}	Number of IPS in a given span
p	Parameter
P	Compression force
P_{Euler}	Euler buckling load
s	Coordinate in direction of the rod axis
S	Span length
$S/2$	Length of the sub-span
t	Time
t_{sc}	Duration of short circuit
T	Conductor tension
T_o	Horizontal conductor tension
U_b	Strain energy of an IPS
U_s	Strain energy of a conductor
U_{sc}	Kinetic energy of the conductor during short circuit
W	Total weight

W_c, w_c	Conductor weight
W_i, w_i	Ice load
x	Length coordinate of the IPS
xa	Length reduction of the rod due to compression
x, y	Cartesian coordinates for galloping
ya	Maximal deflection in the center of the rod
$y(t)$	Conductor deflection curve during vibration
z	Length of the IPS

5.1 Introduction

Line compaction is a proven method not only for reducing line width, but also for increasing the transmission capability of the line due to lower surge impedance. In addition, the electric and magnetic field at the edges of the line corridor are reduced to such an extent that they are not expected to have any negative effects on the environment.

As already explained in the chapter on compact lines, line compaction is primarily based on the principle of minimising interphase spacing. As a result of this, and also since the interphase spacings are somewhat smaller than those encountered in the case of normal overhead transmission lines, there is greater risk of interphase flashover with a compact design when two conductors are moved close to one another, for example as a result of wind, ice shedding or short circuit; this may lead to an interruption in service. Interphase spacers (IPS) are used to avoid this (Fig. 5.1).

In the last few years, increased use of interphase spacers of composite design has been made possible due to the generally massive advances made in the development and acceptance of composite insulators. In this application, it has thus been possible to utilise the specific properties of composite technology, such as low weight, high flexibility and excellent hydrophobicity to name just a few [1].

5.2 CIGRE Survey

5.2.1 Assessment of Survey Responses

As early as 1990, CIGRE began conducting a survey on the global use of interphase spacers [2]. It was found that, at that time, 32 electric power utilities from 13 countries (6 of which within Europe) had used approximately 13,000 interphase spacers over a period of 20 years at practically all voltage levels, from 11 to 420 kV (Fig. 5.2).

Fig. 5.1 400 kV Interphase spacer in the transmission line at the Vorab glacier

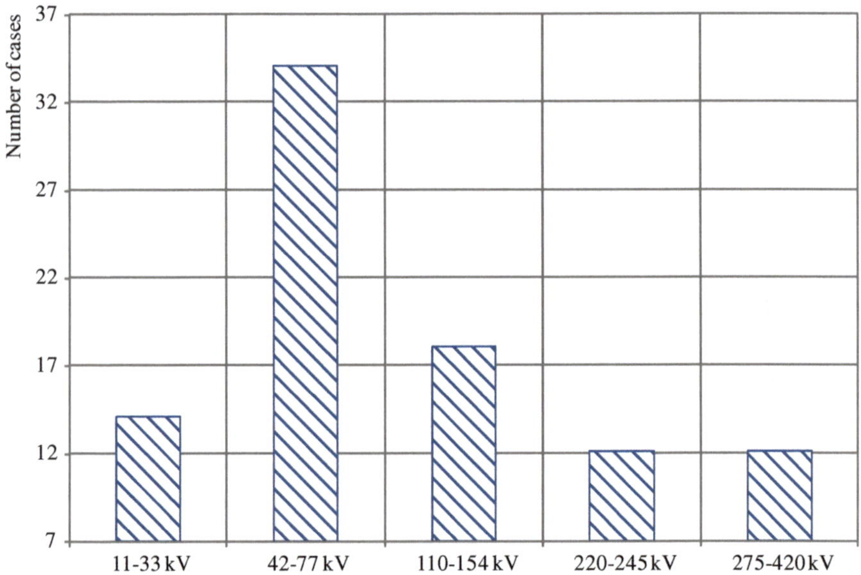

Fig. 5.2 Number of IPS installed worldwide according to voltage level

Table 5.1 Details regarding the used interphase spacers

Total length (mm)	Diameter(mm)	Total mass (kg)	Material	Number of Installations
670–1,999	65	40–44	porcelain	2
	12–30	3–20	composite	9
2,000–2,999	60–65	40–49	porcelain	25
	15–42	6–15	composite	4
3,000–3,999	65	40–154	porcelain	6
	16–43	15–70	composite	4
4,000–4,999	60–90	74–328	porcelain	3
	24–37	36	composite	7
5,000–5,999	60–75	81–142	porcelain	4
	24–63	26–179	composite	9
6,000–8,500	32–47	31–35	composite	4
12,400	–	–	composite	1

The vast majority (92 %) of the EVUs surveyed cited the main reasons for installation of interphase spacers as providing a solution to technical problems, such as:

- Clashing conductors, in particular over long spans in the event of high winds and ice load or ice shedding. Resultant short circuits, which in particular damage the strands in the outer layer of the conductor and also lead to an increase in radio interference. Loose strands which then hang freely in the wind give rise to further short circuits.
- Clashing of the conductors during galloping, which occurs at high wind and with non-uniform icing of the conductors.
- Avoiding flashover between the conductors in the span, since this is where the most intense galloping amplitudes occur.
- Maintaining the necessary (electrical) clearances to earthed structures (for example towers).
- Problems in obtaining rights of way for new line routes.

At the time, of those questioned, 4 % gave line compaction as the reason for installing interphase spacers, which would presumably be considerably higher nowadays as a result of the almost exclusive use of interphase spacers of composite design.

Table 5.1 shows the key parameters of interphase spacers installed at the time of the survey:

In 54 installations, interphase spacers were fixed to the conductor using armour rod fittings (Fig. 5.3). It can be seen from Fig. 5.4 that there is a clear preference for use of pivoted conductor coupling of interphase spacers, irrespective of voltage level; this is advantageous in terms of the large rotations often occurring at this point, which may damage the conductor.

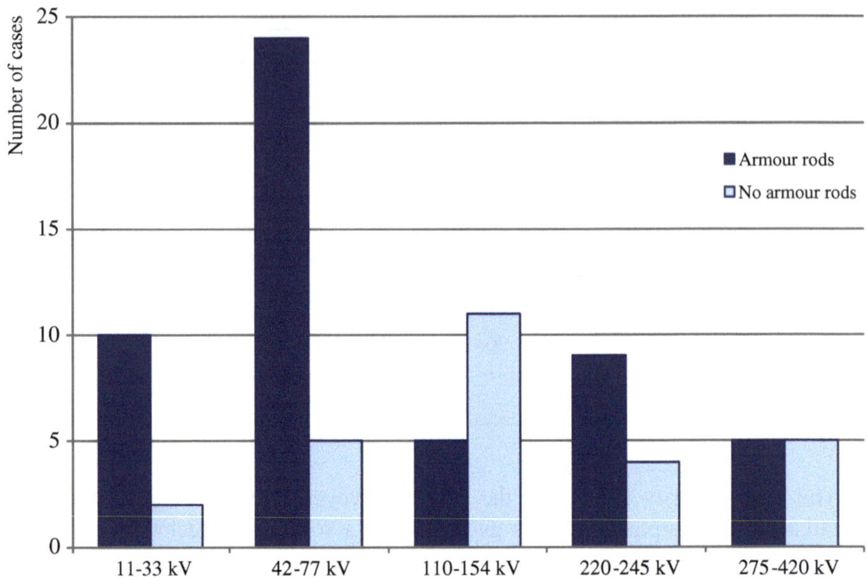

Fig. 5.3 Use of armour rods in IPS according to voltage level

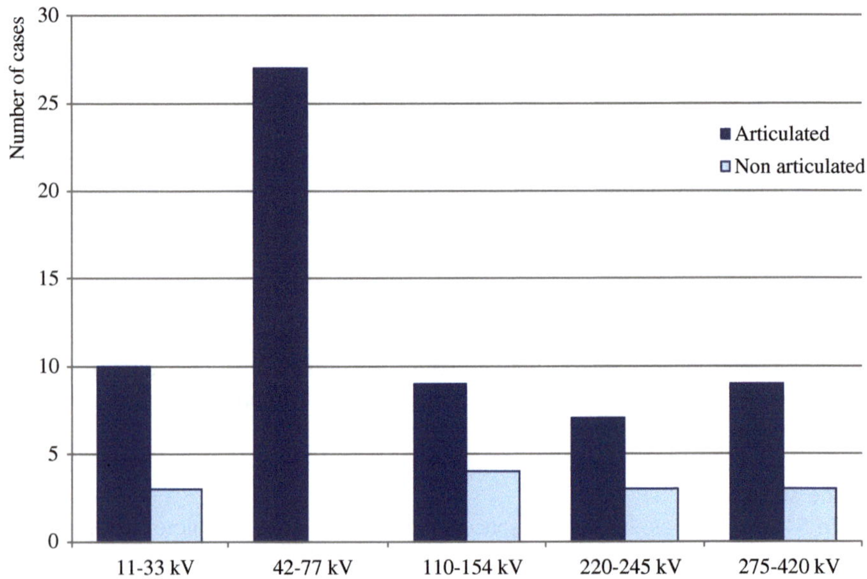

Fig. 5.4 Design of the conductor coupling of IPS according to voltage level

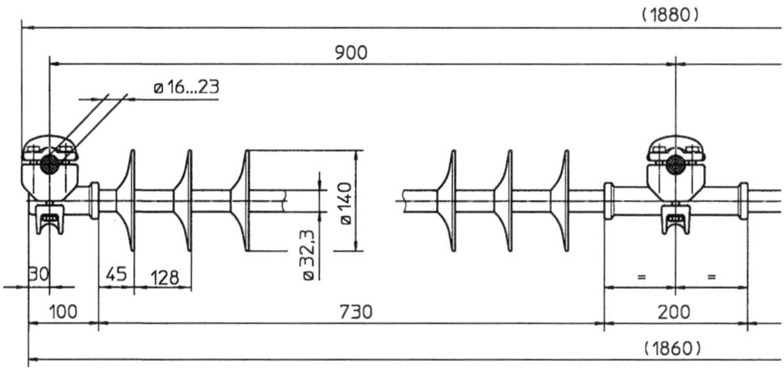

Fig. 5.5 Rigid conductor connection (medium voltage)

5.2.2 Operational Experience

The CIGRE survey proved that user experience of interphase spacers was largely positive, as shown by the following extracts from the respective report [2] :

- Conductor clashing caused by wind and non-uniform ice load were not encountered following installation of interphase spacers.
- Interphase spacers proved to be a cost-effective alternative to installation of additional towers.
- Interphase spacers installed initially and made of porcelain were replaced with IPS of composite design since the former had experienced mechanical failure.
- The rigid attachment of IPS to the conductor failed under shock load (caused by galloping); this rigid design has been replaced by a flexible attachment technique.
- Galloping could not be controlled completely, but conductor clashing and the associated conductor damage could be prevented. In addition, abrasion at the insulator coupling points and at the towers caused by galloping was able to be avoided.
- No further flashover between the phases was observed following installation of IPS in critical spans.

5.3 Attachment Techniques

As already mentioned, it is the technique used for attachment of interphase spacers to the conductor which is of key importance for long-term, reliable operation of the assembly. For medium-voltage overhead transmission lines, a rigid connection (Fig. 5.5) is often sufficient. For higher voltages, an articulated fitting is to be provided (Fig. 5.6), since then neither the sensitive conductor nor the interphase spacer will be overloaded by the torque generated by the above-mentioned loads.

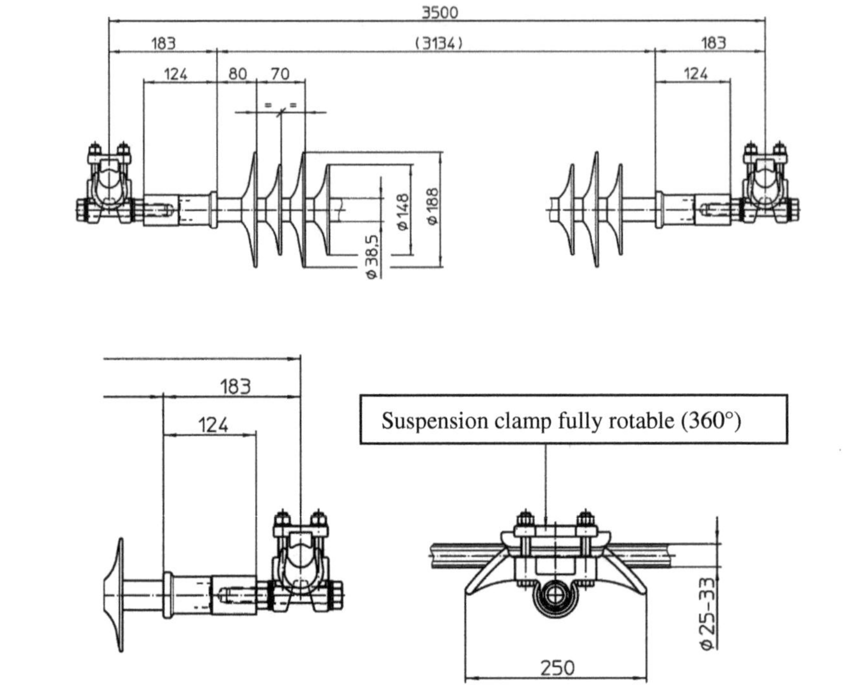

Fig. 5.6 Pivoted conductor connection (high voltage)

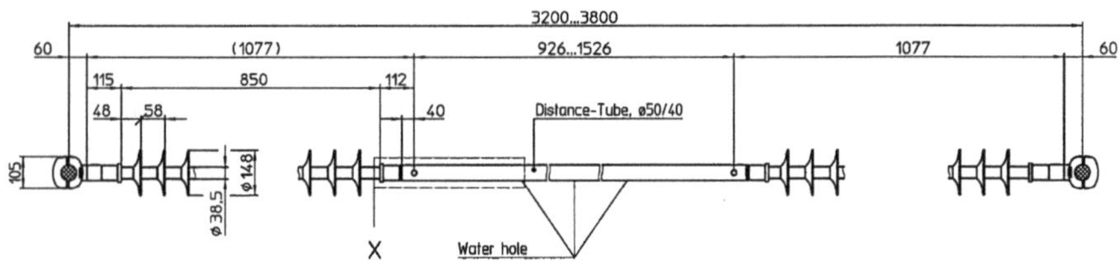

Fig. 5.7 IPS with option to vary coupling length

It is often advantageous to vary the length of the interphase spacer slightly on-site (Fig. 5.7). Any unforeseen assembly problems or last-minute changes to tower location, for example in hilly terrain, can thus be overcome quickly.

Sometimes, rather complicated constructions are used, in particular at high voltages, for example in the case of bundle conductors (Fig. 5.8), and this requires specific know-how on the part of the manufacturer.

In this case, specific attention must also be paid to proper integration of corona protection fittings, which are often necessary (Fig. 5.9).

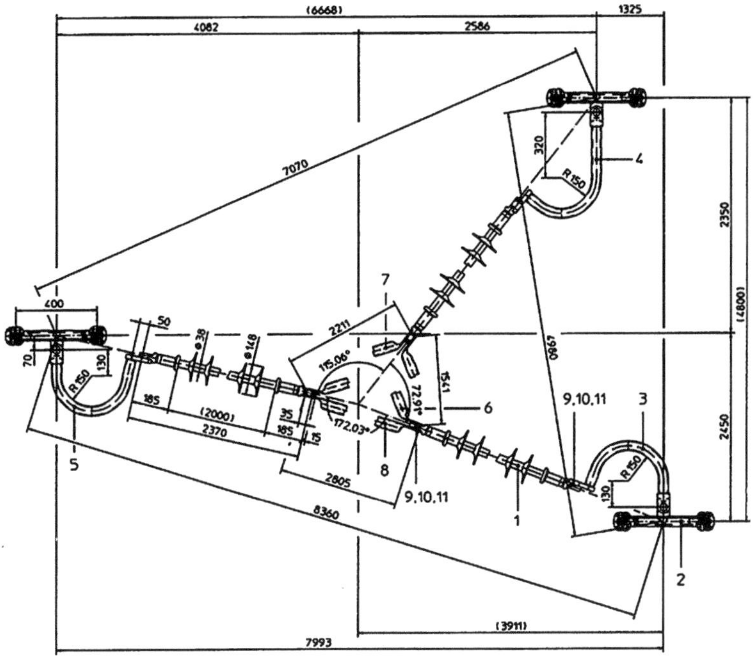

Fig. 5.8 IPS for star-like connection of the three phases

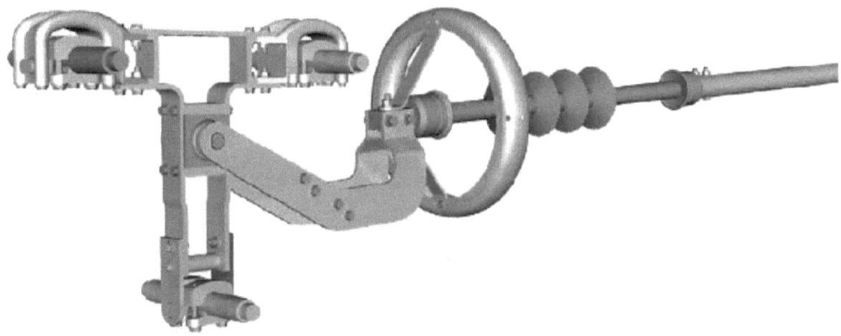

Fig. 5.9 Successful integration of the protection fitting in a 400 kV IPS for a triple bundle

5.4 Mechanical Design of Interphase Spacers

Interphase spacers are subjected to rather different loads compared to the insulators discussed in previous chapters. More specifically, IPS are loaded dynamically primarily by short circuit, ice shedding and galloping. The respective forces have to be taken into account accordingly when dimensioning the IPS.

5.4.1 *Galloping*

Galloping is a low-frequency, high-amplitude, wind-induced vibration which occurs with both single and bundle conductors (often even in single loop). Frequencies range from 0.1 to 1 Hz and amplitudes are 0.1–1 times the conductor sag of the span in question. In the case of medium-voltage lines, galloping amplitudes up to four times the conductor sag have been observed. Galloping is caused by moderate, steady crosswinds acting on an asymmetrically iced conductor surface. In some cases, galloping is also observed in the absence of ice.

Galloping plays an important role in the design of high-voltage towers, both in terms of the necessary distances between towers and in terms of tower load. More specifically, there may be large differences in load between phases, which may lead to increased horizontal and vertical loading, and sometimes even torsional loading. Mechanical failure of conductor coupling fittings and tower connection elements have already been observed in the case of very large galloping amplitudes. It is dangerous if cross-sectional weaknesses of insulator bolts or U-bolts caused by abrasion remain unnoticed as these could cause catastrophic damage, for example during a subsequent storm.

5.4.1.1 Galloping Amplitudes

Numerous tests, observations and calculations have resulted in the following formulas for amplitudes (over conductor diameter) occurring during galloping [3].

For single conductors,

$$\frac{A_{S-S}}{\phi} = 80 \cdot \ln\left(\frac{8 \cdot f}{50 \cdot \phi}\right)$$

For bundle conductors, (up to a wind speed of 10 m/s):

$$\frac{A_{S-S}}{\phi} = 170 \cdot \ln\left(\frac{8 \cdot f}{500 \cdot \phi}\right)$$

$A_{S\text{-}S}$ is the galloping amplitude peak-to-peak [m], ϕ is conductor (or sub-conductor) diameter [m] and f is conductor sag [m] of the span in question at 0 °C.

5.4.1.2 Designs

Interphase spacers are the component most commonly used for galloping control. For example, in North America a wide range of types of interphase spacer have been used since approximately 1970 in overhead transmission lines up to more than 500 kV [4]. The first generation of interphase spacers consisted of porcelain insulators which were interconnected by an aluminium tube. These IPS were

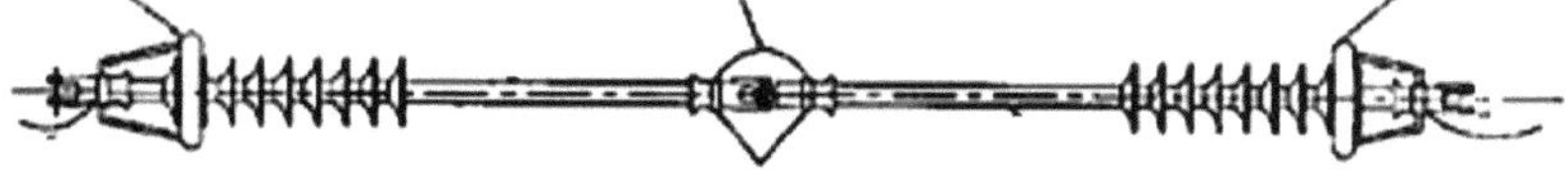

Fig. 5.10 IPS with composite insulators in 230 kV lines used in Canada [5]

heavy and therefore difficult to handle. In addition, some of these IPS failed as a result of high compression loads, such as those which occur during galloping. IPS of composite design were thus introduced as a replacement.

The IPS are attached to the conductors using armour grip suspension (AGS) so as to reduce the local stress at the conductors at the coupling points of the IPS. Later developments provided greater flexibility of the IPS by articulated fittings, wherein the in-between metal tube was initially retained. Ultimately, this was also replaced in favour of a silicone-coated FRP rod. Such IPS in 230 kV and 500 kV lines in operation in Canada can be seen in Fig. 5.10 and in Fig. 5.11 [5]. It can be seen that the connection points are bridged using flexible straps for equipotential bonding. In addition, corona protection rings are used at the high-voltage ends of the IPS so as to reduce the high field gradients occurring here and to prevent erosion of the silicone housing.

Although most IPS are installed in an approximately perpendicular position, they have also been mounted horizontally between the phases at locations where galloping has occurred. Such an IPS in a 420 kV line in North Norway, which always used to become damaged during winter, is shown in Fig. 5.12. It is a rigid IPS with two "thick" composite insulators and a connecting tube made of steel.

The special coupling to the twin bundle means that the original position of the conductors is always retained.

5.4.1.3 Installation Recommendations

Correct positioning of the IPS in the free span is of key importance in order to avoid flashovers caused when the conductors approach one another. Many IPS are installed at the mid-point of the span, where the greatest galloping amplitudes occur. This may help in the case of single-loop galloping, but is less effective in the case of two-loop galloping, since there is already a nodal point of vibration.

An example from Germany is shown in Fig. 5.13, in which there are horizontal, vertical and also so-called delta phase arrangements [6]. For the latter, the IPS are distributed symmetrically over the span length in more than 100 spans of a twin bundle line in South Germany, with, as far as known with positive results.

By contrast, Fig. 5.14 shows two alternatives with two and four interphase spacers per span length for a vertical phase arrangement [4]. In this case, the IPS are distributed in a non-uniform manner over the span for the purposes of

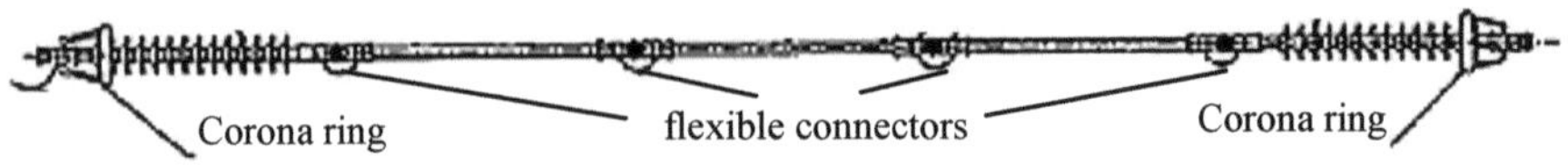

Fig. 5.11 IPS with composite insulators in 500 kV lines used in Canada [5]

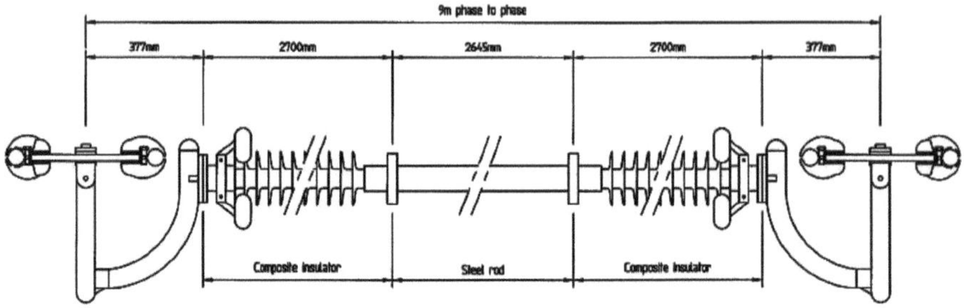

Fig. 5.12 IPS for a 420 kV twin bundle in Norway

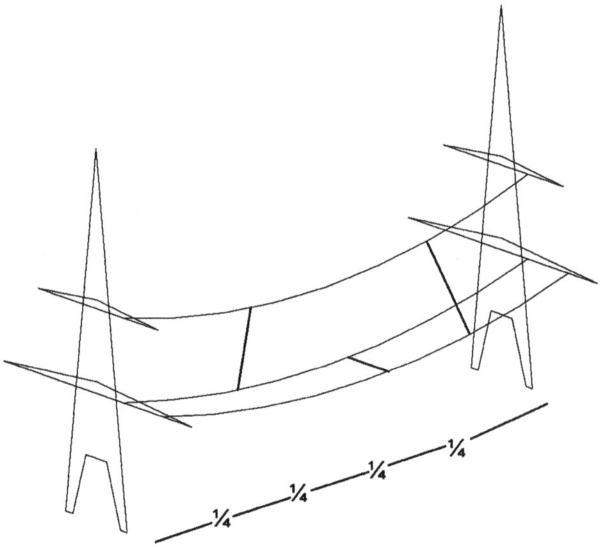

Fig. 5.13 Installation of IPS with triangular phase arrangement (Germany)

"disturbing" the vibratory system. Field tests have shown that these spacers are effective, both in the case of single-loop and two-loop galloping, although the arrangement with four IPS was selected since it prevents better a clashing of the phase conductors at the quarter points of the span length in the event of mixed mode galloping.

Such a case of mixed mode galloping, as was observed at Ontario Hydro, is shown schematically in Fig. 5.15. It can be seen that with two IPS a dangerous

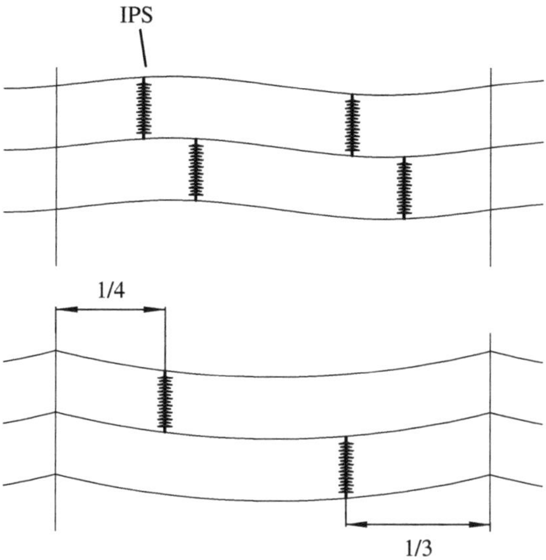

Fig. 5.14 Alternative installation of IPS with vertical phase arrangement

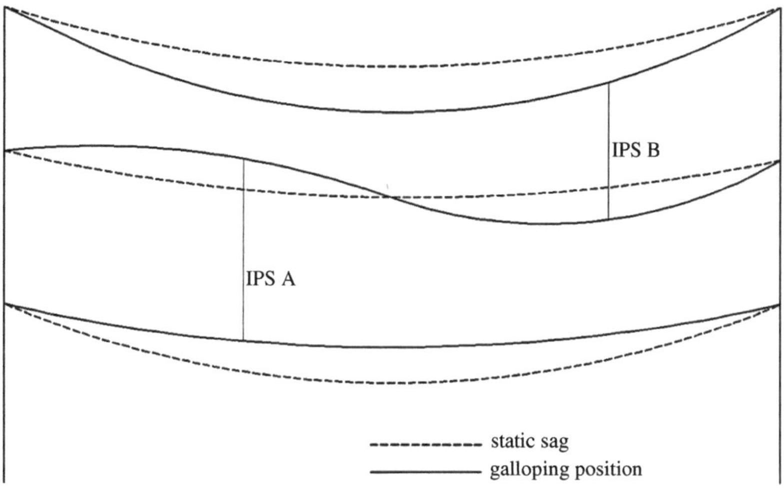

Fig. 5.15 Approach of the phase conductors with two IPS and mixed galloping

approach of the conductors may take place, more specifically at the points at which the IPS have been installed.

In the 1970s, large-scale field tests with IPS provided by different manufacturers were carried out at Ontario Hydro, primarily on single conductor lines [7].

The measurement results are presented in Fig. 5.16 as galloping amplitude peak-to-peak over the number of events observed. A general reduction in amplitude can be seen, although high amplitudes may sometimes occur, even with use of IPS.

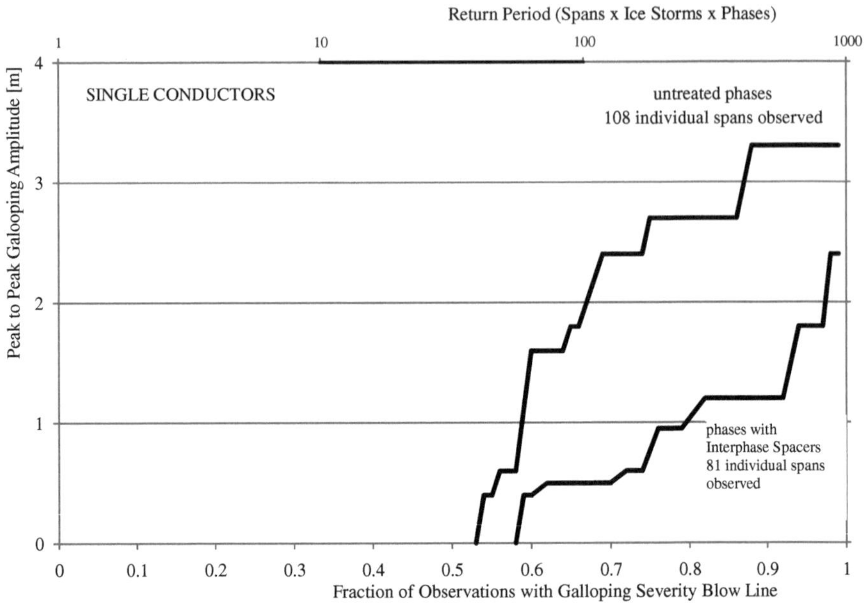

Fig. 5.16 Reduction in galloping amplitude peak-to-peak by IPS [7]

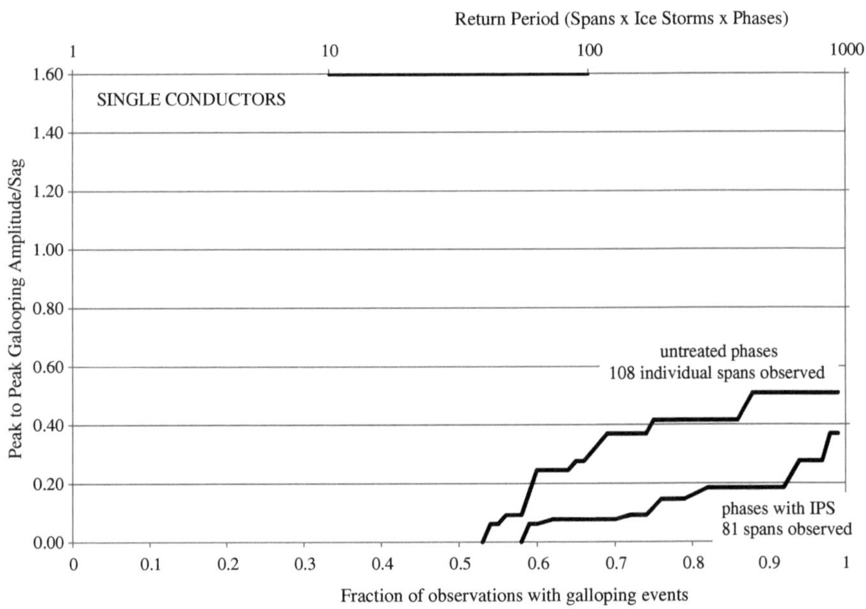

Fig. 5.17 Reduction by IPS in galloping amplitude peak-to-peak/conductor sag [7]

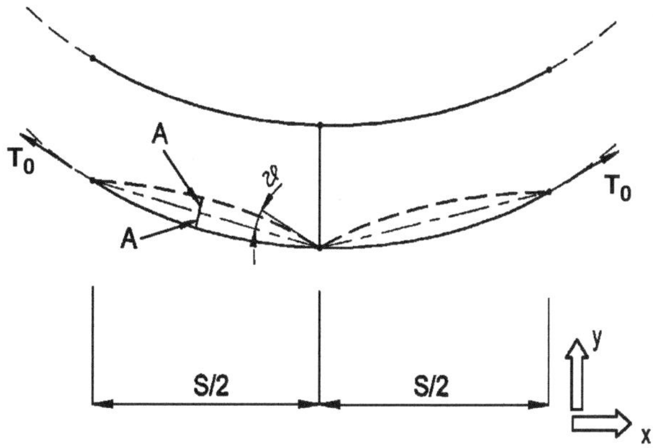

Fig. 5.18 Determination of the compression loads on the IPS during galloping

The same field data are illustrated in Fig. 5.17, as a plot of peak-to-peak amplitude divided by the conductor sag of the span in question. In 90 % of cases, the maximum values were able to be reduced by means of IPS to 0.2 times conductor sag, whilst values up to 0.52 times conductor sag were observed in spans without IPS.

5.4.1.4 Loading of IPS Caused by Galloping

As already mentioned, galloping is a complex aerodynamic process which can rarely be controlled using analytical methods alone. The situation is worsened when the conductors vibrating under the effect of galloping interact mechanically with the IPS, since in this case there is dynamic coupling of two vibratory systems. Due to the development in recent years of efficient computer systems, simulation methods have been created, in particular the FEM, in order to carry out such calculations [3]. On the other hand, it is useful for line engineers to have simple approximation methods available in order to carry out initial rough calculations in the early phases of a project. Such a calculation will be presented hereinafter [8].

The mechanical load of the IPS is examined based on the following assumptions:

- An IPS is installed at the mid-point of the span
- Both sub-spans to the left and right of the IPS vibrate symmetrically with single-loop galloping (Fig. 5.18)
- The maximum galloping amplitude corresponds to the conductor sag of the respective sub-span between IPS and tower

For movement of the conductor during galloping, the known differential equation of the vibrating string can be assumed in a first approximation:

$$\frac{W}{g} \cdot \frac{\partial^2 y}{\partial t^2} - T_o \cdot \frac{\partial^2 y}{\partial x^2} = 0$$

where

W conductor weight

T_0 horizontal conductor tension

g acceleration of gravity

The shape of the vibrating conductor $y(t)$ is given by using $S/2$ as the span length of the sub-span and A as maximum galloping amplitude in said sub-span, thus giving:

$$y = A \cdot \sin\left(\frac{2\,\pi x}{S}\right) \cdot \cos(\omega\, t)$$

$$\omega = \frac{2\,\pi}{S} \cdot \sqrt{\frac{g \cdot T_0}{W}}$$

From this, the maximum slope of the tangent of the vibration curve at the point of installation $(t = 0,\ x = 0)$ of the IPS is obtained, giving:

$$\tan(\vartheta) = \frac{\partial y}{\partial x}(x = 0, t = 0) = A \cdot \frac{2\pi}{S}$$

On the assumption that, in this case, $\vartheta \approx \tan\,\vartheta$ and taking into account the known formula for the sag A of the sub-span (span length $S/2$), that is to say:

$$A = \frac{(W_c + W_i) \cdot (S/2)^2}{8 \cdot T_0}$$

it is possible to obtain the compression load acting on the interphase spacer during galloping as follows:

$$P = 2 \cdot T_0 \cdot \sin(\vartheta) = 2 \cdot T_0 \cdot \frac{\tan(\vartheta)}{\sqrt{1 + \tan(\vartheta)}} \approx \frac{\pi}{8} \cdot (W_C + W_i) \cdot S$$

5.4.2 Shedding of Ice Loads

Ice or snow shedding can lead to serious damage of overhead transmission lines [9–11]. When the ice load deposited on the conductor suddenly drops, the conductor jumps upwards and, if the clearances between the conductors arranged above or the earth wire is exceeded, may lead to clashing or else to short circuit. In some cases, the insulator string may even loop around the cross-arm of the tower, which leads to permanent short circuit. In addition to these electrical problems, ice or ice shedding may cause serious damage to the individual components of the line (conductors, insulators, towers) owing to the high forces which prevail.

Ice or snow which is deposited on the conductors leads to an increase in conductor sag, and therefore also in the strain energy stored in the conductor. If the ice or snow load drops, the conductor jumps upwards. A distinction is made between two phases: In the first phase, the strain energy stored in the conductor is converted into kinetic energy; in the second phase, this kinetic energy is converted into potential energy until the conductor reaches its maximum jump height. Vertical conductor amplitude, rotation of the insulator strings, and the increase in conductor tension all reach their respective maximum values within the first phase of vibration and then decrease relatively quickly in the subsequent phases due to the inherent damping of the system.

Compared to other methods used to avoid such a scenario, such as the melting or removal of ice or the rather uneconomical increase in interphase spacing, the use of interphase spacers of composite design has proven to be the most successful in practice.

5.4.2.1 Shock Loads on the IPS

As in the case of galloping, complex dynamic processes which can only be fully understood using numerical simulation methods (usually FEM) take place in this instance [12].

Also is the case with galloping, it is necessary to create a simple analytical method for carrying out an initial estimation of the loading situation and therefore for dimensioning of the IPS. This is achieved by considering the energy stored in the conductor and in the IPS, as will be explained hereinafter.

The elastic deformation energy U_s in a conductor of cross-section A, length L, modulus of elasticity E and loaded by its own weight w_c and by ice load w_i (which ultimately leads to a tensile force T in the conductor) is:

$$U_s = \frac{1}{2}\frac{T^2 \cdot L}{E \cdot A}$$

with the assumption that this energy, which is released at the moment when the ice drops, is used in a first instance for deformation (bending) of the interphase spacer. The change in potential energy during conductor movement, inherent conductor damping, movement of the towers and insulators, and original deformation of the IPS as a result of its own weight and ice load are disregarded in this case.

On the assumption that the bending moment extends in a sinusoidal-like manner along the IPS and adopts a value of zero at both ends thereof, which are assumed to be pivoted, the deformation energy U_b of the IPS can thus be calculated as follows:

$$M(x) = M_{\max} \sin\left(\frac{x}{L_{sp}}\pi\right)$$

In this case x is the axial coordinate of the ISP, E_{sp} is the modulus of elasticity of the IPS, L_{sp} is the length of the IPS and M_{max} is the maximum bending moment at the centre of the IPS (x/2).

Using the known formula for the bending energy U_b and after integration, it is possible to obtain:

$$U_b = \frac{1}{2}\int_0^{L_{sp}} \frac{M(x)^2}{E_{sp}I_{sp}} = \frac{M_{\max}^2}{2E_{sp}I_{sp}}\int_0^{L_{sp}} \left(\sin\left(\frac{x}{L_{sp}}\pi\right)\right)^2 = \frac{M_{\max}^2}{4E_{sp}I_{sp}}\int_0^{L_{sp}} \left(1-\cos\left(\frac{2x}{L_{sp}}\pi\right)\right) dx$$
$$= \frac{1}{4}\frac{M_{\max}^2 L_{sp}}{E_{sp}I_{sp}}$$

By equating this energy with the strain energy U_s of the tensioned conductor, the following can be solved for the maximum bending moment M_{max} of the IPS:

$$M_{\max} = \sqrt{2}T\left(\frac{L}{L_{sp}}\frac{E_{sp}}{E}\frac{I_{sp}}{A}\right)^{1/2}$$

With this, the maximum bending stress σ_{max} at the centre (x/2) of the IPS, which has a rod diameter d_{sp}, can be calculated as follows:

$$\sigma_{\max} = \frac{M_{\max}}{\pi \cdot d_{sp}^3/32}$$

This bending stress has to be smaller than the breaking stress pf the FRP rod in the IPS (approximately 900 N/mm^2) by an appropriate safety factor. If this is not the case, further IPS must be installed, whereby the conductor energy U_s is then distributed over these, and M_{max} and σ_{max} are reduced accordingly.

The greatest load on the IPS is caused when the entire ice load drops suddenly in only one phase. Field tests in which these conditions were reproduced showed that, similarly to galloping, the compression load on the IPS is approximately proportional to the sum of conductor weight and ice load on the one hand, and to span length on the other [8]. Figure 5.19 shows the results of such a test with a simulated ice load of 44 N/m over a 182 m long span with a 240 mm^2 ACSR conductor (aluminium conductor steel-reinforced) with a conductor weight of 1.11 kg/m. The maximum dynamic compression load on the IPS measured in this case was 4.4 kN.

5.4.3 Electrodynamic Short-Circuit Loads

Interphase spacers are also used with success to maintain the necessary electrical clearances in the event of a short circuit. It is important that the IPS is positioned correctly in the span in question.

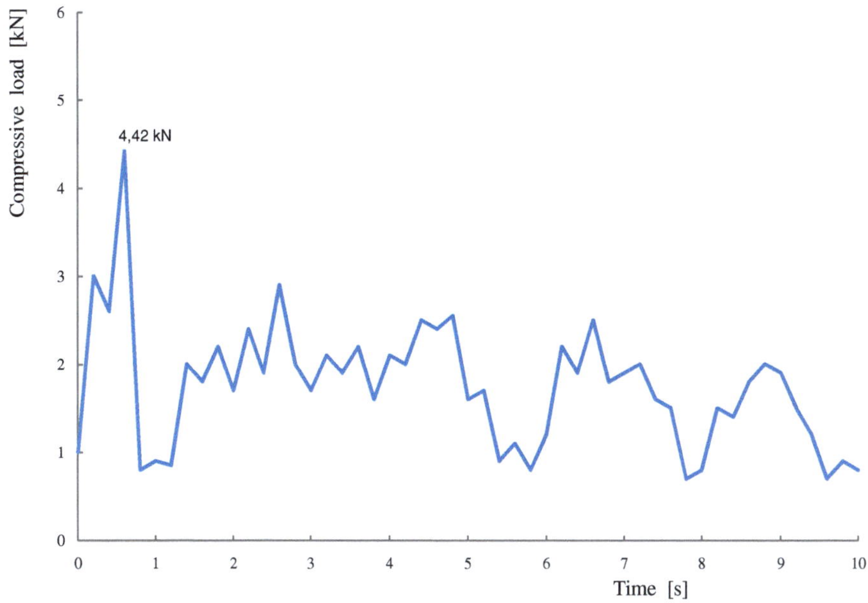

Fig. 5.19 Time history of compression load of an IPS in the event of sudden ice shedding [8]

In the event of a short circuit, mechanical vibrations which cause both compression and tensile loading of the IPS, are produced after the first and maximal force amplitudes, which are dependent on current direction. The IPS must therefore withstand both types of load. In addition, the IPS may be bent by the abrupt wave-like movement. Since this is a complex electromechanical dynamic process, it is not easy to establish the forces acting on the IPS. Normally tests and numerical simulation calculations are used as an aid. On the other hand, it is also advantageous in this instance to have on hand an approximative estimation method [13].

The following calculation is based on two-phase short circuit, which is almost equivalent to three-phase short circuit in terms of the electrodynamic forces, in particular for a horizontal phase arrangement [14]. The velocity of the moving conductor at the end of the short circuit is established by integration of the conductor acceleration caused by the short circuit force. For the simple arrangement of two parallel conductors which are placed at a distance d [m] from one another and through which a current $i(t)$ [kA] flows, the short-circuit force $F(t)$ [N/m] per unit length can be calculated on the basis of the following numerical value equation:

$$F = \frac{0.2 \cdot i(t)^2}{d} [\mathrm{N/m}]$$

in which the short-circuit current $i(t)$ in a first approximation is:

$$i(t) = \sqrt{2} \cdot I_{eff} \sqrt{\frac{1}{2} + e^{-\frac{2 \cdot t}{\tau}} \cdot \sin^2 \phi}$$

where τ [s] is the time constant of the short-circuit path.

On the other hand, according to Newton, the short-circuit force *F(t)* on the conductor of mass per unit length M [kg/m] causes an acceleration b [m/s^2]:

$$b = \frac{0.2 \cdot i(t)^2}{d \cdot M}$$

At the end of the duration of short circuit t_{sc} [s], the conductor, with the assumption of a fully asymmetrical short circuit, that is to say $\sin\varphi = 1$, will have reached the velocity v[m/s]:

$$\upsilon = \int_0^{t_{sc}} b \cdot dt \approx \frac{0.2 \cdot I_{eff}^2 \cdot (t_{sc} + \tau)}{d \cdot M}$$

On the assumption that the total energy of the short circuit is converted fully into the kinetic energy of the conductor, that is to say if conductor elastic strain, conductor heating, and a change in the potential energy of the conductor are disregarded, the total kinetic energy of the conductor U_{sc} [Joules] within a span at the end of the short circuit is given from the integration of $Mv^2/2$ over the span length L [m]. For this step constant conductor velocity is assumed over the entire span, possibly with a reduction factor of 0.75, in order to take into account end effects:

$$U_{sc} = \frac{1}{2} \cdot M \cdot \left[\frac{0.2 \cdot I_{eff}^2 \cdot (t_{sc} + \tau)}{d \cdot M}\right]^2 \cdot \frac{3 \cdot L}{4}$$

As in the previous section, the strain energy *U*b of n_{sp} spacers installed within a span is calculated as follows, on the assumption of sinusoidal distribution of the bending moment over the IPS and pivoted mounting of the IPS at either end:

$$U_b = \frac{n_{sp}}{4} \frac{M_{max}^2 L_{sp}}{E_{sp} I_{sp}}$$

where

U_b	strain energy of an IPS [Joules]
n_{sp}	number of IPS in a given span
E_{sp}	modulus of elasticity (Young's modulus) of the IPS [N/m2]
M_{max}	maximal bending moment of the IPS [Nm]
I_{sp}	moment of inertia of the IPS [m4]

Table 5.2 Test results of the deflection of the IPS during short circuit

Duration of short circuit (ms)	Max. horizontal deflection (m)				Phase approach (m)		IPS deflection (m)	
	Lower conductor		Upper conductor					
	without IPS	with IPS	without IPS	with IPS	without IPS	with IPS	without IPS	with IPS
105	2.30	0.60	2.70	0.80	−0.50	0.40	0.50	0.30
305	3.5	1.30	6.60	1.70	−3	−0.10	1	0.60
505	3.90	1.30	8.20	1.80	−1.80	1.20	2.2	1

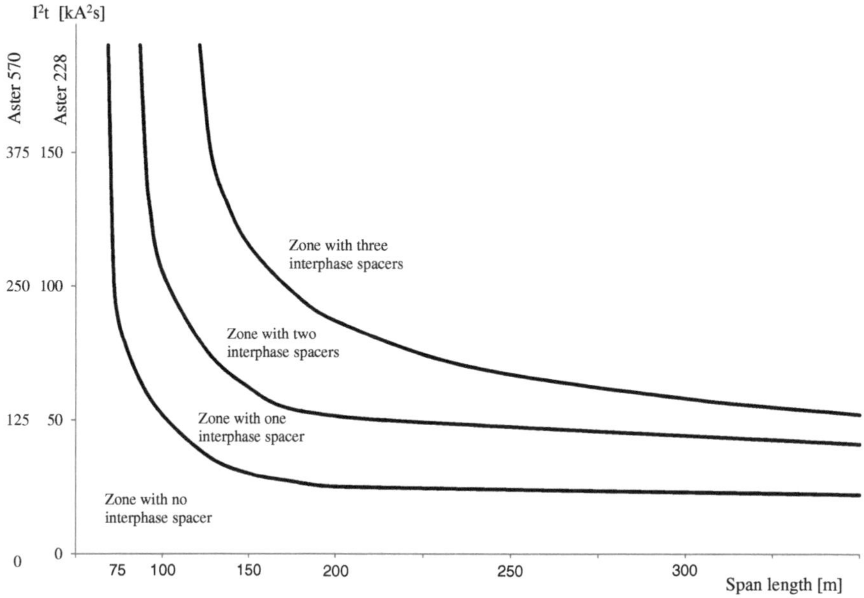

Fig. 5.20 Diagram for determining the number of necessary IPS in the event of short circuit [15]

By equating U_{sc} and U_b, it is possible to obtain the actual dimensioning parameter for the IPS in the case of short circuit—the maximal bending moment $M_{\max}$. The necessary rod diameter of the respective composite insulators is thus determined, as explained in the section above on ice shedding.

Generally, there is little published information regarding the short-circuit behaviour of IPS installed in overhead transmission lines. One exception is the publication [15], where the movements of the conductors and IPS installed in an experimental span have been monitored. This span measuring 150 m was part of a 68–90 kV overhead transmission line having 228 mm^2 Aldrey conductors (ASTER) with a vertical phase spacing of 3 m and with an IPS at the mid-point of the span. The continuous short-circuit current was 20 kA with a duration of short circuit of 105, 305 and 505 ms. The test results are summarised in Table 5.2.

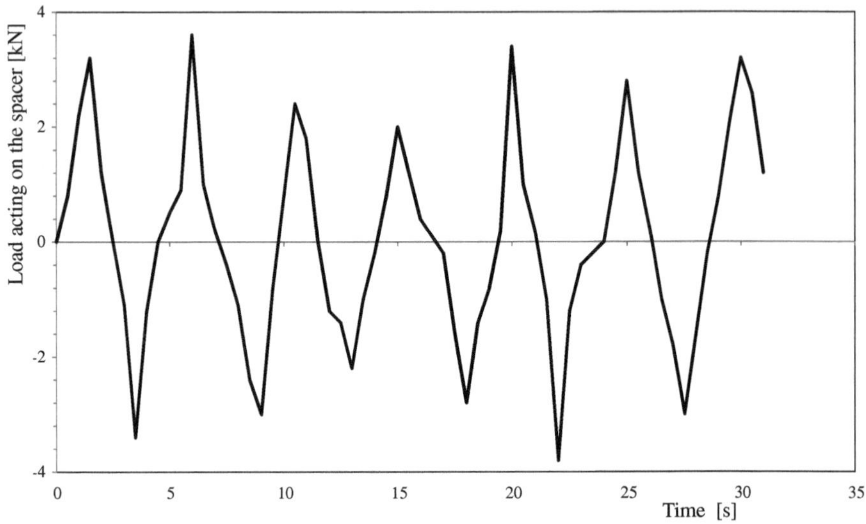

Fig. 5.21 Time history of the load of the IPS during short circuit (numerical simulation) [19]

The conductor movements are clearly reduced significantly by the installation of IPS. In addition, the necessary number of IPS for two conductor types (ASTER 228 and 570, interphase spacing 3 m) and for different span lengths have been calculated on the basis of energy considerations (Fig. 5.20).

There is also some information available regarding the behaviour of IPS during short circuit when installed in outdoor substations with flexible bus bars. In this case, the short-circuit currents (and the respective short-circuit forces) adopt particularly high values, which leads to considerable loading of the IPS.

Such a case is described in a report compiled by CIGRE Task Force ESCC (*Effects of Short-Circuit Currents*) [16]. This report concerns a 40 m long span, with an initial short circuit current of 28.5 kA and a duration of 103 ms. The first 2–3 s of this dynamic process were examined.

The calculation was first carried out without IPS. The FE calculation provided a value of 1.12 m for maximum conductor deflection at the span mid-point. This correlates well with both the measurements and the calculation according to IEC 60865-1, which provided 1.27 m.

On the other hand, in the case of short circuit the minimum phase spacing in a 110 kV system is 0.55 m [17]. If conductor deflection of 1.27 m is expected, this means that the phases will clash, which may lead to a two-phase short circuit.

This problem can be overcome as follows:

- Increase interphase spacing from 2.0 m to 2.75 m
- Reduce span length from 40 to 23 m (for example by installing intermediate support structures)
- Reduce the short-circuit current I_{k2} from 28.5 to 18.2 kA

Fig. 5.22 Static system and load of a slender IPS

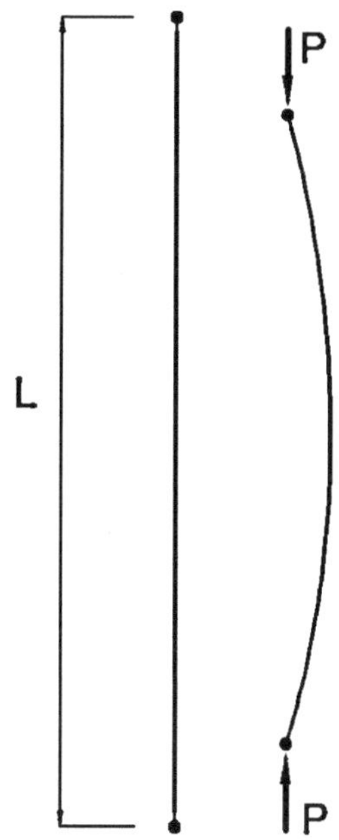

All three measures require considerable modifications and therefore higher costs of the substation. A cost-effective and simple alternative is the installation of IPS formed by composite insulators at the mid-point of the span. The following insulator was considered in the new calculation:

- Coupling length 1,300 mm
- Rod diameter 36 mm
- Modulus of elasticity 34,000 N/mm2
- Weight 12.4 kg

All other parameters remained unchanged. Due to the installation of IPS, the maximum conductor deflection in the mid-point of the span (since this is where the reciprocal conductor movement is controlled by the IPS) was shifted to the ¼ and ¾ points of the span and was reduced by approximately half to the acceptable value of 0.5 m.

Figure 5.21 shows the forces acting on the IPS and caused by conductor dynamics. In spite of a relatively high harmonic content, these forces remain well below the Euler buckling load of 16 kN of this spacer, even with a short circuit current of $I_{k2} = 40$ kA.

5.4.4 Buckling Behaviour of Interphase Spacers

Interphase spacers are often subject to compression loads. For the earlier IPS made of porcelain, this means that they fail catastrophically under buckling as soon as

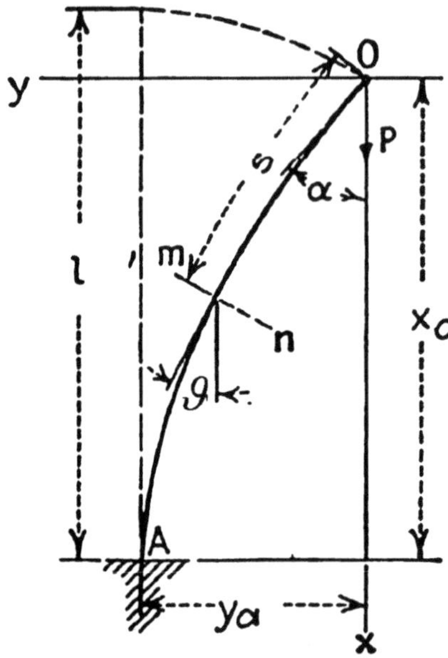

Fig. 5.23 Parameter definition of the compressed rod [20] Abb. 5.23 Parameter definition des Druckstabes

the compression load exceeds the Euler load, as is the case for materials of low breaking strain such as porcelain..

By contrast, modern interphase spacers with composite insulators can withstand compression loads somewhat greater than the critical Euler loads because their FRP rod has a rather high breaking strain. It is therefore important to understand the behaviour of the IPS once the Euler load has been exceeded.

Figure 5.22 shows the initial, starting situation. An originally straight, stiff rod of length L, modulus of elasticity E and moment of inertia I is subjected to a compression force P. The rod is mounted in such a way that both its ends are forced to move in the direction of the originally undeformed, straight rod axis. The deflection of the rod was sought. Leonard Euler, one of the pioneers in this field, dealt with the same problem [18], although, because of symmetry considerations, only one half of the buckling rod was examined, (Fig. 5.23).

It is recalled that the Euler buckling load is determined by using for rod curvature $\kappa(x)$, the conventional Euler–Bernoulli approximation, namely:

$$\kappa(x) = \frac{d^2y(x)}{dx^2}$$

If, by contrast, the exact expression for the curvature is used, it is possible the determine also rod deflection during buckling. With the notations from Fig. 5.23, the "exact" curvature is:

$$\kappa(s) = \frac{d\vartheta}{ds}$$

The "exact" differential equation of the compressed rod becomes:

$$EI \cdot \frac{d\vartheta}{ds} = -P \cdot y$$

After introduction of the parameter $k^2 = \frac{P}{EI}$ and taking into account $\frac{dy}{ds} = \sin \vartheta$:

$$EI \cdot \frac{d^2\vartheta}{ds^2} = -P \cdot \sin \vartheta$$

The solution of this differential equation provides the Euler angle , named after the great Swiss mathematician Leonhard Euler, who was involved in the study of this problem as early as 1739. This is the same differential equation as the exact description of a pendulum oscillation, which is known as the Kirchhoff analogy. The solution requires the use of elliptic integrals. The calculation process is described in detail in [18], which is why it is only summarised here in order to provide for the first time a "rule of thumb" for the selection of interphase spacers.

A further parameter p is initially introduced:

$$p = \sin \frac{\alpha}{2}$$

where α is the maximal Euler angle at load application point. For each predefined angle α, k and thus the force P belonging to α can be determined using the following equation:

$$\ell = \frac{1}{k} K(p)$$

where *K(p)* is the complete elliptic integral of the first kind (the values of which can only be given in table form as a function of p).

It should be noted that the rod length L of the interphase spacer pivoted at either end is related to ℓ from Fig. 5.23 as follows:

$$L = 2 \cdot \ell$$

The maximal deflection in the centre of the rod y_a is then obtained from the equation:

$$y_a = \frac{2 \cdot p}{k}$$

The resultant length reduction of the rod due to compression x_a is given from:

$$x_a = \frac{2}{k} E(p) - \ell$$

Table 5.3 Compression load and deformations of a buckling rod as a function of α

α	0°	20°	40°	60°	80°
P/P_{Euler}	1	1.015	1.063	1.152	1.293
x_α/l	1	0.970	0.881	0.741	0.560
y_α/l	0	0.220	0.422	0.593	0.719

Fig. 5.24 Highly deformed IPS

where $E(p)$ is the complete elliptic integral of the second kind, likewise only available in table form.

Table 5.3 from [18] shows, for some angles α, the values of y_a in relation to ℓ, and the values of P in relation to the respective Euler buckling load:

$$P_{Euler} = \frac{\pi^2 \cdot EI}{4 \cdot \ell^2}$$

On the other hand, the maximum bending stress $\sigma_{b,max}$ must be smaller than the bending strength $\sigma_{b,zul}$ of the FRP core, that is to say:

$$\sigma_{b,max} = \frac{M_{max}}{W} = \frac{P \cdot y_a}{I/(d/2)}$$

If, in this last equation, P and y_a as well as average material values for E and $\sigma_{b,zul}$ (for example $E = 35{,}000$ N/mm^2 and $\sigma_{b,zul} = 800$ N/mm^2) are given and the angle α is limited to $\alpha_{max} = 80°$ for structural reasons (see Fig. 5.24), the following rule of thumb is given for the selection of a "sensible" core diameter d at predetermined IPS length L (predetermined from electrical considerations as explained in Sect. 5.5.1):

$$d \geq L/100$$

The IPS thus dimensioned can withstand compression loads which lie up to approximately 30 % above its respective Euler buckling load without difficulty, that is to say without permanent deformation (see Table 5.3: $P/P_{Euler} = 1.293$ for $\alpha = 80°$). Although most IPS are dimensioned conservatively with the Euler

Fig. 5.25 Damaged end fitting of an IPS

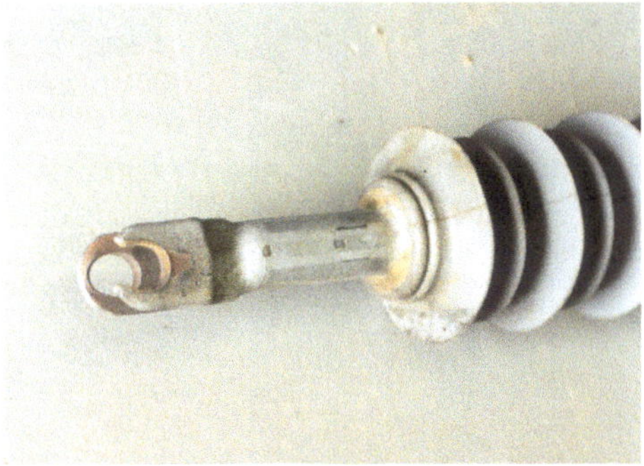

buckling load as a limit load, it is important to know that there is still a load reserve according to rod diameter and, above all, that the post-buckling behaviour of an IPS is stable.

The limitation of the angle α—and therefore the limitation of the deformation of the PAH which can still be adopted—is necessary for structural reasons. Even if the FRP core would allow greater deformations, such extreme deformations would highly load the fittings coupling the IPS to the conductor and would damage them. This has been observed during field tests carried out by IREQ on a test line in Canada. More specifically, suitable measures were used to excite the conductors to strong galloping, which subjected the IPS to extreme loads (Fig. 5.24) damaged the end fittings (Fig. 5.25). It is even conceivable that the silicone housing could become detached with such strong curvature of the rod, particularly on the compressed side of the rod.

The post-buckling behaviour of IPS was also confirmed by our own tests. In these tests, a 3,124 mm long FRP rod of 31.5 mm diameter was provided with conventional end fittings and was subject to vertical load, as is shown in Fig. 5.26. The rod behaved well, even once its Euler buckling load of approximately 2 kN had been exceeded by more than 30 %, and returned to its original straight shape once the load was removed.

Forces of up to 2,670 N were measured with maximal deflection in the centre of the rod of approximately 500 mm and maximum length reduction of the rod due to compression of approximately 200 mm.

5.5 Electrical Design of Interphase Spacers

5.5.1 Minimum Length

The minimum length of an IPS is essentially determined by the voltage of the line in question and by the sag of the span in which the IPS is installed. In the Swiss Regulations on Overhead Transmission Lines [19], IPS are mentioned specifically

Fig. 5.26 Length reduction of a 31.5 mm rod due to compression at approximately 2,000 N

and the following formula is provided therein for determination of minimum IPS length:

$$z = k \cdot \sqrt{d_f} + C_3 \geq C_1$$

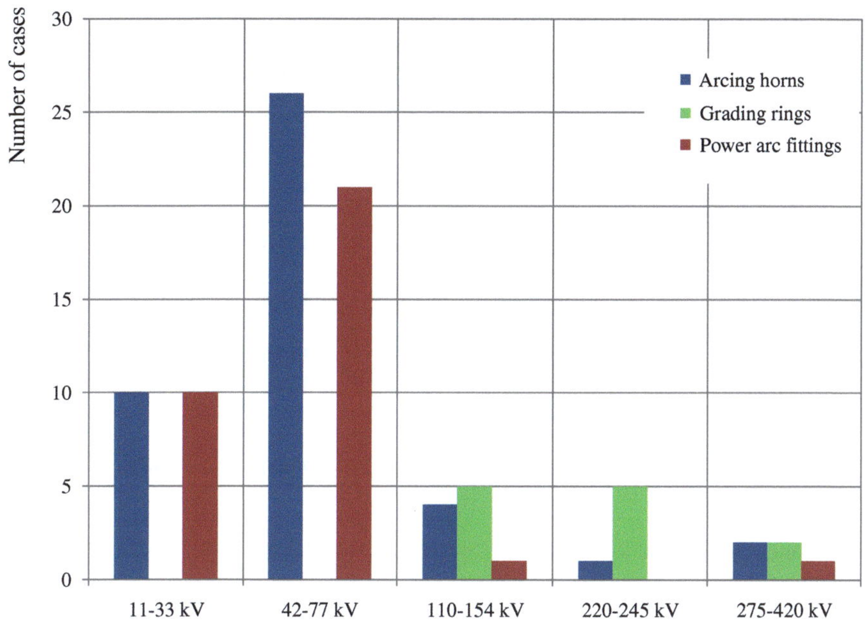

Fig. 5.27 Use of different protection fittings in IPS according to voltage level

In which:

z length of the IPS, at least 0.9 m

d_f maximum sag in the sub-span between two IPS or between one IPS and the next tower

k Correction factor depending on conductor type, conductor diameter and wind pressure

C_i constant (dependent on line voltage)

These factors k can be found in diagrams 2.4.2 and 2.4.3, and the constant C can be found in Table 2.4.1 of the cited Regulation [19].

5.5.2 Corona Effects

Reduced interphase spacings owing to the installation of IPS increases the electric field strength at the conductor surface and therefore increases the likelihood of corona discharges, in particular in the case of line compaction. In addition, the IPS are subjected to the higher phase-to-phase voltage. Numerical field simulations have shown that the field distribution along an IPS is different from that along a comparable suspension string. In the case of an IPS, the field strength runs symmetrical to the centre of the IPS and also has higher values in the vicinity of the conductor fittings.

Table 5.4 Creepage distances in different IPS during operation

Network operator	Highest operational medium voltage (kV)	Coupling length (mm)	Creepage distance (mm)	Spec. creepage distance (mm/kV)
Tokyo Electric Power Co. Inc.	72.5	2,200	4,580	63.2
Tokyo Electric Power Co. Inc.	72.5	2,400	5,080	70.1
Tokyo Electric Power Co. Inc.	72.5	3,000	4,080	56.3
Hokuriku electric Power Co. Inc.	85	2,240	3,500	41.2
VEW Dortmund	123	4,315	5,820	47.30
Österreichische El.wirtschafts AG.	123	5,446	5,700	46.30
Österreichische El.wirtschafts AG	123	6,800	7,000	56.9
Hokuriku electric Power Co. Inc.	170	3,760	7,940	46.7
VEW Dortmund	245	5,600	10,002	40.8
Trans power	245	5,300	5,760	23.5
E.D.F.	245	5,432	6,811	27.8
Lech El.wirtschafts AG	245	3,100	9,020	36.8
BKW AG	420	8,500	9,500	20.2
E.D.F.	420	12,400	14,595	34.8
Norwegian state power board	420	5,384	7,544	18.0

Since this favours an undesired phase flashover, it is recommended to install corona protection rings at both ends of the IPS, even in the case of lower line voltages. A summary of the types of protection fitting (protection horns, protection rings, short-circuit protection fittings) used in interphase spacers in accordance with line voltage, as emerged from the CIGRE investigation [1], are shown in Fig. 5.27.

5.5.3 Pollution

The creepage distance of insulators which are subject to phase-to-earth voltage is often provided with a sufficient safety margin of 20 to 30 % according to installation conditions based on practical design considerations. Accordingly, the creepage distance of IPS, which are subject to the phase-to-phase voltage which is

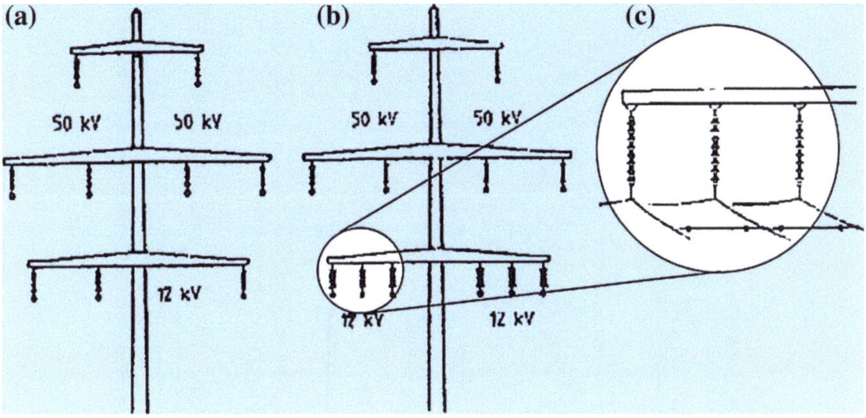

Fig. 5.28 IPS for medium-voltage lines in Switzerland (schematic view)

higher by the factor $\sqrt{3}$, is provided with a similar margin. The coupling length, creepage distance and specific creepage distance for selected IPS of different network operators, again as established by the CIGRE investigation, are summarised in Table 5.4.

5.6 Applications

The following examples are provided on the basis of the authors'practical experience and are intended to demonstrate a few typical uses of interphase spacers.

5.6.1 Compact Line for Medium Voltage

As in many other industrialised countries, in Switzerland, too, it is difficult to obtain way leaves for new line routes. One solution is to increase the transmission capability of an existing line, for example by installing a second circuit on the same line support. In the case presented here, although a concrete pole and concrete cross-arms as shown in Fig. 5.28a were able to accommodate the second 12 kV circuit due to their high mechanical strength, the necessary interphase spacings at the mid-point of the span were not met. Due to the installation of suitable IPS using silicone composite insulators every 40 m (Fig. 5.28c), this problem was able to be solved and a conductor arrangement according to Fig. 5.28b was implemented. In the meantime, thousands of such IPS have been operated successfully for many years in Switzerland (Fig. 5.29).

Fig. 5.29 IPS from Fig. 5.28 in successful operation

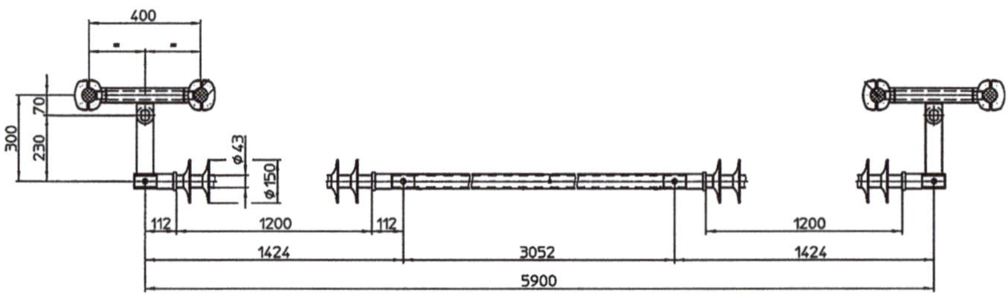

Fig. 5.30 IPS for a traction supply line of DB (Deutschen Bahn) (details)

In a similar case, IPS were used in a somewhat overcrowded route owned by Deutschen Bahn (German Railways) to maintain the necessary clearances (Figs. 5.30 and 5.31).

Fig. 5.31 IPS from Fig. 5.30 in successful operation

Fig. 5.32 Span of a 420 kV line at risk of damage caused by ice shedding in winter

Fig. 5.33 The 420 kV tennis racket tower in West Switzerland

5.6.2 Interphase Spacers in the Event of Galloping Caused by Ice Shedding

Most of the 420 kV lines in Europe at high altitudes can be found in Switzerland. Owing to the unusually harsh weather conditions which prevail during the winter months (Fig. 5.32), there would be interruptions in operation time and time again, particularly within a specific span. These interruptions were caused by two-phase short circuits during clashing of the phase conductors after ice shedding. In addition, the outer layer wires would become damaged, sometimes severely, as a result of the physical contact and also as a result of the high short-circuit currents, which led to the need for short-notice repairs in particularly challenging environmental conditions. For this reason, it was decided to install interphase spacers and to design them (from a mechanical point of view) in such a way that they could maintain the phase geometry, even under extreme dynamic load. For this task, it was obvious to use composite insulators since their relatively low weight compared to porcelain means that the conductor is not subjected to excessively high bending loads, and their high level of elasticity, as mentioned above, should be able to overcome the strong dynamic loads and associated deformations without difficulty. It should be noted in this regard that numerical simulations confirm field reports, in accordance with which the conductors would jump upwards by up to 20 m in the event of ice shedding before installation of the IPS. Figure 5.1 shows the installation of such IPS in the critical span of the 400 kV line in the Vorab

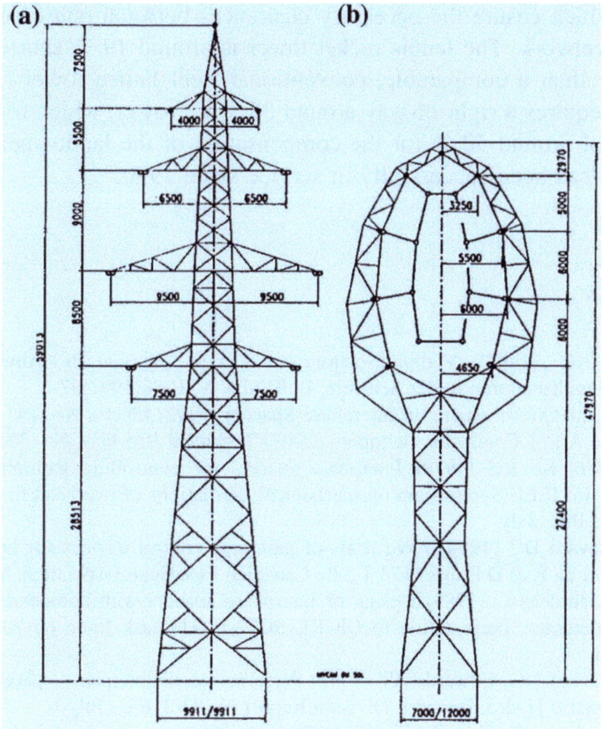

Fig. 5.34 Comparison of the dimensions and spatial requirement of a conventional 420 kV steel lattice tower (**a**) and a tennis racket tower (**b**)

glacier. The individual IPS between two phases are between 10 and 12 m long and weigh approximately 100 kg, including their special fittings which allow full articulated coupling to the phase conductors and thus prevent damage in the event of sudden movements. Each IPS consists of four to five silicone composite insulators having a rod diameter of approximately 80 mm. These IPS have been used with success since 1995, that is to say there has been no outage caused by ice shedding since their original installation.

5.6.3 The Tennis Racket Tower

When installing a major 420 kV line in West Switzerland, it was of importance to develop a special tower which not only had a specific shape, namely the shape of a tennis racket, but which could also be implemented with minimal spatial requirement. This led to the unique concept of the world's only "tennis racket tower" (*pylône raquette* in French) [20] (Fig. 5.33). For aesthetic reasons, but also for reasons regarding weight, the tower top was formed using silicone composite insulators, which ensure the necessary clearances between phase conductors and the tower steelwork. The tennis racket tower is around 19 % shorter and around 10 % lighter than a comparable, conventional steel lattice tower (Fig. 5.34). In addition, it requires a right-of-way around 37 % narrower, which translates into a cost saving of around 50 % for the compensation of the landowners. The tennis racket tower has been successfully in service since 1990.

References

1. Papailiou KO (1997) Verbundisolatoren auf dem Vormarsch—Anwendungen von Silikonverbundisolatoren in der Schweiz, Bull. SEV/VSE 88(1997)07
2. Results on the Questionnaire on Interphase Spacers (1992) Electra No. 143, CIGRE, Paris
3. State of the Art of Conductor Galloping (2007) Technical Brochure No. 322, CIGRE, Paris
4. Edwards AT, Ko RG (1979) Interphase spacers for controlling galloping of overhead conductors In: IEEE Symposium on mechanical oscillations of overhead transmission lines. Vancouver, B.C. July
5. Pon CJ, Havard DG (1994) Field trials of galloping control devices for bundle conductor lines. Report on R & D Project 133 T 386, Canadian Electrical Association. Montréal. March
6. Schmidt J, Jürdens C (1989) Design of interphase spacers with composite insulators and service experience, presentation to CIGRE SC22-WG11 task force on galloping. Rijeka, Yugoslavia
7. Pon CJ, Havard DG, Edwards AT (1982) Performance of interphase spacers for galloping control. Ontario Hydro Research Division Report No.82-216-K. July 6
8. Kito K, Imakoma T, Shinoda K, (1975) Phase-to-phase spacers for transmission lines. In: IEEE PES summer meeting, San Francisco, Calif., July 20–25
9. Morgan VT, Swift DA (1964) Jump height of overhead-line conductors after the sudden release of ice loads. In: Proceedings of IEE, Vol. 111, No. 10, pp 1736–17461964
10. Roshan Fekr M, McClure G, Hartmann D (1998) Investigation of transmission line failure due to ice shedding using dynamic analysis. In: Proceedings of 8th international workshop on atmospheric icing of structures, Reykjavik, Iceland
11. Su F, Hu S (1988) Icing on overhead transmission lines in cold mountainous district of southwest China and its protection. In: Proceedings of 4th International workshop on atmospheric icing of structures, Paris, France
12. Wang J, (1996) Large vibrations of overhead electrical lines: A Full 3-DOF model for galloping studies. Ph.D. Thesis, Collection des Publications de la Faculte des Sciences de l'Universit de Liege No. 151
13. Lilien JL, Dal Maso F (1990) Contribution to line design by accurate predetermination of severe but occasional stresses. Paper 22-103, CIGRE session, Paris
14. Papailiou KO (1989) Mechanische Beanspruchung von Kabelsteckern bei Kurzschluss. SEV_Bulletin Jg. 80
15. Dalle B (1985) Espaceurs pour lignes aériennes, Zeitschrift D.E.R., EDF; Paris
16. The mechanical effects of short-circuit currents in open air substations (2002) Part II. Companion brochure to No. 105, CIGRE, Paris
17. Declercq G (1998) Tests with droppers and interphase spacers. In: Proceedings of the 8th international symposium on short-circuit currents in power systems. Brussels (Belgium). 8–10th Oct, pp 143–148
18. Timoshenko S, Gere J (1961) Theory of elastic stability. McGraw-Hill, NY

19. Verordnung vom 30. März 1994 über elektrische Leitungen (Leitungsverordnung, LeV), Schweizerische Eidgenossenschaft, Bern, 1994
20. Ammann M, de Weck P, Faggiano P (1991) Pylone raquette. Paper 400-04, CIGRE Symposium. Leningrad

Chapter 6
Composite Hollow Core Insulators

Symbols and Abbreviations

2D	Two-dimensional
3D	Three-dimensional
CIGRE	Conseil International des Grands Réseaux Électriques (International Council on Large Electric Systems)
DMS	Strain gauge
FEM	Finite element method
FRP	Fibre reinforced plastic
IEC	International electrotechnical commission
LVDT	Linear variable differential transformer
MML	Maximum mechanical load
VHI	Composite hollow core insulator
MSP	Maximum specified pressure
O-ring	O-ring (seal/gasket)
SF_6	Sulphur hexafluoride (insulating gas)
SIP	Specified internal pressure
SML	Specified mechanical load
VPE	Cross-linked polyethylene (insulating material for cables)
ZnO	Zinc oxide (material of the varistors for lightning arresters)
1, 2, 3	Subscripts related to the fibre direction and to the two orthogonal directions thereto
$v_{i,j}$	Poisson number (for FRP tube, aluminium, glue, epoxy resin)
σ_b	Bending stress
σ_{rz}	Shear stress at the interface between tube and end fitting
d	Tube inner diameter
D	Tube outer diameter
E	Average modulus of elasticity of the FRP tube
E_i	Modulus of elasticity (Young's modulus) (for FRP tube, aluminium, glue, epoxy resin)
$G_{i,j}$	Shear modulus (for FRP tube, aluminium, glue, epoxy resin)
F, F_{b}	Bending load

K. O. Papailiou and F. Schmuck, *Silicone Composite Insulators*, Power Systems,
DOI: 10.1007/978-3-642-15320-4_6, © Springer-Verlag Berlin Heidelberg 2013

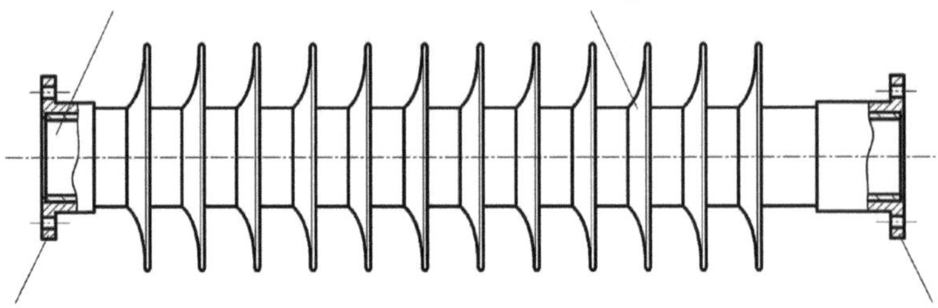

Fig. 6.1 Schematic view of a composite hollow core insulator and its components

l	Lever arm from the point of the load application to the mouth of the metal end fitting
M_b	External bending moment
W	Section modulus of the tube
t	Time
z	Coordinate in the axial direction of the FRP tube

6.1 Main Properties of Composite Hollow Core Insulators

Composite hollow core insulators (referred to hereinafter as HCI for short) are used primarily as housings for various high-voltage equipment and apparatus. Owing to their conditions for use, they primarily have to withstand bending and compression loads and since a large number of the electrical equipment are gas-insulated (with SF_6 and SF_6 gas mixtures), they are also subject to internal pressure. In composite design, composite hollow core insulators are virtually explosion-proof and, in the event of failure, are practically harmless to people or objects. Due to their low weight (10–30 % of a corresponding porcelain insulator) and the flexibility of the insulator housing, they are less prone to damage during assembly and transport, compared to porcelain hollow core insulators. Because they are not sensitive to dynamic loads, they are preferably used in outdoor substations in areas where there is a high risk of earthquake. A further advantage in particular in he EHV and UHV range, is their flexibility in terms of mechanical and geometrical design (single pieces of 10 m or more may be manufactured) to meet a wide range of requirements with short design cycle time. They facilitate innovation in terms of new solutions to known applications and have been widely employed where new solutions such as integration of devices have been developed. Their typical design is shown in Fig. 6.1.

Fig. 6.2 Composite insulator bushings (dry design) on a 420/220 kV grid transformer

6.2 Composite Insulators in Outdoor Substations

Substations are the nerve centres of every power grid and so the users expect and demand a correspondingly high degree of operational safety. It is therefore not surprising that with the growing faith in composite insulators—particularly due to the good experience made in their application in overhead lines worldwide—great interest has developed in recent years in their application also in outdoor substations. Today, if the user so desires, it is possible to design complete substations in silicone composite technology [1] (Fig. 5.1).

6.2.1 Bushings

Increasingly, the design of high-voltage bushings is being influenced by greater demands on operational safety and reliability, damage risk minimisation (to people and property), and not least by greatly increased environmental awareness. The consideration of these factors has led to a new design of these key components on the basis of composite materials. It has been possible to satisfy the above-mentioned demands on bushings by using modern components made of top-quality materials and also by mastering the manufacturing process. Figure 6.2 shows 420 and 220 kV transformer bushings, whilst Fig. 6.3 shows wall bushings for 123 kV, and Fig. 6.4 shows an 800 kV wall bushing in composite design.

Fig. 6.3 Wall bushings (123 kV)

6.2.2 Surge Arresters

Surge arresters were among the first electrical devices designed with a silicone housing, more specifically for the obvious reason of protecting against the risk of explosion in the event of overload [2, 3]. The advances made in ZnO technology in arrester design, which replaced the technology of spark-gap arresters, made it possible to develop new manufacturing concepts. Today, ZnO arresters are manufactured either by applying the silicone housing directly onto the active part, or by using a fibre reinforced silicone coated composite tube as an insulating housing for the arrester, which is advantageous, if a combined post insulator function is required for the arrester, Fig. 6.5.

6.2.3 Outdoor Terminations for Cables with Extruded Insulation

The many users and advocates of high-voltage cables with extruded, solid insulation were soon calling for terminations properly suited to the properties and features of this type of cable. Slip-on silicone terminations were thus developed in Switzerland as early as 1974 and were used successfully at home and abroad (Fig. 6.6). Similarly to general electrical apparatus and components having a

Fig. 6.4 800 kV wall bushing, gas-insulated

silicone housing, outdoor terminations of this type poses high thermal stability and also resistance to ozone, corona discharge, ionising radiation, moisture and exposure to flame, even in the presence of industrial pollution, in railway tunnels, near busy roads, and in high alpine and earthquake-prone areas, and are therefore characterised by low maintenance, high operational safety, reliability, and long service life.

6.2.4 Current and Voltage Transformers

Composite hollow core insulators have been used as housings for current and voltage transformers since around the mid-1970s (Fig. 6.7). This started in locations which placed particular demands on the mechanical strength and flexibility of the housings, for example in areas there is a risk of explosion or where high mechanical stresses such as earthquake, vandalism or high short-circuit forces are likely. Faults, despite rare, could lead to transformer explosion and thus pose a considerable risk to people and property have in the last years led to an increased use of current and voltage transformers in composite technology.

Fig. 6.5 Transformer protection using arresters with silicone housing

6.2.5 Outdoor Circuit Breakers

For the various reasons already mentioned above, outdoor circuit breakers, including their respective control capacitors, and more recently high-voltage load-break switches, have also been equipped increasingly with composite hollow core insulators (Fig. 6.8). The possibility of fitting a fibre optic cable in the composite tube for the transfer of measurement and control signals is regarded as an additional advantage, particularly in circuit breakers.

6.3 Service Experience

Experience of polymer housings started at industrial scale in the early 1980s [4, 5]. Since then the total number of hollow core insulators in service is about half a million in the voltage range above 145 kV (2006 data). The market is in excess of 50,000 insulators/year with a yearly growth rate in the order of 10–20 %. If

Fig. 6.6 Terminations for VPE-insulated cables (110 kV)

Fig. 6.7 Current transformer with composite insulators (123 kV)

directly moulded apparatus at voltages greater than 60 kV are considered, e.g. for surge arresters, cable terminations and bushings, there are probably another million of units in service mainly for applications below 145 kV. Since there are now

Fig. 6.8 420/63 kA circuit breaker and control capacitors with composite hollow core insulators

Table 6.1 Polymer housing penetration in % for HV apparatus [6]

Polymer penetration in HV apparatus			
HV apparatus type	<170 kV (%)	245 kV (%)	>360 kV (%)
Live tank circuit-breakers	<5–10	<5–10	<5–10
Dead tank circuit-breakers bushings	>30	>30	>70
Gas insulated station bushings	>30	>50	>50
Combined circuit-breakers bushings	>90	>90	>90
RIP bushings (wall/transformer)	>50	>50	>50
Gas bushings (wall/transformer)	>50	>50	>50
OIP Bushings (wall/transformer)	10–20	10–20	10–20
Oil Instrument transformers	10–20	10–20	10–20
Gas Instrument transformers	>50	>50	>50
Cable terminations[a]	>50	>60	>80
Surge arresters[a]	>10	>10	>10

[a] Polymer housing with hollow insulators only. If all directly moulded components are considered the polymer penetration is close to 100 % for voltages <170 kV and 40–50 % for >245 kV

significant numbers of composite insulators which have been in service for a number of years it is appropriate to review their performance in the service environment. To this end, CIGRE WG A3.21 has undertaken a survey of experience and collected data from a wide range of source. This information has been reviewed and analysed and the results are presented in CIGRE Technical Brochure No. 455 [6] and hereunder quoted.

The penetration estimated by the CIGRE WG A3.21 from data provided by manufacturers for the year 2007 is reported in Table 6.1.

The CIGRE WG A3.21 employed three main sources of service experience data:

- Data collected by WG members using a questionnaire distributed to power companies
- Publications including test station experience
- Data provided by manufacturers based on their delivery lists and publications

According to the IEC/TS 60815-1, for service experience in naturally polluted conditions a period of satisfactory operation from five to ten years can be considered as acceptable and for test station experience a period of satisfactory operation from two to five years can be considered as acceptable. These indications can be used for the evaluation of service experience of polymeric insulators.

As a conclusion, the reported service experience is positive, with minor degradation in only few cases observed. A summary of above data collection for power companies is presented in Table 6.2.

6.4 Main Components of Composite Hollow Core Insulators

6.4.1 FRP Tube

The main component of a hollow core insulator is the FRP tube. Typical tube diameters range between 50 and 600 mm, wherein tubes measuring up to 800 mm and more have also been manufactured in some instances for apparatus insulators. As suggested by the name, FRP tubes are fibre reinforced plastic tubes which withstand mechanical loads and ensure the mechanical integrity of the composite hollow core insulators. To this end, the reinforcing glass fibres are applied in thin layers, sometimes at different winding angles. Different materials can be additionally integrated for specific applications, for example corrosion-resistant materials or materials which are not attacked by the decomposition product of the insulating gas SF_6. These tubes are produced in practically all instances by the wet-winding process (filament winding). More details regarding this process are described in Sect. 7.2.2.2.

6.4.2 End Fittings

The end or terminal fittings of composite hollow core insulators are flange-like cast or machined parts, normally made of aluminium. Depending on the application, they are custom-designed, for example for customised integration in a circuit breaker or transformer, or are produced in bulk having a standardised bolt circle.

Table 6.2 Summary on service experience collected from power utilities [6]

Source	Apparatus in total	Voltage class range (kV)	Maximum service record (years)	Typical specific creepage distance (mm/kV) phase to phase	Service experience
Hydro-Québec	No data	25–315	17	20–30	In general positive experience. Some loss of hydrophobicity, some deterioration for distribution class only
Czech transmission company	304	220–400	16	30	In general positive experience. Some deterioration only for apparatus installed before 1997
TenneT	195	50–380	15	25–31	In general positive experience. Some visual corona, some deterioration
German power company	230	123–420	18	25	In general positive experience. Some biological growth
Japanese and other power companies	298	55–550	13	24–47	In general positive experience. Some loss of hydrophobicity, biological growth
Svenska Kraftnät	150	245–420	5	16–25	Positive experience
Statnett	642	72.5–420	15	20–25	In general positive experience. Some biological growth
Furnas cenrais electricas	Surge arresters Instrument transformers	13.8–800 800	50 24	20–30 20–30	No visual corona; No loss of hydrophobicity; No flashovers; Some biological growth

Primarily, the end fittings have to transfer the mechanical loads from the tube to the supporting sub-structure and are also partly responsible for the sealing of the hollow core insulator. These flanges thus play a key role in the reliability of the insulator during operation. On the other hand, due to their design, they are much bulkier and heavier than the end fittings of long rod insulators (Chap. 2), and sometimes account for almost 25–30 % of the total weight of a hollow core insulator.

The pre-treatment of FRP tube and fitting and the assembly thereof not only play a key role for safe and reliable operation of composite hollow core insulators, but are also crucial for cost-effective manufacture and therefore for the manufacturing costs of the product.

These procedures are often the result of systematic tests and years of experience. It is therefore no wonder that the manufacturers of composite insulators are not forthcoming in the disclosure of such information. In spite of this, the assembly of end fittings on an FRP tube is discussed to some extent in Sect. 7.3.1.2.

6.5 Mechanical Behaviour of Composite Hollow Core Insulators Under Bending

It was explained in the introduction to this chapter that composite hollow core insulators are primarily subjected to bending loads and, to a lesser extent, internal pressure loads. The mechanical behaviour of composite hollow core insulators under such loads will be examined hereinafter, wherein particular attention will be paid to damage mechanisms and to the respective tests as stipulated in the international standards. The analytical methods for calculating and designing composite hollow core insulators will also be discussed so that the numerical simulation tests can be presented in detail. These are more demanding than the respective numerical simulation tests for solid core post insulators (Chap. 3) owing to the more complex structure of hollow core insulators.

6.5.1 Failure Mechanisms of Composite Hollow Core Insulators and Diagnostic Methods

The failure mechanisms of hollow core insulators are somewhat more complex than those of their porcelain counterparts. There are no simple mathematical models to adequately describe the mechanical behaviour of the FRP tubes used. For example, the impregnating resin matrix is therefore able to withstand pressure loads, but not significant tensile loads. By contrast, the glass fibres can only withstand tensile, but not pressure loads. In the event of torsional loading, the associated stresses are not distributed as evenly in the FRP tube as those in a homogeneous material, since the glass fibres are laid in alternate directions and therefore some of them are subjected to compression, whilst others are subjected to tension, and the resin matrix experiences both compressive and shear stresses (a winding angle ± 45° would be optimal in the case of torsion alone). This is further complicated by the lamination of the glass fibres, particularly in the event of failure. The mechanical properties of an FRP tube are therefore dependent on the properties of the resin matrix, the glass fibres, tube thickness, winding angles and the number of layers.

Delamination (first ply failure) is what commonly occurs in the event of failure of FRP tubes. The bending strength of the tube is then reduced and may lead to progressive damage and ultimately to the total failure of the insulator if it continues to be subjected to loading.

Further failure mechanisms include micro cracks, partial de-bonding between the glass fibre and the resin matrix, and glass fibre breakage. Flaws may also occur during the manufacture of the tubes, such as voids in the resin, glass fibre disorientation, and non-uniform distribution of the glass fibres in the matrix. However, most of these flaws can be detected using visual inspection methods for example.

6.5.2 The Concept of the "Damage Limit Load" in Composite Hollow Core Insulators

It is known that, in the case of tensile load for example, the time to failure of a composite insulator is dependent on the magnitude of this load, and that the higher the applied load, the more the time to failure is reduced. On the other hand, extensive tests and many years of operational experience have shown that, in the case of composite insulators, there is a load which can be withstood by the insulator indefinitely (Chap. 2). When subjected to this load, the insulator will practically never fail, because no internal damage occurs.

This load is called the "damage limit load".

In the case of the composite hollow core insulators detailed in this chapter, this damage limit load is established by applying strain gauges to the FRP tube (Fig. 6.9). Based on experience and specified in the relevant international standard IEC 61462 [7], a residual strain, i.e. a strain after unloading, of ± 5 % at most without exceeding the damage limit load of the tube is permissible (Fig. 6.10a). If the residual strain is greater (Fig. 6.10b), it is assumed that permanent damage will be caused, either on the surface or inside the tube.

6.6 Testing of Composite Hollow Core Insulators

6.6.1 Definitions of the Loads

Hollow core insulators are designed and tested in accordance with the above-mentioned IEC standard. Inter alia, the following loads are named and defined in this standard:

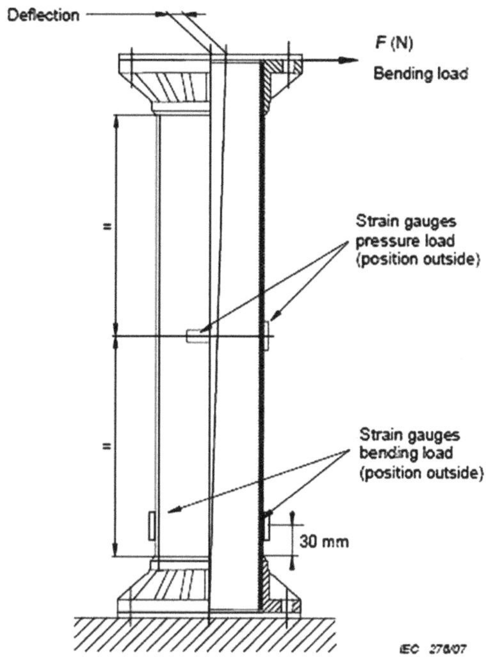

Fig. 6.9 Position of the strain gauges for pressure and bending load (according to IEC 61462: Ed. 1 2007)

Specified mechanical load (SML): Load specified by the manufacturer and applied during mechanical tests.

Note 1: This load is normally applied as bending at normal ambient temperature.
Note 2: The SML forms the basis for the selection of composite hollow core insulators with regard to external loads.

Maximum mechanical load (MML): Highest mechanical load which is expected to be applied on the hollow core insulator in service and also on the equipment in which it is used.

Note: This load is specified by the equipment manufacturer.

Failing load: Maximum load which can be reached when the insulator is tested under the prescribed conditions (valid to bending or pressure tests).

Note: Damage to the core is likely to occur at loads lower than the insulator failing load.

Permanent deflection: Difference between the initial deflection of a hollow core insulator before load application and the final deflection after removal of the load.

Note: The permanent deflection is used for comparison with strain gauges.

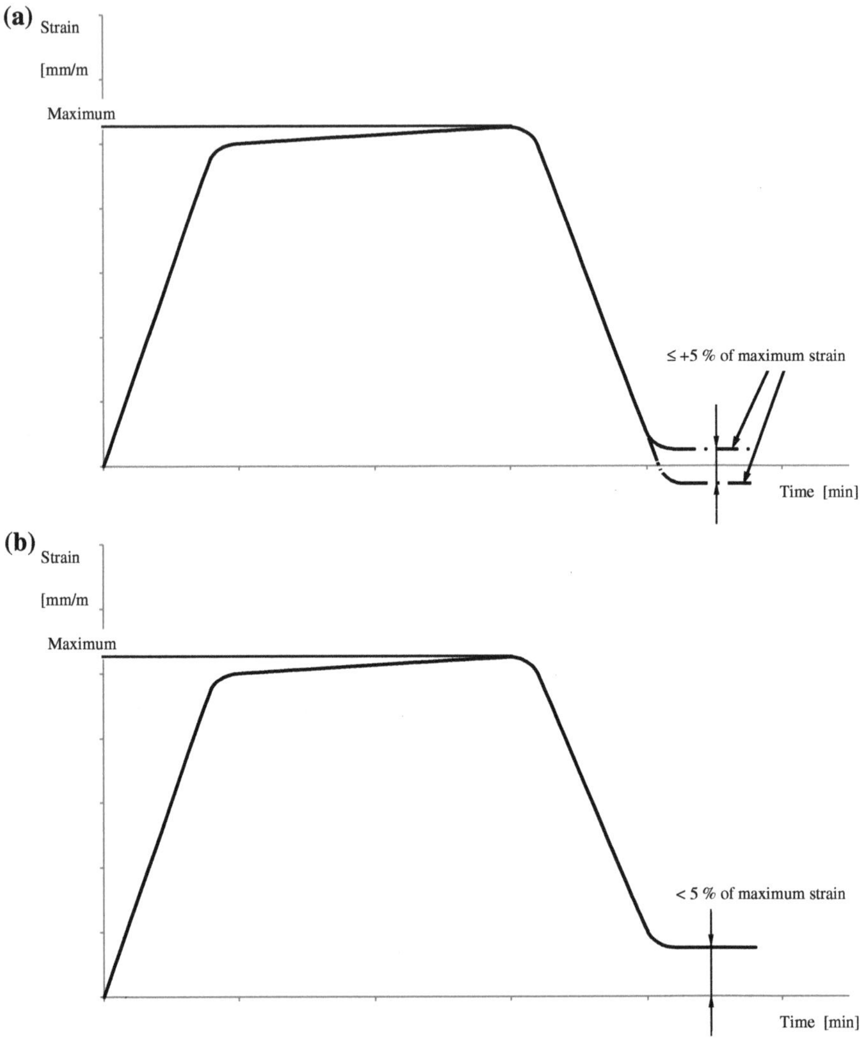

Fig. 6.10 Strain/time curve. **a** reversible elastic phase, **b** irreversible plastic phase, damage limit load (according to IEC 61462: Ed. 1 2007)

Specified internal pressure (SIP): Internal pressure specified by the manufacturer which is verified during a type test at normal ambient temperature.

Note: The SIP forms the basis for the selection of composite hollow core insulators with regard to internal pressure.

Maximum specified pressure (MSP): Difference between the maximum absolute internal pressure, when the equipment (of which the hollow core insulator

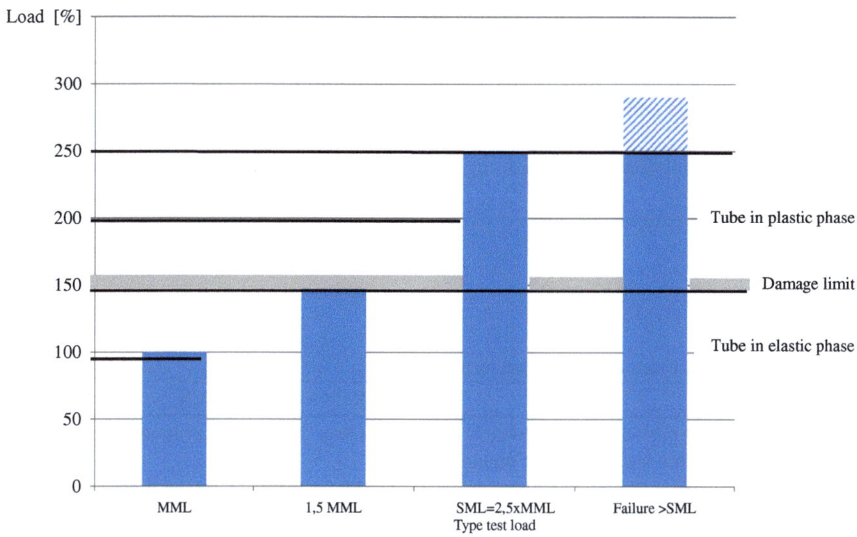

Fig. 6.11 Relationship between the bending loads

is a part) is carrying its rated normal current at maximum operational temperature, and the normal external pressure.

Note 1: The MSP of the hollow core insulator is specified by the equipment manufacturer.

Note 2: The MSP is equivalent to "design pressure" as used for ceramic hollow core insulators (see IEC 62155)

The following tables and diagrams from IEC 61462 summarise the different loads and their relationships to one another.

6.6.2 Bending Loads

These loads are also illustrated in a graph in Fig. 6.11, Table 6.3.

6.6.3 Pressure Loads

The pressure loads are illustrated in a graph in Fig. 6.12, Table 6.4.

Lastly, Table 6.5 shows examples of typical values for the bending and pressure loads, as are often encountered in practice.

Table 6.3 Relationship between the bending loads

Load	Relationship	Tube is in the
Maximum mechanical load (MML), which is the design load for the equipment manufacturer	1.0 × MML	Reversible elastic phase
Damage limit load	>1.5 × MML	Reversible elastic phase
SML—Failing bending load for the bending test	2.5 × MML	Irreversible elastic phase
Failing load	>2.5 × MML	Irreversible elastic phase

Table 6.4 Pressure loads acting on the insulator

Pressure	Relationship	Tube is in the
Maximum specified pressure (MSP), which is the design pressure for the equipment manufacturer	1.0 × MSP	Reversible elastic phase
Routine test pressure	2.0 × MSP	Reversible elastic phase
Damage limit load	>2.0 × MSP	Reversible elastic phase
Type test pressure	4.0 × MSP	Irreversible plastic phase
Specified internal pressure (SIP)	>4.0 × MSP	Irreversible plastic phase

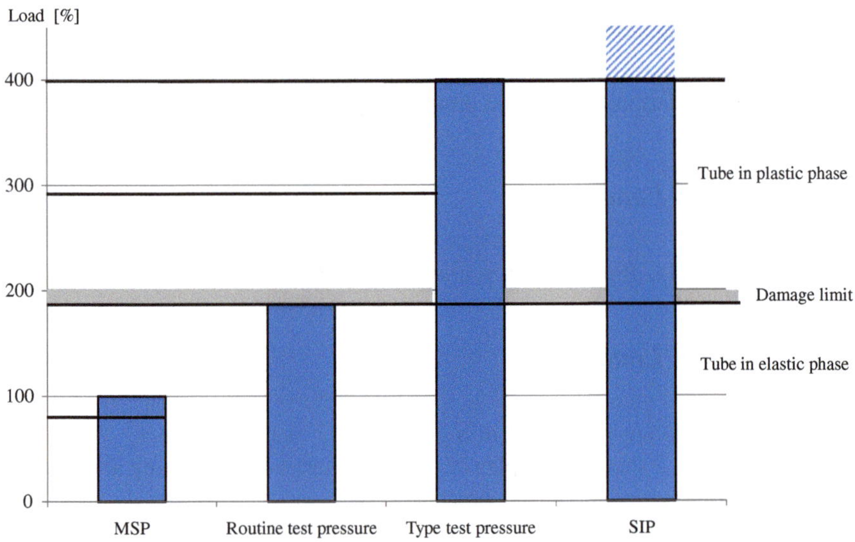

Fig. 6.12 Relationship between the pressure loads

Table 6.5 Examples of pressure/bending values—practical relationships between the values

			Tube phase
Exemplary values			
Maximum specified pressure MSP = 1.00 MPa (design pressure)			Elastic
Specified internal pressure SIP = 4.5 MPa			Plastic
Maximum mechanical load MML = 2,000 N (design load)			Elastic
Specified mechanical load SML = 5,000 N			Plastic
Design tests			
Pressure			
0.25 × SIP gas leakage (detected with He)	= 0.25 × 4.5	= 1.12 MPa	Elastic
1.0 × SIP water tightness	= 1.0 × 4.5	= 4.5 MPa	Plastic
Bending			
0.5 × SML, four directions, $\Delta\theta = 85$ K	= 0.5 × 5,000	= 2,500 N	Elastic
Type tests			
Pressure			
2.0 × MSP	= 2.0 × 1.0	= 2.0 MPa	Elastic
4.0 × *MSP*	= 4.0 × 1.0	= 4.0 MPa	Plastic
1.0 × SIP	= 1.0 × 4.5	= 4.5 MPa	Plastic
Bending			
1.0 × MML	= 1.0 × 2,000	= 2,000 N	Elastic
1.5 × MML	= 1.5 × 2,000	= 3,000 N	Elastic
2.5 × MML	= 2.5 × 2,000	= 5,000 N	Plastic
Sample tests			
Pressure			
2.0 × MSP	= 2.0 × 1.0	= 2.0 MPa	Elastic
Bending			
1.0 × MML, four directions	= 1.0 × 2,000	= 2,000 N	Elastic
1.5 × MML, four directions	= 1.5 × 2,000	= 3,000 N	Elastic
Routine tests			
Pressure			
2.0 × MSP	= 2.0 × 1.0	= 2.0 MPa	Elastic
1.0 × MSP gas tightness	= 1.0 × 1.0	= 1.0 MPa	Elastic

Fig. 6.13 Composite hollow core insulator (without shed) for a surge arrester during the bending test

6.6.4 Examples of Practical Tests According to IEC 61462

6.6.4.1 Hollow Core Insulator for a 145 kV Surge Arrester

Tubes having a length of 1500 mm and diameters of 93/83 mm had to withstand bending loads MML = 1.4 kN and SML = 3.4 kN and reach a pressure tightness up to a pressure SIP = 40 bar.

Bending Test

During the bending test (Fig. 6.13), a failing load of 4.5 kN was reached, which corresponds to a reference ultimate stress of 226 N/mm^2. An average modulus of elasticity of approximately 20 kN/mm^2 has been determined from the corresponding load/path diagram.

The residual strains (measured by DMS, Fig. 6.14) were: Table 6.6

It can be seen that the insulator behaves in a linear elastic manner up to a bending load of at least 1.5 MML, corresponding to a reference bending stress of 105 N/mm^2.

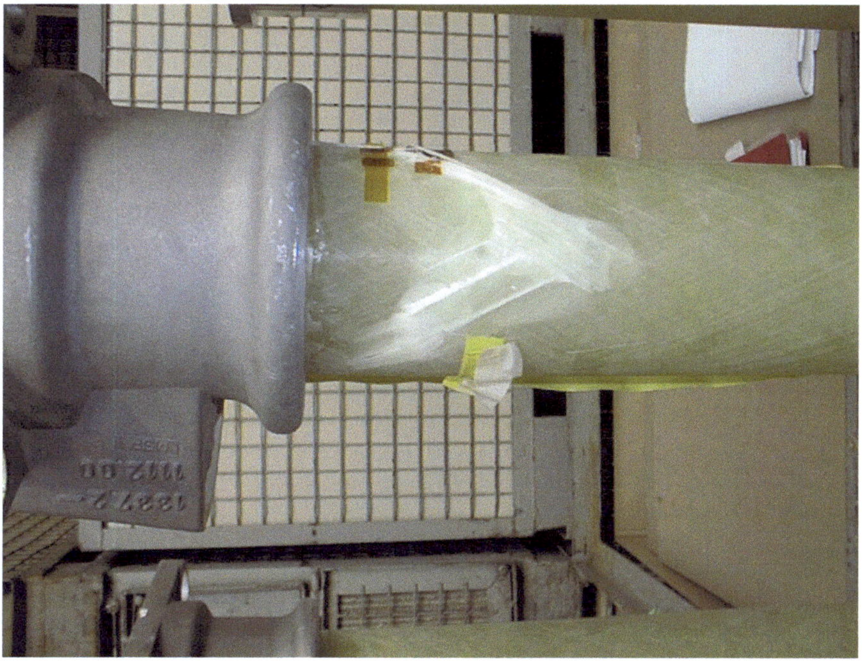

Fig. 6.14 Position of the DMS during the bending test; damage pattern in the vicinity of the fitting at 4.5 kN

Table 6.6 Measured value of residual strains

MML	Percentages
At	1.8
At 1.5	2.8
At 2.5	5.6
At 3	5.7

Pressure Test

Another sample tube was tested using the following values:

- 30 bar for 5 min
- 60 bar for 5 min
- 100 bar for 5 min

At 130 bar, the deformation in the vicinity of the fitting was so great that the O-ring sprang out of the groove. Up to this point, the tube demonstrated visible "sweating" at some points (Fig. 6.15). Individual fibre cracks in between were also

Fig. 6.15 Composite hollow core insulator "sweats" at 130 bar during the pressure test

Fig. 6.16 Composite hollow core insulator (without silicone housing) for a bushing during the bending test

clearly visible, and it is therefore to be expected that 130 bar would not have been maintained over the full 5 min. The required SIP of 40 bar, however, was achieved satisfactorily without difficulty.

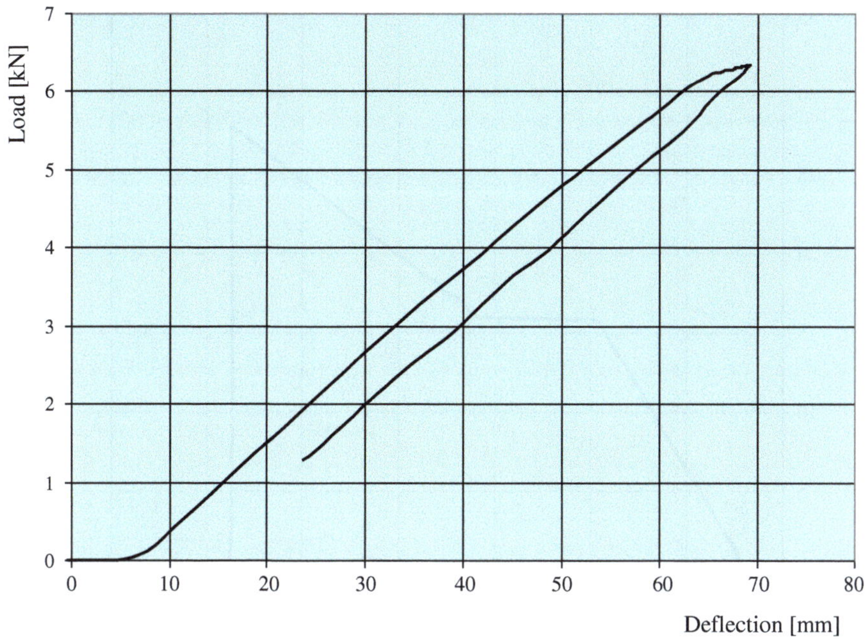

Fig. 6.17 Load/deflection curve during the bending test with an MML of 6.4 kN

6.6.4.2 Hollow Core Insulator for a 500 kV Bushing

This insulator had an FRP tube ø 440/ø 420 mm with a length of L = 5,019 mm and was tested in accordance with IEC 61462 (Sect. 8.4.2) with MML = 6.4 kN and SML = 16.0 kN (Fig. 6.16).

The force/deflection curve of the insulator up to the maximum bending load MML of 6.4 kN can be seen in Fig. 6.17. Once relieved, the insulator demonstrated a residual strain well below the 5 % stipulated by the standard.

Figure 6.18 shows the development of the bending load over time during the test for the SML and the failing load, which occurred at 28.1 kN as a result of failure at the tube/fitting interface (Fig. 6.19), albeit with a satisfactory safety margin with respect to the required SML of 16 kN.

6.7 Mechanical Design of Composite Hollow Core Insulators

As already mentioned elsewhere in this book (Chaps. 2 and 3), the mechanical calculation and dimensioning of a metal composite joint is far from trivial. The calculation has to take into account not only the rather different material properties of the metal and the composite, but also has to model realistically the interfaces

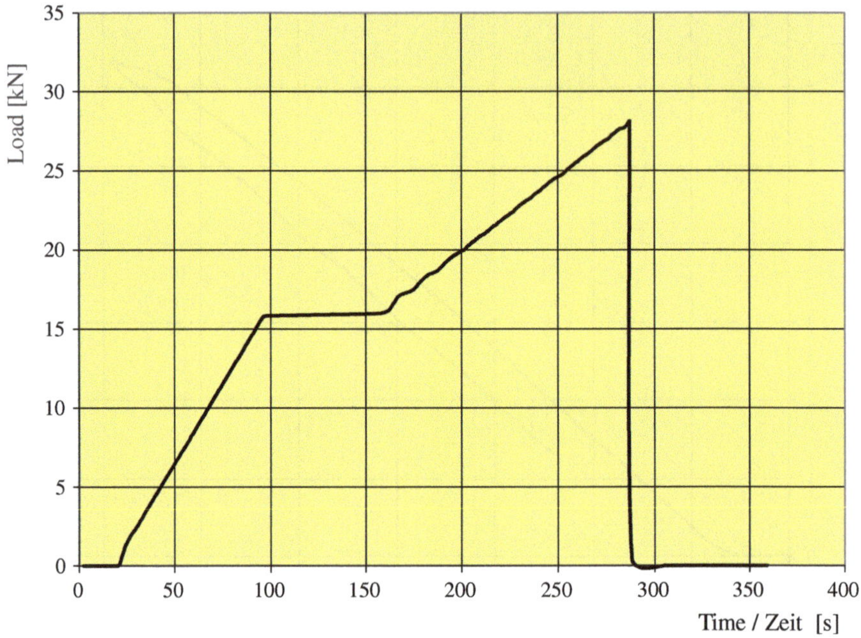

Fig. 6.18 Evolution of the bending load over time with an SML of 16 kN until failure at 28.1 kN

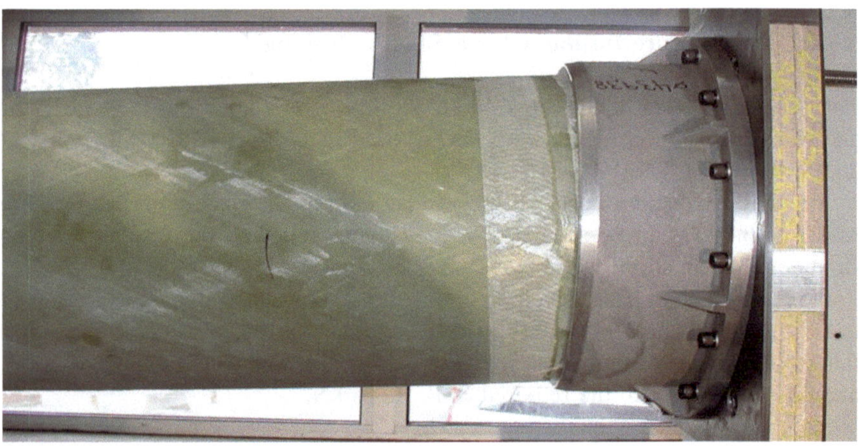

Fig. 6.19 Failure of the bonded surfaces at the lower fitting at a bending load of 28.1 kN

between these two components, since they play a key role in the functionality and mechanical strength of the joint.

This is also true in particular for composite hollow core insulators, since in this case the composite component is a multi-layer FRP tube, the calculation of which is more demanding than that in the solid core FRP rods discussed in the previous

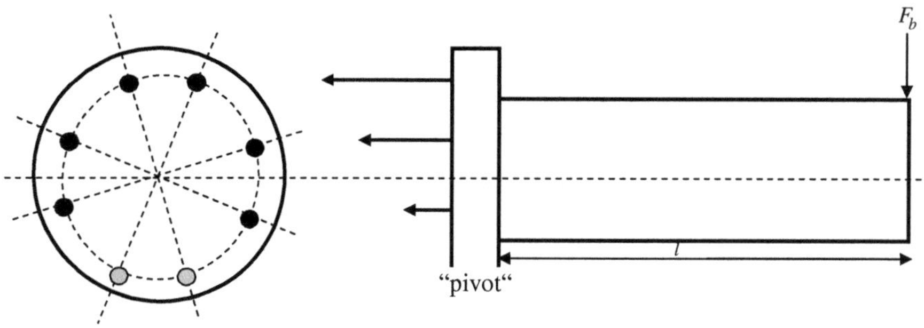

Fig. 6.20 Geometry for analytical tube and screw calculation

chapters, and also because of the adhesive bonding commonly used as the joining method. Such calculations are carried out almost exclusively with the aid of numerical simulations using the finite element method (FEM), as presented in Sect. 6.2. It is nevertheless of interest for practical day-to-day design calculations to briefly examine hereinafter a simple analytical method for dimensioning composite hollow core insulators and their end fittings.

6.7.1 Simple Analytical Method

The following simple analytical method is often used in practice in order:

(a) To select the necessary tube outer diameter and the necessary tube wall thickness (the tube inner diameter is generally determined by the conditions of the respective electrical apparatus), so that the composite hollow core insulator can reliably withstand the external loads (these are often the bending loads) and pass the tests as stipulated in IEC 61462 without difficulty.
(b) To determine the dimensions of the connection elements (these are generally bolted connections) for connection to the sub-structure (this is often a steel lattice structure).

6.7.1.1 Dimensioning of the FRP Tube

The FRP tube is dimensioned by known relationships for the section modulus W of a tube

$$W = \frac{(D^4 - d^4) \cdot \pi}{32 \cdot D}$$

and bending stress

$$\sigma_b = \frac{M_b}{W} = \frac{F_b \cdot l}{W}$$

M_b is the external bending moment and F_b is the bending load. The lever arm l is measured from the point of the load application to the mouth of the metal fitting (Fig. 6.20). The tube outer diameter D can be determined from this if the maximal permissible bending stress (see below) and F_b, l and d, as defined above, are known.

As already mentioned elsewhere, the bending stress thus calculated is merely a reference stress, since the above calculation is only valid for homogeneous, isotropic materials (such as aluminium or steel). However, it is nevertheless useful for comparative purposes, for example in order to compare the quality of the technology regarding application and assembly of the metal fitting on the FRP tube, or else as an easily calculated, rough estimate when carrying out a bending test. The numerical value of this stress is largely dependent on the properties of the FRP tube and also on the dimensions thereof, and is also highly dependent on the bonded length and inner shaping of the bottom metal fitting. It is established systematically by practically all manufacturers by laborious practical tests. A study of manufacturers' brochures gives values between 100 and 250 N/mm^2 for the maximal permissible bending stress, the lower values applying to large tube diameters.

6.7.1.2 Design of the Elements for Connection to the Sub-Structure

A few basic observations with regard to the design of the bolts for fastening the bottom flange of a composite hollow core insulator to the respective sub-structure, for example a concrete or steel structure, will now be listed briefly (see also Sect. 6.6).

When designing bolts for a flange, it is often assumed that the sum of the screw forces times the distances to the assumed point of rotation corresponds to the torque caused by the bending load (Fig. 6.20). The two bolts closest to the point of rotation are assumed to be free from load (shown in grey in Fig. 6.20). The loading of the bolts increases in an approximately linear manner with increasing distance from the point of rotation. The maximum bolt load is established on the basis of this assumed load.

The forces caused by the breakaway torque and any other force present caused by the internal pressure in the hollow core insulator are also taken into account, in addition to the max. bolt force, when dimensioning the bolts. For the breakaway torque, values are established as for clamps of outdoor substations. The necessary bolt cross-section is then determined on the basis of the formulas well known from machine elements.

6.7.2 Numerical Simulation Methods

The state of stress of a composite hollow core insulator under bending load is a complex problem which has not been examined previously in detail. There are

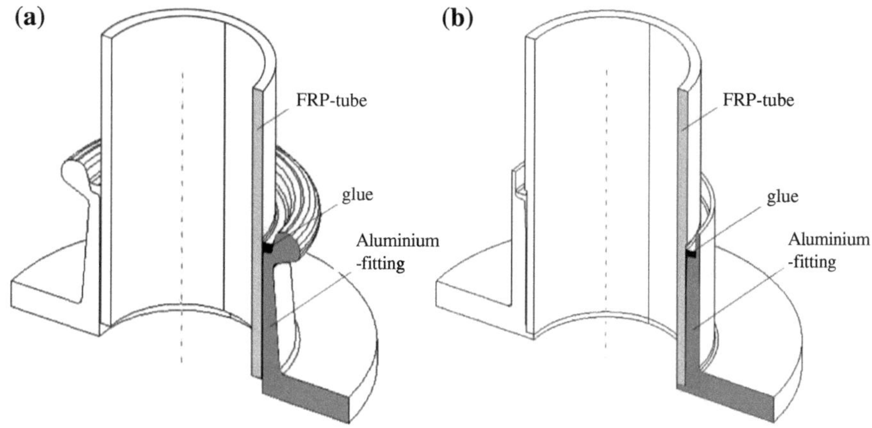

Fig. 6.21 The two metal fittings examined, *left* with rim, *right* without rim

Table 6.7 Elasticity constants of the materials used

Material	E_1[a] (GPa)	E_2, E_3 (GPa)	G_{23} (GPa)	G_{31}, G_{12} (GPa)	ν_{23} (—)	ν_{31}, ν_{12} (—)
Glue	3.2	–	–	–	0.34	–
Aluminium	70	–	–	–	0.30	–
Composite	44.6	15.7	6.6	7.2	0.46	0.28
Resin	6.7	–	–	–	0.37	–

[a] Subscripts 1, 2 and 3 relate to the fibre direction and to the two orthogonal directions thereto

hardly any published tests regarding composite hollow core insulators, with the exception of a research project in which the authors were involved and which specifically examined the metal composite joint of composite hollow core insulators for electrical power networks, the key results of which will be presented hereinafter [8].

6.7.2.1 Finite Element Modelling of the Metal Composite Joint

Silcosil® composite hollow core insulators having an FRP tube made of E-glass and a silicone shed were examined. The two metal fittings made of aluminium were fastened to the FRP tube using a special glue.

The FRP tubes were produced by the wet-winding process (filament winding) and had 10 layers with different winding angles between 20 and 60°. The proportion of the E-glass fibres in the resin matrix was 58 % by volume. Composite hollow core insulators having an outer diameter of 93 mm, wall thicknesses of 5.8 and 6.8 mm, and a length of 1 m were examined. Two geometries of aluminium end fittings (Fig. 6.21) were used: the first is slightly conical inwardly and has a

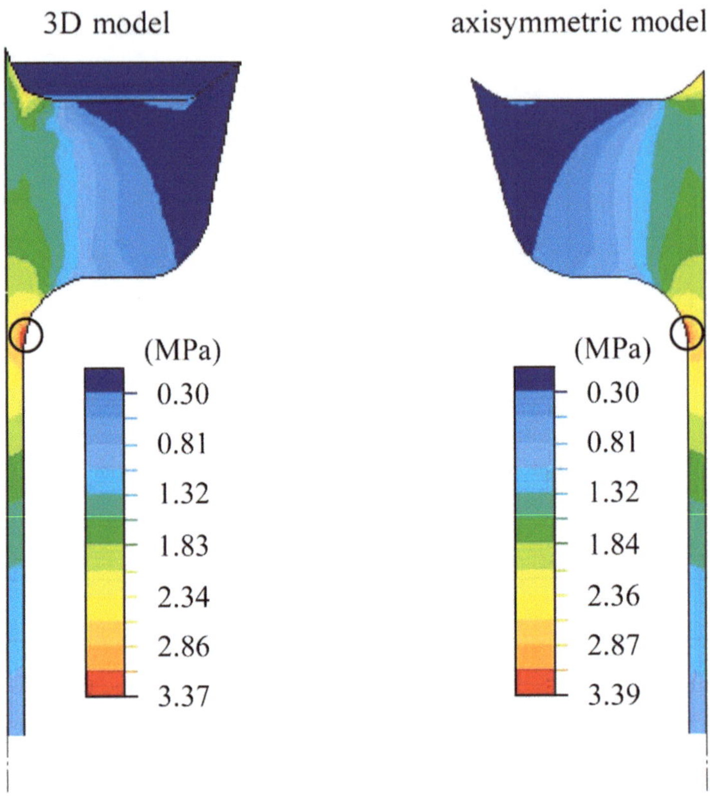

Fig. 6.22 Comparison of the calculated von Mises criterion in the glued joint under tensile load with 3D (*left*) and with 2D rotationally symmetrical (*right*) modelling; the position of the maximal stress is circled in each case

rim, the second is cylindrical; the glued lengths for the inserted resin glue were 75 mm in both cases.

All material properties are shown in Table 6.7, wherein linear-elastic material behaviour was assumed for the calculations.

Three different FE models were implemented for the numerical simulation:

(a) A rotationally symmetrical 2D model loaded purely axially; each layer of the FRP tube was simulated using up to three elements depending on layer thickness. The glue and the end fitting were also modelled rather precisely using a fine mesh of elements, both in the longitudinal and transverse directions. This gave 33,147 elements with a total of 205,008 degree of freedom.
(b) A simplified rotationally symmetrical 2D model loaded purely axially, but with slightly rougher FE discretisation compared to (a), in which the layers having similar geometrical properties (layer thickness, winding angle) were combined in groups and were modelled by one element in each case in the transverse direction. This gave 3,861 elements with 23,904 degree of freedom.

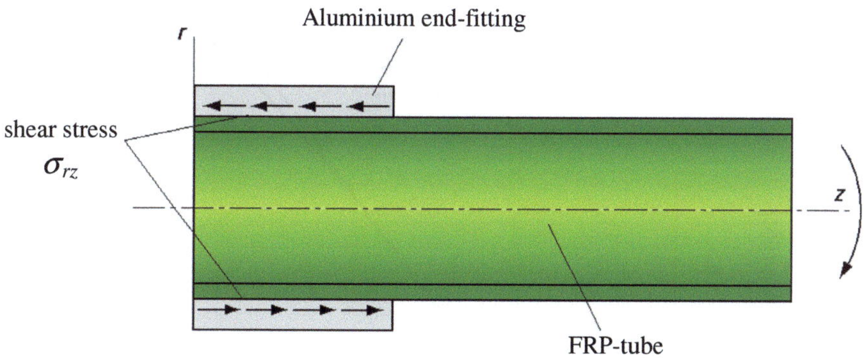

Fig. 6.23 Approach for determining the equivalent tensile load

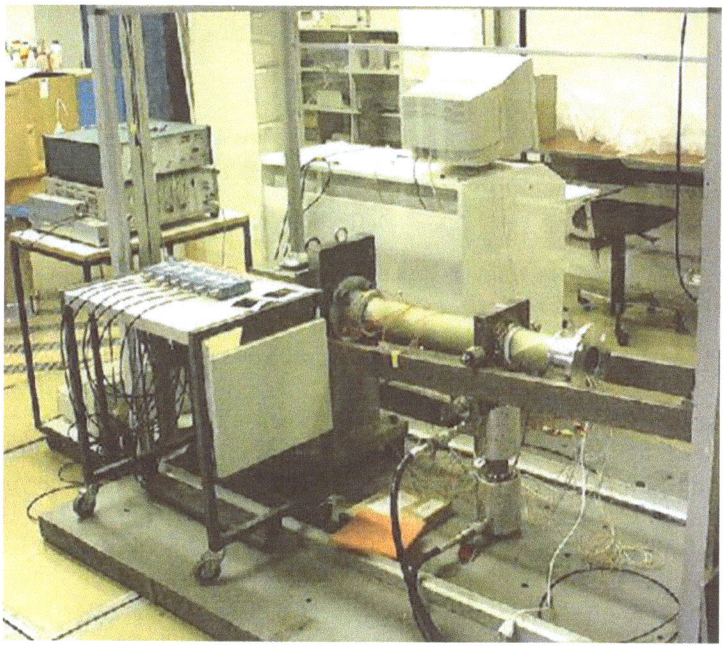

Fig. 6.24 Test rig with test specimen during the bending test

(c) A full 3D model subjected to a bending load, but with the same simplification in modelling as (b), that is to say in this case, too, similar layers were combined in groups and modelled by one element in transverse direction in each case. Accordingly, the glue was also simulated roughly using only one element, and the aluminium fitting was simulated roughly using a few elements, which led to a total of 35,722 3D elements with 468,999 degrees of freedom.

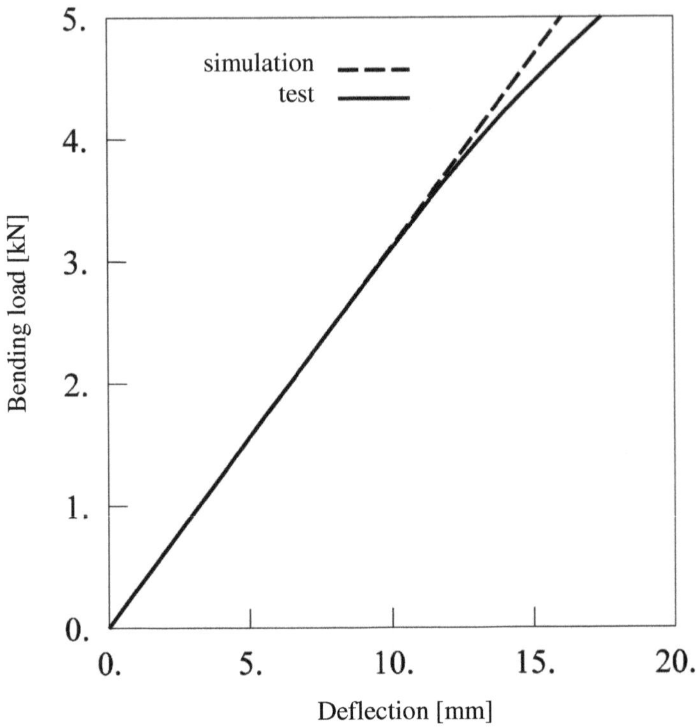

Fig. 6.25 Comparison of FE calculation and measurement in load/deflection curve

6.7.2.2 Numerical Simulation of the Internal Stressed State

In a first step, calculations were made in which all three above-described FE models were loaded by 10 kN tensile force for both tube types. The differences between the different modelling strategies could then be determined; the von Mises yield criterion was selected as a parameter for comparison (Fig. 6.22). There was very good correlation between all three models (a), (b) and (c), which led to the assumption that the more approximate 3D model (c) would provide viable results, even in the case of bending, which is of greater interest.

This important piece of knowledge was then applied in the case of a composite hollow core insulator fixed rigidly at one end and subjected to a bending load of 1 kN at a distance of 650 mm from the point of insertion of the FRP tube into the metal fitting (the fitting mouth). Similarly to when the insulator is subjected to purely tensile load, with bending there are also stress peaks where the glued joint ends, just before the point at which the tube exits from the fitting.

It is quite important to note that it was established that the stress distributions at the point of gluing between the tube and fitting progress very similarly, both under tensile load and under bending load. The respective stresses have a practically constant ratio over the entire glued length, this ratio being given by the ratio of the

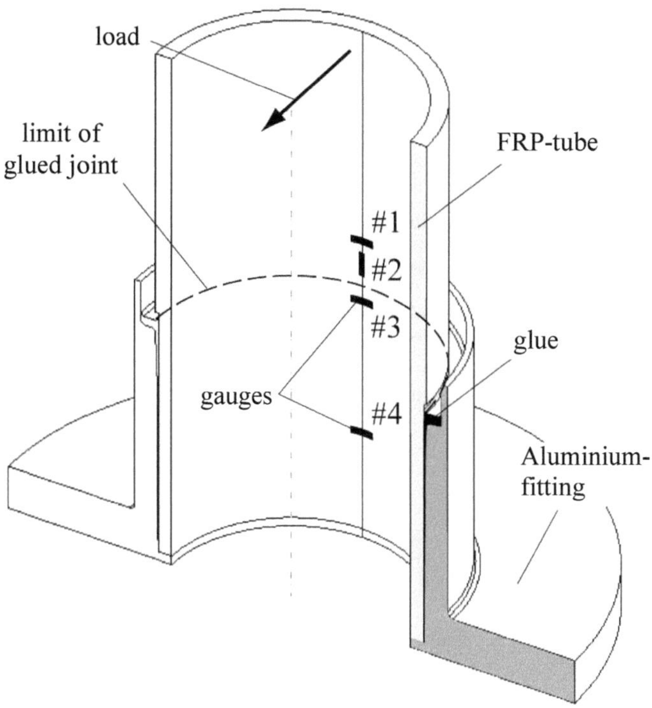

Fig. 6.26 Positioning of the DMS for the bending tests

cross-section of the tube and its moment of inertia. However, this presupposes that an equivalent tensile load is established from the bending moment of the bending load (bending load multiplied by the lever arm), more specifically by equating the bending moment of this tensile load multiplied by the mean tube diameter on the one hand, and by the shear stress σ_{rz} and mean tube cross-section on the other hand (Fig. 6.23).

6.7.2.3 Experimental Verification of the Simulation

Corresponding tests were carried out to verify the numerical simulation. The stresses were measured at selected points, both under tensile load and under bending load, and were compared to the mathematically calculated stresses.

Figure 6.24 shows the test rig constructed specially for the bending load. The load (up to 100 kN at most) is applied using two hydraulic cylinders, and the deflection of the tube is determined using a LVDT sensor. Eight strain gauges (DMS), which were attached at different points to the outer and inner surfaces of the composite tube and to the aluminium fitting, supplied information regarding the stress and strain distributions as a function of the bending load, which was applied at 650 mm from the point of gluing.

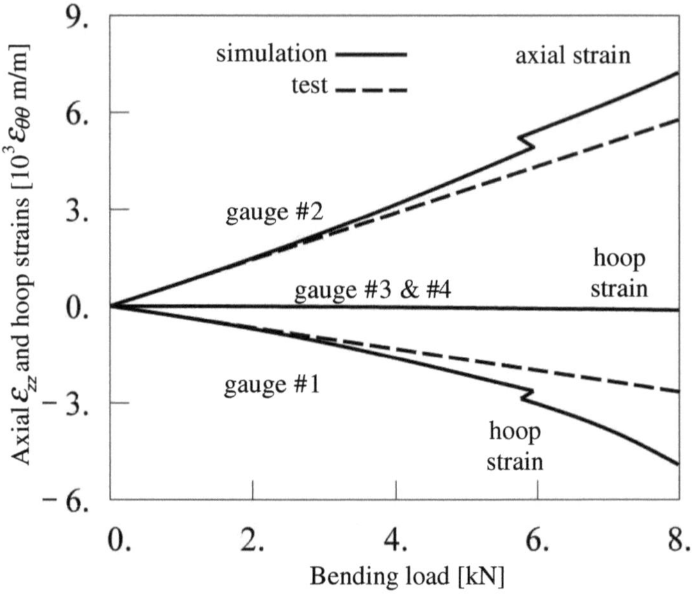

Fig. 6.27 Comparison of FE calculation and measurement of the strains

In order to confirm the measurement results over a broader range and to better estimate their scattering on the basis of the production process, seven test specimens were tested, three of which were taken directly from factory production, and the remaining four of which were manufactured under laboratory conditions.

There were hardly any differences in the measurements between the different test specimens. A typical load–deflection curve is illustrated in Fig. 6.5 for one of the test specimens. Very good correlation can be seen between measurement and simulation, up to a load of 3 kN. In addition, a slight deviation from the linear behaviour has been observed (Fig. 6.25).

Similar behaviour is observed with the stresses (strains): the load/strain curves for the axial strain and hoop strain, which were recorded by DMS on the FRP tube and at the bottom fitting (Fig. 6.26), correlate rather well with the numerical simulation in the linear region (Fig. 6.27).

References

1. Papailiou KO (1997) Verbundisolatoren auf dem Vormarsch—Anwendungen von Silikonverbundisolatoren in der Schweiz. Bull. SEV/VSE 88(1997)07
2. Gutman I, Stenström L, Gustavsson D, Windmar D, Vosloo WL (2004) Optimized use of HV composite apparatus insulators: field experience from coastal and inland test stations, CIGRE Session paper A3-104, Paris 2004

3. Gutman I, Ohnstad T, Wallin L, Windmar D, Åkesson U (2008) Field experience and test verification of HV composite apparatus insulators, CIGRE session paper A3-302, Paris 2008
4. RichterB, Schmidt W, Kannus Lahti K, Hinrichsen V, Neumann C, Petrusch W, Steinfeld K (2004) Long Term performance of polymer housed MO-Surge Arresters, CIGRE session paper A3-110, Paris 2004
5. Hinrichsen V, Petrusch W, Richter B, Steinfeld K (2003) Prüfung des Langzeitverhaltens von silikonisolierten Überspannungsableitern, ETG-Workshop "Silikonelastomere", Berlin, 20.3.2003, ETG-Fachbericht No .93, pp 119–128
6. CIGRE Technical Brochure No 455, WG A3.21 (2011) Aspects for the Application of Composite Insultors for High Voltage (>72 kV) Apparatus, Paris 2011
7. IEC 61462 Ed. 1.0 (2007) Composite hollow insulators—Pressurized and unpressurized insulators for use ine electrical equipment with rated voltages greater than 1000 V—Definitions, test methods, acceptance criteria and design recommendations
8. Bonhôte P, Gmür T, Botsis J, Papailiou KO (2003) Stress and damage analysis of composite-aluminium joints used in electrical insulators subject to traction and bending. Compos Struct 64(2004):359–367

Chapter 7
Material Selection and Manufacturing Processes for Composite Insulators with Silicone Rubber Housing

Symbols and Abbreviation

A_A	Area, when test solution is applied to surface
A_E	Area, when test solution is extracted from surface
AC	Alternating current (voltage)
ACIM	Automated continuous injection moulding
AE	Acoustic emission
AQL	Acceptable quality level
ATH	Aluminium trihydrate, Aluminium hydroxide
C0…C7	Silicone compound description code
CIGRE	Conseil International des Grands Reseaux Électriques (International Council for Large Electric Systems)
D	Start point of an indexed variable
D_A	Diameter, when test solution is applied to surface
D_E	Diameter, when test solution is extracted from surface
DSC	Differential scanning calorimetry
d	Rod diameter
d_F	Fibre diameter
E	Electrical field stress
E-CR or ECR-Glass	Corrosion-resistant glass fibre for electrotechnical applications
E-	SCA electron spectroscopy for chemical analysis
FMEA	Failure mode and effects analysis
FIFO	First in, first out principle
GFK	Glass-fibre reinforced plastic
HDK	Fine-particle silica
HV	High voltage
HTV	High temperature vulcanising
IEC	International Electrotechnical Commission
LMW	Low molecular weight (chains)
LSR	Liquid silicone rubber
M1…MX	Material description code

K. O. Papailiou and F. Schmuck, *Silicone Composite Insulators*, Power Systems,
DOI: 10.1007/978-3-642-15320-4_7, © Springer-Verlag Berlin Heidelberg 2013

M_b	Bending moment
MDA	Mix and dosing equipment
m_L	Mass loss
N	Number
RCC	Company name
SEM	Scanning electron microscope
RTV	Room temperature vulcanising
RTV-1	Room temperature vulcanising 1-component system
RTV-2	Room temperature vulcanising 2-component system
SR	Silicone rubber
tex	Tex number (length mass of a fibre)
TG	Glass transition temperature
TGA	Thermo-gravimetric analysis
t_F	Time to failure
t_{Re}	Rest time
t_T	Test time
t_{Rec}	Recovery time
t_{Tr}	Time of transfer
t_L	Time to loss of a specific property
v_P	Speed of pultrusion
α	Coefficient of thermal expansion
Δ	Relative displacement
γ	Surface tension
θ	Contact angle
θ_R	Receding contact angle
θ_S	Static contact angle
θ_A	Advancing contact angle

7.1 Introduction

This chapter reports on the latest materials and production processes used in the manufacture of composite insulators. It also looks back at the development of rod materials as well as experiences relating to the in-house manufacture of glass-fibre reinforced semi-finished parts. PFISTERER SEFAG AG has more than thirty years of experience in processing low viscosity silicone rubber (namely room temperature vulcanising (RTV) rubber and, more recently, liquid silicone rubber (LSR) as well as high viscosity solid silicone rubber (namely high temperature vulcanising (HTV) rubber). Given this level of experience and silicone rubber's acknowledged importance as a reliable housing material, this chapter will focus on providing a detailed account of this group of materials. As a result of the close interaction between material formulations, applicable manufacturing processes and operation-relevant properties, the results of key material analyses will also be presented here.

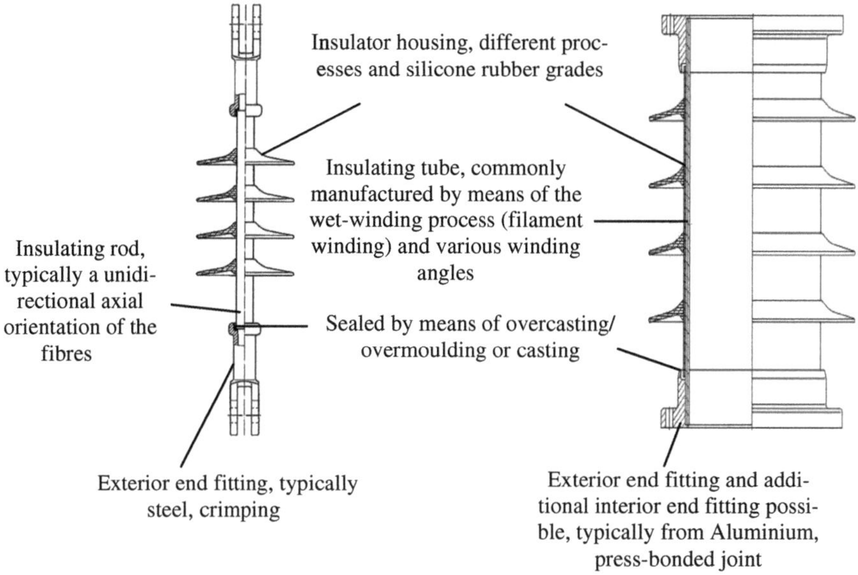

Fig. 7.1 Terms used in this chapter for solid core and hollow core insulators

Composite insulators can be manufactured as composite solid core or hollow core insulators. Both share similarities but also exhibit differences (Fig. 7.1). In the case of the load-transmitting components: The rod is typically reinforced unidirectionally with glass fibres, while the tube has different winding angles depending on the intended use. A crimped steel fitting is used to apply the load to the rod in a composite solid core insulator, while a press-bonded joint is generally used in hollow core insulators. To increase the mechanical strength of a hollow core insulator, an interior fitting can also be provided. If both are joined together (e.g. pinned), high loads can be attained during bending, in particular. Since the housing material can be manufactured from various silicone rubber grades, there are many different application processes.

A distinctive characteristic of composite insulators is its many interface areas, which can be classified as microscopic or macroscopic interface areas, and need to be considered in accordance with the relevant applicable test philosophies in each case [1–4]. Macroscopic interface areas form between the rod and the insulator housing or between the rod and the end fitting, for example, while microscopic interface areas form, for example, between the glass fibres, their sizes and the resin matrix when manufacturing rods or between the axial and radial vulcanising points when manufacturing the insulator housing. All of these interface areas are critical for the reliable operation of a composite insulator.

Since this chapter will discuss operation-relevant material properties as well as test methods to verify these properties, it is worth referring to IEC 62039 [5], which was published in 2007 and describes, for the very first time, key material properties and the minimum requirements for polymeric insulating materials for outdoor insulation systems. This standard is based on the contributions of CIGRE Working Group D1.14, which published the Technical Brochure TB 255 [6] in 2004. The comparison below in Table 7.1 is limited to the properties of composite insulators (whereas the standard also describes the material properties of 1-component insulating (structural) systems) and references the relevant sections of the chapter in which the corresponding property is evaluated.

A note on the terminology used: The terms *ageing* and *damage* describe changes to the surface of the insulating material or changes within the insulating material itself. When compared against a polymeric material in its new state, there is a reduction in insulation strength. However, unlike the established ageing processes for life time testing, this does not cause the insulator to fail immediately, but it may have an adverse effect on its long-term performance. In [21], this is categorised as follows: Irreversible changes to surface and volume properties (e.g. as a result of chemical processes such as oxidation, hydrolysis, etc. or as a result of weathering) are associated with the term “ageing”, while damage triggered by certain phenomena such as bulk erosion or prolonged tracking are associated with the term “damage” and are characterised by transgressions of certain limit loads (depending on the material and geometry).

The English literature referenced in this book uses the terms differently: In 1990, a CIGRE report [22] defined the term *degradation* as a temporary “reduction of a property” that disappears as soon as the stress factor(s) have been resolved. The original profile of the property is then restored. In contrast, the term *deterioration* means that the “reduction of a property” remains permanently even after the cause(s) has (have) been removed. Often, the term *degradation* is used to describe ageing mechanisms or their outcome [23]. In [24], the terms *deterioration* and *damage* are used with a practical reference, which makes it easier for the line inspector performing the visual inspections to decide whether or not to replace or to continue using the insulators (Figs. 7.2, 7.3).

7.2 Components/Semi-Finished Parts for Composite Insulators

7.2.1 (End) Fittings

Steel fittings are typically used for *composite solid core insulators*. Depending on the application and load, either steel casting (e.g. EN-GJS 400) or steel forging (CK 45) is used. Forged steel is the material of choice for insulators that are predominantly subjected to tensile loads and specified mechanical loads greater than 70 kN.

Because of the need for substitution compatibility, the connection dimensions for composite long rod insulators are based on the standards for conventional

Table 7.1 Minimum properties of polymeric insulating materials for outdoor applications

Property	Minimum requirement		Standard	Section of chapter	Remark
	Housing material	Core			
Resistance to erosion and tracking	1 A 3.5	–	IEC 60587 [7]	Greater Erosion and Tracking Resistance	
Resistance to ozone and corona	x	–	In progress	8.5	E. g. IEC 60343 [8], but an evaluation of more influencing factors (elongation) is recommended.
Resistance to chemical and physical degradation by water	Immersion in boiling water for 100 hours followed by 1-minute voltage withstand test at 12 kV_{rms} (AC) without a breakdown or flashover, current during voltage test ≤ 1 mA		IEC 62217 [9]	–	The second edition of IEC 62217 will be published in 2012. This test is called the water diffusion test and provides sufficient information about material suit-ability.
	Optionally, measurement of dissipation factor tan δ after 50 days of water immersion at 50 °C-threshold tan $\delta < 0.2$		IEC 60250 [10]	–	This test gives additional information about possible ageing mechanisms.
Tear strength	>6 N/mm^2	–	ISO 34 [11]	11.6.3.5	DIN 53504 [12] is applied as well
Volume resistivity	>10^{10} Ωm				IEC 60093 [13]
Breakdown strength (electrical)	10 kV/mm		IEC 60243-1 [14]	–	IEC 60243-3 [15] for Impulse voltage test
Resistance to chemical attack		96-hour test involving simultaneous mechanical (340 MPa) and chemical (1N HNO_3) stresses without failure	IEC 62039	Choosing the Glass Fibre and Resin System	Only applicable for acid-resistant core material
Resistance to weathering and UV	x		IEC 62217 with reference to ISO 4892-2 [16] and 4892-3 [17]		The second edition of IEC 62217 will only have the stress with Xenon lamp radiation (ISO 4892-2).

(continued)

Table 7.1 (continued)

Property	Minimum requirement		Standard	Section of chapter	Remark
	Housing material	Core			
Resistance to flammability	x		IEC 62217, with reference to IEC 60695-11-10 [18]	9.5.1	V0, V1 or HB40-25mm is to be proven depending on the application and nominal voltage; V0 is the highest class.
Arc resistance	>180 s		IEC 61621 [19]	–	
Glass transition temperature		$T_g > T_{max} + 15$ K	IEC 61006 [20]	–	T_{max} = maximum temperature in service
Resistance of hydrophobicity	x		In progress	Effect on the Resistance of Hydrophobicity	Only for polymeric materials, which have permanent intrinsic and dynamic hydrophobic properties
Transfer of hydrophobicity	x		In progress	Effect on the Hydrophobicity Transfer	

Fig. 7.2 Chalking as an example of deterioration [24]

Fig. 7.3 Peeling as an example of damage [24]

insulators and are described extensively in the following standards: IEC 60120 [25], IEC 60471 [26] or IEC 61466 [27, 28]. The thickness of the zinc layer is set in accordance with the recommendations in IEC 60383 [29], a thickness >85 μm is the established norm. Greater layer thicknesses (e.g. 150 μm) are possible for highly corrosive operating conditions (e.g. deserts, tunnel operation and coastal routes) or direct current applications. If the layer thickness increases, it is important to keep in mind that delaminations resulting from partial fractures (splintering) do not occur during crimping.

High-tensile chill (coquille) casting aluminium is often used in special applications (including catenary insulators).

Due to the high degree of standardisation, batches of 10,000 units or more are quite common (Fig. 7.4). Quality assurance is based on the Acceptable Quality Level (AQL) philosophy, which is a method for determining the required sample size in quality management. The following important aspects of fittings are tested (Table 7.2).

In *hollow core insulators* and in most cases, aluminium is used. Depending on the degree of standardisation for fittings, sand casting or chill casting (gravity casting) is used, or pultruded semi-finished parts or semi-finished parts that have been pressed are machined. To ensure wall impermeability or zero porosity in the area of O-ring seal(s), it is imperative that air pockets are prevented from being formed during casting. To all intents and purposes, there is no particular degree of

Fig. 7.4 Steel fittings, considered to be a "bulk material", require AQL methods; a no-go gauge is shown on the *bottom right*

Table 7.2 Quality assurance procedures for steel fittings

Measurement category	Measuring device	Standard	Destructive	Non-destructive
External dimensions	Gauge dynanometer for split pins	IEC 61466		x
Material composition	Spectroscopy		x	
Zinc layer thickness	Magnetic measurement	ISO 2178 [30]		x
Bore diameter and its inner contour	Gauge	Manufacturer-specific		x
Hardness	Eddy current measurement,	E. g. [31] DIN EN ISO 6506		x
	penetration test		x	

standardisation for connection dimensions (the current transformer head, crimping head for cable terminations, etc.). Instead, an agreement is usually reached with the manufacturer of the device. Due to the relatively low number of units and the functional importance of coordinating the press fit diameter on the outside of the tube with the press fit diameter on the inside of the fitting, dimensional accuracy is tested routinely.

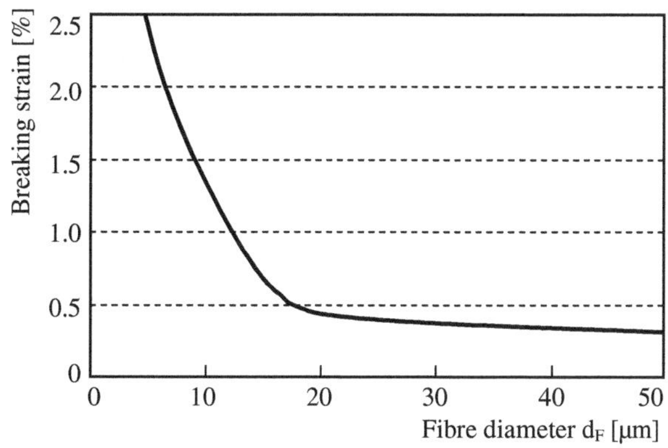

Fig. 7.5 Breaking strain in proportion to the fibre diameter [32]

7.2.2 Internal Insulation

7.2.2.1 Rod–Materials

Glass Fibre

The glass fibres used in the pultrusion process for insulators are pulled from the glass melt as individual glass filaments and, without any additional intermediate step (assembly), they are wound directly in parallel, without twisting, onto bobbins. This is known as "direct roving" [32]. Since rovings can be pulled from the outside or inside of the bobbin, it is necessary to choose between these two forms of delivery. When pulled from the outside of the bobbin, the roving is not twisted around its longitudinal axis; the filaments are all evenly aligned. However, the thread force also needs to be regulated here. Such propulsion systems are comparatively cost-intensive and time-intensive (because a certain amount of time must be devoted to bobbin changes). For this reason, rovings pulled from the outside are used only for exact laminate properties with a low tolerance deviation. Rovings pulled from the inside (i.e. from the centre of the bobbin) are most commonly used. The difference whereby the roving rotates once around its longitudinal axis for each inner circumference can be ignored when the rod is used in composite insulators.

Rovings used in rod pultrusion are available in various *fibre diameters*, *sizings* (coatings with protective layers, or combined protective layers with adhesive agents), *levels of corrosion resistance* and *tex numbers*.

The basic effect of the *fibre diameter* of a glass fibre on its breaking strain is shown in Fig. 7.5. The (tensile) strength of a glass fibre structure is always higher in comparison to an equivalent solid glass structure.

This is due to the following effects:

- When the fibre is pulled out from the glass melt, the outside of the fibre cools faster than the inside; internal stresses that occur in the fibre and increase its

tensile strength must be overcome in addition to the fracture. This principle is also used in the manufacture of tempered glass, for example.
- Statistically, the glass fibre is accredited with a lower imperfection density, which also increases its tensile strength.
- If a crack occurs in a fibre, it does not propagate to adjacent fibres.

Direct rovings are surface-coated; this *size* has the dual function of being both a protective layer and an adhesive agent:

- Protection against abrasion.
- Better sliding ability of filaments in order to prevent crippling.
- A reduction in electrostatic charges.
- Improved spreading behaviour of resin on the fibre surface in the impregnation bath.
- A defined chemical bond between the resin matrix and glass fibre.

In most cases, the size comprises silane compounds that make it possible to obtain a good chemical bond between the inorganic glass surface and the organic resin matrix applied to the surface. There have been a number of studies on the effect of this interface between the resin matrix and the size/glass surface, which can have a thickness of 200 nm. However, this relatively small number must be set in proportion to the small diameter of the glass fibres; if the fibre diameter is 10 μm, the resulting amount is approx. 8 % of the entire fibre composite. If the resin system is not correctly adjusted to the size, or if the system is stored incorrectly (exposure to UV-radiation) etc., the properties of a pultruded rod (e.g. its longitudinal impermeability) may be reduced considerably.

With regard to *corrosion resistance* (to acid, in particular), a distinction is nowadays made between two types of glass fibres, namely E-Glass, which was developed in 1938, and the more corrosion-resistant E-CR-Glass. Both names are generic; in other words, they represent product lines that have certain properties in common, but may nevertheless differ in detail from manufacturer to manufacturer.

Glass fibre properties may exhibit the differences outlined in Table 7.3 [33–39].

At the start of the 1980s, an acid-resistant E-CR-Glass was patented under the name ECRGlas© [38, 39]. Its glass composition was borne out of a commercial consideration whereby more stringent environmental regulations demanded a reduction in fluorine and boron emissions during the glass melting process. The availability of ECRGlas© established a high resistance within a pH range of ≤ 3; within the pH range 4–12, there is effectively no measurable differences between E-CR-Glass and E-Glass [39, 40]. Since measurements taken after corona or partial discharges have recorded pH values of 0 (see also Sect. 8.5.2), the use of corrosion-resistant glass fibres can reduce the risk of a brittle fracture occurring. The E-CR-Glass from the early years had, when compared to E-Glass, the following functional disadvantage: The modified composition meant that increased quantities of undissolved gases could accumulate in the melt. If, as a result of viscosity, these gas bubbles did not reach the surface of the melt and then they are

Table 7.3 Differences between E-Glass and E-CR-Glass [33–39]

Composition [%]	E-Glass	E-CR-Glass	Effect
SiO_2	53–57	54–62	Basic matrix, glass former (network former)
Al_2O_3	12–15	9–15	Better thermal, mechanical and chemical properties (amphoteric oxide = network former or network modifier)
$CaO + MgO$	22–26	17.29	Stabilisers CaO: reduces the melt viscosity, but it is a devitrification catalyser[a] MgO: counteracts devitrification
B_2O_3	5–8	0	Reduces fibre brittleness, increases chemical resistance and thermal shock resistance, and reduces acid resistance
F_2	0–0.6	–	Reduces the melt viscosity
$Na_2O + K_2O$	<1	0–2	Simplifies processing, accelerates the melting process, refining agent[b]
Fe_2O_3	0.5	0–0.8	Mechanical
ZnO	–	2–2.5	Mechanical (increases thermal conductivity, reduces thermal expansion)
TiO_2	–	0–4	Fluxing agent, instead of F_2 and B_2O_3
Property			
Density [g/cm^3]	2.6	2.72	
Tensile Strength [N/mm^2]	3,400	3,400	
Breaking strain [%]	2.2–2.5	4.8	
Coefficient of linear expansion [1/K]	5.3×10^{-6}	5.9×10^{-6}	

Explanatory note [a] Devitrification catalyser: Glass is also known as a "frozen, super-cooled liquid" that does not crystallise. If germs occur in the melt, the glass may crystallise. This is also known as devitrification

[b] Refining agent: Alkali oxides that have a degassing effect. They decompose in the glass melt when gas bubbles form and combine with other bubbles on their way to the surface, thus increasing their size and rising more quickly

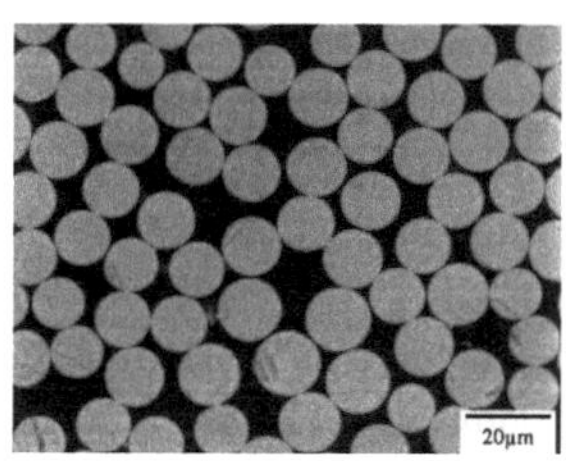

Fig. 7.6 SEM exposure of E-Glass [41]

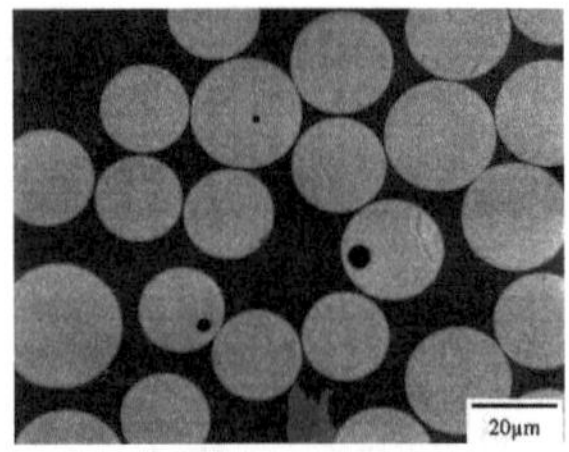

Fig. 7.7 SEM exposure of "normal" E-CR-Glass [41]

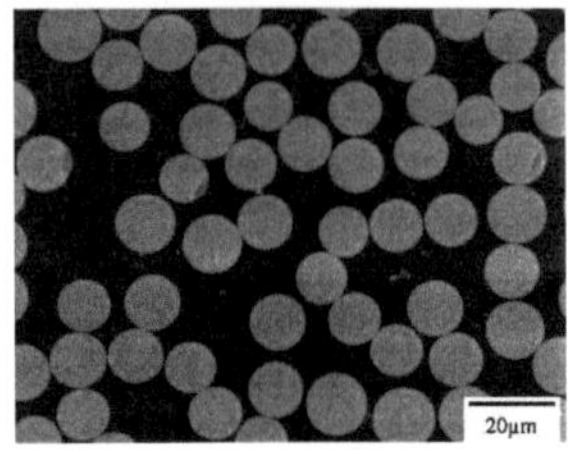

Fig. 7.8 SEM exposure of seed-free E-CR-Glass [41]

extracted in the spinnerets; the filaments can contain relatively long seeds as a result of this.

Figures 7.6 and 7.7 [41] show scanning electron microscope (SEM) exposures of rods with E-Glass and E-CR-Glass fibres. The diameters of the E-CR-Glass fibres are greater and have hollow spaces (capillaries). Studies on rods that have an identical resin system, but either E-Glass or E-CR-Glass fibres with seeds, have shown that the dielectric strength is already reduced significantly without further stressing (e.g. immersion in water), if seeds are present [42]. This effect is intensified when capillary condensation occurs or when the rods come into contact with water [36, 43].

Various design philosophies emerged as a result of this situation (i.e. increasing the acid resistance of the glass fibre while possibly reducing the electrical strength of the rod). Only one manufacturer used rods with these fibres [44, 45]. The process of "vertical impregnation while using the capillary effect" was used in this case (see also section Rod Manufacturing Process). This comparatively slow process clearly resulted in a degree of impregnation that reduced the operational effect of the unwanted capillary effect in the individual fibre. Other manufacturers have solely used rods with E-Glass and focused on constructive measures to prevent an ingress of moisture or acid into the rod. Both design philosophies were successful in the past; the effect of the resin system used in each case and the set-up of the entire insulator string or set from the perspective of preventing corona are other important factors. Thanks to a rise in demand for E-CR-Glass fibres and a growing market acceptance for composite insulators, the glass composition and glass melting processes were further optimised [46] and today, it is possible to manufacture corrosion-resistant glass fibres with a very low seed count. Such glass fibres are also known as boron-free low-seed E-Glass fibres [47]. Figure 7.8 shows an example. It is interesting to note that the diameter of the seed-free E-CR-Glass is in the order of magnitude of E-Glass.

Table 7.4 Comparison of resin systems [48]

Resin system	Density [g/cm^3]	Breaking strain [%]	Cure temperature °C	Continuous operation temperature [°C]	Shrinkage [%]	Storage life [months]	Price
Epoxy resin	1.1–1.25	4–8	15–230	55–230	1–3	12	High
Vinyl ester resin	1.07	4–8	15–70	100	1	6	Medium
Unsaturated polyester resin	1.12–1.25	<3	15–180	50–160	6–10	3–6	Low

Like the number of filaments, the *tex number* is a unit of measurement for the thread count and describes the length mass. The tex number 9,600 means a roving with a weight of 9.6 g/m. Tex numbers between 4,400 and 9,600 are common for rod pultrusion.

Resin

Nowadays, the following thermosetting plastics, shown as generic names, are used as resin systems:

- Epoxy resin.
- Vinyl ester resin.
- Unsaturated polyester resin.
- Blends of the resin systems mentioned.

The following comparison is referenced in the literature (Table 7.4) [48]:

Epoxy resin is cured by polyaddition [49]; the combination of biphenyl A resin and anhydride hardeners is widely used in insulation technology. Polyester resin and vinyl ester resin are cured by radical polymerisation [50].

When the three resin systems are compared against each other, epoxy resin is deemed to be the highest-grade resin and displays the best properties in terms of strength, rigidity, temperature resistance and chemical resistance.

Polyester resin has the lowest purchase price and facilitates higher pull-out speeds during the pultrusion process. Both produce a cost-optimised rod. The following negative factors are associated with polyester resin: A lower mechanical strength and a danger of cracks forming as a result of a high level of shrinkage during the curing process. Consequently, epoxy resin is compulsory in many composite insulator specifications.

Other additives are added to the resin system, namely:

- Inorganic fillers

These may be added for functional reasons in order to reduce brittleness and increase viscosity and thermal conductivity. An increase in resin viscosity must be taken into consideration because it can have an adverse effect on fibre wettability. Calcium carbonate is an example of a filler for electrical applications.

- Internal mould release agent

  This additive has an important effect on product quality. Because of the low level of shrinkage associated with epoxy resin during the curing process, the mould release agent reduces the peel strength and improves surface smoothness. However, the mechanical properties and the glass transition temperature may be reduced if an inadequate quantity is added. There are various materials such as zinc stearate or complex systems comprising synthetic resins and glycerides. It is also important to keep in mind that the mould release agent does not hamper the subsequent adhesion of the housing material.

- Wetting agents

  Wetting agents reduce the surface tension of the resin and improve the resin's ability to be spread across the glass fibre. Effects similar to those associated with the mould release agent must be considered when determining the quantity to be added.

- Catalysts

  Catalysts can be added to increase reaction kinetics. Examples of catalysts include amines or imidazoles.

- Pigments

  These are added as required in order to visually distinguish between the various material qualities of semi-finished parts, for example.

Choosing the Glass Fibre and Resin System

As described above, corrosion-resistant glass fibres with a low seed count are the glass fibre of choice, while epoxy resin systems are the highest grade of impregnation. Service experiences and systematic analyses confirm this [41, 50–55].

In the United States, more than 30 failures occurred between 1992 and 2004 as a result of a brittle fracture within a voltage level of 69–500 kV [53]. Approximately 90 % of these failures concerned rods manufactured from E-Glass and impregnated with polyester resin. This trend was also confirmed in an experiment where rods impregnated with E-Glass fibres and the three resin systems (polyester, vinyl ester and epoxy) were exposed to various concentrations of nitric acid (HNO_3) and subjected to a 4-point bending load. The rods were tested in their as-received state without any further conditioning changes. The results shown are valid for the chosen type of process parameter, the E-Glass fibre used and the corresponding resin system. The acoustic emission (AE) measurement was used here, which has a specific pattern for destruction resulting from a brittle fracture,

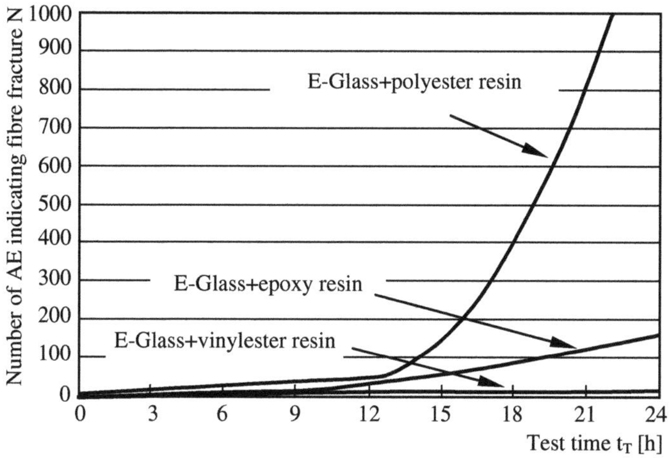

Fig. 7.9 Glass fibre fractures resulting from an acid attack, measured in terms of AE, different resin systems [51, 53]

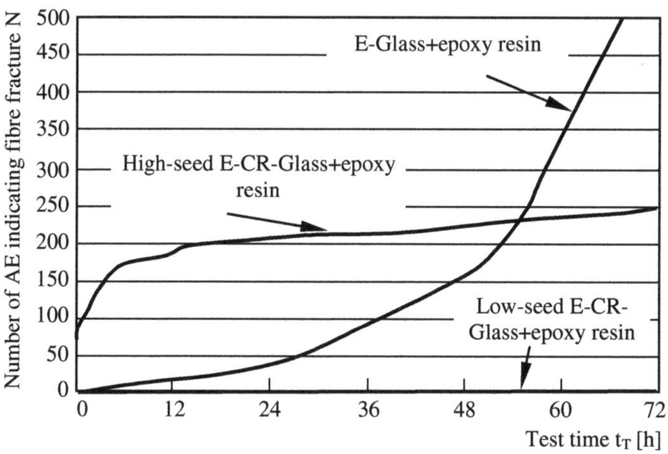

Fig. 7.10 Glass fibre fractures resulting from an acid attack, measured in terms of AE, solely epoxy resin system [51]

namely a small amplitude, low energy and a large number of acoustic emissions [56]. The progression of the AE-measurements is shown in Fig. 7.9. The E-Glass and polyester resin composite has the strongest emissions (i.e. the least durability under these conditions). This corresponds to the failures that occur while in service. Interestingly, the material combination of E-Glass and vinyl ester resin copes particularly well under these test conditions. At the time of publication in 2005, the authors did not know of any brittle fracture failures involving this composite.

Table 7.5 Comparison of the number of AEs in different resin glass systems during an acid attack [51]

Composite	Mean value for the number of acoustic emissions after 72 h	Number of test specimens
E-Glass + polyester resin	44,000	6
E-Glass + epoxy resin	1,500	5
E-Glass + vinyl ester resin	34	5
High-seed E-CR-Glass + polyester resin	330	3
High-seed E-CR-Glass + epoxy resin	140	3
High-seed E-CR-Glass + vinyl ester resin	170	3
Low-seed E-CR-Glass + epoxy resin	4	3
Low-seed E-CR-Glass + vinyl ester resin	15	2

Fig. 7.11 "Classic" brittle fracture test whereby an acid attack and tensile load occur simultaneously

This test was also used to evaluate the behaviour of E-CR-Glass with different seed counts (Fig. 7.10). If one compares the progression of the E-Glass curves in Figs. 7.9 and 7.10, there are some differences that may also be due to the variation among the test specimens (e.g. different glass melts). Furthermore, scanning electron microscopic studies [51] have shown that a pure resin layer is not necessarily formed on the rod surface. Rather, some fibres may be exposed. Neither E-CR-Glass composite failed the test. At the beginning of the test, however, brittle fractures were evident in individual fibres in the high-seed E-CR-Glass, the number of which reduced to a lower level after approxi. 12 h. The low-seed E-CR-Glass displayed virtually no brittle fractures under these test conditions.

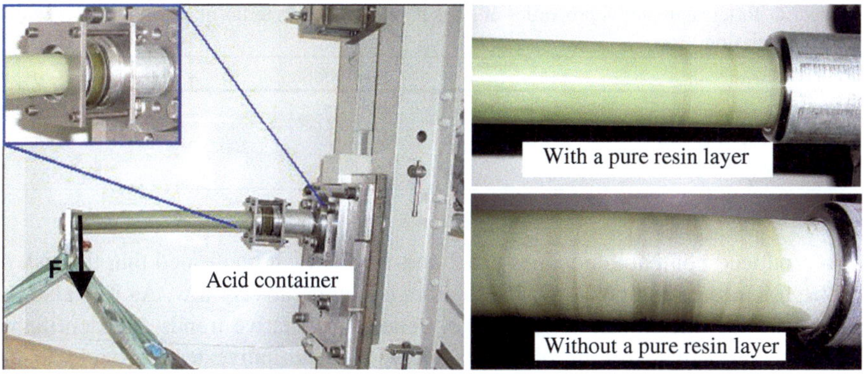

Fig. 7.12 Modified brittle fracture test for a bending load

The results of these studies on various composites are summarised in Table 7.5. The use of low-seed E-CR-Glass and epoxy resin or vinyl ester resin can significantly reduce the risk of a brittle fracture occurring. Many power utility specifications now prescribe E-CR-Glass as the glass fibre of choice for use in composite insulators that are subjected to tensile loading.

The results of the 4-point bending load were compared against the results of the "classic" brittle fracture test [44] whose set-up is shown in Fig. 7.11. The study revealed that, under these test conditions, all E-Glass fibre composites, irrespective of the resin matrix used, failed within a period of only 2 h (!). The E-CR-Glass composites passed the test conducted over a period of 96 h, with the low-seed E-CR-Glass once again displaying the lowest number of acoustic emissions.

The following question is often posed: Are composite post insulators also at risk of brittle fracture? Reference [52] reports of only one known case that is classified as a combined failure owing to a mechanical failure (poor crimping of the end fitting) and a possible acid attack in small subareas of the rod.

This finding is particularly noteworthy because a large number of suspension strings in the United States are designed as composite post insulators, and a tendency towards brittle fractures would translate into a high failure rate, similar to that associated with composite long rods. Geometrically, the larger diameter of the post insulator fitting leads to reduced electrical stressing, thus lowering the probability of occurrence of corona or water droplet corona.

Furthermore, post insulators are subjected to a bending load that is a combination of tension and compression, which is likely to impede the damage mechanism of the brittle fracture on the compressed side. In [57], the set-up for the classic brittle fracture test was modified and a bending load was applied to a 63 mm E-Glass rod, which corresponded to 1.1 times the maximum design cantilever load (1.1 $\times$ MDCL). The test duration was 96 h. In the first test, the pure resin layer of the rod surface exposed to the acid attack was not removed. This was done in a second test on a new test specimen. Both tests were successful without any damage being caused (Fig. 7.12). Moreover, the subsequent ultimate load test

Table 7.6 Relevant material properties of glass fibre and resin in terms of a composite for tensile load transmission

Property	Glass fibre	Epoxy resin
E-Modulus [N/mm^2]	72,000	3,000–4,500
Tensile strength [N/mm^2]	3,400	80
Breaking strain [%]	2.2–4.8	4–8

also produced comparable values (16.2 and 16.8 kN). It confirmed that the risk of brittle fracture in composite post insulators is comparatively low. As it is also the case with the studies mentioned above, general qualitative trends are identifiable here, the immediate results of which apply, in quantitative terms, solely to the material combinations tested. Reference [51, 56] also analysed the effect of sand blasting on brittle fracture behaviour. Sand blasting is often used to remove internal mould release agents and to improve interfacial adhesion between the rod surface and the insulator housing. It confirmed that E-CR-Glass fibres, when combined with epoxy resin or vinyl ester resin, do not suffer any reduction in their resistance to brittle fractures. When using E-Glass, it is recommended to limit the surface roughness to 5 μm [56].

Since the brittle fracture failures in the United States are due to the poor string design of certain manufacturers, a certain degree of uncertainty has arisen on this subject. Reference should be made in 2010 published IEC 62662 [58], which uses failure mode and effects analysis (FMEA) to describe influencing parameters that are relevant for insulators, and can provide support during a risk assessment.

Coordination of Properties Between Glass Fibre and Resin

Once the glass fibre and the resin system have been chosen, it is necessary to coordinate the following properties, which are universally valid for fibre-reinforced plastics:

- The modulus of elasticity for the glass fibre must be greater than that for the epoxy resin matrix.
- The tensile strength of the glass fibre in the direction of the load applied must be greater than that of the epoxy resin matrix.
- The breaking strain of the epoxy resin matrix must be greater than that of the glass fibre.

If one compares the typical values for glass fibres and epoxy resin, the results are as follows (Table 7.6):

Due to major differences in the modulus of elasticity and the tensile strength, it is necessary to coordinate the breaking strain as follows:

1. If the breaking strain of the glass fibre and the epoxy resin matrix is equal, both materials determine the tensile strength of the composite.

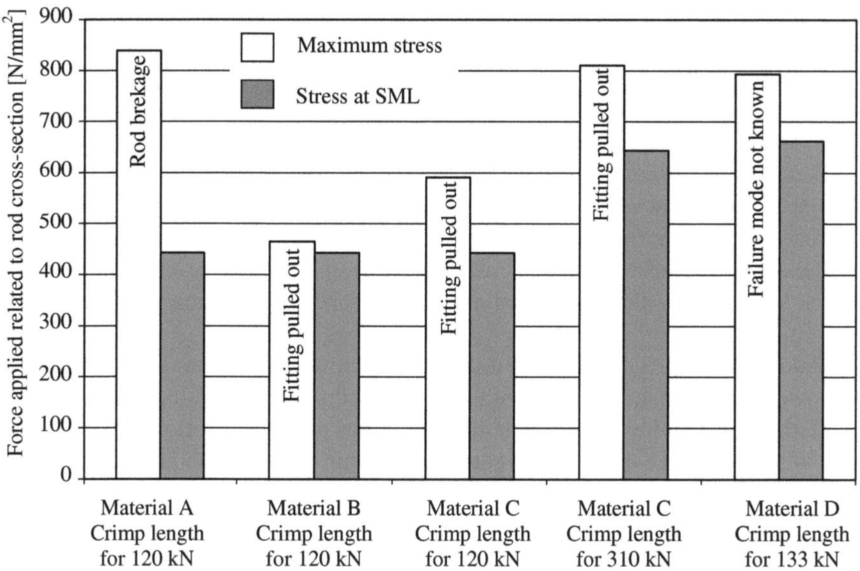

Fig. 7.13 Effect of manufacturing on the maximum stress determined in the tensile test

2. If the breaking strain of the epoxy resin is lower than that of the glass fibre, the composite fails prematurely; the tensile strength of the glass fibre is not used.
3. If the breaking strain of the resin is greater, the glass fibre determines the tensile strength of the composite.

As noted above, the breaking strain is coordinated on the basis of the last-mentioned combination. The contribution that the resin matrix makes to the tensile strength of the composite is low as a result of its lower intrinsic value and a percentage by weight of approx. 25 %. This also minimises the risk of the second scenario occurring for the same target breaking strain, but as a result of process variation.

Rod Manufacturing Process

The following different processes are deployed in the manufacture of rods for composite solid core insulators:

- Vertical impregnation while using the capillary effect.
- Special lay-in techniques.
- Pultrusion process.
- Combined pultrusion and winding process (pull-winding).

The pultrusion process is the most widely used process because it is economical and ensures high levels of productivity. However, is this at the expense at quality?

Figure 7.13 shows a comparison of the maximum tensions achieved, the design size SML for different materials and the use of crimping technology.

Material A was manufactured using special lay-in techniques and has the highest glass content. A high maximum tension is obtained with a comparatively short crimp length for the fitting. However, the failure mode is always a rod breakage. Due to the high manufacturing costs, this material is used solely for special applications.

Material B is a rod manufactured using the pultrusion process; due to the brittleness of the resin matrix, the degree of crimping in the crimp process could not be set in a way that facilitated a sufficient load transmission between the rod and fitting.

Material C is also manufactured using the pultrusion process; in comparison to B, however, it is manufactured with an optimised resin system that, in addition to exhibiting great strength at room temperature, has a glass transition temperature of 160 °C. The greater the crimp length, the greater the loads obtained without the failure mode changing from "Fitting pulled out" to "Rod breakage".

Material D represents the example from IEC 61109 [59], which is also manufactured using the pultrusion process.

From the above comparison, it is clear that the pultrusion process has achieved a high degree of maturity and can manufacture very high-grade semi-finished parts. In the case of the "composite insulator", not only the intrinsic properties of the rod but also the complex interactive design of the end fitting and crimp process are key factors for the attainable maximum load, its variation and the type of failure mode.

A brief description of the principles of the pultrusion process is shown in Fig. 7.14.

The number of glass fibre bobbins depends on their tex number and the rod diameter. As noted above, a roving pulled from the inside of the bobbin (i.e. from the centre of the bobbin) is the most frequent pulling type. If, for example, a 9,600 tex fibre is used for a 63 mm rod, this produces approx. 550 glass bobbins. Approximately 20 kg produces 11 t of glass, which must be placed on racks (Fig. 7.14, bottom left). For the thread guide, it is important to ensure that the individual rovings of the 550 individual bobbins cited in the example do not become entangled.

The resin bath (Fig. 7.14, top left) is typically a manufacturer-specific solution because the set-up, temperature and its distribution need to be adjusted to the resin blend used in each case. Today, resin blends are created using computer-aided weighing systems and logged so that they can be traced. After the resin bath, the impregnated rovings pass through the pre-feeder where they are pre-oriented so that they can be evenly fed into the heating sections and the surplus resin can be stripped away.

The exothermic reaction of resin polymerisation is initiated in the heating section (Fig. 7.14, top right). It is also a manufacturer-specific solution and can

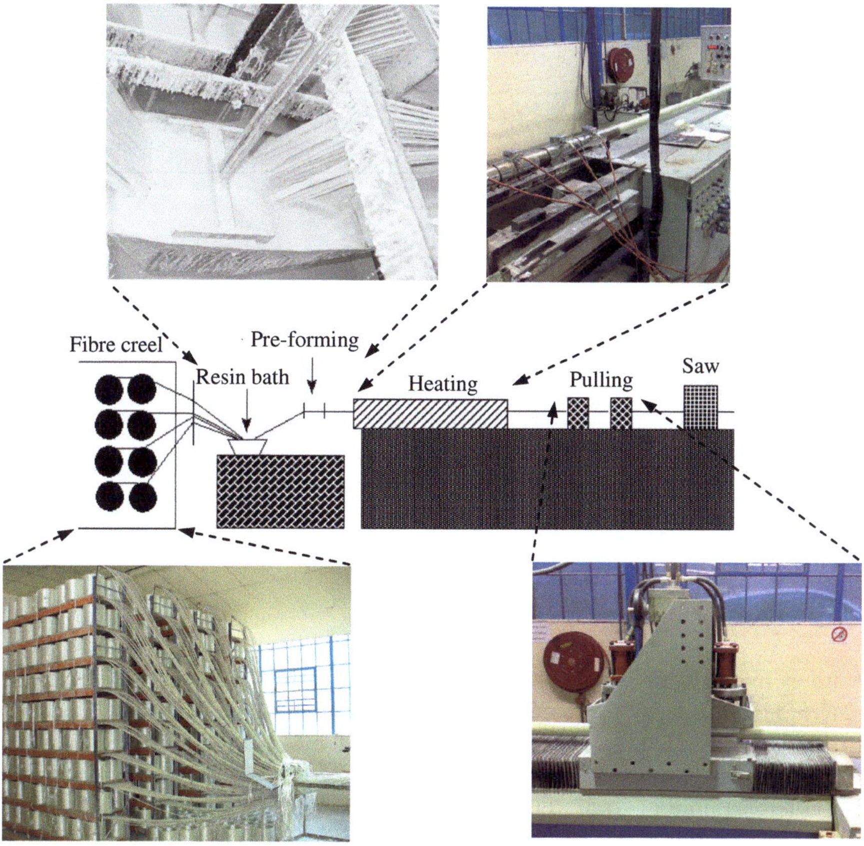

Fig. 7.14 Brief description of a rod pultrusion

comprise multiple heating zones of different temperatures. The temperature can be generated using heating elements, microwaves or a combination of the two. The advantage of microwave heating, especially in the case of rods that have larger diameters, is that the heat does not have to be transported into the rod core by means of thermal conduction. Since epoxy resin shrinks slightly, the surface quality of the heating section is important for a low pull-out friction and for low-scratch rod surfaces.

The speed of pultrusion depends greatly on the diameter of the rod. Figure 7.15 shows the example of a pultrusion process with an epoxy resin impregnation and a thermal heating by means of heating elements. The effective speed of pultrusion is often greater in the case of smaller diameters because parallel heating sections are used for the simultaneous manufacture of three or more rods.

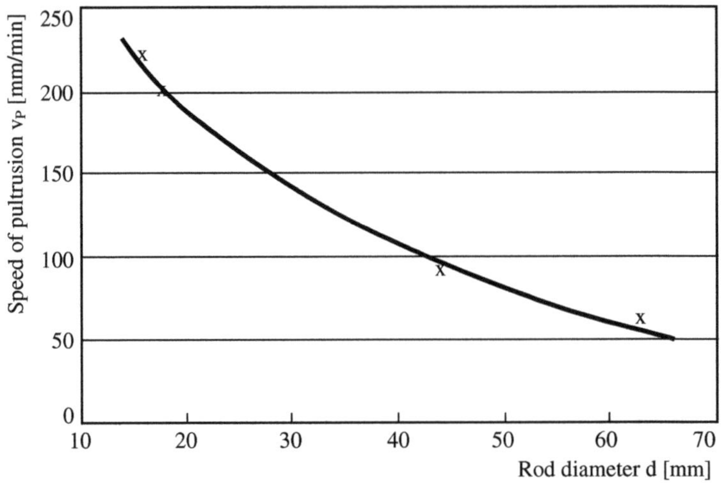

Fig. 7.15 Speed of pultrusion depending on the rod diameter, manufacture with heating elements

Quality Control

Various methods are used to control the quality of pultruded rods. These include determining the density whose value corresponds to the glass content and modulus of elasticity for a given rod composition, and measuring the glass transition temperature by means of differential scanning calorimetry (DSC). A glass transition temperature (TG) of 160 °C, as shown in Fig. 7.13 is not functionally relevant for operations if the insulator string/set design is such that the fault current of the power arc does not flow over the insulator fitting. When high temperature vulcanising silicone elastomers are used, a high rod TG is beneficial for preventing deformities when applying the insulator housing and therefore preventing rod eccentricity. The trend whereby epoxy resin systems with a high TG tend to suffer from a certain brittleness must be considered when determining the resin formulation.

A dye penetration test is used as a design test; due to its simplicity, it is also used as a routine test. Here, capillary forces are used to uncover fine cracks, seeds and pores in the material. After conducting a series of round-robin tests, CIGRE Working Group B2.21 presented its recommendations [60] for the revised version of IEC 62217, because the testing accuracy in the previous version differed greatly as a result of the wide variety of chemicals permitted. In future, this test will be standardised in such a way that a 1 % solution of Astrazon BR 200 and methanol will be defined as the dye penetrant. This composition is somewhat more stringent in comparison to the dye penetrant for IEC 61109 in 1992 (0.99 % Fuchsine solution in ethanol). The effectiveness of the dye penetrant is shown in comparison to the spreading behaviour of a water droplet (Fig. 7.16).

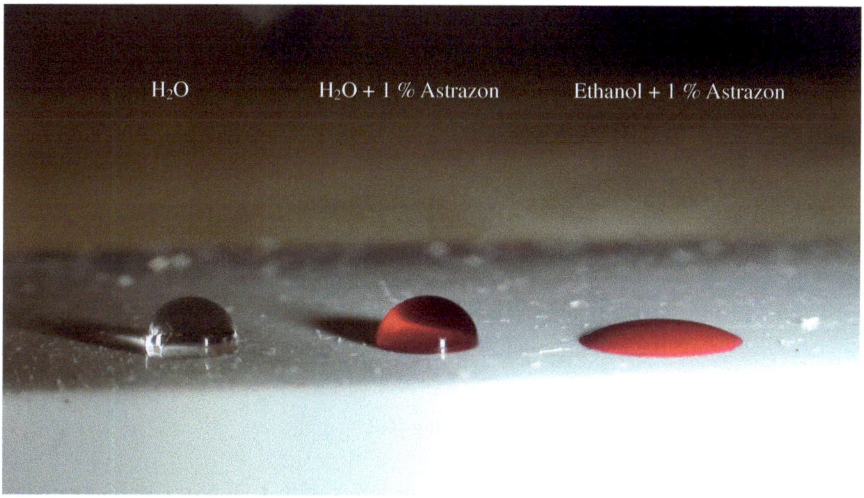

Fig. 7.16 Spreading behaviour of the test solution components of the dye penetration test

Fig. 7.17 Partial penetration

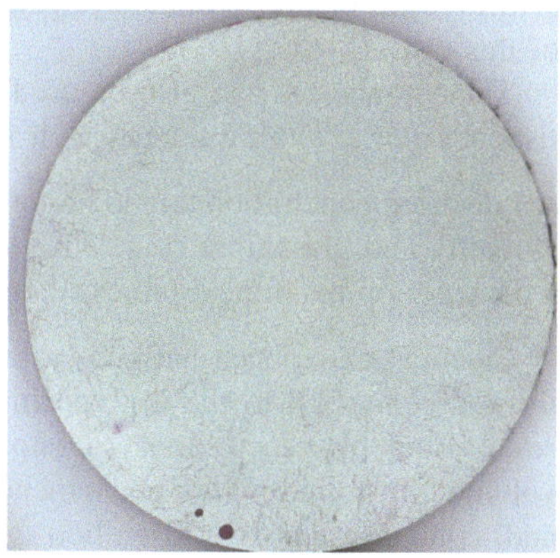

Fig. 7.18 Penetration due to an exothermal crack

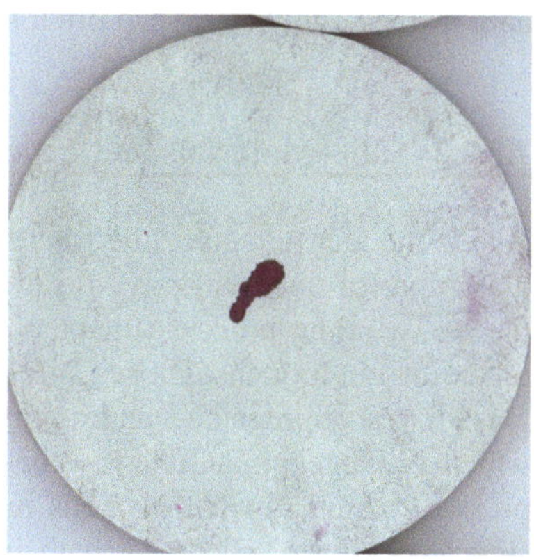

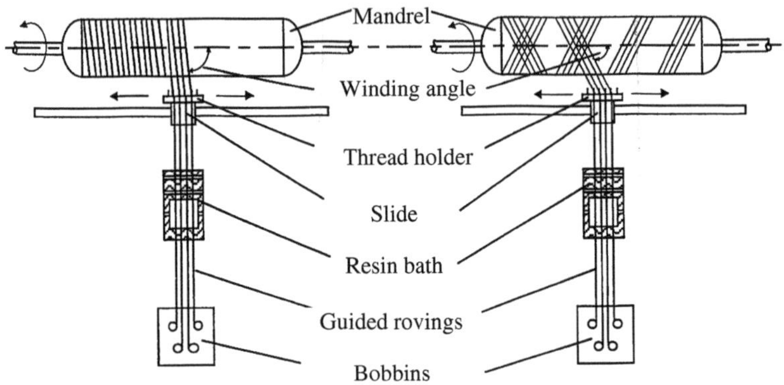

Fig. 7.19 Principle of parallel and cross winding [48]

Equal volumes of water droplets, water droplets with 1 % Astrazon, and ethanol droplets with 1 % Astrazon were placed on a hydrophobic surface. It is apparent that, as a result of the wetting effect, the methanol/ethanol and Astrazon blend will always result in a raise of the dye penetrant in capillaries (seeds), which is the goal of this test.

This simple test can also be used to evaluate batches in accordance with statistical criteria. During diagnosis, it is important to distinguish between:

- Localised penetration due to hollow fibres and/or no adhesion between the resin matrix and glass fibre (Fig. 7.17) and
- Penetration resulting from exothermic cracks (Fig. 7.18).

Practical experience has shown that testing additional sections detached from the rod connected to the test specimen in accordance with Fig. 7.17 achieves a result whereby penetrations can no longer be detected. It is obvious that the volume effect causes this phenomenon; the rod shown has a diameter of 88 mm and contains approx. 20 million glass fibres. Exothermic cracks (Fig. 7.18) resulting from thermal stresses cannot be tolerated. These may affect the entire rod and, owing to their size, can often be detected even without a dye penetration test.

7.2.2.2 Tube–Materials and Processes

Due to the fact that no brittle fractures have been reported for tubes, E-Glass is the glass fibre of choice. Epoxy resin is used for impregnation purposes. Nowadays, the wet-winding process (filament winding) is mostly used in the manufacture of tubes for outdoor insulators. This process is similar to pultrusion. However, the glass fibres impregnated under atmospheric conditions are wound around a core, which is named mandrel. In other words, the impregnated fibres are not compressed, which is what happens in the heating section of the pultrusion process (Fig. 7.19, [48]). The resin bath is therefore also a key know-how element for

Strength type		Trend	
Internal pressure	High	←	Low
Bending	Low	→	High
Rigidity	Low	→	High

Fig. 7.20 Basic effect of the winding angle on the mechanical behaviour of a tube

manufacturing: Furthermore, it is important to control the amount of resin bonded to the fibres in order to be able to reproducibly manufacture the glass to resin ratio (see also the function of the pre-feeder during the pultrusion process). This can be achieved using rolls or rollers in the resin bath, after winding the mandrel or a combination of the two. The resin matrix is usually polymerised over several stages in a convection oven. Until the gelling phase is complete, the mandrel must be rotated slowly to ensure that the surplus resin is drained uniformly. To increase productivity, the mandrels can be heated from the inside. The use of a peel ply is one possible option for compressing the composite further before polymerisation. This is wound around the outer surface, when the resin is still in a liquid or gel state and removed again after polymerisation.

Dry winding is used less for outdoor insulators than for components in insulating media such as oil or SF_6. The fibres are dry when applied to the mandrel and later impregnated in moulds or in an autoclave. Impregnation occurs by means of vacuum or compression. Polyester fibres and aramid fibres are also used for the switch components within a device, for example. The choice of winding angle over a wide area can influence the mechanical properties of the tube. The behaviour then tends to change from that of a maximum compression design to a bending design as shown in the simplified diagram in Fig. 7.20 [61, 62].

A suitable FEM process can be used to achieve the optimal winding angle [63]. The example provided in Fig. 7.21 shows the effect of the winding structure on displacement as a measure of rigidity [61, 62]. The values for M2–M5 relate to the displacement of material M1, which has the lowest modulus of elasticity as a result of a winding structure of $\pm 53°$. However, material M3 also has the lowest compression strength. M1, M2, M3 are wet-winding tubes, while M4 and M5 are dry-winding and vacuum-impregnated tubes respectively.

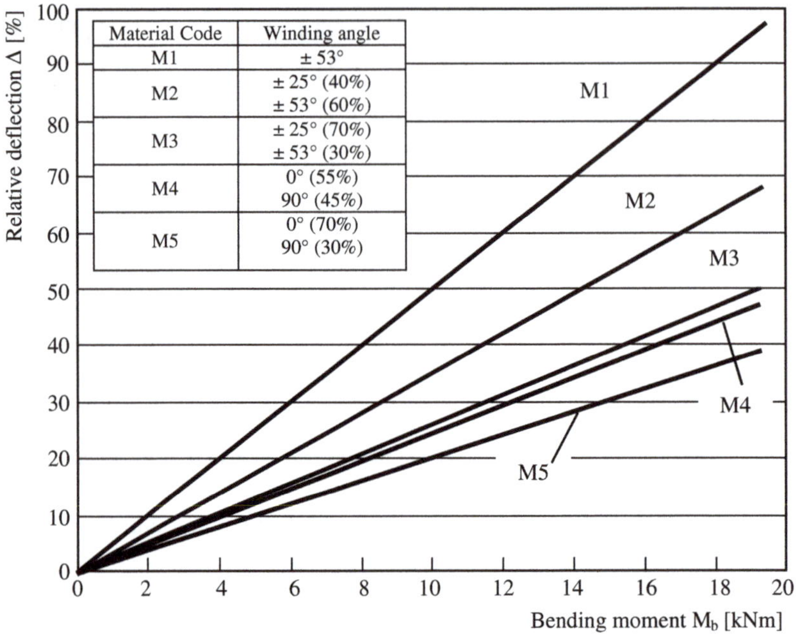

Fig. 7.21 Comparison of relative rigidity (deflection) of various tube winding structures

Fig. 7.22 Use of conical hollow cores as bushings of a 3-phase dead tank breaker [63]

Conical hollow core insulators are also used to reduce the volume of the inner insulating medium or for applications in which the centre-to-centre distances of the phases are constructively an issue (Fig. 7.22 [63]).

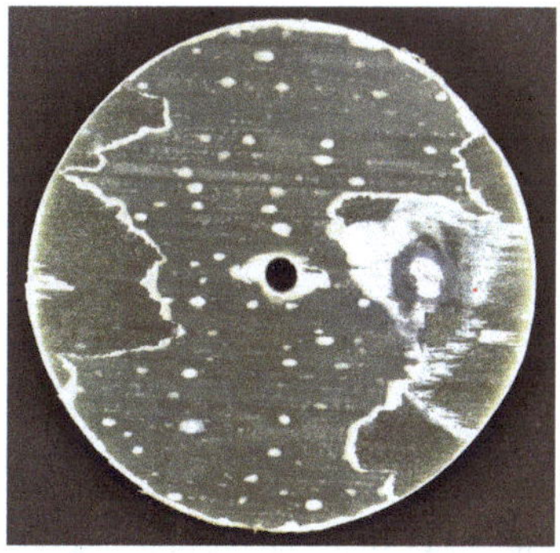

Fig. 7.23 Effect of hydrofluoric acid on the tube's inner wall after repeated switching operations [61, 62]

In the case of switching operations under SF_6, hydrofluoric acid (HF) may be produced when water molecules are present [64–66]. Hydrofluoric acid can attack the E-Glass of the tube's inner wall and cause destructive delaminations (Fig. 7.23 [61, 62]).

The following is a simplified representation of the effects on the inner interface of the FRP-tube (reinforced with glass fibres):

SF_6 decomposition during the switching operation

$$SF_6 + Me \rightarrow SF_4 + MeF_2 \, (Me = \text{bivalent metal}) \tag{7.1}$$

Formation of gaseous HF with high penetration ability (size approx. 0.3 nm *only)*

$$SF_4 + H_2O \rightarrow 2\,HF + SOF_2 \tag{7.2}$$

Attack on the glass fibre

$$SiO_2 + 4\,HF \rightarrow 2\,H_2O + SiF_4 \tag{7.3}$$

Formation of conductivity-enhancing electrolytes in the damaged area

$$SiF_4 + 2\,HF \rightarrow 2\,H^+ + SiF_6^- \tag{7.4}$$

It is clear to see that the first layer of the tube segment has already been destroyed. Appropriate counter-measures protect the glass fibres with material barriers that involve, for example, the addition of metal oxide to the resin on the first winding layer or the use of polyester fibre or polyester fleece as an inner tube liner [61, 62].

Recently concluded studies have shown that the low moisture penetration rates estimated in the past for the tube wall [61, 62] are also confirmed after cyclic mechanical stresses if the press-bonded joint was designed correctly [67].

As further reading literature, the Technical Brochure 455 [68] provides a wide range of information in regard to the use of composite hollow core insulators.

$$SiO_2 + 2\,C \text{ --------> } Si + 2\,CO \qquad (7\text{-}5)$$

			Yield	
$2CH_3Cl + Si$ --Copper-->	CH_3SiCl_3	Methyl trichlorsilane	5...10%	(7-6)
	$(CH_3)_2SiCl_2$	Dimethyl dichlorsilane	85...90%	
	$(CH_3)_3SiCl$	Trimethyl chlorsilane	4...5%	
	$(CH_3)_2SiHCl_2$	Methyl hydrogen dichlorsilane	2...3%	
	+ more Silanes		4...5%	

Fig. 7.24 Silane production as a preliminary stage for silicone rubber [49, 69]

7.2.2.3 Quality Control

The quality control performed here corresponds to the quality control associated with a rod test. Thanks to the inner hollow space, it is easy to visually evaluate the tube wall by means of an optical procedure.

7.2.3 Outer Insulation–Silicone Rubber Grades for the Insulator Housing

There are numerous ways to distinguish between silicone rubber grades. In a few cases, the user must rely on his or her own knowledge in order to deal with the non-conformities that arise in specifications as a result of differently used terminology. One example is when liquid silicone rubber is offered as high temperature vulcanising silicone rubber even though the user expects a highly erosion-resistant solid silicone rubber with a comparatively high filler. For this reason, the following section also presents key properties in terms of whether they are relevant to a particular process of housing manufacture.

7.2.3.1 Manufacture of Silicone Rubber

Silicone rubber is manufactured in accordance with the direct Müller-Rochow synthesis [69]. This double-barrelled name is used because, in the period between 1940 and 1941, the American chemist, Eugene George Rochow, and the German chemist, Richard Gustav Müller, discovered the catalytic synthesis of methylchorosilane at almost the same time. The two simplified main equations for this process are shown in Fig. 7.24 [49, 69].

The quality of the reaction products greatly depends on the composition and purity of the initial materials. SiO_2 is in powder form with a particle size of between 45 and 250 μm. Copper has a catalysing effect. The reaction occurs between 250 and 300 °C. The silane blend obtained with the main product, dimethyldichlorosilane, is separated by means of distillation and the dimethyldichlorosilane is converted into polysiloxane, the pre-product of silicone rubber, by means of hydrolysis. "Silicone rubber semi-finished parts" are created by adding silica, other fillers, additives, etc. as well as the peroxides, catalysts, etc. required for vulcanisation.

The possible main components of silicone rubber are:

- Polysiloxane with different functional groups (–OH, –C=CH_2, –H and others).
- Fillers (pyrogenic or precipitated silica, quartz powder and others).
- Vulcanisers (peroxides, H-siloxane).
- Catalysts (platinum compounds, tin compounds and others).
- Additives (adhesive agents, softening agents, pigments and others).
- Carbon blacks (pigment types, conductivity types and special types).
- Aluminium trihydrate, aluminium hydroxide (ATH).

7.2.3.2 Description and Classification of Silicone Rubber

An overview of the silicone rubber grades used as housing material can be compiled in accordance with Table 7.7.

From this overview, it is evident that there is an importance difference in terms of processing, namely the process pressure required to fill a mould cavity with silicone rubber of a certain viscosity. In the case of solid silicone rubber, "classic" processes from the rubber industry are used, which, owing to the simultaneous presence of high pressure (up to 1,500 bar) and a high temperature (approx. 200 °C when manufacturing composite insulators), are very challenging from a process technology perspective. In the case of low viscosity silicone rubber, more simplified mix and dosing equipment is used in a low pressure process whereby the viscosity pre-defined by the equipment must not be exceeded. This significantly determines the composition of the usable silicone rubber: While high pressure injection moulding machines can, with some adjustments, also process liquid silicone, it is not possible for mix and dosing equipment to process solid silicone rubber in low pressure processes. The addition of active fillers (e.g. the addition of ATH [70]) can significantly increase the ageing resistance of silicone rubber to thermal decomposition as a result of thermal-ionised partial discharges such as pre-arcs (stabilised by a pollution layer) over dry band zones. For the ATH quantity required, however, the resulting increase in viscosity can only be processed by high pressure injection moulding machines designed to process highly viscous materials from the outset.

In the literature [71], hydrophobicity and a low propensity to form conductive tracking are cited and acknowledged as being a dual protective mechanism for silicone rubber. If, however, there is a loss of hydrophobicity, and stable pre-arcs across dry bands and fed by the pollution layer form in the mA range, thermal

Table 7.7 Silicone rubber grades used for composite insulators

	Silicone rubber				
Property	RTV-1	RTV-2	Low viscosity LSR	Paste-like LSR	HTV
Vulcanising temperature [°C]	20	20–100	80–130	100–200	100–200
Viscosity [mPa s]	10,000–1,000,000 castable–firm	1,000–100,000 castable	30,000 castable	1,000,000 paste-like	5–20 × 10^6 solid
Number of components, blend ratio	1	2/9 : 1	2/1 : 1	2/1 : 1	1
Vulcanisation chemistry	Condensation	Addition, rare condensation	Addition	Addition	Peroxidic, Addition
Pot life	A few minutes/hours, depending on the air humidity	Hours, depending on the temperature	Days	Days	Months
Use[a]	Vulcanisation of pre-fabricated elements, e.g. in the modular process (shed/shank extrudate), seals	Insulator housing, seals	Insulator housing, vulcanisation of pre-fabricated elements, e.g. in the modular process (shed/shank extrudate)	Insulator housing	Insulator housing
Processes[a]	Casting, coating	Low pressure cavity casting, casting	Low pressure cavity casting	Low pressure cavity casting, injection moulding	Extrusion, compression moulding, transfer moulding, injection moulding, modular process

Explanatory note [a] The various uses and processes will be discussed in greater detail in Sect. 7.3.2

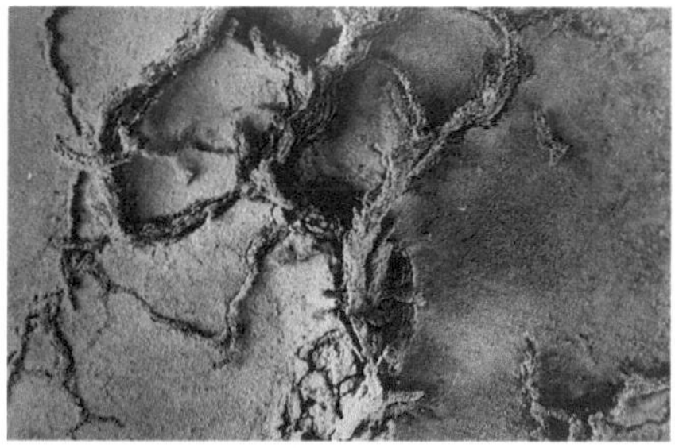

Fig. 7.25 Decomposition products on a silicone rubber surface (SR without ATH-filler)

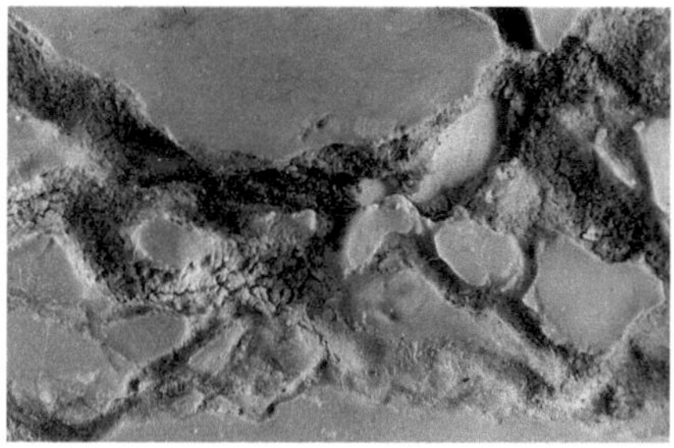

Fig. 7.26 Damage pattern caused by bulk erosion

decomposition of silicone rubber may commence, which often has the pattern of damage associated with bulk erosion [72].

Decomposition products have a high SiO_2 content and form a brittle structure (Fig. 7.25) that looses adhesion from the surface of the insulating material and easily detaches under mechanical force. The remaining "moat" (Fig. 7.26) can result in preferential adsorption of the pollution layer or, in the next humidification phase, in a pollution layer that is higher than on the surrounding smooth surface. In the worst case, both result in further cumulative damage; the bulk erosion worsens despite the low intrinsic conductivity of the direct decomposition products. Moreover, it is possible to expose the core material (rod, tube, etc.) whose surface is to be classified as being non-resistant to outdoor conditions.

For the above reasons, solely hydrophobicity is inadequate as a unique characteristic to ensure successful operation without intensive damage to the silicone surface, especially if the environment calls for a high pollution class, which is often accompanied by multiple, temporary losses of hydrophobicity.

In addition to the "classical" pollution, further severities of the environmental conditions can accelerate critically the loss of hydrophobicity. These include, for example, mold growth (especially when humid storage conditions promote this growth), algae growth, hygroscopic pollution layers [73–81]. This is evidenced in the following example of heavy pollution: Fig. 7.27 shows two composite insulators with RTV housing, which were used as suspension and tension insulators in the Lötschberg-Scheitel-Tunnel in a 15 kV network of the Swiss railway operator BLS (Bern-Lötschberg-Simplon). While these insulators have proven themselves with a creepage distance of 680 mm in an outdoor environment, a loss of

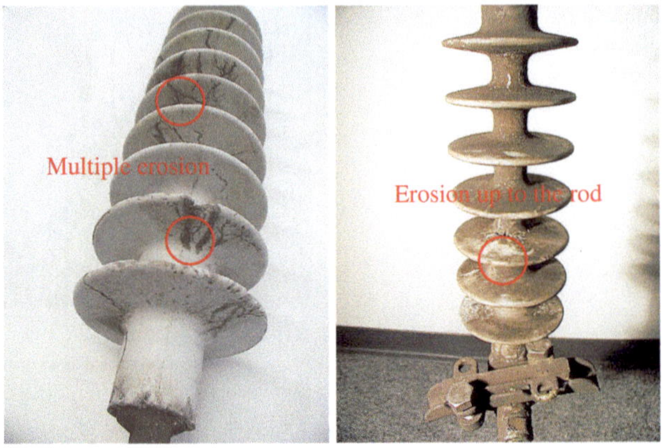

Fig. 7.27 Erosion tracks after service in tunnel

Fig. 7.28 Shed erosion on a 420 kV current transformer after short period in service

Fig. 7.29 Result after a 2,000 h clean fog test—insulator with an HTV housing filled with ATH [83]

hydrophobicity was experienced in the tunnel due to the accumulation of a pollution layer resulting from a lack of self-cleaning, and partial discharges that thermally attack the surface of the insulating material were formed because they favour the intrinsic conductivity of the pollution layer as a result of brake and pantograph abrasion. These insulators were identified as individual cases during line inspections and their mechanical strength was tested in the laboratory. The maximum load was not affected by this erosion; the rod did not show any evidence of erosive damage [82]. When operated under the same conditions, cantilever insulators with RTV silicone housing (also known as post insulators, see also Sect.

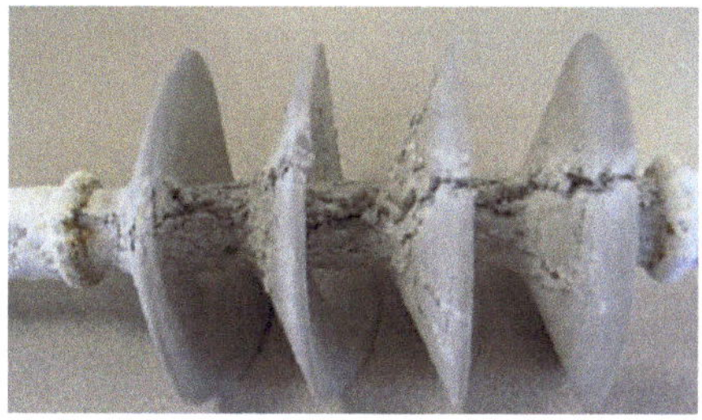

Fig. 7.30 Result after a 2,000 h clean fog test—insulator with an LSR housing [83]

10.5.3) have a creepage distance of 1,115 mm and, since the start of their installation in 1979, none of the approx. 4,000 insulators have needed to be replaced as a result of a similar pattern of damage.

Figure 7.28 shows the housing of a 420 kV current transformer that exhibited signs of erosive destruction after only six months in service at a coastal location. It is clear that the ambient conditions were so severe that they resulted in a frequent loss of hydrophobicity or delayed the recovery of hydrophobicity or the hydrophobicity transfer. The erosion resistance of the LSR used was too low under these conditions.

This difference can also be demonstrated in the laboratory by means of accelerated ageing tests (Figs. 7.29, 7.30 [83]). Both insulators were aged over a period of 2,000 h in a clean fog test. At the same creepage distance and a slightly different shed profile, the HTV test specimen filled with ATH has the edge over the LSR test specimen not filled with ATH. The destruction corresponds to the bulk erosion shown in Fig. 7.26.

It is also worth mentioning that hydrophobicity does not have any effect on the potential design weaknesses of a composite insulator if, for example, the sealing system fails and, in the case of line insulators, no acid-resistant glass was used for the rod.

As mentioned above, active fillers are used in solid silicone rubber to increase erosion resistance. The most popular filler is aluminium trihydrate (ATH), which is one of the world's most important mineral-based flame retardants and was patented for the first time in 1909 by Fritz Hofmann. ATH can undergo single-stage or two-stage decomposition.

Single-stage decomposition

$$2\ Al(OH)_3 \rightarrow\ Al_2O_3 +\ 3\ H_2O \qquad \text{as of } 180 - 200°C \tag{7.7}$$

Two-stage decomposition

$$Al(OH_3) \rightarrow AlO(OH) + H_2O \qquad \text{as of } 180 - 200°C \tag{7.8}$$

$$2\ AlO(OH) \rightarrow Al_2O_3 + H_2O \qquad \text{as of } 450 - 560°C \tag{7.9}$$

Under standard pressure, and as of approx. 180–200 °C, ATH is converted directly into Al_2O_3 without the intermediate stage of boehmite formation AlO(OH)

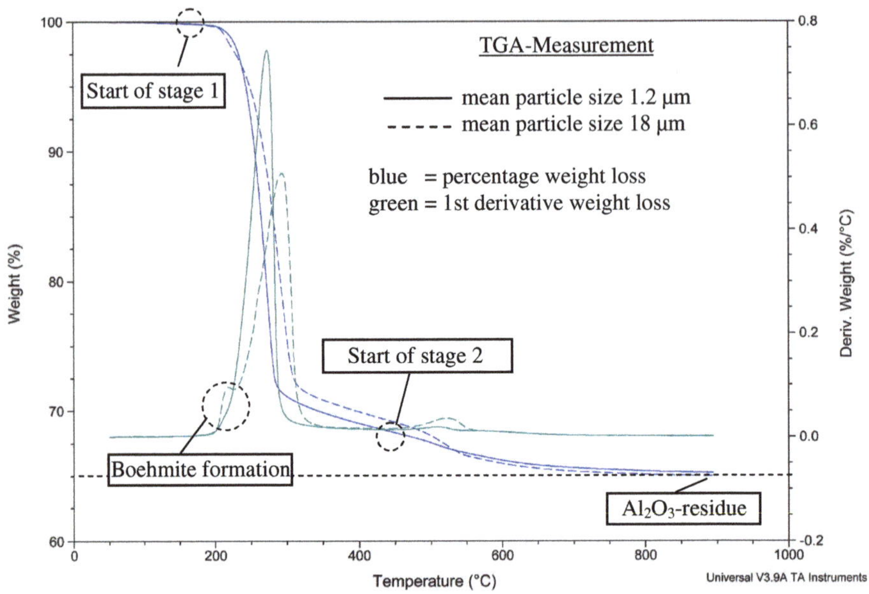

Fig. 7.31 TGA of two-stage thermal decomposition of ATH

[84] (single-stage decomposition). A two-stage decomposition occurs if a high temperature and high pressure occur simultaneously. This situation may arise during the thermal decomposition of particles if the surface is spontaneously converted into Al_2O_3. The internal $Al(OH_3)$ then experiences conditions conducive to the formation of AlO(OH), which does not enter the second decomposition stage until higher temperatures are reached. The particle size determines the quantity of boehmite formed. Larger particles have a more enclosed volume, while smaller particles have a larger surface to volume ratio. In other words, larger particles form more boehmite.

These stages can also be demonstrated in thermo-gravimetric analysis (TGA). In Fig. 7.31 only different-sized ATH particles without silicone rubber were analysed. It is easy to see where both stages start for the two ATHs. As of Stage 1, the volume of water discharged is always greater than the water release after reaching Stage 2. The intermediate stage of boehmite formation is less pronounced in the case of the small particle size, which corresponds to a small volume of water released in the second decomposition stage. The behaviour identified here corresponds to that of a fine-precipitated ATH. The ATH with the larger particle size has a clearly visible shoulder in the first derivative. The large quantity of boehmite formed in comparison to the small particle size gives rise to a second pronounced shoulder in the first derivative at approx. 450 °C. The subsequent water release can also be recognised by the percentage weight loss; the ATH with the larger particle size does not achieve the weight loss of the ATH with the smaller particle size until the boehmite has decomposed.

If one considers the overall reaction of ATH, there is a weight loss of approx. 34 %, which is achieved asymptomatically in Fig. 7.31.

$$100 \text{ g Al(OH)}_3 \rightarrow 65.4 \text{ g Al}_2\text{O}_3 + 34.6 \text{ g H}_2\text{O} \quad (7.10)$$

The effect of fire retardancy as a flame retardant for increasing resistance to partial discharges or power arcs in a filled elastomer is more complex. The following mechanisms are effective [85–87]:

- Substitution of polymer as a thermally decomposable material fraction.
- Use of ATH to increase the thermal capacity (i.e. more thermal energy must be supplied to facilitate the thermal decomposition of the polymer).
- Endothermic ATH decomposition (i.e. discharge of energy from the surface).
- Use of nascent steam to cool the surface.
- Use of steam to "dilute" flammable gases.
- Use of steam to displace oxygen required for oxidation purposes.
- Reduction in the formation of flammable cyclic oligomers with a low flash point (e.g. $(SiO(CH_3)_2)_3$, which has a flash point of 135 °C) in silicone rubber [85].

The aforementioned "particle size effect" can cause a more intensive release of H_2O in the case of smaller particle sizes. At lower temperatures, the insulating material is already protected by the accompanying processes. This behaviour correlates with the importance of the ATH particle size described in the literature [88, 89].

Moreover, the effect of increased thermal conductivity as a result of adding ATH cannot be ignored: A commercially available HTV system filled with ATH (overall weight: approx. 50 % ATH) has a thermal conductivity of 0.5 W/mK, while an RTV system that has not been filled with ATH has only 0.2 W/mK. This property is relevant from an ageing perspective in order to dissipate the temperature rise at the PD roots [90, 91].

On the basis of the tried and tested use of ATH to increase erosion resistance, the effect of ATH on the following operation-relevant properties will be evaluated below:

- Greater erosion and tracking resistance.
- Resistance of hydrophobicity.
- Recovery of hydrophobicity behaviour.
- Hydrophobicity transfer behaviour.
- Acid resistance.

The results presented here are a selection of those results provided in the literature and based on the own investigations. Qualitatively, they show the development trend associated with a particular material modification; quantitatively, they apply only to the chosen test conditions and test specimens. The results also confirm that solely adding a certain quantity of ATH does not serve any purpose. Rather, the entire system (polymer, fillers, the treatment of fillers, etc.) [90–92] needs to be optimised in a balanced manner.

Table 7.8 "Kärner matrix" tailored to the link between materials and insulators manufactured using these materials

Material from the perspective of outdoor resistance	Insulator geometry[a] encourages ageing/damage	Insulator geometry[a] does not contribute to ageing/damage
Inadequate	Very high probability of failure	Cannot achieve reliable long-term behaviour
Suitable, optimised	Reduced probability of failure, but frequent diagnosis recommended, long-term behaviour is critical	Reliable–confirmed after almost 40 years of outdoor experience involving different silicone rubber grades

[a] In the case of nominal voltages that require external field control by means of corona rings, the string design or device design must be considered accordingly

For the most part, the results presented in this chapter come from material investigations/tests. From an insulation perspective, however, their long-term behaviour is determined from the complex interaction between the materials used, the geometric design of the insulator housing (e.g. shed distances and shed overhang) as well as the manufacturing process (parting line or mould line). Consequently, the use of suitable materials that have been optimised for the intended purpose is a necessary but by no means a sufficient condition for ensuring reliable operational behaviour. This is illustrated in the "Kärner matrix" [93, 94], whose basic information has been tailored to the subject matter of this chapter (Table 7.8). For further reading on complete insulation systems, please refer to the laboratory and outdoor experiences of medium-voltage cable terminations involving different materials and geometries [95–97] and to the systematic analyses involving HTV silicone rubber as a dedicated housing material under laboratory and outdoor conditions but with a different geometry (variations that conform to IEC 60815-3) [21, 98–100].

7.2.3.3 Use of ATH as a Filler

Greater Erosion and Tracking Resistance

Greater erosion and tracking resistance is the most tested property because comparatively simple processes in the form of the Inclined Plane Test, salt fog test or wheel test are available to test it. Due to the high pressure method required to process HTV solid silicone rubber, ATH in the required quantity (depending on the formulation) can almost always be added to the housing material. The Inclined Plane Test also reveals the differences shown in Fig. 7.32 [101, 102]. The erosion rate of each specimen is calculated as quotient of its erosive weight loss (a possible mass increase by water uptake during test should be corrected) divided by its real exposure to PD-activity. This procedure considers that in the case of no PD (e. g. caused by a diversion of the electrolyte), a specimen is not stressed and can pass the test more easily.

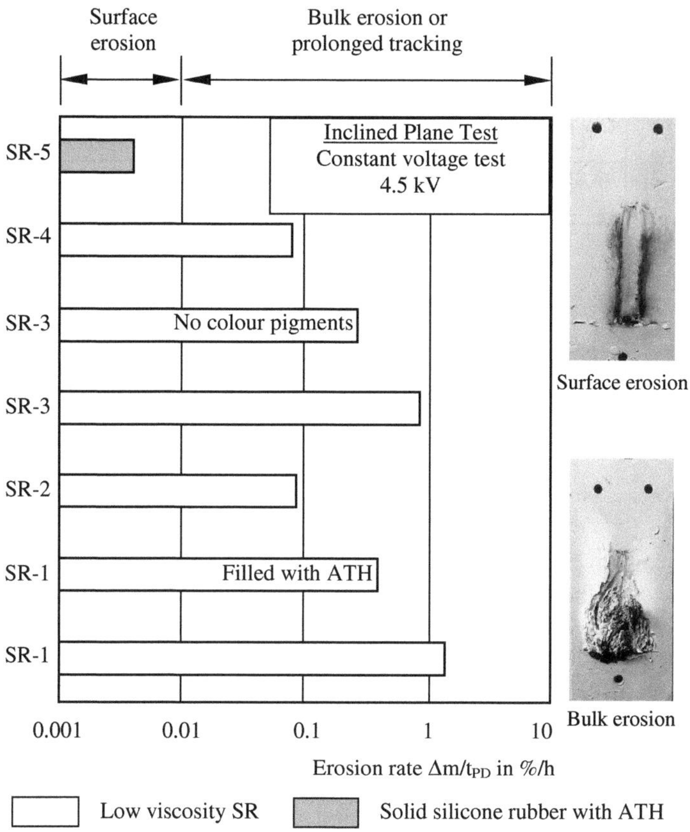

Fig. 7.32 Erosion rate of different SR-Systems [101, 102]

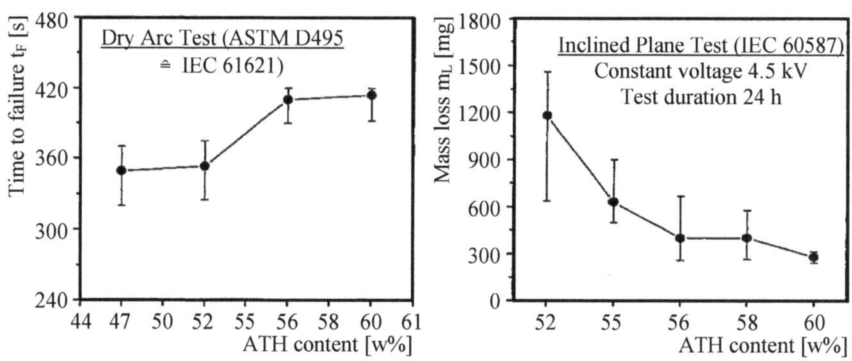

Fig. 7.33 Greater erosion resistance as a result of adding ATH, SR-system 1 [103]

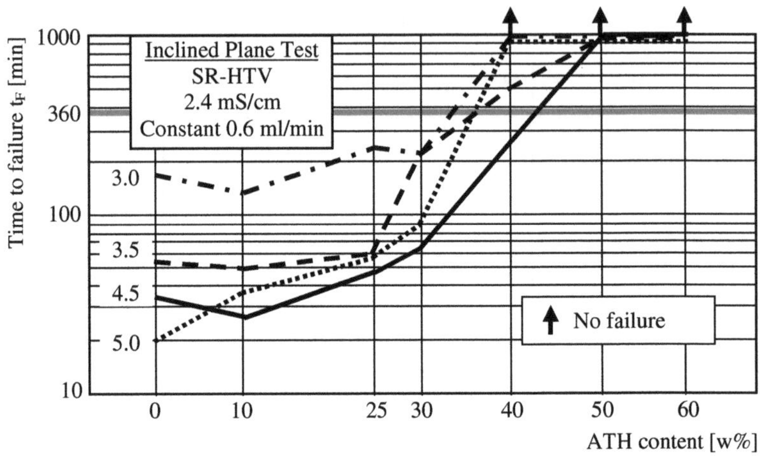

Fig. 7.34 Greater erosion resistance as a result of adding ATH, SR system-2 [85]

In the case of SR-1 (RTV grade), the erosion rate can be lowered by adding ATH. However, the material is so highly viscous that processing in the form of low pressure cavity casting is no longer possible. SR-3 is an LSR grade and illustrates an example whereby the addition of an unsuitable colour pigment can increase the erosion rate. SR-2 (RTV grade) and SR-4 (LSR grade) are commercially available systems that are castable without the addition of ATH and, with a suitable formulation, achieve a lower erosion rate than, for example, SR-1 with ATH. This can occur, for example, by adding suitable additives such as TiO_2 or a surplus platinum catalyst. The low erosion rate associated with SR-5, a commercially available HTV with approx. 48 % ATH, is not achieved with the low viscosity systems tested under these conditions. The quantity of ATH needed to deploy the flame-retardant effect must always be tailored to the carrier material and coordinated with the contribution of other additives. In Fig. 7.33 [103], a HTV system with different quantities of ATH is evaluated in a dry arc test and an Inclined Plane Test (24 h). For this system, the tests established that erosion strength increases significantly when 56 % ATH by weight is added.

The effect of the polymer is confirmed in Fig. 7.34 [85]. The Inclined Plane Test is performed using four different voltage levels and a constant flow rate for the electrolyte that varies the ATH content. Depending on the voltage level, erosion resistance increases as of 40 % ATH by weight; 50 % ATH by weight is required to ensure that the system definitely passes the test. In each case, the test was concluded after 360 min and an expected time to failure was derived by means of a probability evaluation based on the extent of the damage caused [105]. This produced the sections of the curve at 1,000 min. It is interesting to note that, under these test conditions, 40 % ATH by weight is insufficient for 3.5 and 4.5 kV. The burn time of the PD is longer as for the voltages levels 3.0 and 5.0 kV and can cause more severe damage to the material [72, 104].

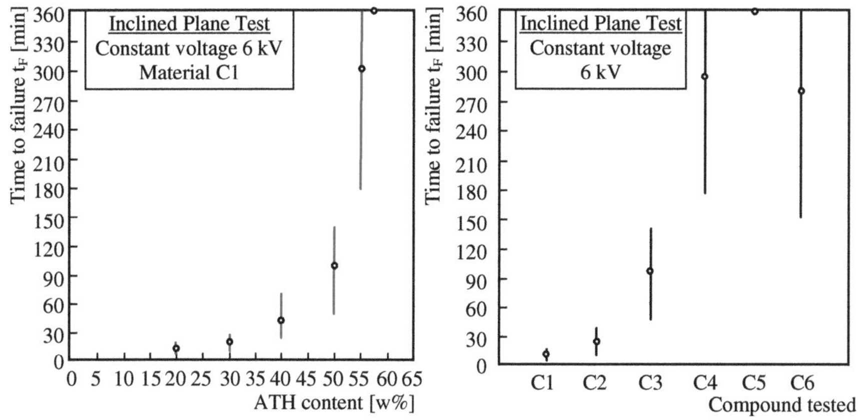

Code	Material
C1	Base recipe
C2	C1+ 50 w% unmodified ATH
C3	C1+ 50 w% modified ATH
C4	C1+ 55 w% modified ATH
C5	C1+ 57 w% modified ATH
C6	C1+ 57 w% modified Silica

Fig. 7.35 Effect of ATH content and modified ATH in the Inclined Plane Test, SR system 3

Due to continuous material developments and growing demands for greater erosion strength (e.g. 6 kV as a testing voltage level in the Inclined Plane Test), systematic analyses were performed to increase erosion strength while simultaneously reducing the processing viscosity [89, 106]—a combined goal that cannot be achieved solely by increasing the ATH-content. 6.0 kV was chosen as a reference voltage in the Inclined Plane Test. It is well recognised that an increase of the test voltage does not cause necessarily more severe damages [107, 108], because higher voltage can have an opposite tendency for leakage current (increase) and PD-burn time (decrease) [72, 104]. However, for the own investigations it was found that the damage at lower voltages was no more intensive under the test conditions chosen, the testing technology used and the materials evaluated.

The base formulation (Fig. 7.35), comprising two different base blends with silica and a peroxidic vulcaniser, is tested with different quantities of ATH. As expected, the more ATH that is added, the longer the period until the time to failure occurs; the variation is high between 40 and 55 %, but the increase in erosion resistance is unstable. The HTV system does not definitely pass the testing voltage level of 6 kV until it has an ATH content of 57 %. The effect of filler treatments was also analysed (Fig. 7.35–right). "Modified" is generally understood to mean that the ATH surface is enriched with certain groups in order to improve (by means of silanisation, for example) how ATH is distributed and embedded in silicone rubber (here, the base formulation). The positive effect

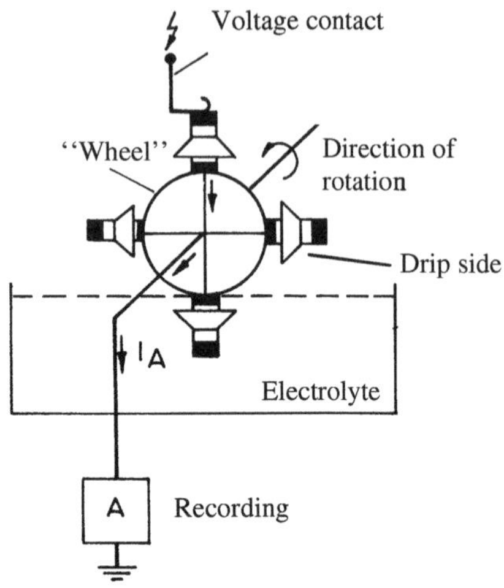

Fig. 7.36 Principle of the wheel test (Modified Rotating Wheel Dip Test)

associated with modifying the ATH content is shown in the comparison between C2 and C3. However, the test involving these two systems failed because the overall quantity of ATH was too low. The literature presents results that confirm that the combined or sole addition of fillers other than ATH also achieves high erosion resistance [90–92]. For this reason, only modified silica was added to C6 in the quantity required for the base formulation and ATH to pass the test. The erosion resistance to C1–C3 also increases, but it does not achieve the stable fill value of 57 % ATH.

The results shown infer that specifically adding ATH can also achieve a significant increase in erosion resistance up to 6 kV in the Inclined Plane Test. The required quantity of ATH depends on how the entire system is formulated. Preferably, it should be determined through systematic tests on an adequate sample size [89]. Modified ATH has a positive effect on erosion resistance.

Effect on the Resistance of Hydrophobicity

The wheel test is a proven method for evaluating the resistance of hydrophobicity (Fig. 7.36 [109–115]). In contrast to its use as tracking and erosion test [116–118], the testing parameters are chosen in such a way that it is possible to distinguish between the early ageing phases (identification of hydrophobicity loss and not tracking/erosion) of various materials. For this reason, the term *Modified Rotating Wheel Dip Test* (*MRWDT*)) was introduced [111]. The test specimens, which are generally cylindrical or rotationally symmetrical, must be manufactured in such a way that there is no (axial) mould line. This can work as a "mechanical water carrier" and shorten the time to failure. Are different materials tested for the purpose of comparison, the surface roughness of the specimens must be in the same order of magnitude. Because the hydrophobicity is an interface property

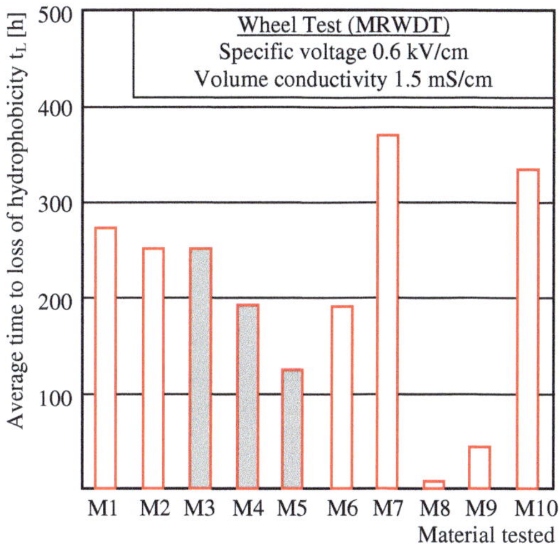

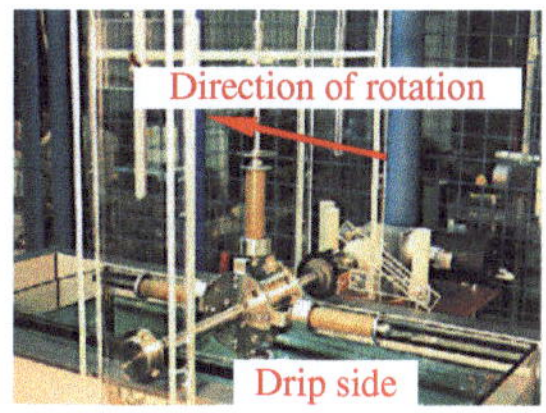

Code	Material/thickness
M1	RTV-2, 10 mm
M2	RTV-2, 5 mm
M3	HTV with silanised ATH, 3 mm
M4	HTV with silanised ATH, 10 mm
M5	HTV with untreated ATH, 10 mm
M6	LSR-1, 10 mm
M7	LSR-2, 10 mm
M8	RTV-1, Coating-1 0.1 mm
M9	RTV-1, Coating-1 0.3 mm, $R_z = 8.8\ \mu m$
M10	RTV-1, Coating-2 0.3 mm, $R_z = 3.7\ \mu m$

Fig. 7.37 Loss of hydrophobicity of various silicone rubber grades in the MRWDT [109]

depending on the intrinsic hydrophobic effect and the surface finish of the specimen tested [112–114].

Under the test conditions chosen, the stable occurrence of a leakage current of 1 mA was defined as a criterion for the loss of hydrophobicity. The number of test specimens should, at the very least, be between 4 and 8 units. Furthermore, adequate measures must be deployed to protect the test solution against a microbial attack because a loss of hydrophobicity can also occur as a result of biogenous settlement without electrical stressing [72–74]. It is then not possible to reproduce the test.

The results reveal interesting trends (Fig. 7.37 [109]). M1, M2 and M6 are commercially available low viscosity silicone rubber grades without ATH, while M3, M4 and M5 (coloured grey) are commercially available HTVs with and without modified ATH. M7 is an LSR laboratory prototype without ATH, while M8, M9 and M10 are various RTV-1 coatings. If one compares the result for M1, M2 and M6 with the result for M3, it is evident that a high resistance of hydrophobicity, comparable to that in systems without ATH, can be achieved with ATH silanisation. M3 and M4 differ despite the greater material thickness of M4; however, the test specimens were manufactured under different processes. In the case of the M8 and M9 coatings, the material thickness has a major effect in the range tested. Small coating thicknesses have a very low resistance of hydrophobicity and, when used, are also sensitive to mechanical stresses. The comparison between M9 and M10 cannot be conducted unequivocally because, even though the coating thickness is the same, the materials are different. On the basis of further investigations [109], it can be assumed that roughness is the dominant effect under these test conditions.

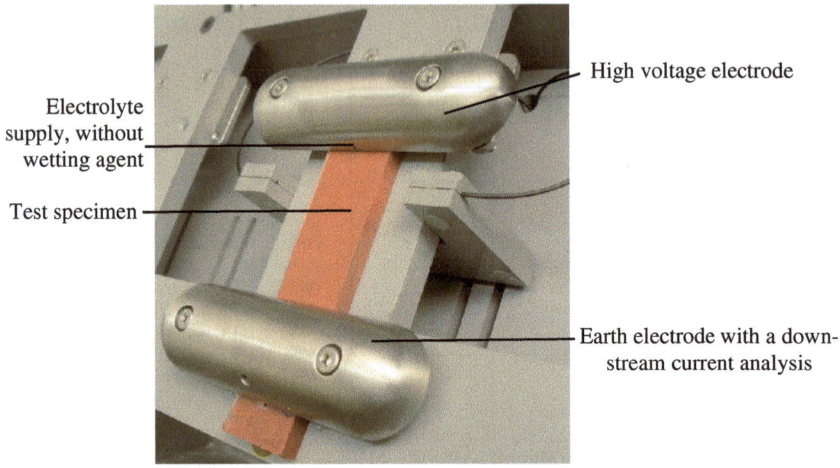

Fig. 7.38 Set-up of the dynamic drop test for evaluating the resistance of hydrophobicity

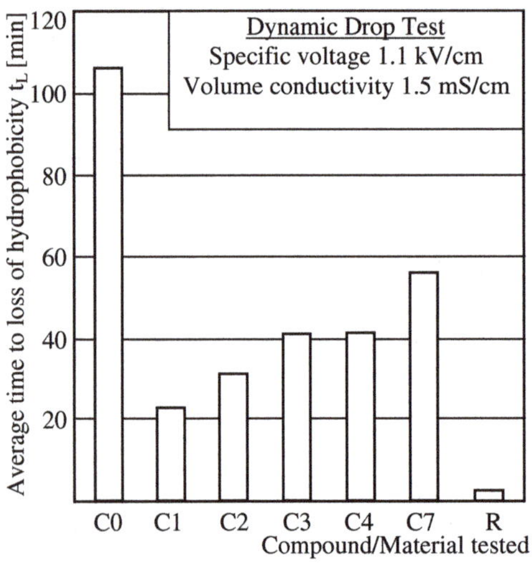

Code	Material
C0	Base polymer (gum)
C1	Base recipe
C2	C1+ 50 w% unmodified ATH
C3	C1+ 50 w% modified ATH
C4	C1+ 55 w% modified ATH
C7	RTV-2
R	Glass specimen

Fig. 7.39 Results of the Dynamic Drop Test

As it was the case with erosion resistance, modified ATH also has a positive effect here.

As a method for evaluating the resistance of hydrophobicity, the wheel test has a drip edge in a unidirectional direction of rotation (Fig. 7.36). If this drip edge coincides with random increases in roughness, there is a greater variation among the measurement results.

In order to use more simple geometric test specimens and to automate the test within the framework of quality control, a test method called *Dynamic Drop Test*

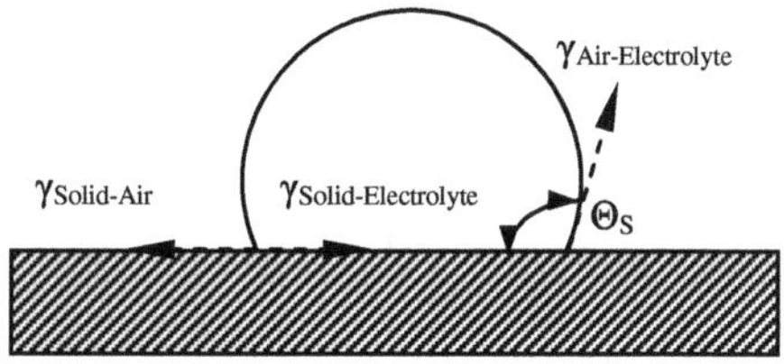

Fig. 7.40 Model description of the different interfacial tensions

[119–121] that can, for example, be integrated into the set-up of the Inclined Plane Test device was proposed (Fig. 7.38).

The testing parameters for the results [89, 106] concerning the selected materials from the Inclined Plane Test (Cx) were defined with a test specimen gradient of 45°, an electrolyte flow rate of 1 ml/min, electrolyte conductivity of 1.5 mS/cm and a testing voltage of 5.5 kV based on systematic preliminary tests.

The results show that this method is an interesting alternative to the relative comparison of various materials (Fig. 7.39). A glass plate is used for calibration and reference purposes. The base polymer C0 consists of polydimethylsiloxanes of varying density and the peroxidic vulcaniser. It does not contain any silica. The material has effectively no erosion resistance, but exhibits the highest resistance of hydrophobicity. Material C1 corresponds to the base formulation from the erosion tests in Fig. 7.35. The low resistance of hydrophobicity, when compared to systems filled with ATH, required further tests for clarification purposes. The results for C2, C3 and C4 confirm that the effect of modified ATH is greater than the ATH content in the range tested (50–55 %). The difference between C4 and C7 is due to surface roughness because the test specimens were manufactured using different tools.

Effect on the Recovery Behaviour

The contact angle measurement is often used to describe dynamic processes of hydrophobicity. A droplet on the surface of an insulating material is an example of effective interfacial tension (also known as surface tension in systems between gas and liquid) at equilibrium, which can be described for an ideally smooth surface using Young's equation [see also (Fig. 7.40)].

$$\gamma_{\text{Solid-Air}} - \gamma_{\text{Solid-Electrolyte}} = \gamma_{\text{Air-Electrolyte}} \cos\theta \tag{7.11}$$

There exist different methods to quantify the interfacial tension. For insulating materials, IEC 62073 [122] mentions contact angle method, surface tension method with different ratios of two reference solutions mixed (ethylene–glycol-monoethyl-ether, formamid) and spray method with visual comparison to reference photos. In the case of contact angle measurements, a droplet of a defined size is applied to the surface of the insulating material and, depending on the process, the static or dynamic contact angle is measured using a goniometer or a digital

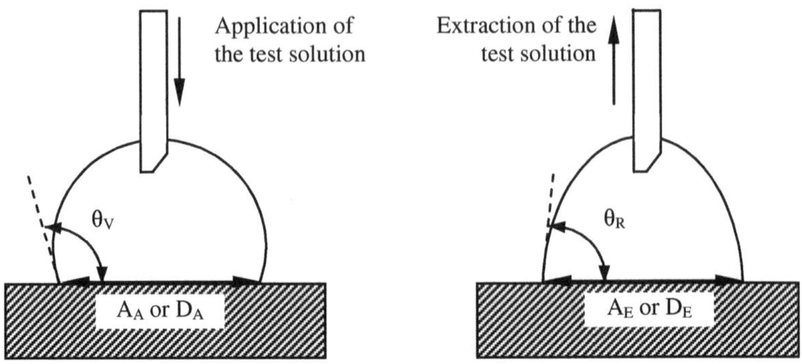

Fig. 7.41 Dynamic contact angle measurement

video system. The drop analysis for contact angle measurement can be in diverse ways: The tangent method is the easiest meanwhile the theoretically most exact Young-Laplace-method considers the droplet deformation caused by its own weight [123, 124]. The method of measurement used should be mentioned in the protocol, because the different methods can lead to different values of the contact angle.

The static angle corresponds to the angle Θ_S in Fig. 7.40. The definition of the dynamic angle can be obtained from the diagram in Fig. 7.41: The test solution is applied to the surface by means of a cannula, which should remain in place in order to improve measurement reproducibility in the droplet during application and extraction. When taking the measurement, it is important that surfaces A_A and A_E or, in the case of a two-dimensional consideration, diameters D_A and D_E remain constant between the dynamic processes for a change in volume. Generally, the relationship between the angles is as follows: $\theta_R < \theta_S < \theta_V$. Since various factors can influence the measurement (e.g. surface roughness, the cleanliness of the surface, the droplet volume, etc.), it is generally recommended, depending on the variation among the individual measurement values, to take several measurements at different points on the surface of the insulating material.

Recovery behaviour means that groups that have a hydrophobic effect on the insulating material orientate towards the surface again and the loss of hydrophobicity after corona treatment, for example, is counter-balanced. The recovery can also be accelerated by desorption of H_2O-molecules or the migration of low molecular silicone rubber parts to the surface [109, 111, 125]. Systematic analyses on recovery behaviour in association with different quantities of ATH are not extensively documented in the literature. This is also due to the fact that, in order to prevent the need for cost-intensive cleaning, the property of hydrophobicity transfer is more operation-relevant for housing materials that have hydrophobic properties. In [126], tests were performed on an LSR with different quantities of ATH after 30 min corona treatments (Fig. 7.42). When evaluating the dynamic contact angle, the tests reveal that very high advancing and receding contact angles are achieved again after a long rest time. The damage to the surface of the

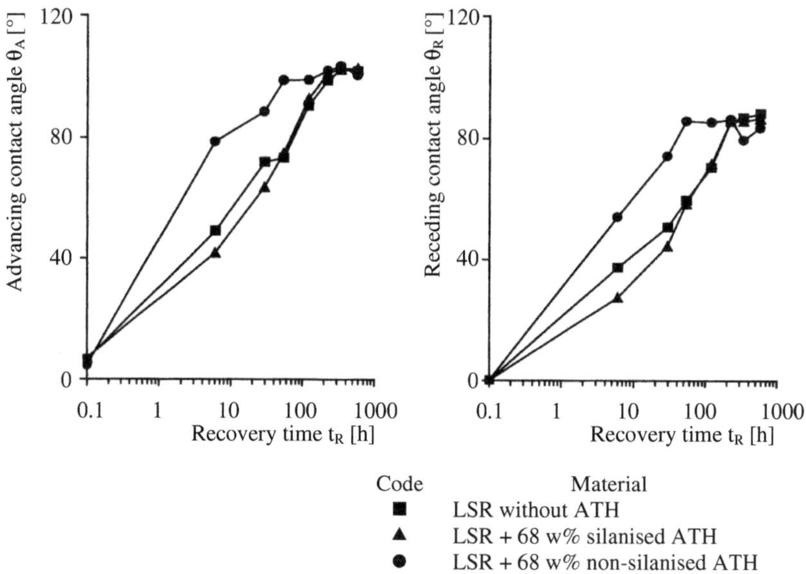

Fig. 7.42 Increase in the dynamic contact angle after corona treatment [126]

insulating material appears to be less than in the case of the loads in Figs. 8.27, 8.28, 8.29, and 8.30 [127]. Despite the various ATH contents, the differences between the materials tested are negligible and do not reveal a clear trend.

It can therefore be concluded that the addition of ATH does not have an adverse effect ("in the sense of delaying") on the recovery behaviour.

The measurement of the static contact angle on horizontally arranged test specimen surfaces, which have an RTV coating and different quantities of ATH, seems to confirm this trend (Fig. 7.43) [128]. It should be noted that the static contact angle can show a certain correlation with the advancing contact angle, but generally not with the receding contact angle. The discussion in relation to Fig. 7.36 already made reference to the effect of insulating material thicknesses; the test specimens in these analyses have an insulating material thickness of 0.7 ± 0.08 mm.

The stress applied to the cylindrical test specimens occurred in a fog test. The contact angles were measured after cleaning. The different increases in contact angles at various positions (HV electrode, centre of the test specimen and nearby earth electrode) do not correlate with the visual observation of a formation of dry zones and, for this reason, cannot be interpreted further. As main conclusion—the recovery behaviour is also not adversely affected by the quantity of ATH.

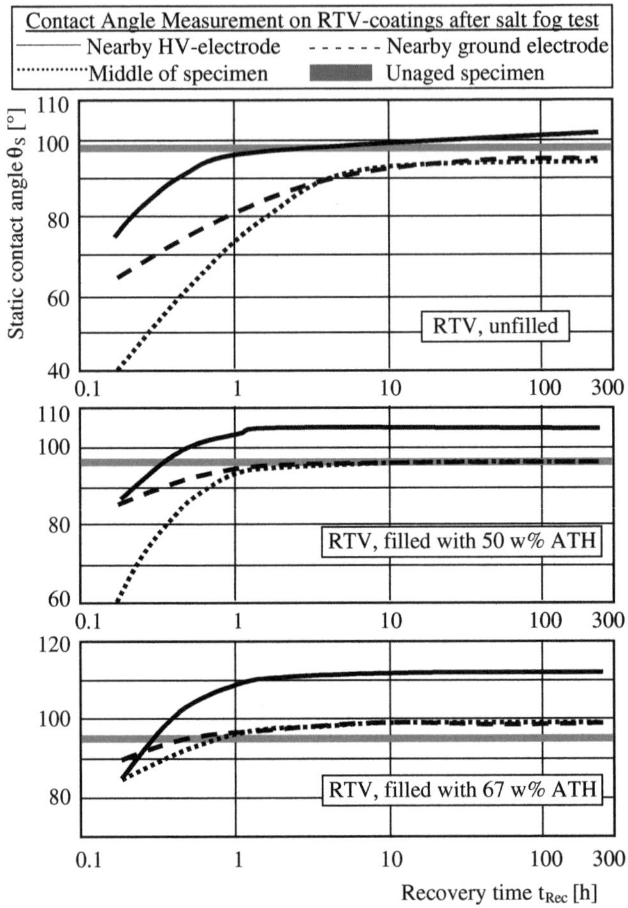

Fig. 7.43 Increase in the static contact angle of various RTV-coatings after salt fog test [128]

Effect on the Hydrophobicity Transfer

The "hydrophobing" of pollution layers on the surface of the insulating material is a proven and operation-relevant unique characteristic of composite insulators with silicone rubber housing and is due to the migration of low molecular weight (LMW) chains. According to [129], this property was described in 1981 as a possible explanation for the return of hydrophobicity [130], whereby, at that time, there was still no distinction, within recovery, between the reorientation of methyl groups and an LMW transfer. In the years that followed, this property was specifically analysed using, among other things, electron spectroscopy for chemical analysis (ESCA) [131, 132], the result of which was that, after a series of ageing tests and a specific rest phase, the quantity of silicone in the carbon pollution layers

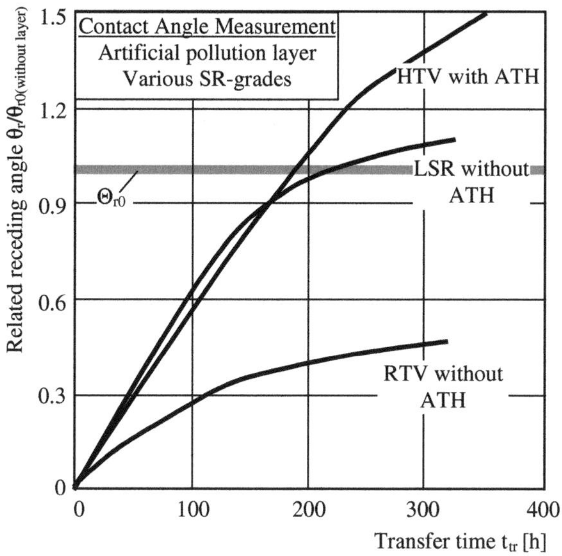

Fig. 7.44 Increase in the related receding angle of various silicone rubber grades [102, 143]

applied for diagnostic purposes increased again. This property was also extensively analysed for RTV coatings [133].

The hydrophobicity transfer causes more adhesion of the pollution layer to the silicone rubber surface. In an ideal scenario, however, this can exhibit better hydrophobicity than the surface without a pollution layer [3]. If the hydrophobicity transfer can take effect, it is not necessary to clean the insulators while they are in service, even under critical operating conditions (e.g. in tunnels [134] or in arid conditions [135, 136]). The first systematic analyses on this phenomenon involving laboratory pollution layers of different thicknesses were published in 1989 [137]. They revealed that, in order to reproducibly apply an artificial pollution layer to hydrophobic and rubber-elastic surfaces, it is necessary to comply with certain rules, which were published in Technical Brochure CIGRE TB 442 [121] in 2010 following many years of work.

Not only the material itself but also the environment influence the intensity and speed of the hydrophobicity transfer. The base polymer, content and type of filler are key factors for the material [126]. The low molecular weight chains are present in silicone rubber as linear and cyclic diffusible parts. When silicone rubber is used for housing materials, these parts typically occupy 3–5 % by weight [126, 138]. Since migration processes are involved here, the linear parts can result in a faster hydrophobicity transfer [139].

The environment influences the hydrophobicity transfer through the content of hygroscopic salts in the pollution layer [109], the temperature [126] and the duration of the humidification phases. Low temperatures, hygroscopic salts and long humidification phases slow down the hydrophobicity transfer. It was also experienced that a hydrophobicity transfer fails to appear, if biogenic pollution layers are formed on the surface. They have often a hygroscopic behaviour and a

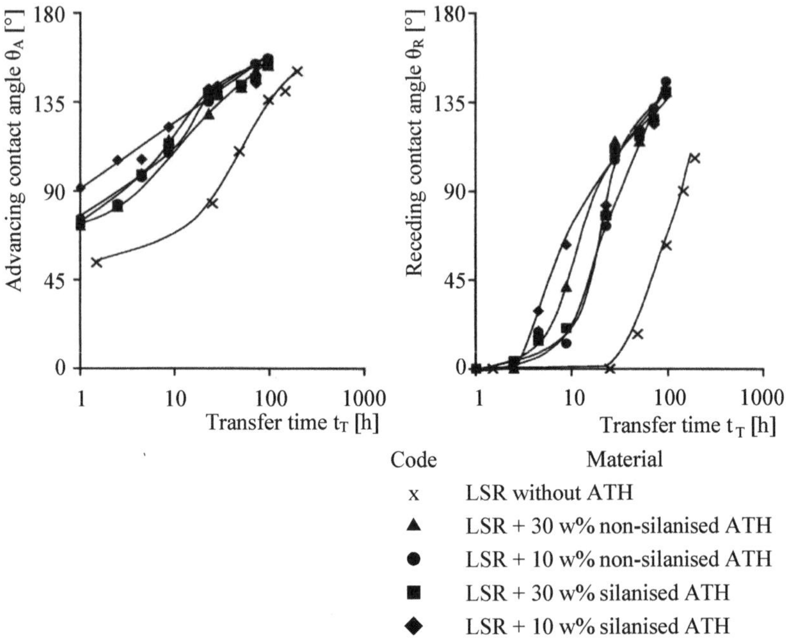

Fig. 7.45 Increase in the dynamic contact angle after the application of an artificial pollution layer [126]

permanent high moisture content in the biogenic pollution layer can slow down or interfere the migration processes of the LMWs [80, 140–142]. Does the addition of ATH adversely affect the hydrophobicity transfer? Current studies involving commercially available silicone rubber grades that have been in use for many years show that this is not the case. The two examples below demonstrate this point.

Plate-shaped silicone test specimens were immersed in an aqueous solution containing 5 % non-silanised silica and the pollution layer was dried at room temperature. The dynamic receding contact angle was used as a measurement (Fig. 7.44 [102, 143]) because it not only shows the hydrophobicity transfer that has taken place but it also has a good correlation with behaviour during simultaneous humidification and electrical stressing [125]. It is surprising that the material with ATH shows a rapid increase and significantly outperforms the unfilled materials in terms of the end value it achieves. This result is significant because low viscosity silicone rubber grades should have a fast hydrophobicity transfer as its property profile so that they can "compensate" for their lower erosion resistance (when compared to HTV grades filled with ATH) under critical operating conditions. It can be assumed that the slow hydrophobicity transfer and the low erosion resistance caused the damage pattern shown in Fig. 7.27.

In [126], the effect of ATH content on a hydrophobicity transfer with various liquid silicones (LSR) was analysed. The results are shown as advancing and

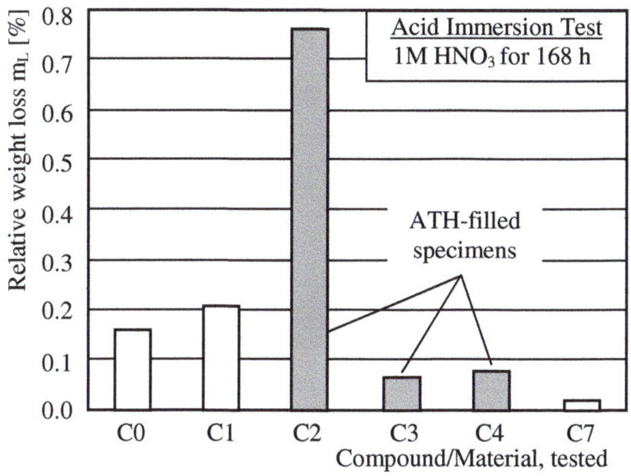

Code	Material
C0	Base polymer (gum)
C1	Base recipe
C2	C1+ 50 w% unmodified ATH
C3	C1+ 50 w% modified ATH
C4	C1+ 55 w% modified ATH
C7	RTV-2

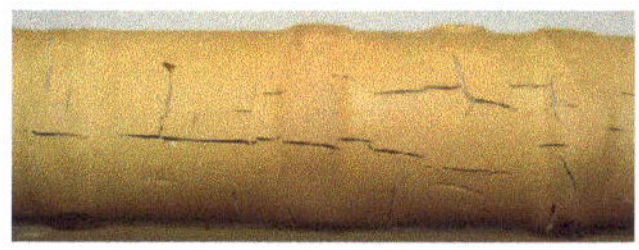

Fig. 7.46 Weight loss of various materials after immersion in 1 M nitric acid

receding contact angles (Fig. 7.45). The major trend differences are due to the advancing contact angle's stronger reaction to the roughness of the artificial pollution layer. This analysis also reveals a trend whereby the addition of ATH appears to accelerate the hydrophobicity transfer. The minor differences between the materials added are neither significant nor unique from the perspective of the effect of the ATH-content and modified ATH. The reasons for the positive effect of ATH are not yet fully understood.

As is also the case with the recovery behaviour, there is no evidence to suggest that the addition of ATH adversely effects the hydrophobicity transfer. Rather, the addition of ATH seems to accelerate the hydrophobicity transfer.

Effect on Acid Resistance

The resistance of polymeric housing materials to acid is an important topic in the context of acid formation due to corona discharges. As a preventive measure, it is recommended to avoid continuous corona discharges at hardware components and to comply with the threshold values for the electrical field stress at the insulator shank (see also Sect. 8.7). In certain areas, acid concentrations may result from industrial pollution, exhaust emissions, etc., which result in low pH values if there is no rain but dew occurs. If the housing material experiences a loss of hydrophobicity and corona activities result in the additional formation of acid, certain fillers may react with the nitric acid, which is formed in most cases, and cause embrittlement (alligatoring) of the housing material. Examples for such fillers include $CaCO_3$ [144] or non-silanised ATH [83], which, to some extent, are used for commercial reasons.

There is no standard for testing the acid resistance of polymeric housing materials. CIGRE Working Group D1.27 is currently addressing this topic. For this reason, the following exploratory test was conducted [106, 145]. Different test specimens in the compounds used in the erosion resistance tests were stored

Property	HTV silicone rubber	RTV/LSR silicone rubber
Viscosity	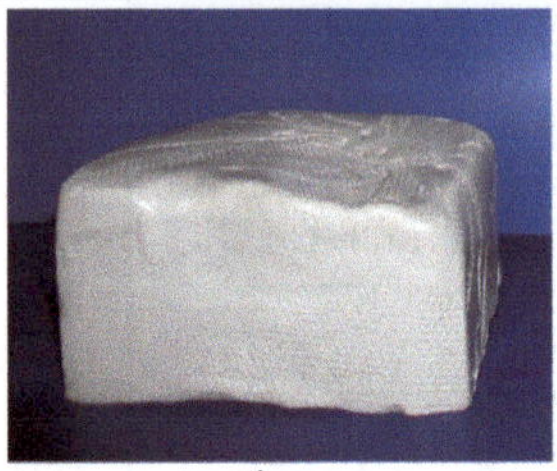 5…20 x 10^6 mPa s (solid)	 10,000…30,000 mPa s
Ageing resistance		
Tracking/Erosion	High	Lower
UV-Resistance	High	High
Flammability	High	Lower.... High
Acid resistance	Lower	High
Hydrophobicity		
Recovery	Fast	Fast
Hydrophobicity transfer	Fast	Fast

Fig. 7.47 System comparison of silicone rubber grades

separately in a solution of 1 M nitric acid (which corresponds to a pH value of 0) for seven days. To prevent neutralisation, the acid volume was 500 ml in each case (the pH value rose by only 0.1 while in storage). The mass loss after storage was determined by weighing the test specimens (Fig. 7.46). When C0, C1 and C7 are compared against each other, C7 shows the lowest loss. The difference between C0 and C1 is the absence of silica in C1; both contain peroxidic vulcanisers. This is the main difference between C1 and C7, whereby the latter is a platinum-catalysed vulcanisation system. These differences in mass loss were not analysed further in this exploratory test. The acid attack on ATH, which has an alkaline effect, is a key factor here. If this is not protected by silanisation (as is the case with C2), considerable neutralisation of ATH, which may cause cracking/embrittlement, occurs under these harsh test conditions. One possible reaction is shown below (7.12), which produces the highly water-soluble aluminium nitrate as salt:

$$3\ HNO_3 + Al(OH)_3 \rightarrow Al(NO_3)_3 + 3\ H_2O \qquad (7.12)$$

Figure 7.46, bottom right, shows the shank section of an HTV-SR insulator with silanised ATH after being stored in 1M nitric acid for approx. 5 months (!). Despite this unrealistically harsh test, the cracks never reached the rod.

The chemical resistance to strong acids is a required property that is limited to a few cases of application. The alkalinity of ATH can trigger a neutralisation reaction that causes cracking/embrittlement of the housing material. ATH silanisation is one way to delay an acid attack in such a way that no chemical attack occurs as a result of rain-induced acid dilution under outdoor conditions.

Fig. 7.48 View of a crimping machine with eight crimping jaws and crimping dies with identification numbers

7.2.3.4 Summarising Comparison

The properties can be contrasted as follows (Fig. 7.47):

7.3 Processes

After discussing the three material components of a composite insulator (namely the fitting, rod/tube and silicone rubber), an overview of the processes used to apply the fitting, the housing material and the seal between the fitting, insulator housing and rod is provided.

7.3.1 Fitting Assembly

7.3.1.1 Crimping

Crimping is a complex process that is important for transmitting the load from the fitting to the rod. Nowadays, crimping is an established state-of-the-art technology. During this joining process, the rod and fitting are joined together through plastic deformation of the fitting.

Procedure

Nowadays, controlled crimping machines with, for example, eight crimping jaws (Fig. 7.48) are commercially available. The transmission of the load to the crimping jaws can occur individually or centrally. The following two functionally independent and integrated control mechanisms are common:

- Crimping pressure.
- Motion distance of crimping jaws.

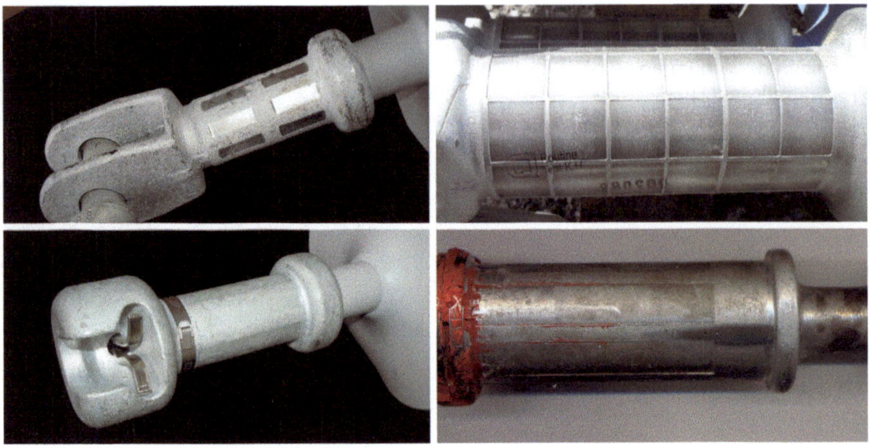

Fig. 7.49 Different crimp patterns on steel end fittings

An empirically established set of crimp parameters is used for each metal grade (steel, aluminium, etc.), rod type (glass content, resin matrix, etc.) and end fitting/rod diameter combination. When commissioning new crimping machines or applying maintenance to existing ones, the machines need to be calibrated in order to ensure that empirically established crimp parameters, introduced over a period of several years, can be used.

Insulator manufacturers have special expertise in terms of the crimping pattern, the application of the load to the end of the open fitting, and so on. Examples of different crimping patterns are shown in Fig. 7.49. It is important to note that the crimping pattern cannot be used to draw conclusions about the mechanical behaviour. In each case, a design test must be conducted over a period of 96 h in order to verify the suitability of each crimping pattern. The variation of the maximum load is an important criterion for the stability of the chosen crimping parameters, coordinating the materials used in the rod and fitting, the shape of the fitting, and so on. Here, the preferable failure mode is for the fitting to pull out of the rod without the rod breaking. The following rule has proven to be feasible: The smallest value for the maximum load (e.g. for the 3-times standard deviation 3 s = 99.7 %) should be at least 20 % above the specified mechanical load (SML).

Quality Assurance: Acoustic Emission (AE)

Routine testing of at least 50 % of the specified mechanical load is mandatory for composite long rod insulators [59]. For historic reasons, some power utilities specify a routing testing load of 80 % of the specified mechanical load, as is the case with porcelain long rods. This requirement may well be justified. A scenario whereby a continuous break in the resin matrix occurred relatively deep within the

Fig. 7.50 Example of a rod breakage as result of over crimping, but routine test with 50 % of SML passed successfully

fitting during the crimp process was simulated in [83] (Fig. 7.50). However, this composite long rod held 50 % of the specified mechanical load value.

For this reason, acoustic emission (AE) analysis is an interesting tool for detecting such rod breakages. AE has two basic areas of application:

- The use of the Kaiser and Felicity effects [146] to evaluate cyclic stresses or defined increasing stresses [63].
- Quality control through the comparison of reference measurements.

Although it is not directly relevant to this topic, it is worth providing a brief description of the Kaiser effect (Kaiser 1950) here because, for structured plastics (including glass-fibre reinforced plastics), it postulates the following: It is not possible to replicate an acoustic emission image for the same structured object subjected to the same stress.

Figure 7.51 shows an example of a compression load whose very first load causes a high level of acoustic emissions, which is characteristic of the set-up of the moulded part to be analysed. Only if the value for this first load is exceeded or another damage mechanism is adjusted are AE-signals of a certain level generated again [147]. In the case of AE, the very first load is used as a crimping control measure for insulators.

In [148], a quality control example shows how AE can be used for a specific system (Figs. 7.52, 7.53, all diagrams are displayed with a test time in s). Ranges I–III are described as follows:

Range I: Movement of crimping jaws.
Range II: Contact between the fitting and the crimping jaws.
Range III: Crimp process.

It is clear that there is a permitted range of AE-signal intensities in which crimping is implemented correctly. It is important that the detection frequency of the measurement system is high enough to detect peaks that indicate a breakage.

In this context, "specific" means that it is necessary to establish a link between the materials, crimping machine and measuring technology used. The procedure described in Figs. 7.54–7.60 has proven to work in practice.

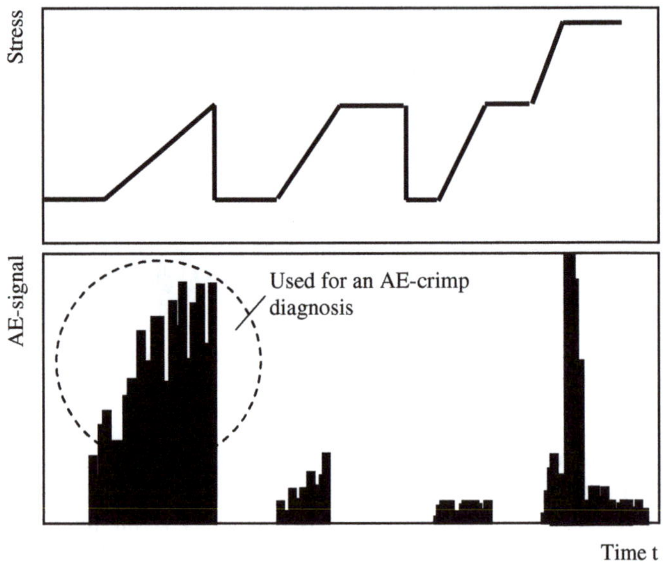

Fig. 7.51 Simplified diagram of the Kaiser-Effect

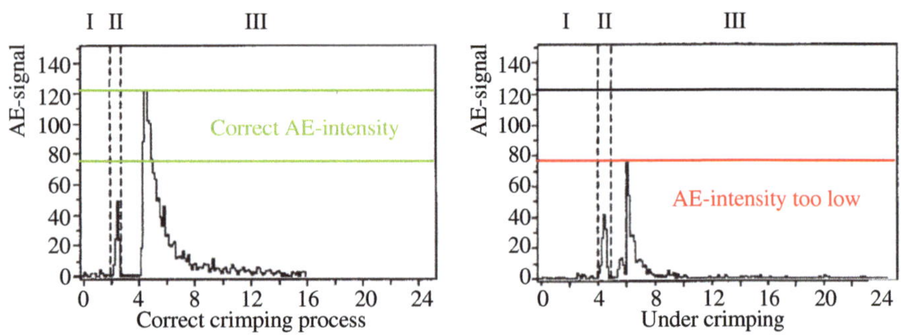

Fig. 7.52 AE-diagnosis for detecting under crimping [148]

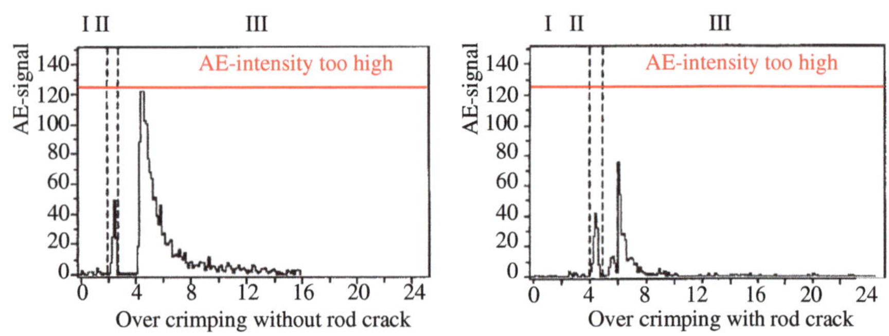

Fig. 7.53 AE-diagnosis for detecting over crimping [148]

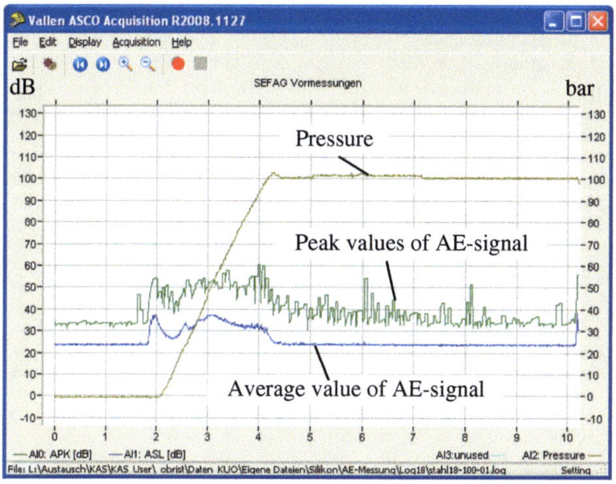

Fig. 7.54 AE-signal sequence for crimping steel onto steel

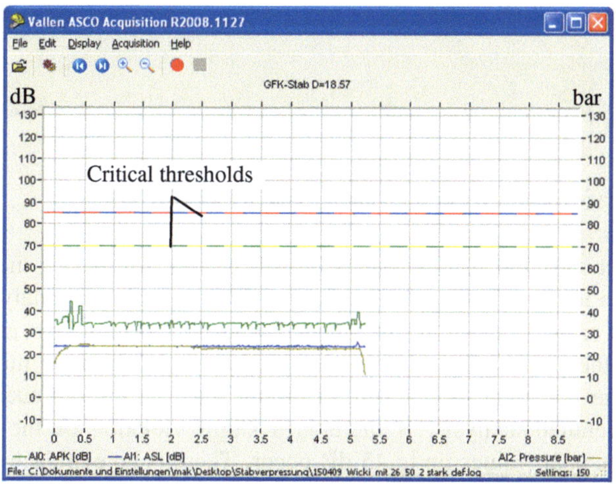

Fig. 7.55 AE-signal sequence for direct crimping onto a FRP-rod, crimp pattern #1 (Fig. 7.57)

The measuring equipment is calibrated using a reproducible crimp process. For this purpose, the steel fitting is crimped onto a steel rod, for example (Fig. 7.54). The measurement system used outputs the peak value and average value of the AE-signal. Averaging occurs as low pass filtering. In other words, the lower the frequency of the AE-signal, the greater the difference between both value representations [149]. The peak value is used for crimping quality. However, the average value is also recorded. The peak values before the rise in pressure can be traced back to process noise. The crimp process remains at values $\leq$60 dB.

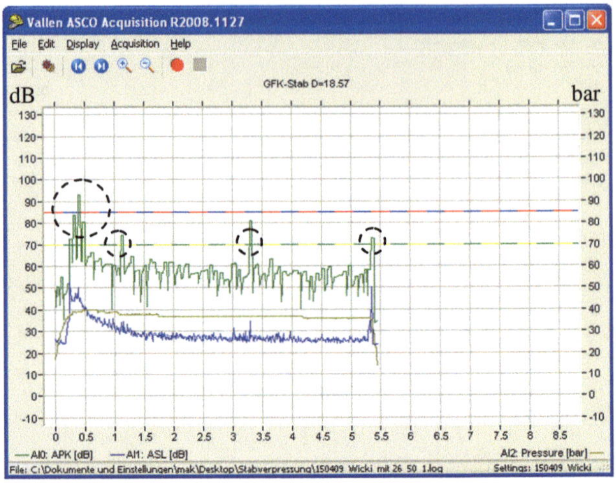

Fig. 7.56 AE-signal sequence for direct crimping onto a FRP-rod, crimp pattern #2 (Fig. 7.57)

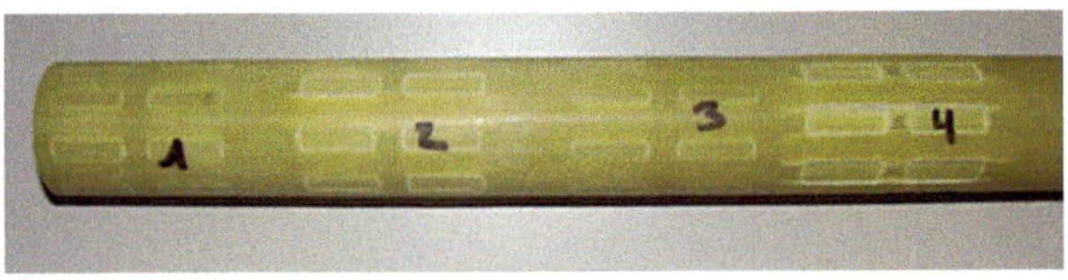

Fig. 7.57 Damage patterns for direct crimping onto a FRP-rod

In a further step, the specific behaviour of the rod is tested by means of direct crimping onto its surface (Figs. 7.55, 7.56, 7.57). To make the analysis easier for the user, two critical ranges are introduced, highlighted in yellow and red. If the AE-level of the peak value exceeds the yellow range, the user must check the settings. If the red range is reached, crimping must be discontinued in order to facilitate a detailed analysis. If there is a slight violation of the resin matrix (crimping #1), peak values up to 45 dB occur. There is no evidence of glass fibre delamination within the violated range. If the pressure is increased (crimping #2), the first glass fibre fractures occur and the critical ranges at 70 or 85 dB are demonstrably exceeded.

For each material combination involving a rod and fitting (and their design), it is necessary to determine the pressure at which the first rod breakages occur. This occurs iteratively; a pressure of 105 bar was determined for the example.

The results confirm its "threshold valency"; crimping can take place without (Fig. 7.58) or with (Figs. 7.59, 7.60) rod breakages. It is also interesting to note the different AE-signal for the two rod breakages Figs. 7.59 and 7.60. While the sequence in the upper diagram shows an intensive emission with numerous transgressions of the red range, there are only two transgressions in the sequence shown in the lower diagram. A reduction in pressure is another characteristic of

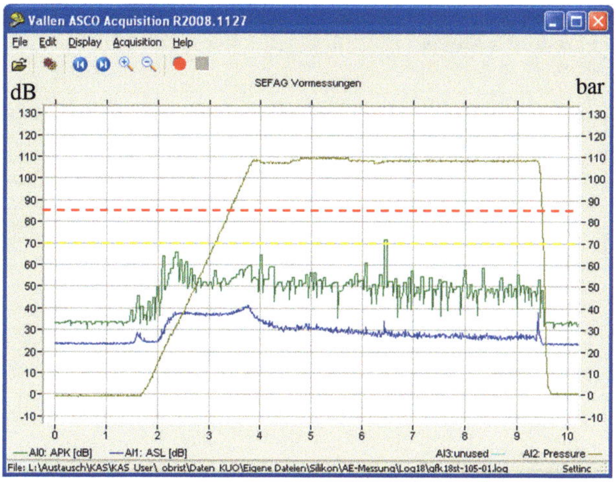

Fig. 7.58 Crimping a steel end fitting onto a FRP-rod, pressure: 105 bar, no rod breakage

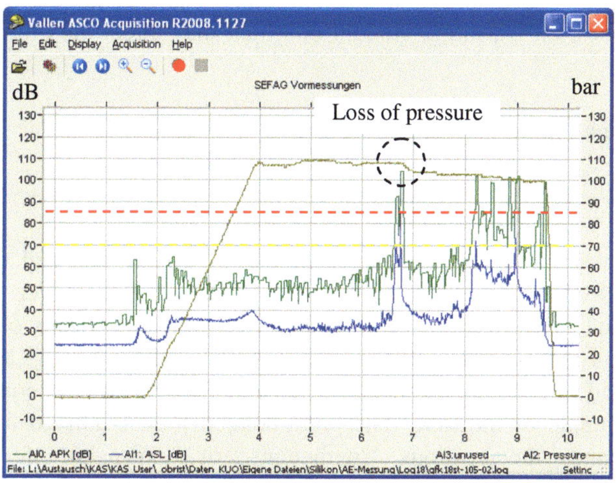

Fig. 7.59 Crimping a steel end fitting onto a FRP-rod, same material batches as in Fig. 7.60, pressure: 105 bar, example 1 of rod breakage

over crimping; the analysis of this sequence as an additional quality characteristic is still being examined statistically.

The base interference level of the crimping machine is often filtered out. If AE-sensors that measure within a certain frequency range are used, both signals (i.e. the crimp process and the crimping machine emissions) can be recorded without any mutually disruptive interference. The advantage of this dual detection is its

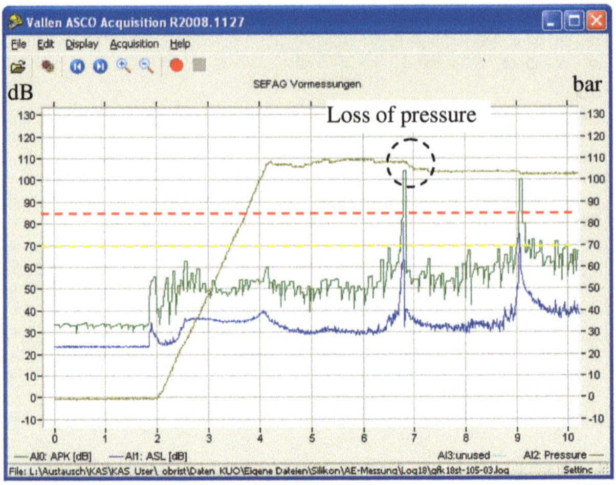

Fig. 7.60 Crimping a steel end fitting onto a FRP-rod, same material batches as in Fig. 7.59, pressure: 105 bar, example 2 of rod breakage

ability to detect any changes in machine emissions that indicate, for example, frictional losses, oil warming, etc., all of which may affect the crimping quality.

7.3.1.2 Flange Application to the Tube

Procedure

In the early days, glueing of a gap (clearance) between tube and flange or of a thread of both as joint technology were used. Nowadays, however, press-bonded joints are used to bond aluminium flanges with tubes by means of form-fit and adhesive bonds. The flange's load application is configured using a radius at the tube-facing side and must be adjusted to the modulus of elasticity configured using the winding angle of the tube. The press-bonded joint is affixed using hot-melt adhesives, and the various coefficients of expansion for aluminium ($\alpha = 23.8 \times 10^{-6}$) and the tube ($\alpha = 8\text{–}16 \times 10^{-6}$) are used to reduce the swaging friction at an increased temperature. In order to prevent deformities, the glass transition temperature of the tube must be taken into account while curing the adhesive.

Quality Assurance

IEC 61462 [150] describes mandatory routine tests. In the case of gas-pressurised insulator applications, it is necessary to perform an impermeability test at an operating pressure. This test must be performed as a cumulative measurement (IEC 60068-2-17, Test Qm, Test Method 1 [151]). The use of helium leak

Fig. 7.61 Atmospheric single shed casting with RTV

detection has proven to work in practice; in vacuum, it can have a sensitivity of up to 10^{-11} mbar l/s [152]. When used in the presence of atmospheric air, the sensitivity falls to 2×10^{-6} mbar l/s because the 5 ppm of helium present in the air causes ambient noise interference [153].

7.3.2 Applying the Insulator Housing and Seal

7.3.2.1 Overview of Insulator Housing

Low Pressure Cavity Casting Process for RTV-2 and LSR

Low pressure cavity casting was first used to process RTV-2 silicone rubber in the 1970s. In view of the absence of product standards and the difficulty of predicting how well the new "composite insulator" product family would establish itself, it became necessary to find a compromise between great flexibility and comparatively low investment costs. This led to the use of low pressure or gravitational moulding. Sheds and shank sections were vulcanised step-by-step onto the internal tube (known as RTV-stepping–Fig. 7.61). This made use of silicone rubber's ability to develop full homogenous vulcanisation between a previously vulcanised section of the insulator housing and the subsequently moulded section by bonding silica groups to a vulcanised section and moulded section (see also Fig. 7.75), without the formation of a macroscopic interface. By adopting the approach of simultaneous housing manufacture (Fig. 7.61) for several composite hollow core insulators, productivity was brought up to a level that reflected demand at that time. The vast majority of these composite hollow core insulators were used for cable terminations (Fig. 7.62).

With the increasing use of composite hollow core insulators and the development of gas-insulated bushings for 525 kV in the early 1990s, which was another important milestone, the demand for more productive methods grew. With the

Fig. 7.62 123 kV cable terminations, installed in 1975

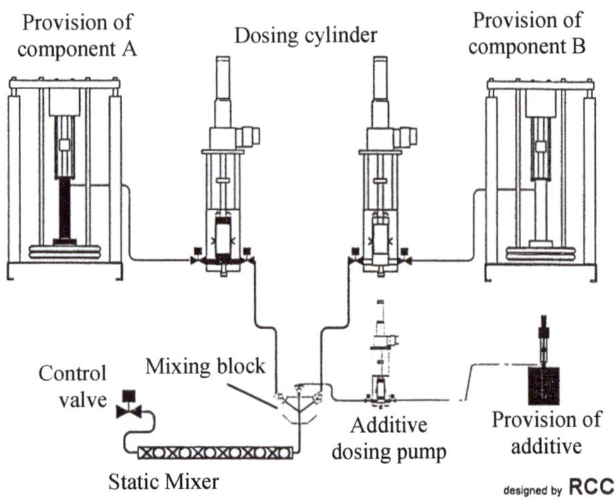

Fig. 7.63 Principle of low pressure mix and dosing equipment

availability of low viscosity LSR systems that underwent vulcanisation in the technically used temperature range for RTV-2 systems and had to be processed in a fault-tolerant 1:1 blend ratio, a new generation of mix and dosing equipment (MDA) was developed for low pressure casting [62, 143, 154]. Nowadays, they permit individual shot volumes of up to approximately 130 l for low viscosity LSR grades (Figs. 7.63, 7.64).

Fig. 7.64 View into a mould used for an external tube diameter of 610 mm [62]

MDA facilitates continuous, precise volumetric dosing and cavity casting in one work process (Fig. 7.63). No screws or boosters are required. The equipment works without a vacuum. However, depending on the history of the material (e.g. intensive movement during transportation), it may be necessary to vacuum pack individual components. The required volume of each component is forced into the dosing cylinders under a defined pressure and then continuously filled into the mould (with the chosen volume per time unit) by means of a control value.

The mixing header and static mixer are extremely important here. Both must be adjusted in terms of viscosity, the blend ratio, the dosing valve (manual or pneumatic) and the flow rate. Especially in the static mixer, with its inner structure, the components are subjected to a shear stress, which reduces the mixing viscosity of thixotropic materials and accelerates mould filling without air pockets. In the case of small-volume parts or parts with a relatively low wall thickness, the vulcanisation process or one of its preliminary stages, depending on the vulcanisation system (peroxidic, addition, UV, etc.), may be triggered when the parts pass through the static mixer.

Today, the use of this principle of low pressure cavity casting for RTV-2 and LSR is predominantly limited to the manufacture of insulator housing for hollow core insulators. Since the mid-1970s [45] HTV technology has been the established technology for overhead transmission line insulators, especially in the case of transmission voltages $\geq$123 kV. The percentage of composite insulators created using HTV-technology is estimated to be more than 90 % for transmission voltages $\geq$123 kV. There were two reasons for changing from RTV-2 to HTV, namely:

- The lower erosion and tracking resistance of low viscosity systems coincides with installations high up on pylons where it is difficult to diagnose problems with the insulator. For this reason, the more redundant material (namely HTV with ATH) has become compulsory in many specifications.
- The mechanical properties of early RTV-systems were inadequate. Even at the installation stage, damage to the shank may cause the rod to be exposed, or shed tears may occur.

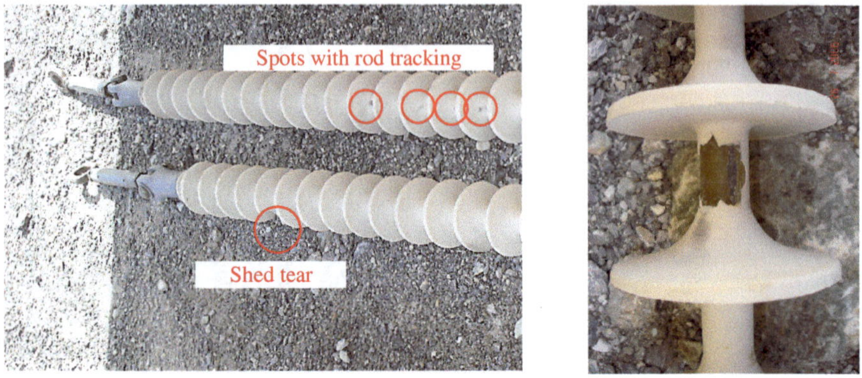

Fig. 7.65 Damage patterns for first-generation RTV insulators

Effect of a mould line (flash)	Preventive measure
Physical obstacle to a smooth surface, leads to better adsorption of the pollution layer.	Mould line height should be ≤ 1 mm
Effect on axial distribution of electrical field by improved pollution layer adsorption.	Offsetting of mould line for voltages > 145 kV (turning of mould line position)
Susceptibility to bulk erosion/tracking in the case of thin flash membrans.	Prevention or removal during routine testing
Susceptibility to bulk erosion/tracking, if a segregation reduces amount of filler.	Inspection of injection parameters and impermeability of the mould

Fig. 7.66 Effects of a mould line and preventive measures

Figure 7.65 shows examples of these first-generation silicone composite insulators, which were used in a zone within a 420 kV network with extremely low levels of pollution. These insulators have most definitely reached their end of service life because tracking in the resin matrix has already been diagnosed in some places.

The matter of the "parting line or mould line" must be mentioned in connection with the injection moulding process because there may be some critical effects if an axial mould line runs along the insulator. These effects and preventive measures can be summarised as shown in Fig. 7.66.

High Pressure Methods for HTV-Processing

Traditional Methods

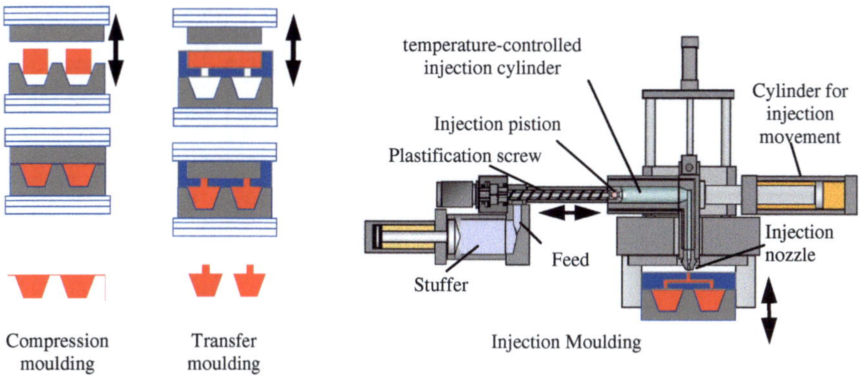

Fig. 7.67 Overview of high pressure processes used traditionally for insulator housing manufacture

The traditional methods for processing silicone rubber for insulator housing materials can be summarised as follows:

- Compression moulding.
- Transfer moulding.
- Injection moulding.

Hybrid forms (e.g. transfer moulding with injection moulding) are possible, as is the use of cold runners to reduce waste and to prevent preliminary vulcanisation in complex tools. Figure 7.67 shows the functional principles of these methods. In comparison to low pressure cavity casting for RTV-2 and LSR, the final solid silicone rubber compounded blend (usually HTV with ATH) is processed (1-component). With pressure levels of up to 1,500 bar and temperatures up to 200 °C, new challenges face this apparent simplification of the process.

These include the following process parameters:

- Mould temperature.
- Injection unit temperature.
- Plasticising profile and dynamic pressure.
- Injection velocity and pressure.
- Filling of cavity.
- Dwell pressure.
- Pre-heating in the injection piston.
- Vulcanisation time.
- Process time for injection unit.
- Velocity profile at the clamping unit.
- Mould clamp force.

Today, compression moulding is used for medium voltage insulators. Just one major manufacturer of insulators deploys it for transmission level purposes. It is the "oldest" method and requires only a "low" level of process expertise. Since

Fig. 7.68 HTV injection moulding for a 420 kV long rod insulator

the tool costs associated with long rod insulators can be high, the investment strategy determines the level of flexibility associated with the shape of sheds on manufactured insulators. The costs associated with series manufacturing depend on the level of rework required.

Today, transfer moulding has largely replaced compression moulding in the case of medium voltage insulators.

The injection moulding process is the most developed process for composite insulators. The example shown in Fig. 7.67 [155] is an injection unit that works in accordance with the first-in-first-out principle (FIFO). The stuffer is filled with the solid silicone rubber. The plasticising screw fills the injection cylinder, which can be heated to reduce the process times. As a result of the forward movement of the injection piston (to the right in Fig. 7.67), the first plasticised material is always injected into the tool, thus achieving a uniform length of stay for the blend in the injection unit at a constant material pressure. A new plasticising process can be performed while the insulator housing is being vulcanised in the tool.

Due to the high vulcanising temperature and the process pressure that arises, special attention should be paid to centreing the rod. Depending on the tool length, rods with a diameter of 88 mm may be deformed incorrectly when using a unidirectional or asymmetrical injection nozzle system.

In addition to the modular process, which will be discussed later, HTV injection moulding is one of the key processes for manufacturing composite insulators. HTV-stepping has been successfully used for long rod insulators since the 1990s (Fig. 7.68) and nowadays is considered to be a safe and established technology. Section 10.5.1 evaluates insulators that were manufactured using this process and were in service in a 420 kV network for a period of ten years.

Due to the effect of the two process parameters, namely pressure and temperature, acting on the tube simultaneously, the use of HTV injection moulding for

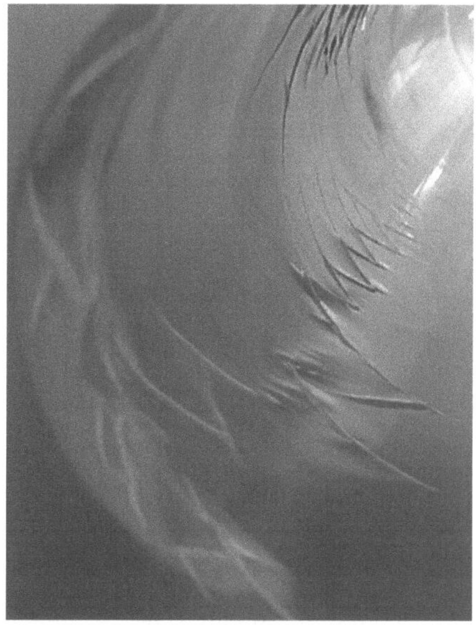

Fig. 7.69 Challenging tube housing manufacture (thermal and pressure cracks)

Fig. 7.70 Successful housing manufacture after process tuning

hollow core insulators is even more sophisticated. If the tube has a large diameter, it is often supported from within in order to prevent delaminations (Fig. 7.69). The tolerances between the inner diameter of the tube and the outer diameter of the supporting tube must be designed as a "fit". If aluminium is the material used for supporting tubes, it can create an inner counter-pressure as a result of the larger coefficient of expansion during (prior) heating. A high number of possible process

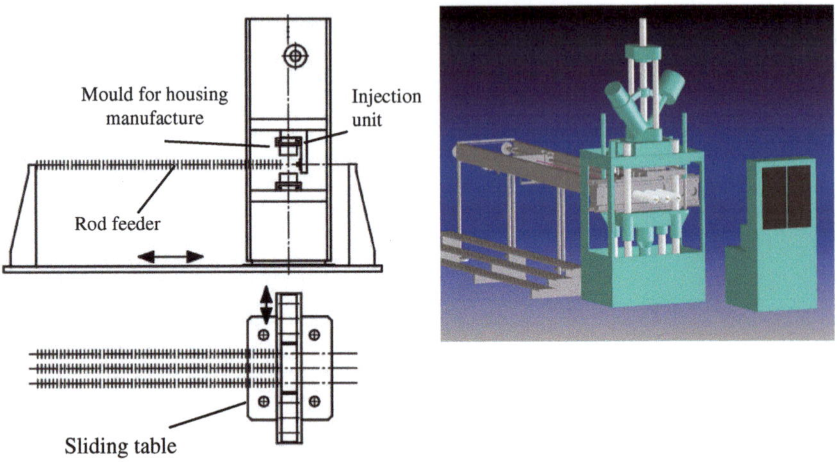

Fig. 7.71 Principle of ACIM technology

parameters is involved in the successful implementation of HTV injection moulding for tubes (Fig. 7.70).

Automated continuous injection moulding (*ACIM*)

Automated continuous injection moulding [156] is a highly developed process for the manufacture of composite insulators with HTV silicone rubber housing. The many years of experience acquired with HTV stepping since the 1990s was an important basis for ACIM. It was essential to combine the proven experiences of outdoor use with an increased level of automation and a high level of flexibility.

The simplified principle is shown in Fig. 7.71: There are four basic components. The mould for housing manufacture determines the geometry of the insulator housing; its length can be shortened or extended in predefined steps by choosing a suitable number of steps. The injection unit consists of an injection moulding machine with a modified control and an automated injection nozzle system. Depending on the diameter, the rod feeder can hold up to three rods at once. The sliding table, which moves at right angles to the rod feeder, holds the different moulds (for example, for the different-shaped shed sections at the start, in the middle and at the end of the insulator) and positions them, in a fully automated manner, on the insulator according to the particular steps (for example, start, middle or end).

The advantages of this process can be summarised as follows:

- Combination of a high degree of flexibility plus manageable tool costs with a high level of automation.

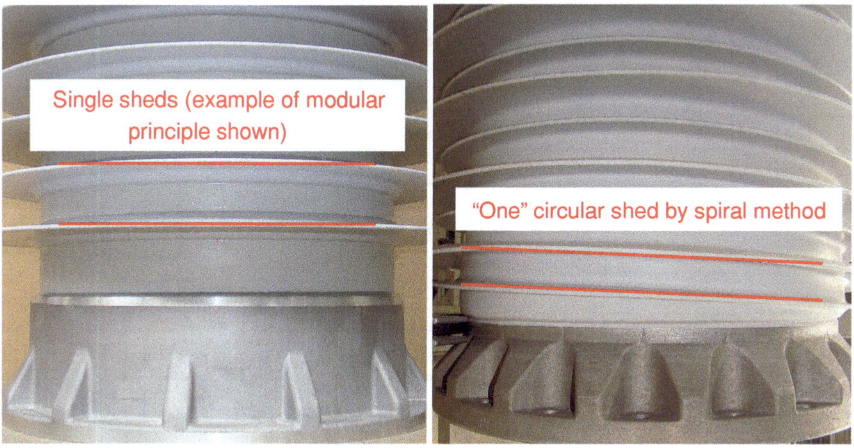

Fig. 7.72 Differences between single shed and spiral-shaped housing

- High level of automation = low probability of human error.
- Tried and tested manufacturing technology since the 1990s.
- Processing of (ATH-filled) HTV silicone rubber with extensive, positive outdoor experiences worldwide for more than 30 years.
- Offsetting of the parting line by potentially turning the insulator after each step in order to increase reliability under heightened pollution conditions by avoiding a continuous axial line (see also Fig. 7.66).
- Effectively no length restriction.
- Can also be used for post insulators and hollow core insulators up to medium diameters.

Extrusion Process

Today, two processes play an important role in extrusion, namely the positioning of an extruded shed profile at right angles to the tube surface (I) and the concentric-axial application of the cylindrical "shedless" insulator housing for the modular process (II) (the last process described in section Modular Process). The tube sheathing as per I requires axial relative movement between the extruder header and the tube; the insulator housing is pressed down by suitable rollers [157, 158] and is spiral-shaped (Fig. 7.72). Due to the required bending stiffness and unpressurised vulcanisation (no outer form), special HTV formulations are necessary. This process has the great advantage of flexibility: The cost of a new extrusion die is comparatively low and unnecessary for a change in tube diameter.

There are two drawbacks to this process: In the case of small tube diameters, the curvature at the outer shed periphery is high. In other words, adequate measures must be taken to ensure that cracks do not form in this peripheral area. Furthermore, as was also the case with earlier spiral-shaped insulators

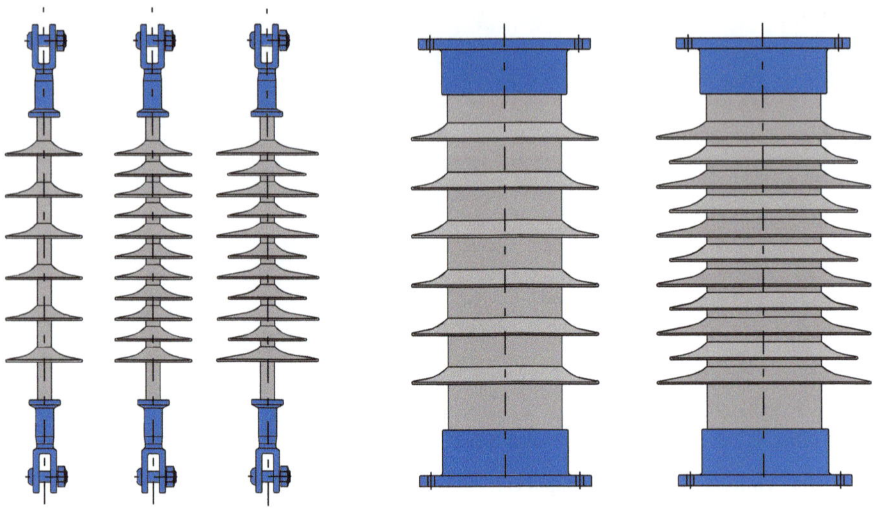

Fig. 7.73 Modular process for solid and hollow core insulators

manufactured using porcelain, the spiral shape in pollution behaviour is inferior to a non-spiral shape in most cases where there is a loss of hydrophobicity [159–161].

The extrusion of long rod composite insulators has been patented several times [162, 163], but is not yet widely used.

Modular Process

Of all the processes associated with the manufacture of insulator housing, the modular process offers the greatest flexibility (Fig. 7.73). Any technically useful creepage distance can be set for a composite solid core or hollow core insulator of a given length. The cylindrical insulator housing is freely extruded from any mould line concentrically onto the rod. Due to the simplicity of the process, special highly ageing-resistant HTVs with a high ATH content can also be used for this purpose. This coordination with the material used for sheds makes sense because, primarily, the shank can typically be damaged by partial discharges.

Sheds are manufactured separately in tools that contain multiple mould cavities. The choice of vulcanisation system and the level of rework required essentially determine the level of productivity.

The sheds can have different diameters and/or underribs, and are applied to the cylindrical housing in a further step.

The use of underribs [164] results in a greater creepage distance per shed, which gives rise to the following two application philosophies:

- For a given creepage distance, the number of sheds in comparison to a flat profile without underribs falls and the distance between each shed rises. It must be noted that each shed also results in a protected creepage distance for the

Fig. 7.74 Production steps within the modular process

shank. Therefore, in this cost-optimised variant, if there are fewer sheds, the creepage distance is less protected and there may be a loss of hydrophobicity, assuming that the operating conditions are adequate.

- The insulator length is limited and a certain creepage distance is required. In these cases, it was possible to successfully use underrib sheds as a technical compromise [165].

Clamping moulds or machines are used to apply the sheds to a rod with extruded silicone rubber housing (Fig. 7.74). It has proven beneficial for the inner diameter of the sheds to be smaller than the outer diameter of the extruded silicone rubber housing. The resulting press fit stabilises the shed position until the silicone rubber used to achieve an adhesive bond between the sheds and shanks has been fully vulcanised (often known as "silicone adhesive").

To secure the interface between two vulcanised silicone elastomers, the specific property of silicone rubber is used to achieve chemical vulcanisation of three physically independent components through the use of an additional silicone elastomer. Figure 7.75 shows the "chemistry" for an acetate system (RTV-1) that reacts when acetic acid is released. The fine-particle silica (HDK) in silicone rubber plays an important role here: Since not all hydroxyl groups of HDK react during vulcanisation of the sheds and the shank material, they are available for subsequent reactions. The silicone adhesive (RTV-1 or RTV-2 system) applied where the shed is positioned or applied liberally over the shank reacts with the HDK and produces an adhesive composite. Bonds are also created without the direct presence of water [166]. To document the interface strength, a colourant was added to the RTV-1 used in this case before it was applied. It shows that an interface fracture occurs in the extrudate when sheds are mechanically separated and detached. No colourant is visible at the positions where sheds have been detached.

The insulation strength of this important interface was also tested using a more stringent boiling test than the current standardisation (maximum in IEC 62217: 100 h) [143]. Simple insulator test specimens with one shed were

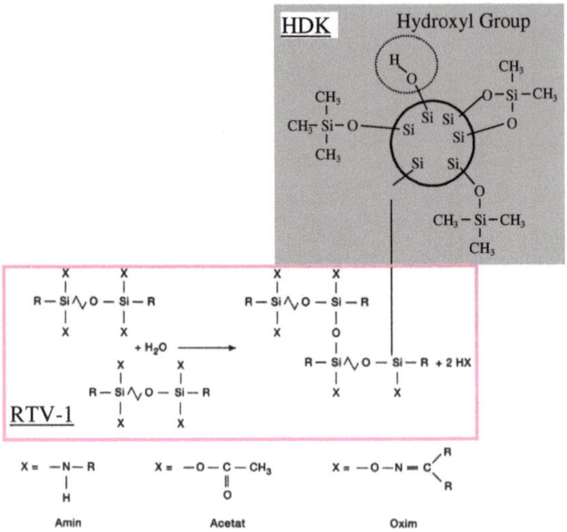

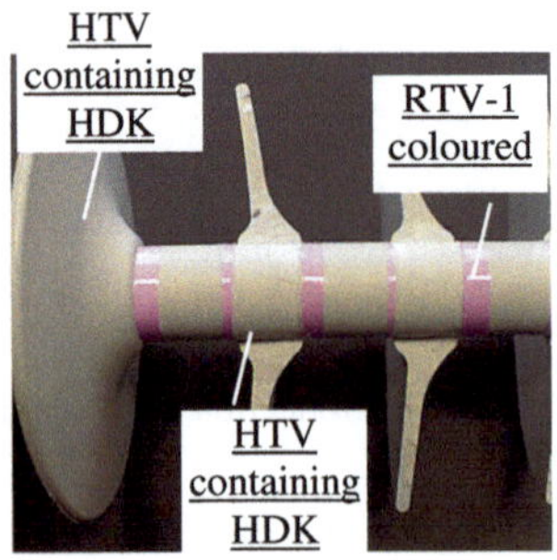

Fig. 7.75 Chemistry for the vulcanisation of pre-fabricated HTV semi-finished parts with an acetate-vulcanising RTV-1

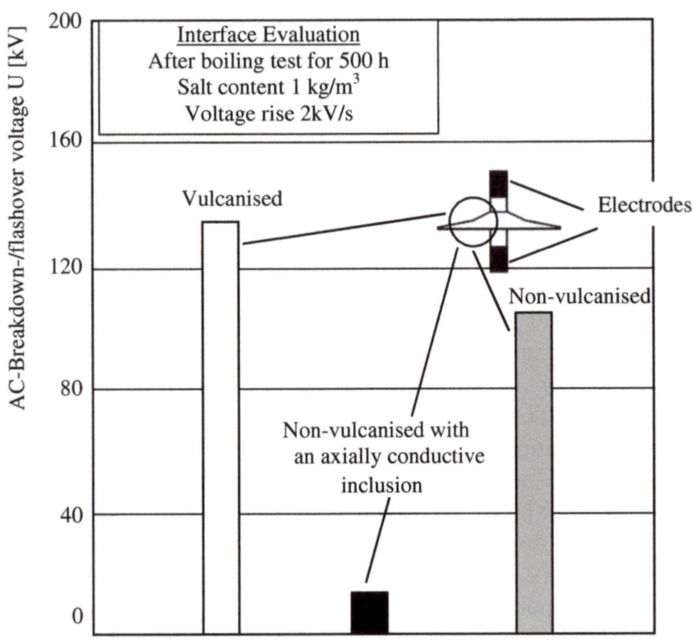

Fig. 7.76 Interface behaviour after boiling test

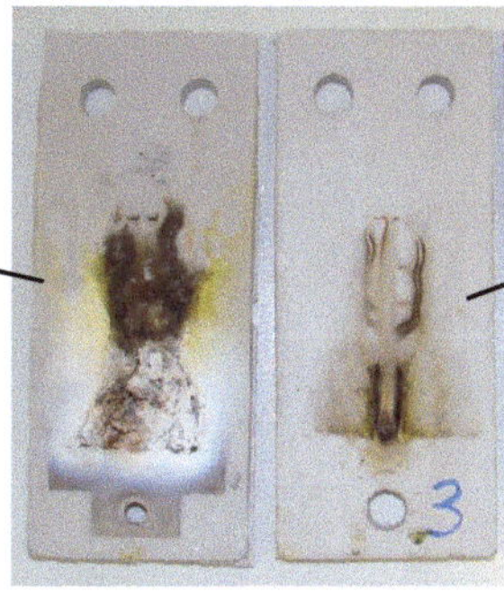

Fig. 7.77 Influence of thickness of RTV-silicone rubber layer used for vulcanisation of pre-manufactured HTV-parts

designed (in terms of the creepage distance to striking distance ratio) in such a way that they were more at risk of a dielectric breakdown in the shank/shed interface than an external flashover. Consequently, the voltage measurement took place under oil. Figure 7.76 shows, by way of an example, the result for three test specimens: The "Vulcanised" test specimen has a press fit and underwent RTV-1 chemical vulcanisation between the shank and shed. The "Non-vulcanised" test specimen only has a press fit on the shank. Only the vulcanised test specimen showed a shed flashover under oil; the test specimen with a non-vulcanised interface experienced a dielectric breakdown in the interface. To verify the "diagnostic ability" of this simple process, a test specimen was prepared with a conductive inclusion (copper wire). As expected, this test specimen fails at a lower value and with an interface flashover. This test can be performed to verify that the chosen material combination of HTV-and RTV-1 silicone rubber, and the coordination of the diameter fits, will not lead to a dielectric breakdown under atmospheric conditions.

The choice and application of the "silicone adhesive" is another important service-relevant detail. In general, these silicone elastomers used have a lower erosion and tracking resistance as the shank material has. Systematic investigations have shown that the silicone adhesive experiences intensive thermal decomposition during the Inclined Plane Test when an excess amount of unsuitable materials is applied. The temperatures that occur can also cause highly-resistant silicone rubber formulations with a high percentage of ATH, which are underneath the adhesive, to fail (Fig. 7.77).

Summary of Processes

A comparison of the individual processes is summarised below (Tables 7.9, 7.10).

Composite Solid Core Insulators
Composite Hollow Cores

Table 7.9 Differences in manufacture for composite solid core insulators

Process	Silicone rubber grade	End fitting sealing method	Advantages	Disadvantages	Investment/level of process expertise required
Compression moulding, transfer moulding, injection moulding 1-shot	HTV with ATH	Casting or overmoulding	Cost-effective	Axial mould line, existing moulds with limited flexibility, rod must be supported for long moulds	High/High
Compression moulding, transfer moulding, injection moulding, multiple shots	HTV with ATH	Casting or overmoulding	Mould line can be offset, cost-effective	Tighter process parameters due to stepping	High/High
ACIM, multiple shots (stepping)	HTV with ATH	Casting or overmoulding	Mould line is offset axially, cost-effective, low rod stress, high degree of automation	Tighter process parameters due to stepping	High/High
Low pressure cavity casting 1-shot	LSR without ATH	Casting or over moulding	Simple mould technology, cost-effective	Axial mould line, existing moulds with limited flexibility	Medium/Low
Low pressure cavity casting, multiple shots (stepping)	LSR without ATH	Casting or over moulding	Simple mould technology, mould line can be offset axially, cost-effective	Step interface more critical as a result of a lower pressure or temperature	Medium/Low
Modular process	HTV with ATH	Casting	No mould line, very flexible	Manufacturing costs due to separate "module production"	High/High

Table 7.10 Differences in manufacture for composite hollow core insulators

Process	Silicone rubber grade	End fitting sealing method	Advantages	Disadvantages	Investment/level of process expertise required
Injection moulding 1-shot	HTV with ATH	Casting or overmoulding	Cost-effective	Axial mould line, existing moulds with limited flexibility, danger of tube delamination in the case of large diameters	High/High
Injection moulding, multiple shots	HTV with ATH	Casting or overmoulding	Mould line can be offset axially, cost-effective	Tighter process parameters due to stepping, danger of tube delamination in the case of large diameters	High/High
ACIM, multiple shots (stepping)	HTV with ATH	Casting or overmoulding	Mould line is rotated, cost-effective, lower tube stress, high degree of automation	Tighter process parameters due to stepping, danger of tube delamination in the case of large diameters	High/High
Low pressure cavity casting, 1-shot	LSR without ATH	Casting or overmoulding	Simple mould technology, cost-effective	Axial mould line, existing moulds with limited flexibility	Medium/Low
Low pressure cavity casting Multiple shots (stepping)	LSR without ATH	Casting or over moulding	Simple mould technology mould line can be offset axially, cost-effective	Step interface more critical as a result of a lower pressure or temperature, axial mould line if large tubes are not rotated	Medium/Low
Modular process	HTV with ATH	Casting	No mould line, very flexible	Manufacturing costs due to separate "module production"	High/High
Extrusion process	HTV with ATH	Casting or overmoulding	Very flexible, no axial mould line	Curvature stress at the outer edge of the shed in the case of small tube diameters	Low/High

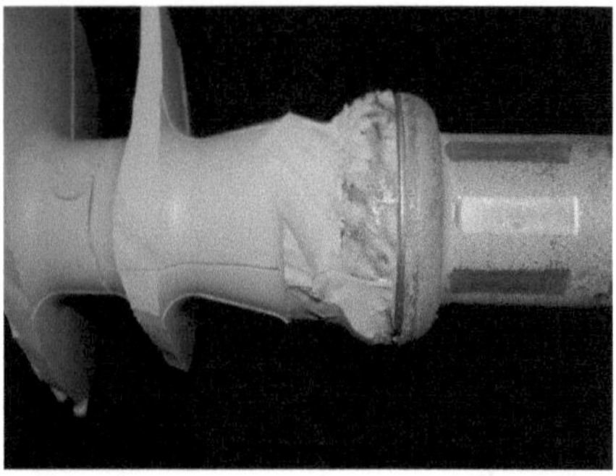

Fig. 7.78 Example of an adhesion test on an over moulded end fitting by peeling

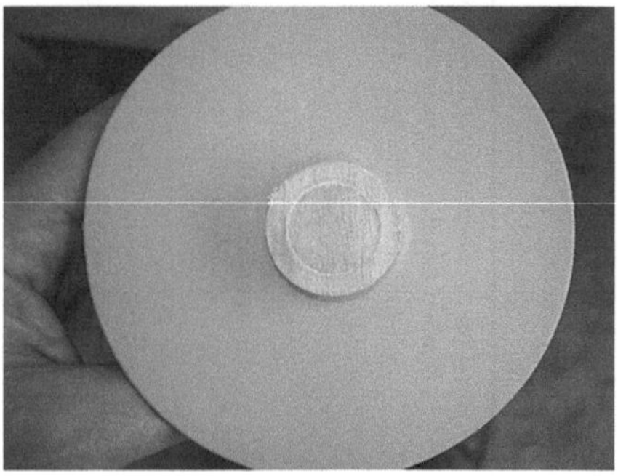

Fig. 7.79 Visual control of rod concentricity

7.3.2.2 Quality Assurance

Numerous quality assurance measures are associated with applying the insulator housing and seal. Such measures can have the following sample structure:

- Checking the characteristic properties of semi-finished parts, for example:
  - Rod roughness.
  - Primer test.
  - Density.
  - TGA.
  - Viscosity.
  - Vulcanisation behaviour.
  - Erosion resistance of silicone rubber.
- Process control.
  - Temperature recording and logging.
  - Type sample in the event of a batch change.
  - Shore A hardness measurement.
  - Adhesive tests (Fig. 7.78).
  - Concentricity test (Fig. 7.79).

Fig. 7.80 (Subsequent) casting

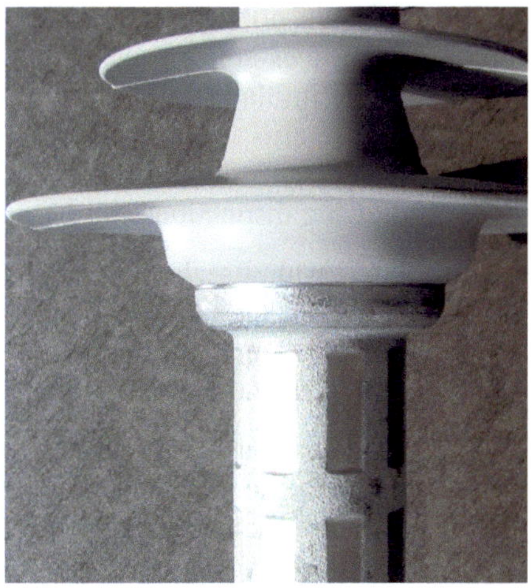

Fig. 7.81 Overmoulding

Ensuring traceability is one important measure; this ensures that the relevant batches of semi-finished parts can be assigned to a particular batch of insulators.

7.3.2.3 Sealing Systems

Process Dependency

Main Principles of Sealing Systems

Years of experience have resulted in "casting" (Fig. 7.80) and "overmoulding" [or "overcasting" (Fig. 7.81)] being the main principles of sealing systems used

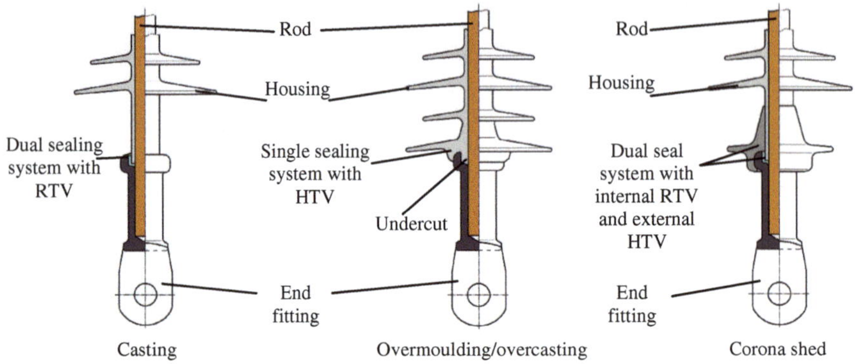

Fig. 7.82 Comparison of the three main sealing concepts

today to apply the insulator housing in RTV/LSR-low pressure cavity casting, in the HTV-injection moulding process and in the HTV-modular process.

The two main principles of sealing systems operate to a reliable level when certain design rules are respected (including corona-free insulator string design). The composite insulator failure reported in the United States in relation to acid penetration (see also section Choosing the Glass Fibre and Resin System) led to further developments that resulted in the various sealing systems available today.

The following distinctions can be made:

- 1-fold [167] with a sealant
- 2-fold [156] sealant and a covering layer
- 3-fold [168] sealant and crimping/pressing of the insulator housing
- 4-fold [169] sealant and multiple crimping/pressing of the insulator housing

If the insulator housing is sealed by means of overmoulding, an adhesive bond is used to bond the housing material with the fitting. The “chemistry” for this type of sealing is more challenging than for casting because the composition of the zinc layer on the steel fitting essentially determines the adhesion. HTV injection moulding is beneficial here because, empirically, it has been established that a higher process pressure leads to better adhesion and a greater fault tolerance than low pressure cavity casting. In accordance with the relevant standards, proof of a safe design is furnished by means of a steep front test after various preliminary loads, including boiling (interface tests).

The Corona Shed as an innovative flexible Sealing Solution

As it will be shown in Sect. 8.6.2, a higher field stress can occur with casting than with overmoulding. This is due to the non-embedded fitting at the triple junction. At the same time, the sealing materials used have a low erosion and tracking resistance due to the required elasticity. This combination is unfavourable. The sectional view of both sealing systems (Fig. 7.82) illustrates this set of problems. Two-fold casting is shown on the left-hand side. The primary sealing function is provided by a silicone gel with a permanently low viscosity. Its gel-like consistency and permanent adhesion make it necessary to have a protection against dirt build-up and particle embedding. This function is fulfilled by a covering layer made from silicone rubber. The vulcanisation systems of the silicone gel and covering layer are coordinated in such a way that there is good adhesion between both.

The erosion and tracking resistance of the covering layer (and silicone gel) is rather low. Furthermore, there have already been examples of "curious" linesmen destroying the sealing system during a "closer" inspection.

This danger does not exist with overmoulding, especially in the case of HTV. The thickness of the insulator housing must be designed in such a way that no dielectric breakdowns occur through the cover during an insulator flashover. The undercut shown in the centre of the figure is also filled with HTV (without air). This produces a material volume that essentially determines the vulcanisation time required to prevent air pockets and to ensure secure adhesion.

Following the development of the corona shed [170], a solution that uses the combined experiences of casting and overmoulding was tested and established:

- The proven primary sealing system (silicone gel) is retained.
- Embedding the fitting reduces the electrical field stress.
- The conical shape also results in a lower field stress in the first shank area.
- The sensitive yet service-relevant sealing interface is protected by HTV silicone rubber, which consists of a housing material with the known high erosion and tracking resistance.
- The shank and fitting are sealed using a form-fit and adhesive bond that will ensure the long-term protection of the primary sealing system.

This solution, which was previously used successfully up to 765 kV AC, can further increase the reliability of composite insulators.

Field Considerations

The three sealing variants were evaluated in terms of electrical field distribution so that the matter of water droplet corona could also be considered (see also Chap. 8). The simulations were performed for identical insulators (length: 1.5 m, 100 kV AC, no modelling of adjacent phases, power arc protective fittings, corona rings, etc.).

The result in Fig. 7.83 confirms the simulations in Sect. 8.6.2; casting experiences the highest field stress. In addition to the different outer contour, the difference between overmoulding and the corona shed can be attributed to the greater wall thickness of the corona shed. This was chosen in order to prevent a material breakdown. The original intention was also to apply a conductive layer to the inner surface of the corona shed in contact with the fitting, which would have resulted in the electrical field in the material always remaining constant and no discharges forming in hollow spaces as a result of the roughness of the galvanised layer.

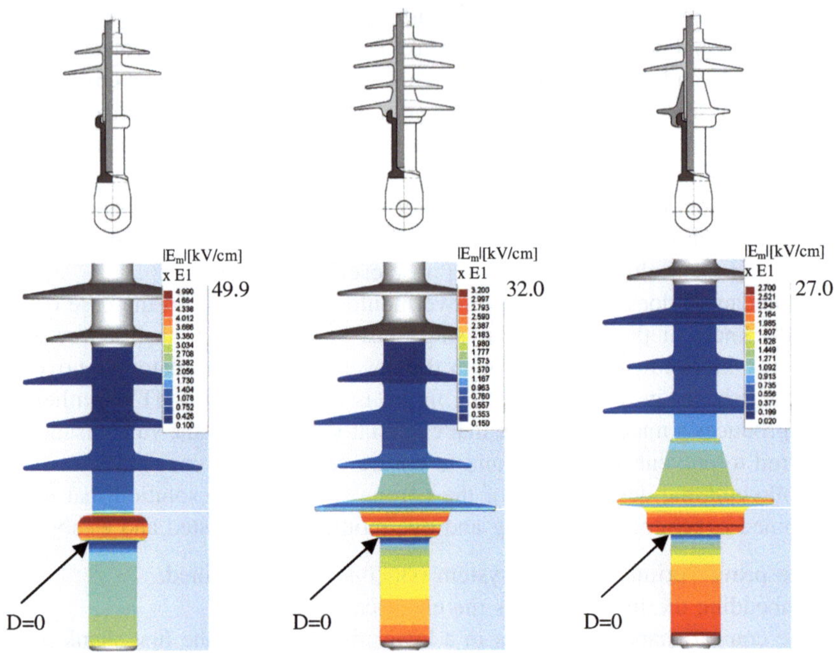

Fig. 7.83 Result of field simulations involving the three sealing systems, D = 0 start point of the field analysis, maximum value shown on the right in kV/cm

However, extensive tests, including a prior boiling test and a subsequent steep front test, have shown that the behaviour is identical with or without a conductive layer. In other words, a flashover always occurred.

In order to assess the field stress from the perspective of using a water droplet corona, it is necessary to analyse the field distribution parallel to the insulator axis; the field stress at the shank sections is important in terms of droplet deformation and corona ignition [171].

This was not possible for this simulation because the conical parts of the over moulded housing or corona shed do not run parallel to the axis. For this reason, the field stress along the surface (Fig. 7.84) was analysed from the starting point D = 0 (Fig. 7.83). In the case of casting, it is clear that a very high field stress occurs at the triple junction (i.e. at the junction between the fitting and the sealing material) and then steadily decreases. Under these conditions, the entire first shank section would have a field stress that is above the recommended limit of 4.5 kV/cm (3.5 kV/cm for seal surface) [172]. In comparison, the shank sections of the other two sealing systems are subjected to less stress.

The effect associated with the relatively high electrical field in the overmoulding and corona shed simulations to D = 15 mm is interesting: After performing the salt fog tests as an ageing test (1,000 h in accordance with IEC

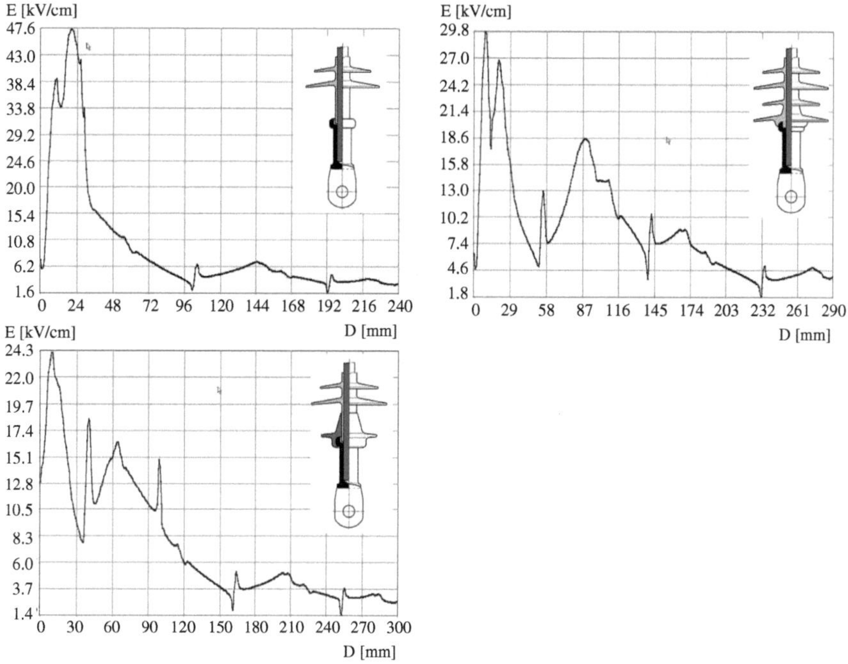

Fig. 7.84 Field stress distribution along the surface of the insulating material used in the three sealing systems, highest value in kV/cm

62217), it was established that this area had virtually no discharge activity and was completely hydrophobic (Fig. 7.85). However, the intensive formation of dry zones on the shank was evident. The conical part of the corona shed also had signs of partial discharge; the grey colour of the layer implied that SiO_x had formed. It is a fact that the testing severity of the standardised ageing tests results in a rapid loss of hydrophobicity and the pollution layer then determines the field distribution. To facilitate an interpretation of the ageing test, it is important to use a corona shed made from HTV silicone rubber in order to effectively protect the sealing system against an erosive attack.

7.4 Summary

This chapter on manufacturing composite insulators has shown the inter-disciplinary nature of this topic; mechanics, high voltage technology, chemistry and process technology are all very closely related and inter-dependent. It is important for the user of composite insulators to know that, after more than 30 years in service, a more interactive exchange of information of experiences between power

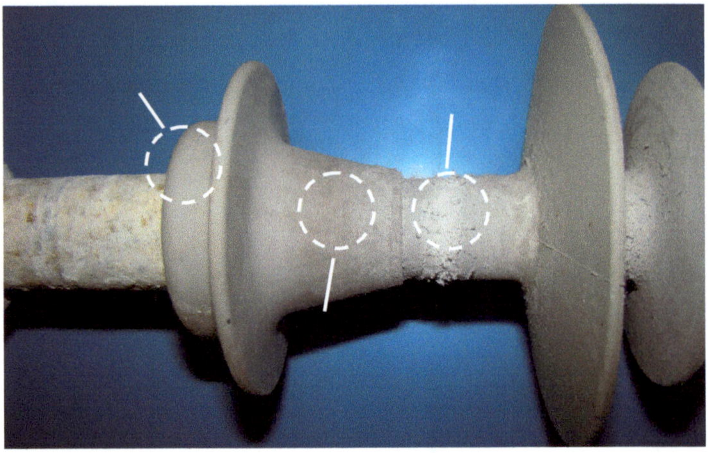

Fig. 7.85 Corona shed surface after salt fog test

utilities and manufacturers, as well as the involvement of academic institutions, have made important contributions to the level of product maturity achieved today. This is especially true for long-time manufacturers whose active participation in the creation of product standards has also resulted in "standardised" components that must fulfil minimum defined requirements.

When composite insulators are used in higher voltage levels, it is important to consider that a composite insulator manufactured from the "best" semi-finished parts will fail if the set-up of the insulator string does not meet the necessary requirements. Due to the wide range of operating conditions, these requirements are difficult to standardise and require the expertise of the manufacturer.

References

1. Gubanski SM, Derfalk A, Andersson J, Hillborg H (2007) Diagnostic methods for outdoor polymeric insulators. IEEE Trans Dielectr Electr Insul 14(5):1065–1080
2. Papailiou KO (1999) Grenzflächen bei Silikon-Verbundisolatoren, Bulletin SEV/VSE 21/ 1999
3. Ansorge S, Camendzind A, Pratsinis SE, Ammann M, Schmuck F, Papailiou KO (2008) Evaluation of silicone rubber housing interfaces after service exposure and performance improvements by nanofillers enriched silicone rubbers. Paper B2-208, CIGRE 2008
4. Ammann M, Papailiou KO, Ansorge S, Schmuck F (2008) Zur Bewertung von Verbundisolatoren und deren relevanten Interfacebereichen nach Freilufteinsatz. ETG Fachtagung Grenzflächen in elektrischen Isoliersystemen, September 2008
5. IEC TR 62039 Ed. 1 (2007) Selection guide for polymeric materials for outdoor use under HV stress
6. CIGRE Working Group D1.14 (2004) Material properties for non-ceramic out-door insulation: state of the art. Technical Brochure 255, 2004

7. IEC 60587 Ed. 3 (2007) Electrical insulating materials used under severe ambient conditions—test methods for evaluating resistance to tracking and erosion
8. IEC 60343 Ed. 2 (1991) Recommended test methods for determining the relative resistance of insulating materials to breakdown by surface di-scharges
9. IEC 62217 Ed 1 (2005) Polymeric insulators for indoor and outdoor use with a nominal voltage >1,000 V—general definitions, test methods and acceptance criteria
10. IEC 60250 Ed. 1 (1969) Recommended methods for the determination of the permittivity and dielectric dissipation factor of electrical insulating materials at power, audio and radio frequencies including metre wavelengths
11. ISO 34-1 (2010) Rubber, vulcanized or thermoplastic—determination of tear strength—part 1: trouser, angle and crescent test pieces
12. DIN 53504 (2009) Prüfung von Kautschuk und Elastomeren—Bestimmung von Reißfestigkeit, Zugfestigkeit, Reißdehnung und Spannungswerten im Zugversuch
13. IEC 60093 Ed. 2 (1980) Methods of test for volume resistivity and surface resistivity of solid electrical insulating materials
14. IEC 60243-1 Ed. 2 (1998) Electrical strength of insulating materials—test methods—part 1: tests at power frequencies
15. IEC 60243-3 Ed 2 (2001) Electric strength of insulating materials–test methods—part 3: additional requirements for 1.2/50 μs impulse tests
16. ISO 4892-2 (2010) Kunststoffe—Künstliches Bestrahlen oder Bewittern in Geräten–Teil 2: Xenonbogenlampen (ISO 4892-2:2006/Amd 1:2009); Änderung A1
17. ISO 4892-3 (2006) Kunststoffe—Künstliches Bestrahlen oder Bewittern in Geräten—Teil 3: UV-Leuchtstofflampen (ISO 4892-3:2006)
18. IEC 60695-11-10 Ed. 1.1 (2003) Fire hazard testing—part 11-10: Test flames–50 W horizontal and vertical flame test methods
19. IEC 61621 Ed. 1 (1997) Dry, solid insulating materials—resistance test to high-voltage, low-current arc discharges
20. P-IEC 61006 Ed. 2 (2004) Electrical insulating materials–methods of test for the determination of the glass transition temperature
21. Kindersberger J, Bärsch R Grenzfläche Feststoff-Gas—Beanspruchungen, Wechselwirkungen, Design, Prüfverfahren, Lebensdauer. ETG Tagung. ETG Fachtagung "Grenzflächen in elektrischen Isoliersystemen" März 2005, Hanau, ETG-Fachbericht 99,VDE-Verlag GmbH Berlin Offenbach, S. 7-26
22. State of the Art Report On behalf of Study Committees 15 and 33 presented by Working Group 33/15.08 (1990) Dielectric diagnosis of electrical equipment for AC applications and its effects on insulation coordination. CIGRE Technical Brochure 059
23. Haas A, Kindersberger J (2009) Determination of ageing of polymeric insulating materials by thermal analysis methods. In: Proceedings of the 16th international symposium on high voltage engineering, 2009 SAIEE. Innes House, Johannesburg, RSA
24. Gutman I (2004) STRI guidelines for diagnostics of composite insulators, for visual inspections; for hydrophobicity; for ir helicopter inspections. In: International conference on suspension and post composite insulators: manufacturing, technical requirements, test methods, service experience, diagnostics. St. Petersburg October 2004
25. IEC 60120 Ed 3 (1984)Dimensions of ball and socket couplings of string insulator units
26. IEC 60471 Ed. 2 (1977) und Amendment 1:1980 (Dimensions of clevis and tongue couplings of string insulator units)
27. IEC 61466-1 Ed 1 (1997) Composite string insulator units for overhead lines with a nominal voltage greater than 1,000 V—part 1: standard strength classes and end fittings
28. IEC 61466-2 Ed 1.1 (2002) Composite string insulator units for overhead lines with a nominal voltage greater than 1,000 V—part 2: dimensional and electrical characteristics
29. IEC 60383-1 Ed 4 (1993) Insulators for overhead lines with a nominal voltage above 1,000 V—part 1: ceramic or glass insulator units for a.c. systems–definitions, test methods and acceptance criteria

30. DIN EN ISO 2178 (1995) Nichtmagnetische Überzüge auf magnetischen Grundmetallen—Messen der Schichtdicke—Magnetverfahren
31. DIN EN ISO 6506-1 (2005) Metallische Werkstoffe–Härteprüfung nach Brinell–Teil 1: Prüfverfahren
32. Schürmann H (2007) Konstruieren mit Faser-Kunststoff-Verbunden. Springer, ISBN 978-3-540-72189-5
33. Culimeta (2010) Isolier- und Abschirmtechnik. Factbook, 9/2010
34. de Tourreil C, Schmuck F (2004) On behalf of CIGRE working group B2.03: brittle fractures of composite insulators—field experience, occurrence and risk assessment. Electra No. 214 June 2004
35. de Tourreil C, Schmuck F (2004) On behalf of CIGRE working group B2.03: brittle fractures of composite insulators—failure mode chemistry, influence of resin variations and search for a simple insulator core evaluation test method. Electra No. 215 August 2004
36. Schijver C (1998) Optimierung der elektrischen Eigenschaften glasfaserverstärkter Kunststoffe. Dissertation 1998, Shaker, ISBN 3-8265-3842-0
37. Lange J (1993) Rohstoffe der Glasindustrie. Deutscher Verlag für Grundstoffindustrie, Leipzig 1993, ISBN 3-342-00663-3
38. Patent US 000004199364A (1980) Glass composition, 22 Apr 1980
39. Owens/Corning Fiberglass (1991) ECRGLAS—technical status
40. Bunker BC, Arnold GW, Day DE, Bray PJ (1986) The effect of molecular structure on borosilicate leaching. J Non-Cryst Solids 87:226–253
41. Armentrout D, Kumosa M, Kumosa L (2004) Water diffusion into and electrical testing of composite insulator GRP rods. IEEE Trans Dielectr Electr Insul 11(3):506–522
42. PPG Industries Fiber Glass (1995) E-Glass and ECR-Glass pultrusions dielectric breakdown voltage. Firmenreport
43. Schramm J (1985) Über den Einfluss von Mikrohohlräumen auf die elektrische Festigkeit glasfaserverstärkter Epoxidharzformstoffe. Dissertation
44. Kuhl M (2001) FRP rods for brittle fracture resistant composite insulators. IEEE Trans Dielectr Electr Insul 8(2), April 2001
45. Schütz A (2000) The history of rodurflex—quite a long story covering more than 3 decades. LAPP Firmenbroschüre
46. Patent US 000004542106A (1985) Glass composition 17, Sept 1985
47. McQuarrieTS (1999) Improved dielectric and brittle fracture resistant core rods for non-ceramic insulators. INMR Congress, Barcelona
48. Neitzel M, Breuer U (1997) Die Verarbeitungstechnik der Faser-Kunststoff-Verbunde. Hanser, ISBN 3-446-19012-0
49. Domininghaus H, Elsner P, Eyerer P, Hirth T (2008) Kunststoffe. Springer, ISBN 978-3-540-72400-1
50. Phillips A, Bologna F, Shaw T (2009) Application of corona rings at 115 and 138 kV. EPRI-Document 1015917 2008, CIGRE WG B2.21 IWD 063-2009
51. Kumosa LS, Kumosa MS, Armentrout DL (2005) Resistance to brittle fracture of glass reinforced polymer composites used in composite (nonceramic) insulators. IEEE Trans Power Delivery 20(4):2657–2666
52. Kumosa M, Kumosa L, Armentrout D (2005) Failure analyses of nonceramic insulators part 1: brittle fracture characteristics. IEEE Electr Insul Mag 21(3):14–27
53. Kumosa M, Kumosa L, Armentrout D (2005) Failure analyses of nonceramic insulators: part II—the brittle fracture model and failure prevention. IEEE Electr Insul Mag 21(4):28–41
54. Chughtai AR, Smith DM, Kumosa LS, Kumosa M (2004) FTIR analysis of non-ceramic compos-ite insulators. IEEE Trans Dielectr Electr Insul 11(4):585–596
55. Liang X, Wang J, Dai J (2010) Surface micro-crack initiated brittle fracture in fiber reinforced plastic rod of composite insulator. IEEE Trans Dielectr Electr Insul 17(2):368–374
56. Liang X, Fan J, Chen L (2000) Research on the brittle fracture of FRP rods and its acoustic emission detection. IEEE PES 2000WM, Paper 16-19-01, January 2000

57. Huiber W, Papailiou KO, Peter M, Schmuck F (2001) Increased installation performance and application solutions using composite insulators—a manufacturerer's philosophy. INMR World Insulator Congress 2001, Shanghai
58. IEC TR 62662 Ed1 (2010) Guidance for production, testing and diagnostics of polymer insulators with respect to brittle fracture of core materials
59. IEC 61109 Ed. 2 (2008) Composite suspension and tension insulators for a.c. overhead lines with a nominal voltage greater than 1,000 V–Definitions, test methods and acceptance criteria
60. Hill B, Sklenicka V, Schmuck F (2010) On behalf of CIGRE WG B2.21: investigation of different liquid solutions for dye penetration tests used in standard IEC 62217 for design and routine testing. ELEKTRA 251_1. 2010
61. Büchner H, Schmuck F, Zanetti A (1997) Cellpack' s cevosil—10 years of technical and marketing experience 3. Symposium on non-ceramic insulators in Miami 1997
62. Schmuck F (1998) Cevosil composite hollow insulator—facts, history, state of the art and design basics. Factbook 1998, Cellpack AG
63. Rocchetti G, Moal E (2005) Computer-simulated design of high performance hollow core insulators for specialized applications. INMR Congress 2005, Hong-Kong
64. Chu FY (1986) SF_6 Decomposition in gas-insulated equipment. IEEE Trans Electr Insul 21:693—725
65. Piemontesi M, Niemeyer L (1996) Surface reactions of SF6 decomposition products. CEIDP S. Francisco 1996
66. Eidinger A, Schaumann R (1979) Schwefelhexaflourid
67. Hinrichsen V, Möhring N, Wietoska T, Haupt H, Bockenheimer A, Heinemann C, Berger C, Gottschalk I, Kurda N, Mikli N, Schmuck F, Seifert J (2010) Resistance to vapour permeation of factory new and of mechanically stressed composite hollow insulators. CIGRE Paris 2010, A3-304_2010
68. CIGRE WG A3.21 (2011) Aspects for the application of composite insulators to high voltage ($\geq$72 kV) apparatus technical brochure 455, April 2011
69. Tomanek A (1991) Silicones and industry. Hanser, ISBN 3-446-17264-5
70. Patent DE 2650363 C2 (1976) Verbundisolator für Hochspannungsfreiluft-Anwendungen, 3 Nov 1976
71. Cherney EA, Burnham JT, Gorur RS (1999) Outdoor insulators, ISBN 0967761107
72. Schmuck F (1992) Zur zeitraffenden Alterungsprüfung von Silikongummi-Oberflächen unter Fremdschichtbelastung und simultaner 50 Hz-Spannungsbeanspruchung. Dissertation 1992, HTWS Zittau, Germany
73. Schmuck F, Bärsch R (1993) Electrochemical and microbiological phenomena during accelerating ageing tests of polymeric insulators 8. ISH Yokohama
74. Schmuck F, Ramm W (1991) Zum Einfluss einer biologischen Wechselwirkung zwischen polymerem Isolierstoff und Prüfmilieu während zeitraffender Alterungsprüfungen mit fließenden Fremdschichten. Elektrie 11/1991
75. Winter HJ, Bärsch R (2006) Oberflächenverhalten von Siliconelastomeren unter Feuchte- und biogenen Belastungen. RCC Konferenz, Berlin 2006
76. Neumann C (2003) Betriebsverhalten von Verbundisolatoren, ETG-Fachbericht 93, VDE-Verlag 2003
77. Gubanski S, Derfalk A, Wallström S, Karlsson S (2005) Biological contamination of insulators: influence on performance and diagnostic techniques to assess the problem. INMR World Insulator Congress 2005 Hongkong
78. Fernando MARM, Gubanski SM (2010) Ageing of silicone rubber insulators in coastal and inland tropical environment. IEEE Trans Dielectr Electr Insul 17(2):326–333
79. Fernando MARM, Rajamantri H, Gubanski SM (2006) Performance of silicone rubber composite insulators in Sri Lanka. First international conference on industrial and information systems, ICIIS 2006, 8–11 Aug 2006, Sri Lanka
80. Gubanski SM, Karlsson S, Fernando MARM (2006) Performance of biologically contaminated high voltage insulators. ICIIS 2006, 8–11 Aug 2006, Sri Lanka

81. Bärsch R, Kuhl M (1999) Betriebserfahrungen und Untersuchungen an Kunststoffisolatoren in eine 20 kV-Leitung auf der Insel Nordstrand. ETG-Fachbericht 76 S:249–257
82. Papailiou KO, Peter M, Fluri W, Schmuck F (2003) A review of material development, recent 420 kV braced line post designs and long-term evaluation of composite insulators in silicone rubber technology. INMR Insulator Symposium 2003 Marbella
83. George J-M (2009) Review of polymer materials and design considerations based on field experience and laboratory testing. INMR Congress 2009, Kreta
84. Korrespondenz mit HJ Winter, Wacker Chemie AG. Burghausen, Germany: Thermisches Verhalten von ATH
85. Kumagai S, Yoshimura N (2001) Tracking and erosion of HTV silicone rubber and suppression mechanism of ATH. IEEE Trans Dielectr Electr Insul 8(2):673–678
86. Schmidt LE, Kornmann X, Krivda A, Hillborg H (2010) Tracking and erosion resistance of high temperature vulcanizing ATH-free silicone rubber. IEEE Trans Dielectr Electr Insul 17(2):533–540
87. Kautschuk Gummi Kunststoffe 10/2006
88. Meyer LH, Jayaram SH, Cherney EA (2005) A novel technique to evaluate the erosion resistance of silicone rubber composites for high voltage outdoor insulation using infrared laser erosion. IEEE Trans Dielectr Electr Insul 12:1201–1208
89. Ansorge S, Schmuck F, Papailiou KO (2011) Improved silicone rubbers for the use as housing material in composite insulators. Trans Dielectr Electr Insul 19 (1)
90. Meyer L, Jayaram S, Cherney EA (2004) Thermal conductivity of filled silicone rubber and its relationship to erosion resistance in the inclined plane test. IEEE Trans Dielectr Electr Insul 11(4):620–630
91. Meyer LH, Cherney EA, Jayaram SH (2004) The role of inorganic fillers in silicone rubber for outdoor insulation—alumina tri-hydrate or silica. IEEE Electr Insul Mag 20(4):13–21
92. Chang RJ, Mazeika L (1996) Electrical activities associated with inclined-plane tracking and erosion test. CEIDP San Francisco 1996
93. Kärner H (1995) Research at braunschweig—studies of interfacial phenomena in composite insulators. Insul News Market Rep 3(3), May/June 1995, pp 9–11
94. Kindersberger J (2003) Stand und Tendenzen der Prüfung von Kunststoffen für Freiluftisolierungen HIGHVOLT Kolloquium, Mai 2003, Dresden, Beitrag 4.2, pp 121–131
95. Gorur RS, Cherney EA, Hackam R (1990) Polymer insulator profiles evaluated in a fog chamber. IEEE Trans Power Delivery 5(2):1078–1085
96. Gorur RS, Bernstein BS (1998) Field and laboratory aging of polymeric. Distribution cable terminations: part 1-field aging. IEEE Trans Power Delivery 13(2):316–322
97. Gorur RS, Bernstein BS (1998) Field and laboratory aging of polymeric cable terminations: part 2—laboratory aging. IEEE Trans Power Delivery 13(2):323–329
98. Seifert J, Bärsch R (2007) Design evaluation of silicone rubber composite insulators under the aspect of surface pollution stress 15. In: International symposium on high voltage engineering, Ljubljana, Slovenia, Aug 2007
99. Seifert JM, Bärsch R, Vosloo WL (2008) Dimensioning of the housing profile of silicone rubber composite insulators for harsh marine pollution conditions. CIGRE Session 2008, Paris, Paper D1–303
100. Bärsch R (2009) Polymere lsolier- und Funktionswerkstoffe für Hochspannungsisolatoren und Kabelgarnituren—Beanspruchungen, anwendungsspezifische Eigenschaften und Prüfverfahren, RCC-Fachtagung “Werkstoffe- Forschung und Entwicklung neuer Technologien zur Anwendung in der elektrischen Energietechnik, Berlin Mai 2009, Tagungsband S. 1–54
101. Büchner H, Schmuck F, Zanetti A, Zingg A, Bärsch R, Jahn H (1996) Description of a complex test programme for the evaluation of outdoor aging relevant properties of silicone rubber, 7. In: IEE conference in bath
102. Büchner H, Schmuck F, Zanetti A, Bärsch R, Jahn H, Lambrecht J (1997) A comprehensive programme for the evaluation of essential material properties of silicone rubber for outdoor high voltage applications. Lokale CIGRE in Cairo 1997

103. Han DH, Cho HG, Han SW (2003) Effects of alumina trihydrate on the electrical insulation properties of HTV silicone rubber. In: Proceedings of the 7th international conference on properties and applications of dielectric materials, 1–5 June 2003
104. Schmuck F, Barsch R, Pilling J (1996) About an improved evaluation of results by the standardized salt-fog procedure for composite insulators. Eur Trans Electr Power 6:237–243
105. Kumagai S, Yoshimura N (1999) Influences single and multiple environmental stresses on tracking and erosion of RTV silicone rubber. IEEE Trans Dielectr Electr Insul 6(2):211–225
106. Ansorge S, Schmuck F, Aitken S, Papailiou KO (2010) Improved performance of silicone rubbers for the use in composite insulators. CIGRE 2010
107. Krivda A, Schmidt LE, Kornmann X, Ghorbani H, Ghorbandaeipour A, Eriksson M, Hillborg H (2009) Inclined-plane tracking and erosion test according to the IEC 60587 standard. IEEE Electri Insul Mag 25(6), Nov/Dec 2009
108. Seifert JM, Bärsch R, Vosloo WL (2008) Dimensioning of the housing profile of silicone rubber composite insulators for harsh marine pollution conditions. CIGRE Session 2008, Paper D1–303
109. Lambrecht J (2001) Über Verfahren zur Bewertung der Hydrophobieeigenschaften von Silikonelastomer-Formstoffen. Dissertation, TU Dresden 2001, Shaker Verlag ISBN 3-8265-9129-1
110. Barsch R, Jahn H, Lambrecht J, Schmuck F (1999) Test methods for polymeric insulating materials for outdoor HV insulation. IEEE Trans Dielectr Electr Insul 6(5):668–675
111. Bärsch R, Lambrecht J, Winter H-J (1995) On the valuation of the early aging period of polymer insulator surfaces during accelerated aging tests. In: 9th ISH, proceedings, vol 3. Pollution Phenomena, Paper 3023, Graz
112. Bärsch R, Lambrecht J, Jahn H (1996) On the evaluation of the hydrophobicity of composite insulator surfaces. In: IEEE conference on electrical insulation and dielectric phenomena IEEE CEIDP 1996, San Francisco, pp 5A–19
113. Bärsch R, Lambrecht J, Winter HJ (1997) On the evaluation of influences on the hydrophobicity of silicon-rubber-surfaces. 10th ISH montreal (1997), proceedings, vol 3, pp 13–16
114. Kaltenborn U, Kindersberger J, Bärsch R, Jahn H (1997) On the electrical performance of different insulating materials in a rotating-wheel-dip-test. IEEE CEIDP Minneapolis Oktober 1997, paper 5A–1 0
115. Jahn H, Bärsch R, Kaltenborn U, Kindersberger J (1998) The evaluation of the early aging period of castings made of epoxy and PUR resins. IEEE CEIDP Atlanta, October 1998, paper 6B–7
116. IEC 61302 Ed. 1 (1995) Electrical insulating materials—method to evaluate the resistance to tracking and erosion, rotating wheel dip test
117. IEC 62730 TR Ed. 1 (2012) HV polymeric insulators for indoor and outdoor use—tracking and erosion testing by wheel test and 5,000 h test (expected for 2012)
118. ANSI C29.18 (2003) Insulators composite—distribution line post type
119. Bärsch R (2003) Bewertung der Hydrophobie sowie des Kriechstromverhaltens von Silikonelastomeren für Hochspannungs-Freiluftisolatoren. ETG-Fachbericht 93:97–108
120. Cervinka R, Bärsch R, Exl F, Kindersberger J, Winter H-J (2008) Untersuchungen zur Beständigkeit der Hydrophobie von polymeren Isolierstoffoberflächen und ihrer Wiederkehr mit dem Dynamischen Tropfen-Prüfverfahren ETG-Tagung
121. CIGRE WG D1.14 (2010) Evaluation of dynamic hydrophobicity properties of polymeric materials for non-ceramic outdoor insulation—retention and transfer of hydrophobicity. Technical Brochure 442, Dec 2010
122. IEC TS 62073 Ed. 1 (2003) Guidance on the measurement of wettability of insulator surfaces
123. KRUESS Messung des Kontaktwinkels (2011) Online-Dokumentation http://www.kruss.de/de/theorie/messungen/kontaktwinkel/messung-des-kontaktwinkels.html
124. Kopczynska A (2008) Oberflächenspannungsphänomene bei Kunststoffen—Bestimmung und Anwendung. Diss. Technische Fakultät der Universität Erlangen-Nürnberg

125. Kindersberger J, Bärsch R (2005) Grenzfläche Feststoff-Gas—Beanspruchungen, Wechselwirkungen, Design, Prüfverfahren, Lebensdauer. Übersichtsvortrag der ETG-Fachtagung 2005 in Hanau, Deutschland
126. Jahn H (2003) Zur Bewertung stofflicher und herstellungsbedingter Einflussgrößen auf das Hydrophobie- und Erosionsverhalten von Silikonelastomeroberflächen. Dissertation, TU Dresden 2003, Shaker, ISBN 3-8322-1963-3
127. Liang X, Li CR, Ding L (2009) Study on the hydrophobicity of HTV SIR treated by different corona intensity. In: Power and energy engineering conference, APPEEC 2009. Asia-Pacific
128. Kim SH, Cherney EA, Hackam R (1990) The loss and recovery of hydrophobicity of RTV silicone rubber insulator coating. Trans Power Delivery 5(3):1491–1500
129. Hackam R (1999) Outdoor HV composite polymeric insulators. IEEE Trans Dielectr Electr Insul 6(5):557–585
130. Lee CL, Homan GR (1981) Silicone elastomer protective coatings for HV insulators. IEEE-CEIDP
131. Gorur RS, Cherney EA, Hackam R (1986) Factors influencing the performance of polymeric insulating materials in contaminated environments. IEEE-CEIDP
132. Gorur RS, Cherney EA, Hackam R, Orbeck T (1988) The electrical performance of polymeric insulating materials under accelerated aging in fog chamber. IEEE Trans Power Delivery, vol 3
133. Kim SH, Cherney EA, Hackam R (1992) Effects of filler level in RTV silicone rubber coatings used in HV insulators. IEEE Trans Electr Insul 27(6):1065–1072
134. Kocher M (1993) Experience with silicone composite insulators in the tunnels of BLS Lötschberg. Railway Technology 1993
135. Al-Hamoudi IY (2008) Field test results of composite silicone rubber insulators at shoaiba of Saudi Arabia. In: GCC power 4th CIGRE conference 2008, Bahrain
136. Munteanu R (1994) Silicone rubber insulators life cycle costs. Transmission and Distribution, International 1994
137. Kindersberger J, Kuhl M (1989) Effect of hydrophobicity on insulator performance 6. ISH New Orleans 1989
138. Janssen H, Herden H, Kärner HC (1999) LMW components in silicone rubbers and epoxy resins 11. ISH London
139. Janssen H, Herden H, Kärner HC (1997) The loss and recovery of hydrophobicity on silicone rubber surfaces 10. ISH Montreal 1997
140. Winter HJ, Bärsch R (2006) Oberflächenverhalten von Siliconelastomeren unter feuchte und biogenen Belastungen. RCC-Fachtagung "Werkstoffe für Isolatoren, Überspannungsableiter, Kabelgarnituren, Schaltgeräte, Berlin 06.-07. April 2006, Tagungsband S. 49–58
141. Gubanski S (2008) Biological growth on non-ceramic insulators. CIGRE 2008 Paris, SC D1 Pref. Subject: 3, Question N°: 3.6
142. Dinesh MN, Vasudev N, Vasudevan Nambudri PV, Suryanarayana K, Ravi KN, Krishnan V (2008) Performance of composite insulators with and without bio contamination 2008. In: International conference on high voltage engineering and application, Chongqing, China, November 9–13, 2008
143. Papailiou KO, Peter M, Schmuck F (1999) SEFAG's Silcosil—20 years of innovative solutions using silicone composite insulators for high voltage electrical systems. INMR World Insulator Congress 1999 Barcelona
144. Koshino Y, Umeda I, Ishiwari M (1998) Deterioration of silicone rubber for polymer insulators by corona discharge and effects of fillers. In: Annual report conference on electrical insulation and dielectric phenomena, vol 1 and 2
145. Schmuck F (2009) Status of test procedures for FRP cores of composite insulators. INMR Magazine 2009
146. Klöpel S (2002) Schallemissionsanalyse zur Untersuchung von Stahlbetontragwerken Institut für Baustatik und Konstruktion. ETH Zürich ISBN 3-7281-2837-6
147. Leistner M (2011) Onlinestudium Schallemissionsanalyse, www.mleistner.de

148. Ju F, Xidong L, Yu Y, Chengsheng W, Ling C (2000) Application of acoustic emission technology on structure design and quality control of composite insulators. In: Proceedings of 6th ICPADM, June, 2000
149. Vallen Systeme (2008) Handbuch ASCO-P
150. IEC 61462 Ed. 1 (2007) Composite hollow insulators—Pressurized and unpressurized insulators for use in electrical equipment with rated voltage greater than 1,000 V—definitions, test methods, acceptance criteria and design recommendations
151. IEC 60068-2-17 Ed. 4 (1994) Basic environmental testing procedures—part 2-17, Tests–Test Q: Sealing
152. Schröder G (2001) Neue Norm zur Auswahl eines geeigneten Verfahrens zur Lecksuche und Dichtheitsprüfung. ZfP-Zeitung 74, Apr 2001
153. Franke H Lecksuchtechnik (2011) www.lecksuchtechnik.de
154. Steinberger H (1997) Products and processing for high voltage applications. In: INMR symposium 1997, Miami
155. Schmid H (2011) Improved manufacturing methods for cable accessories and insulators. In: INMR symposium 2011
156. Technische Fakten—Silcosil© Silikonisolatoren (2007) PFISTERER SEFAG AG
157. Moal E, Fritsch M, Chevalier F, Rocchetti G, Pigini A (2011) 20 years experience with extruded HTV housings in EHV applications. INMR Symposium 2011, Seoul
158. Holmberg A (2011) Shed profile for hollow core composite insulators on HV apparatus. In: INMR symposium 2011, Seoul
159. Kanters AJLM (2008) Experience with composite insulators in 380 kV-substation Maasvlakte. CIGRE Paris, 2008, A3_Question 11
160. Vosloo W (2007) Experience and lessons learned from testing composite insulators in South Africa. In: INMR symposium 2007, Hong-Kong
161. Chrzan KL, Kindersberger J (2005) Pollution behaviour of insulators with spiral shaped sheds. ETG Fachtagung Grenzflächen in elektrischen Isoliersystemen 2005 in Hanau
162. Patent DE000019944513 C1 (1999) Verfahren zur Herstellung eines Hochspannungsisolators, sowie Vorrichtung zur Durchführung eines solchen Verfahrens. 16. 9. 1999
163. Patent EP000001415308 B1 (2001) Herstellung von Verbundisolatoren durch Anspritzen von Einzelschirmen an einen Strunk, 12 Oct 2001
164. Patent EP0774157 B1 (1995) Elektrischer Isolator aus Silikongummi für Hochspannungsanwendungen, 7 July 1995
165. George J, Iqbal M, Papailiou KO, Huiber W, Schmuck F (2007) Dewa's new 420 kV compact line as an example of maximizing capacity of overhead lines in an urban environment. GCC POWER 07 Dubai, UAE 2007
166. Körner G, Goldschmid T, Schulze M, Weiss J (1991) Silicone–chemistry and technology. Vulkan ESS. 1991, ISBN 3-8027-2161-6
167. LAPP: Rodurflex Produktbrochüre (2011)
168. MacLean Power Systems: P-S-T End Fitting Seal—Produktbroschüre (2011)
169. Hubbel: Quadri Sil—Produktbroschüre (2011)
170. Patent EP000002071592A1 (2007) Koronaschirm und Verbundisolator mit Koronaschirm
171. Phillips AJ, Childs DJ, Schneider HM (1999) Aging of non-ceramic insulators due to corona from water drops. IEEE Trans Power Delivery 14(3):1081–1089
172. IEEE Taskforce on Electric Fields and Composite Insulators (2008) Electric fields on AC composite transmission line insulators. IEEE Trans Power Delivery 23(2):823–830

Chapter 8
Composite Insulator Design from the Perspective of Corona Protection

Symbols and Abbreviations

A_R	Distance between ring and insulator end fitting
A_S	Shed overhang
A_{SF}	Distance between shed and insulator end fitting
A_{SS}	Distance between shed and shed
ACSR	Aluminum conductor steel reinforced
ATH	Aluminium trihydrate, Aluminium hydroxide
C	Empirical constant (0.452–0.462)
D_F	Diameter of insulator end fitting
D_R	Outer diameter of ring
D_S	Diameter of seal
E	Electrical field stress
E_{cr}	Critical field stress
E_{max}	Highest field stress
E-CR or ECR-Glass	Corrosion-resistant glass fibre for electrotechnical use
EPDM	Ethylene propylene diene monomer (insulant)
EPRI	Electric Power Research Institute
EVU	Power utility
HV	High voltage
IEC	International Electrotechnical Commission
IEEE	Institute of Electrical and Electronics Engineers
LSR	Liquid silicone rubber
PDMS	Polydimethylsiloxane
PD	Partial discharge
R_{eq}	Equivalent droplet radius
R_F	Radius of end fitting rim
R_S	Radius of shed edge
RIV	Radio interference voltage
RTV	Room temperature vulcanising
SiO_x	Oxidised silicium

K. O. Papailiou and F. Schmuck, *Silicone Composite Insulators*, Power Systems,
DOI: 10.1007/978-3-642-15320-4_8,

TE	Partial discharge
TGA	Thermo-gravimetric analysis
(U_{rms}/d)	RMS value of voltage related to striking distance
U_f/d	RMS value of voltage related to creepage distance
d_R	Diameter of ring profile
i_{peak}	Peak value of PD leakage current
r	Radius of droplet
s	Distance
t_{MS}	Time to morning star inception
ε_0	Dielectric constant
θ	Static contact angle
ε_r	Relative permittivity
γ	Surface tension
κ_S/κ_{S0}	Increase of surface layer conductivity in relation to the value at $t = 0$

8.1 Introduction

Corona discharges, also known as Saint Elmo's fire, have been observed for hundreds of years. Initially, their natural occurrence on ships' masts or towers was extremely mystifying. Credit must therefore be given to the German physicist G.C. Lichtenberg, the first professor of experimental physics in Göttingen in 1770, who built what at that time was considered to be a very large capacitive generator (2.5 m^2) and found, after a series of tests, branching tracks in the dust on the insulating plates, which were to become known as Lichtenberg figures and heralded the beginning of corona physics. At the start of the 19th century, J.S.E. Townsend's pioneering work involved him systematically analysing gaseous discharges, which led to Townsend developing a set of useful equations to describe the corona phenomenon. In later years, L.B. Loeb, J.M. Meek and H.A. Rather continued his good work.

Today, corona discharges are predominantly known as partial discharges (PD). If the electrical field stress within an electrical insulation system exceeds a critical value (electrical strength) within a certain range, charge carriers (e.g. electrons and ions) propagate as a result of the acceleration of existing free electrons. As the voltage increases, the intensity and propagation of partial discharges also increases until a breakdown or a power arc occurs. Composite insulators are surrounded by air as insulating medium. Since the electrical strength of air is significantly lower than the electrical strength of the solid insulating materials in the composite insulator, partial discharges initially occur in the air surrounding those areas subjected to a high field stress (i.e. at the fittings and/or conductor). The processes implemented here are usually pulse-like and involve, to a greater or lesser extent, fundamental processes such as impact ionisation, photoionization and thermal ionisation, as well as dissociation processes for air molecules. Pulse-shaped currents associated with electromagnetic waves are caused by the partial discharges, which can be detected using devices that measure partial discharges. The sounds associated with partial discharges are also used in acoustic PD measuring technology.

Partial discharges on the surface of a composite insulator, known as surface partial discharges, are particularly significant because their energy input resulting from an impact and/or heat may damage the solid insulating materials. Whereas air molecules can regenerate themselves, damage to the solid surface is usually permanent and leads to cumulative damage (surface ageing), which may cause the insulation system to fail (fracture, flashover, etc.). Since the term "corona" has been retained in the context of overhead transmission lines, it will also be used here. While the measurement unit Pico-Coulomb (pC) is used in PD measuring technology to measure apparent charges, dB/1 μV is used, for practical reasons, to measure the radio interference voltage (RIV) in overhead transmission line technology.

Today, the corona phenomenon is deployed in a range of commercial and industrial applications, including:

- Ozone generation
- Water disinfection
- Removal of particles from flowing gaseous media (smoke, electrostatic filter, etc.)
- To increase the wettability of polymeric surfaces in adhesive technology and printing technology
- Photocopiers and printers.

Such broad industrial uses have also given rise to further analyses on the effects of corona, in particular, the interactive effects between ambient atmosphere, humidity, water and polymeric surfaces. The results of such analyses provide an important basis for understanding secondary and/or ageing mechanisms on polymeric insulating material surfaces.

Corona discharges are undesirable side effects in energy transmission. If the transmission voltage increases, it makes sense to limit the occurrence of corona discharges through the use of multi-conductor systems and corona protection devices on insulator strings. The following negative effects are associated with corona:

- Energy loss
- Acoustic interferences (under dry conditions and during rainfall)
- Electromagnetic interferences
- Luminous effects
- Ozone formation
- Damage to string components and insulator housing.

This chapter will discuss the various design criteria that composite insulators, in particular, need to fulfil in order to prevent corona-induced damage to their surfaces.

8.2 Corona as a Design Problem: Recent Examples

The following recent examples reflect the many potential causes of faults that may result in the composite insulators and composite insulator strings of new and long-time users being subjected to continuous corona discharges. They include:

- Cost pressures
- Inadequate tender process
- Inadequate specification for on-site conditions
- Design errors
- Installation faults.

8.2.1 An Example for 525 kV-Double Tension Set on a Lattice Tower

The photographs discussed below were taken by a first-time user of composite insulators who replaced the glass cap and pin insulator strings on a strategically important 525 kV overhead transmission line with equivalent composite insulators (equivalent in terms of their length, creepage distance, insulator fittings, strength class, etc.), while retaining the metallic string components. During a line inspection, a night-time UV detection camera detected intensive corona occurrences. This type of camera works on the principle of image intensification and is configured to receive a mixture of wavelengths (230–450 nm) that are typical for corona discharges. For this reason, measurements can only be taken in darkness. In order to localise the position of the corona, the object to be evaluated is briefly illuminated with an extraneous light source and the camera is focused on the area of interest. The light reflected is very intense, thus making it impossible to detect a corona in this situation (Fig. 8.1). The glass cap and pin insulator set previously installed uses so-called "rackets" as power arc protective fittings, which is the state-of-the-art for this technology. However, for the new installation with composite insulators, their field homogenising effect of limiting corona discharges is too low. And as an installation fault, corona rings were not installed in the insulator set. Thus, corona discharges are clearly visible once the extraneous light source has been switched off (Fig. 8.2). As expected, continuous and locally stable corona discharges occur at the rim of the two insulator end fittings on the high voltage side.

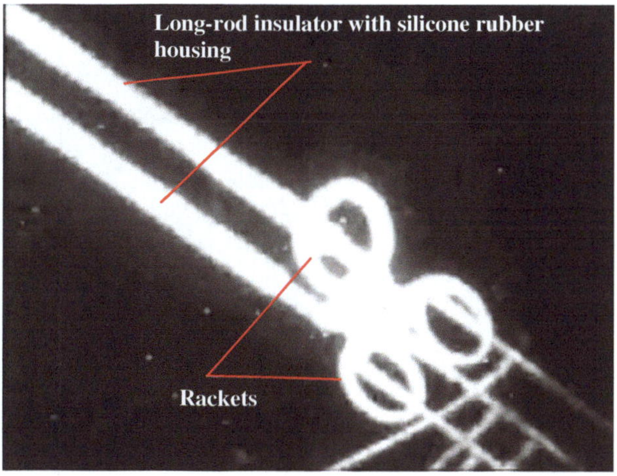

Fig. 8.1 Photograph of an insulator set illuminated with extraneous light

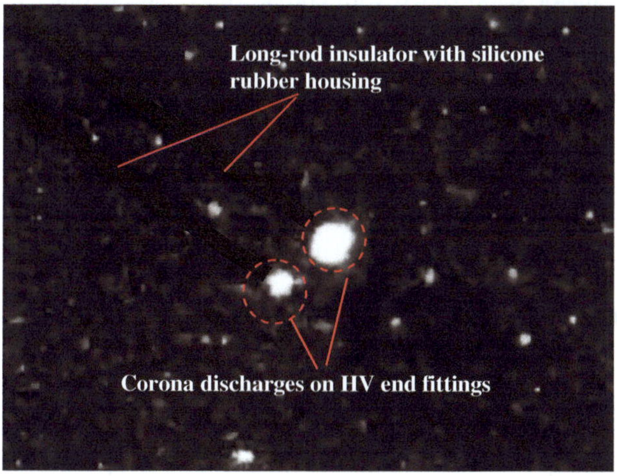

Fig. 8.2 Stable corona discharges on the HV end fitting of 525 kV insulators taken by a night vision UV camera

Today's state-of-the-art solution involves fitting composite insulators with corona protection rings that have been specifically designed to meet the needs of the particular case of application and insulator design, and are included in the scope of delivery. As in the case presented here, problems arise if the corona rings are forgotten during installation because the installation screws are not compatible with the new "composite insulator" technology. In other cases, corona rings are not installed because they do not fit into the existing string configuration. The 525 kV strings shown here have since been retrofitted with corona rings. Furthermore, composite insulators with this particular service history have been withdrawn from the network in order to undergo a detailed inspection.

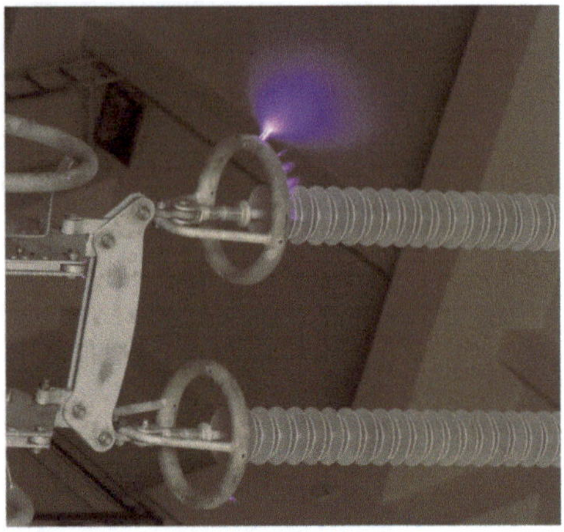

Fig. 8.3 Corona type test of a 525 kV double tension set

The following practice is recommended but not always applied: The complete composite insulator set(s) (fitted with corona rings and the other hardware components) were successfully type-tested in suspension and tension configuration before their shipment to installation site. This test included an evaluation of corona and interference voltages. Corona was incepted at the corona rings (Fig. 8.3), at a value (Fig. 8.4) that was approx. 18 % above the value for the highest phase to earth voltage. The radio interference voltage (RIV) value of 45 dB/1 μV at 333 kV was also compliant (46 dB/1 μV was specified).

8.2.2 An Example for 245 kV-Double Tension String at a Station Entry Point

This example demonstrates the use of a day-time UV-detection camera (Fig. 8.5) in a 245 kV-European substation. This camera system typically displays corona discharges as integral "measurement clouds". Due to the interference of day-time UV-light, it is not possible to resolve the form of the individual discharge. In this case, the string design was not adapted to the voltage level. The upward-oriented racket is clearly visible as a power arc protective fitting, which, due to its position and geometry, is unable to provide complete corona protection. Consequently, after an average service time of just over 1 year, there was considerable damage to the galvanised layer and specifically to the sealing system (Fig. 8.6). Furthermore, despite the low level of pollution in the area of application, an intensive grey colour was visible on the first shank section as a result of oxidation-stimulated cross-linking, which caused SiO_x to form on the silicone rubber surface. The position of the damage on each outward facing area of the fittings on the left and right insulator is due to the fact that the field-controlling effect of the racket is at its lowest at these locations. The insulators were retrofitted with corona rings

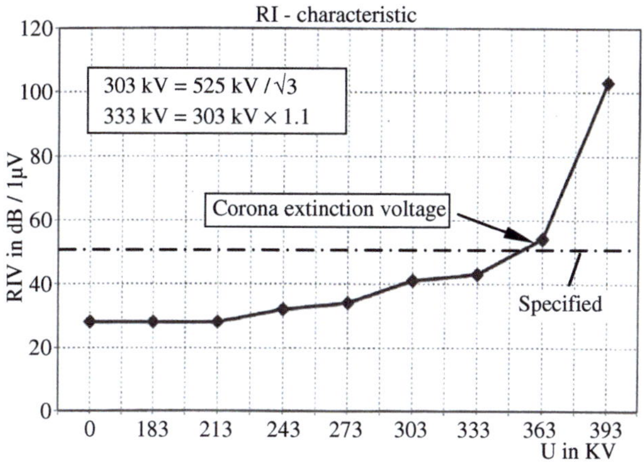

Fig. 8.4 Progression of the radio interference voltage and value for the visible corona extinction

Fig. 8.5 Photograph taken using a day-time UV detection camera

Fig. 8.6 Damage to the galvanizing layer and seal

Fig. 8.7 Retrofitting of corona rings, damage cannot be rectified

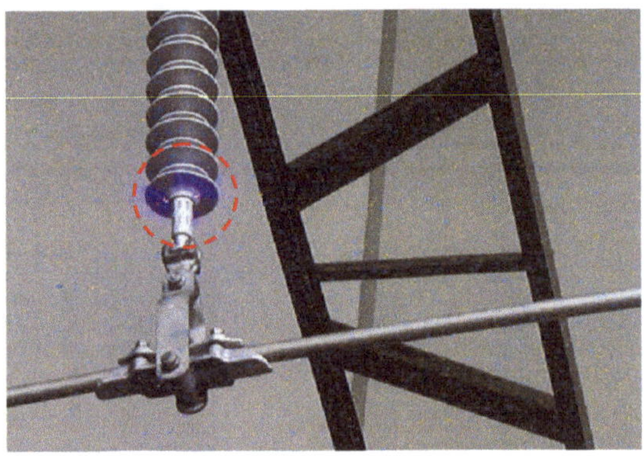

Fig. 8.8 Sharp-edged damage to the galvanising layer as a source of corona

(Fig. 8.7). Following the withdrawal of few damaged insulators and a detailed analysis of the seals affected, one could safely assume that there was a real risk of further cumulative damage unless some additional measures were taken. For this reason, all of the composite insulators were replaced and the string design was adapted.

In this case, the individual insulator set components had not been tendered as a packaged solution, nor was a comprehensive type test performed, which would have easily identified the design errors.

8.2.3 Examples for 115/138/145 kV-Various Insulator Sets

Since the distribution level often concerns service voltages $\leq$145 kV, the cost pressures are particularly high. Figure 8.8 shows a 145 kV composite insulator during a type test. Power arc protective fittings may be compulsory, depending on the specification. If there is no such requirement, corona rings are usually dispensed with at these voltage levels. In this case, it was established that, as a result

Table 8.1 Results of corona studies performed on a network [1]

Application	Service voltage				
	115 kV		138 kV		
Single suspension set	x	x	x	x	0
Single tension set	x	x	x	x	x
Insulated cross-arm	–	x	–	–	–

Explanatory note
x corona discharges detected
0 no corona discharges detected
– no measurement taken

of a fault that occurred during the manufacture of the composite insulator, the sealing knife edge of the injection mould caused sharp-edged damage to the galvanised layer of the metal fitting on the insulator, and the demand for corona freedom at 1.15 times the value of the phase to earth voltage was not fulfilled.

While, in this particular example, it was still possible to eliminate the cause of the fault prior to delivery and installation, it is often costly to identify such faulty designs when the insulator is in service. Furthermore, this process may require special diagnostic and replacement programmes. In light of failures involving certain insulator designs in some applications (dead end towers, in particular), the Electric Power Research Institute (EPRI) in the United States launched a programme that used a day-time UV detection camera to examine various string installations in five EVUs [1]. The results summarised in Table 8.1 were unexpected because the general rule is that all supplied or similarly constructed insulators or insulator strings are type-tested and, if important changes are necessary (e.g. to optimise costs), a further type test is performed to verify their suitability.

During the evaluation, the insulators and/or insulator strings involved in corona activity were also visually inspected, which revealed that, after 6–7 years of a continuous (dry) corona, one particular design was so badly damaged that it was necessary to define its end of service life. This faulty design was replaced with a further development, which also displayed evidence of damage after 3 years in service. The failure mode of the composite insulators in question falls into the category of brittle fracture (see also sections "Choosing the Glass Fibre and Resin System" and 10.3.1) because the permanent corona or the corresponding secondary reactions caused embrittlement/alligatoring of the insulator housing and damage to the sealing system (Fig. 8.9). The rod was therefore effectively exposed to chemically aggressive secondary products. In almost all cases, the first shed, facing the point on the fitting that produced the corona, was also so alligatored that it caused cracks to form upon bending (Fig. 8.10). This situation, which is visually easy to diagnose, is not functionally relevant, but can be used as a basis for making decisions in relation to replacements. This is all the more true if the tower is inspected and it is not practical to use a dye penetration test to assess impermeability or the formation of cracks as a result of alligatoring.

Fig. 8.9 Embrittlement (alligatoring) of housing material [1]

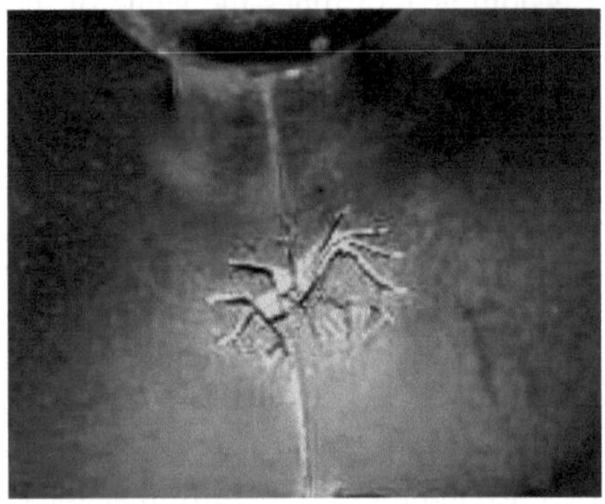

Fig. 8.10 "Litmus test" on a bended shed to identify embrittlement [1]

Based on these experiences, the following rules, which have a high degree of general validity and were also used in the examples cited at the beginning of this chapter, were defined for an essential replacement or retrofit.

A replacement is performed if there are:

- Visible cracks in the insulator housing
- Traces of rust (especially in the seal area)
- Cracks in the first shed (after bending)
- Other types of damage (e.g. to the seal).

A corona ring retrofit is performed if there is:

- A loss of hydrophobicity, visually evaluated using the spraying method
- Increased surface roughness or a change in colour
- Incipient damage to the sealing system (if an experienced inspector deems that a replacement is not necessary).

When such damage occurs, it is recommended to increase the frequency with which the insulators of affected manufacturers are inspected.

Table 8.2 Maximum electrical field stress in kV/cm on the shank, simulated for the string/set designs of various manufacturers [2]

Manufacturer	A		B	C	D		E
Corona ring	–	x	x	x	–	x	x
Service voltage (kV)							
115	3.7–6.3				5.3–5.8		
138	3.6–8.5						
161					3.0–6.8	1.0–5.0	1.9–6.8
235	5.7	2.0–6.6			8.0	2.0–3.8	
345		6.0	3.0–4.5	3.8		3.5	
500		4.2	3.6–7.0	4.3		3.3–4.7	5.4
765		4.0	3.7			8.3	

8.3 Analysis of the Electrical Field of Composite Insulator Sets Being Installed in Service

Given the analysis of such service results and the fact that the relevant standards did not include a definition for permanently permissible electrical field stresses on the surface of polymeric insulation systems as a supplement to the proven threshold values for conventional insulation systems, there is great variation among the electrical field stresses while the insulators are in service.

One such example is a study conducted by the EPRI, the results of which were published in 2008 (Table 8.2 [2]). Various insulator designs, supplied by five manufacturers, were modelled and the highest field stress on the shank was calculated (generally up to a distance of 100 mm from the high voltage fitting). The calculation was performed not only for suspension and tension configurations but also for insulated cross-arms. The two simulation processes, namely finite elements and boundary elements, yielded similar results. These time-consuming simulations were simplified by modelling only one phase, without taking account of adjacent phases.

The results show that manufacturers apply different design rules. In the examples analysed, only the designs of manufacturer C produce field stresses of ≤4.3 kV/cm. In the case of the other manufacturers, the differences between one voltage level and another do not seem to follow any rule. In an unfavourable scenario, a higher field stress can occur at 138 kV than at 765 kV (see also the corona discharge shown in Fig. 8.8), most likely because of cost pressures.

8.4 Current Standardisation

Proof of compliance with the threshold values for insulators and insulator sets is furnished by means of radio interference voltage, corona inception voltage or corona extinction voltage measurements, while taking account of the relevant

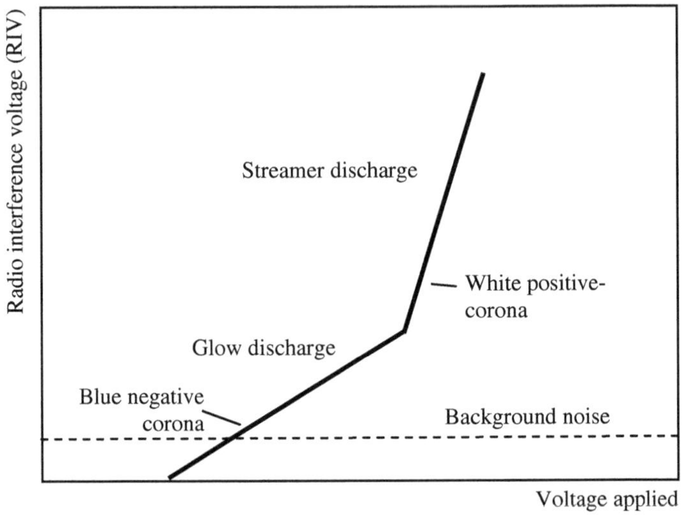

Fig. 8.11 Basic progression of the radio interference voltage [3]

standards (IEC 61284 [3], IEEE 539-2005 [4], IEEE 1829 [5], IEC 60437 [6]). The main link between the corona and radio interference voltage intensity is shown in Fig. 8.11 [3]. It demonstrates, in simplified form, the typical progression of a measurement when the voltage increases. Conveniently, the rise in voltage occurs in steps that correspond to 10 % of the voltage to be verified [6]. A corona discharge can be positive or negative; both types may occur with AC-stressing. In an arrangement, the corona will always occur at the electrode with the higher curvature. The physical properties of a positive or negative corona are different, which is due to the differences in mass between electrons and ions, and which causes a negative corona to occur at lower voltages. Electrons ionise more atoms by means of impact ionisation, and the requisite electron work flow from the negative electrode as a result of field emissions is much lower than it is for (positive) ions. Thus, it is possible to identify the pulse-shaped Trichel discharges generated at the negative peak electrode at first followed by positive corona at higher voltage values.

With regard to the threshold values for radio interference voltages, the standard BS 137 Part 2 [7] was used for many years as an important reference not only for conventional insulators and insulator strings but also composite insulators and composite insulator strings. A selection of values is shown in Table 8.3.

This standard was withdrawn in 1997 and replaced with the standard BS EN 60305. Since then, these threshold values are no longer defined in a standard but in locally applicable laws or by stipulating special requirements. In those countries that do not have any legal regulations governing the erection and operation of overhead transmission lines, the values shown in the table continued to be used as reference values.

The negative/positive example of a composite insulator string for 245 kV, which was designed with a shortened overall length as a result of the ground

Table 8.3 Threshold values for interference voltages (applied for many years) [7]

Nominal voltage (kV)	Test voltage level (kV)	RIV-level (dB over 1 μV)[a]
72.5	53	24
	44	32
145	110	30
	88	34
245	150	40
420	320	40
	260	46

Explanatory note
[a] Over 300 Ω at 1 MHz, (shaded) Increased demands

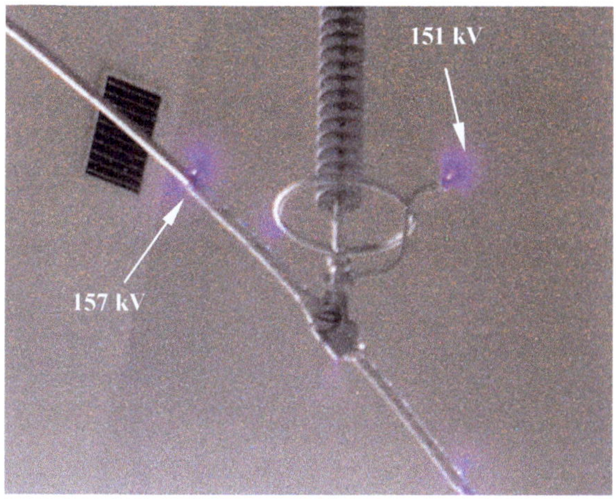

Fig. 8.12 Corona discharges (negative example)

clearances required for the conductor, shows what information can be recorded with just one measurement. Due to a lack of specifications, the recommendations contained in the standard BS 137 were applied. In this project, the composite insulators and hardware components were procured separately on behalf of the construction firm. The type test revealed that the string components were inadequately dimensioned for a service voltage of 245 kV and a single conductor configuration (Figs. 8.12, 8.13). The sphere on the arcing horn is too small to compensate for possible variations in the galvanisation quality and greater roughness as a result of corrosion. The arm rod was also incorrect; single conductors require special rounded end finishes (parrot bill or ball-ended) as of a certain voltage level. Both cases of under sizing resulted in corona inception voltages of 151 or 157 kV. (The corona inception voltage was used here to determine the sequence of the occurrence).

The progression of the radio interference voltage shows that there is a steady increase as of approx. 70 kV and that the threshold value of 40 dB/1 μV is exceeded at approx. 85 kV.

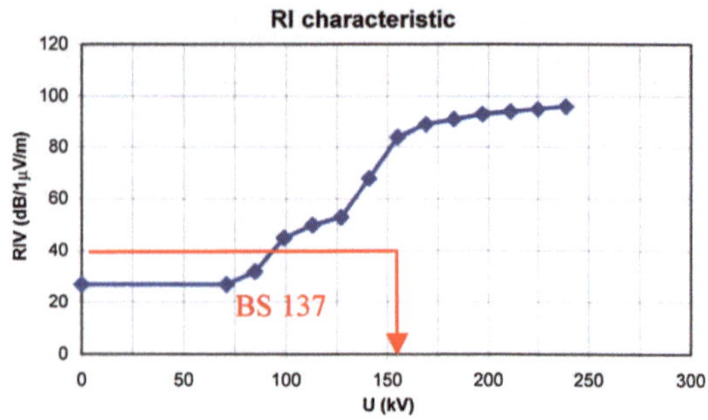

Fig. 8.13 RIV-progression of the negative example

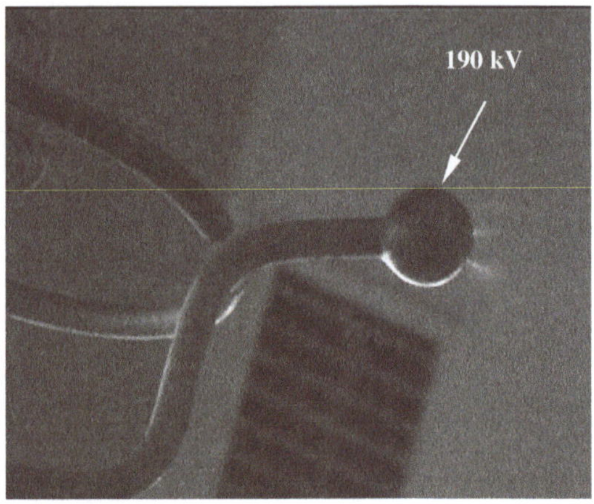

Fig. 8.14 Corona discharge (positive example)

The same set-up of the insulator set, but with an adjusted arm rod and a modified power arc protective fitting on the high voltage side, produces significantly higher values of the inception/extinction voltage (Figs. 8.14, 8.15). The corona extinction voltage is approx. 181 kV; only the exposed sphere is sparkling. The progression of the interference voltage "exceeds" the stipulation in BS 137 by 20 %. This example shows that conducting a type test while taking account of the current installation situation can help to detect particular shortcomings in the string design (continuous corona).

8.5 Water Droplet Corona on Hydrophobic Housing Materials

As already shown, corona discharges can occur at various positions on an insulator or an insulator set (Fig. 8.16). While it is possible to measure the occurrence of corona discharges at hardware components and insulator end fittings under dry conditions in the laboratory, the phenomenon of water droplet coronas occurs at considerably lower field stresses and must be taken into account when designing

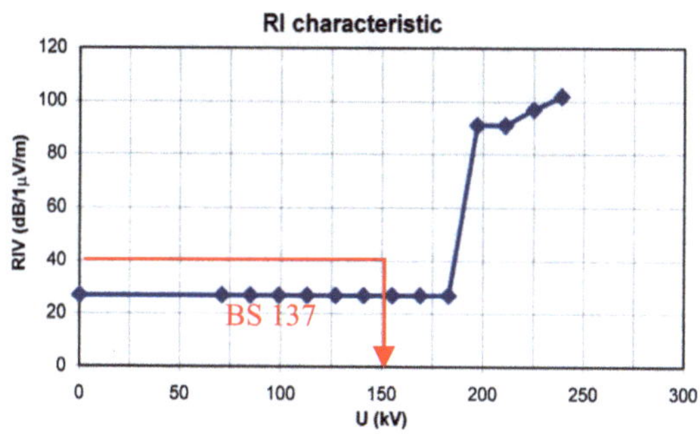

Fig. 8.15 RIV progression of the positive example

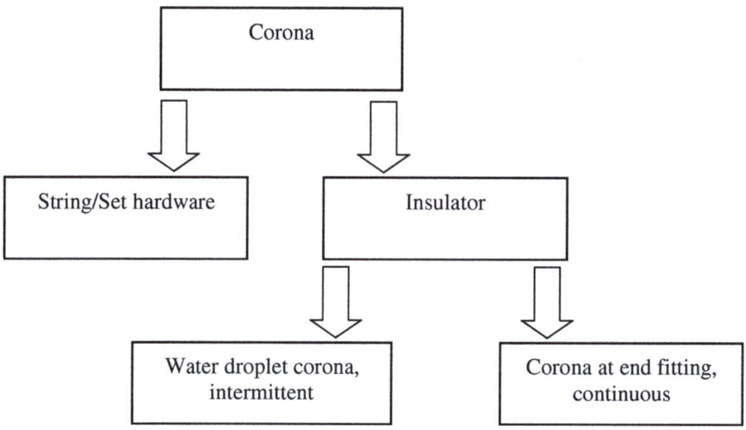

Fig. 8.16 Positions where corona can occur

an insulator or string. From a danger perspective, a continuous corona at the insulator end fitting must be regarded as the most critical case because it has a lasting effect, not only during periods of humidification, and may cause material damage (see also Figs. 8.6, 8.9).

8.5.1 Formation of Water Droplet Corona

A water droplet corona occurs when there is an increase in the field stress at the poles of a water droplet on the insulator surface if the electrical strength of the air is sufficiently exceeded (Fig. 8.17). The field stress increases because the field lines are concentrated at the poles of the medium with high relative permittivity ε_r.

Many studies have been conducted on the behaviour of free water droplets and the results were later applied to the behaviour of water droplet on an insulating material surface. The critical field stress is calculated as follows:

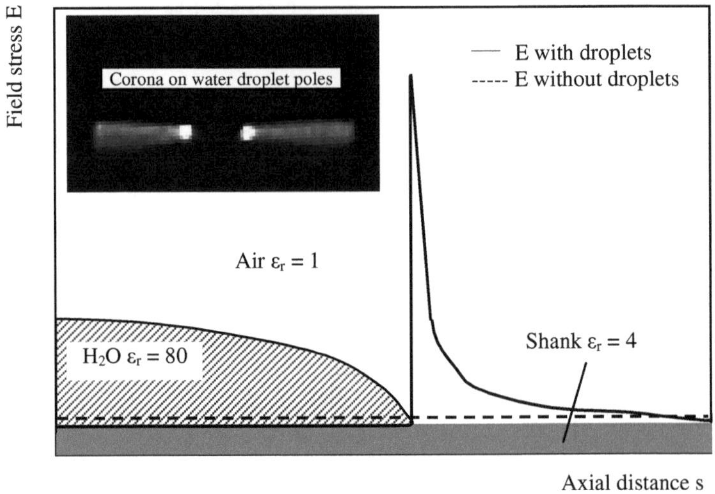

Fig. 8.17 Qualitative diagram of increased field stress at the pole of a water droplet on a cylindrical insulating material surface

$$E_{\mathrm{cr}} = C \times \sqrt{(\gamma \div (\varepsilon_0 \times r))} \tag{8.1}$$

where γ = the surface tension of the liquid, ε_0 = the dielectric constant, r = the radius of the droplet and C = the empirical constant (0.452–0.462). This equation shows that small droplets or a high surface tension result in higher critical values of the field stress. The droplet only becomes unstable and forms new shapes when the critical field stress is exceeded (peaks form at the droplet poles, water threads grow out from these peaks, and smaller droplets form from these water threads; flashover can be triggered see also Fig. 8.18) [8–12]. Depending on the field stress, field distribution, surface roughness, hydrophobicity, etc., these hydrostatic and hydrodynamic processes can be present individually or simultaneously. The air pressure also influences the inception of corona discharges at droplets. In [8], the correlation shown in Fig. 8.19 is reproduced for a peak-plate arrangement with a free water droplet at the peak. In [13], 47 kPa is specified as the threshold value for negative polarity and 60 kPa is specified as the threshold value for positive polarity, which corresponds to altitudes of between 4.5 and 7 km above sea level. In [14], the dependency of corona inception on the air pressure at vertical droplets on conductors is examined both analytically and experimentally. This produces a threshold value of 66 kPa. All of these values indicate that, for the majority of composite insulators used, droplet instability and the aforementioned formation of new shapes precede the ignition of a water droplet corona.

If one adopts the equivalence approach [7, 8, 15], whereby, during electrical stressing, a sessile droplet behaves like a free droplet, the critical field stress can be calculated as a model, depending on the contact angle and the various droplet radii (Fig. 8.20).

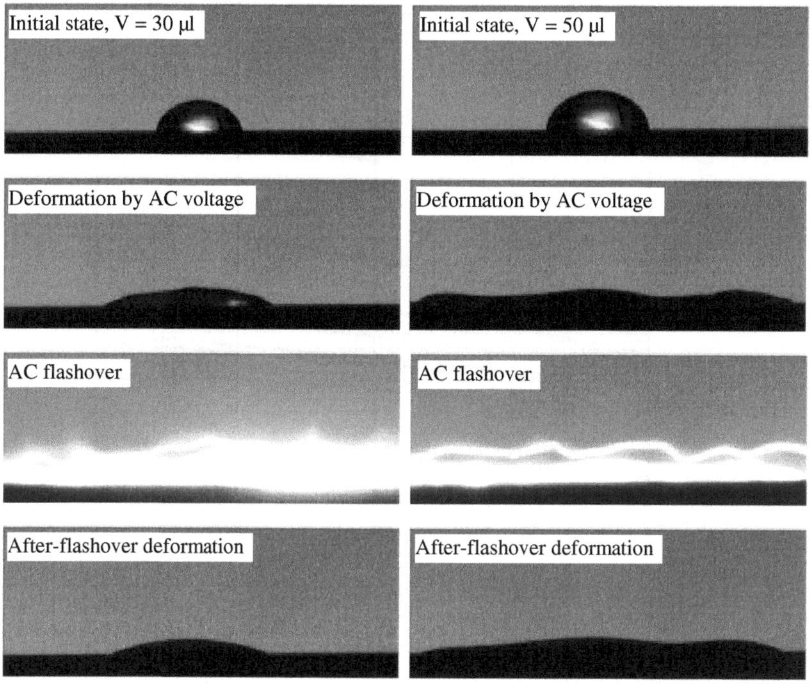

Fig. 8.18 Water droplet deformation due to voltage stress and after flashover [10]

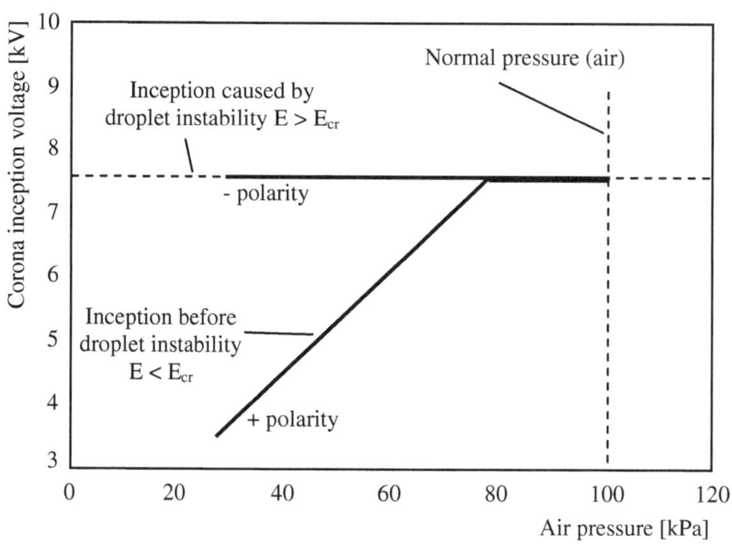

Fig. 8.19 Primary process for a corona inception voltage dependent on air pressure [8]

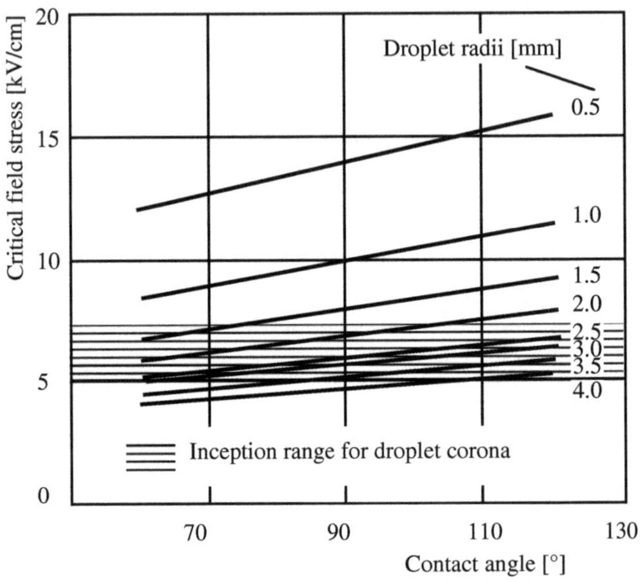

Fig. 8.20 Critical field stress for inception of hydrodynamic instability [15]

$$E_{Cr} = C \times \sqrt{\left(\gamma \div \left(\varepsilon_0 \times R_{eq}\right)\right)} \tag{8.2}$$

$$\mathrm{R_{eq}} = \mathrm{R_{180}} \times \sqrt[3]{(4 \div (3 \times (1 - \cos \Theta)^2 - (1 - \cos \Theta)^3))} \tag{8.3}$$

It is clear that very high field stresses are necessary for droplet radii between 0.5 and 1.5 mm. However, tests on complete 500 kV silicone insulators in an accelerating, combined ageing test have shown that, under these test conditions [16] values of only 3.8 kV/cm are enough to trigger a water droplet corona. The test conditions were a combination of UV, salt fog, clean fog and rain exposure during specific electrical stressing (based on the creepage distance) of various insulators (0.27–0.29 kV/cm). The salt content was low (1 g/l) and the level of rain intensity was very low. In other words, there was a heavy accumulation of decomposition products caused by corona discharges. This led to a reduction in the contact angle and an increase in the droplet size. As a result of this trend, the curves move to smaller critical field stresses, as was confirmed in the test.

In [8], an experiment involving a weak non-homogeneous model arrangement with a hydrophobic insulating material surface satisfactorily verified that the inception field stress decreases when the droplet volume increases (Fig. 8.21).

When compared with the critical field stress calculated for a free droplet of the same volume, the curve slopes are the same, which indicates that the same hydrodynamic mechanisms occur. The absolute values for the insulating material

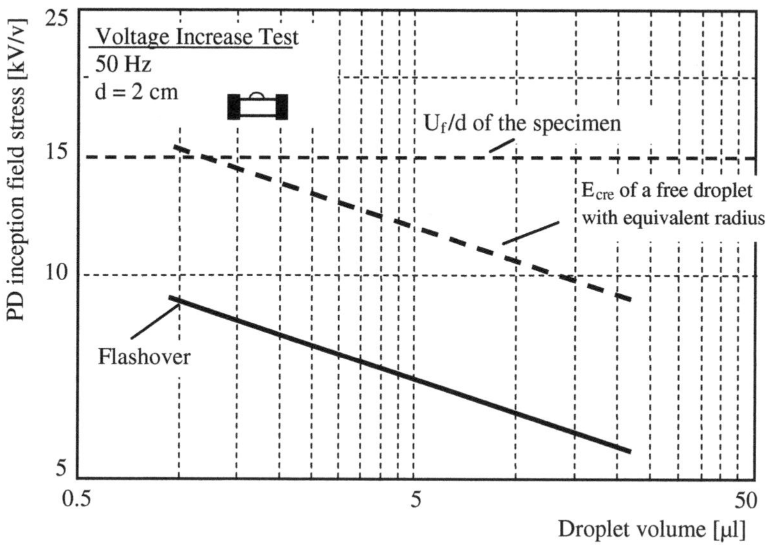

Fig. 8.21 Inception field stress depending on the droplet volume [8]

interface (contact angle = 105°) are moved to smaller values; this is due to the interaction between the insulating material surface [roughness, formation of receding contact angles (see also section "Effect on the Recovery Behaviour")—a small introduction about the contact angle), etc.] and the real shape of the droplet.

In [9], an experiment involving two different materials was also performed in a weak non-homogeneous field to verify the effect of hydrophobicity on the corona inception field stress (Fig. 8.22). Better hydrophobicity leads to smaller droplet diameters for the same droplet volume and a higher curvature strain in the droplet, which requires a higher field stress in order to achieve instability.

The values stated for field stress are to be understood as effective tangential field stresses. Higher values are required for normal field stresses. For a given testing arrangement, the following differences in values were noted: The tangential inception field stress is 5 kV/cm, while the normal inception field stress is 10 kV/cm [17].

These simplified descriptions serve for the understanding of the processes involved and consider only a quasi-static process behaviour. The interactions of dynamic processes are more complex. As example, droplets, which are down-running on an inclined surface of insulating material cause contact discharges (in the case of droplet merging) or tear-off discharges (in the case of droplet separation). This can be evaluated by means of the Dynamic Drop Test (Fig. 7.38) and reference can be made to further reading literature [18–22].

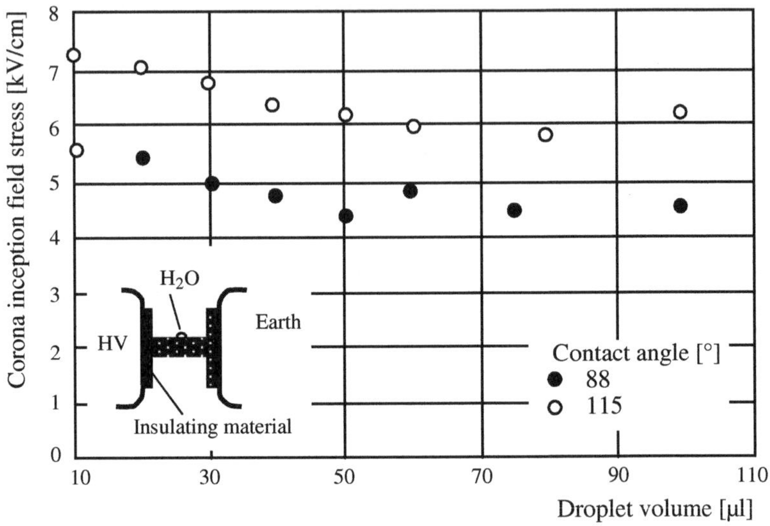

Fig. 8.22 Corona inception field intensity depending on droplet volume and hydrophobicity [9]

8.5.2 Effect of Corona on Polymeric Surface, Especially Silicone Rubber

The effects of corona discharges on polymeric insulating material surfaces are extremely complex because, under outdoor conditions, in particular, there is no inert atmosphere. Instead, additional secondary reactions occur with oxygen, nitrogen and water (Fig. 8.23). The changes to silicone rubber surfaces and the effect on hydrophobicity as a result of UV, corona discharges, immersion in water and dry zone discharges are described extensively in [23–27]. By using the example of the effect of various levels of exposure to UV radiation, the Eqs. (8.4)–(8.17) demonstrate the complex nature of potential changes to the insulating material. Excited oxygen atoms, in particular, have a strong oxidising effect and can result in major changes to the surface of the insulating material.

8.5.2.1 Change Due to UV Radiation

Radiation by means of a low-energy xenon lamp causes methyl groups to decompose due to the formation of free radicals. Hydrogen radicals are highly unstable; they separate hydrogen from the methyl groups and cause the formation of H and Si–CH_2 radicals [Eqs. (8.4), (8.5)], [28].

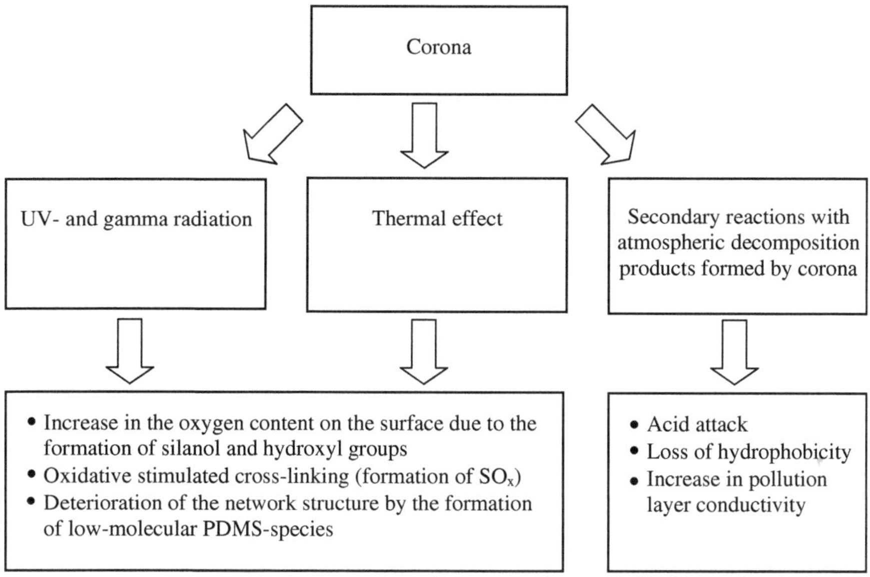

Fig. 8.23 Using the example of silicone rubber to demonstrate the effect of corona discharges

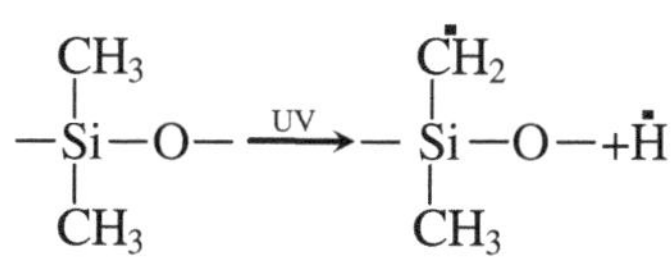
(8.4)

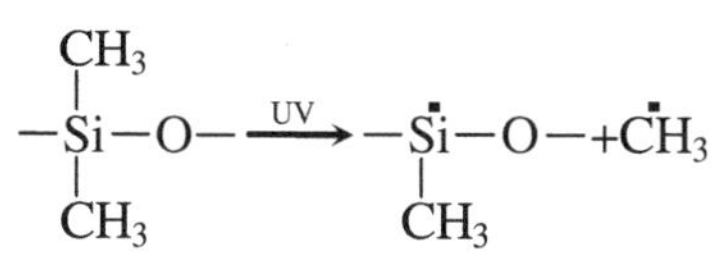
(8.5)

Methyl radicals can separate hydrogen from the main chain and release gaseous methane [Eq. (8.6)], [28].

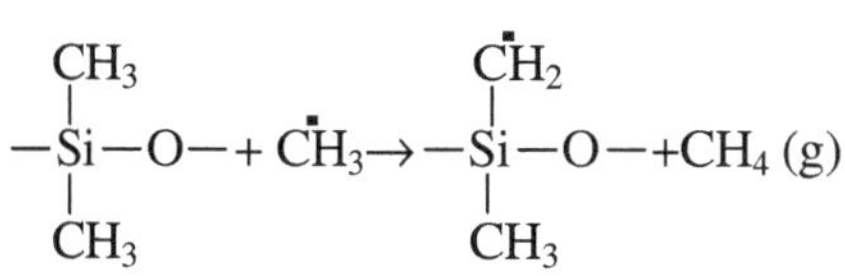
(8.6)

Silicon-containing radicals can form $SiCH_2Si$-species [Eq. (8.7)], [29].

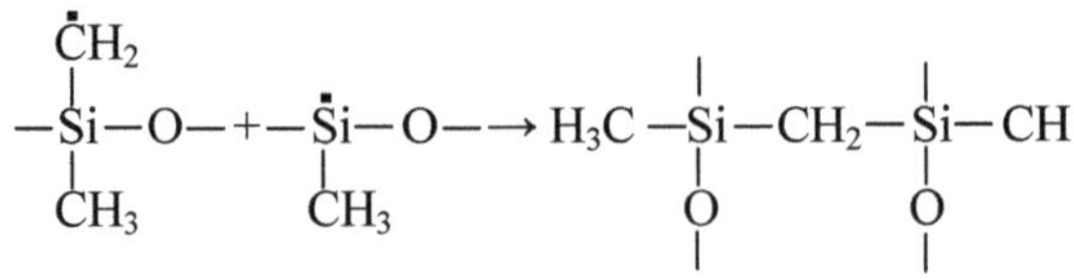

(8.7)

Radiation by means of higher-energy UV lamps (mercury vapour) leads to oxygen excitation, which forms Si–OH compounds (8.8) with active Si radicals from the converted peroxide (in the case of systems that have been vulcanised accordingly), [29].

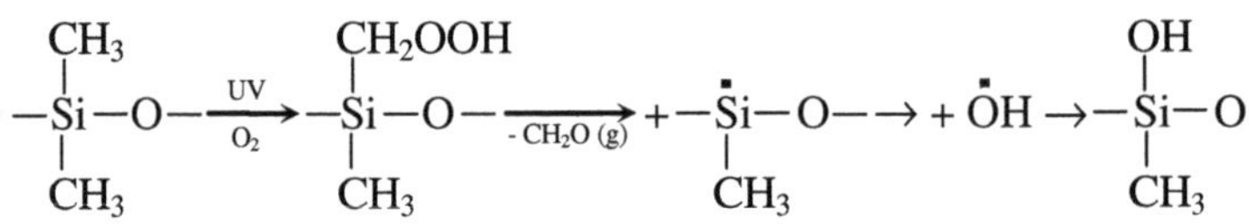

(8.8)

Due to the great potential difference between the positive Si atom and the negative oxygen atom, the formation of Si–O–Si compounds is more likely than the formation of the aforementioned Si–OH compounds [Eqs. (8.9), (8.10)], [29].

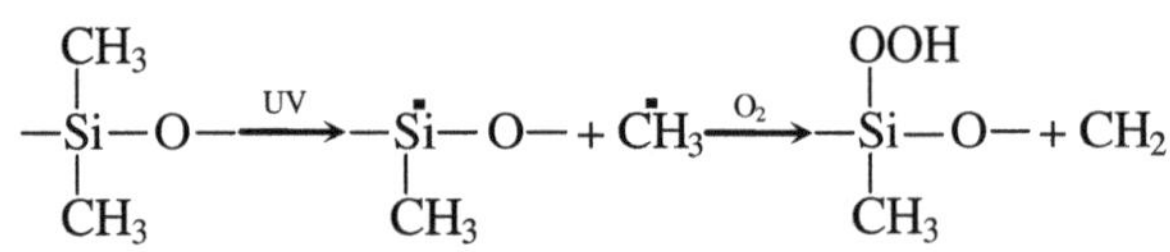

(8.9)

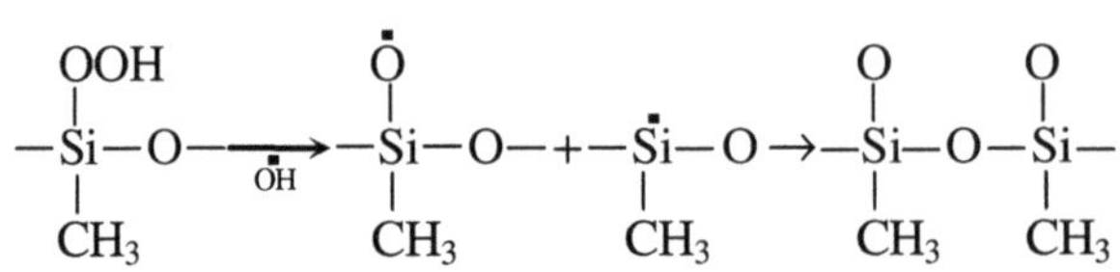

(8.10)

The formation of $SiCH_2CH_2Si$ can occur as a secondary reaction [Eq. (8.11)] corresponds to the vulcanisation process for addition vulcanisation with a platinum catalyst ($Si\text{–}H + CH_2 = CHSi \rightarrow SiCH_2CH_2Si$), [29].

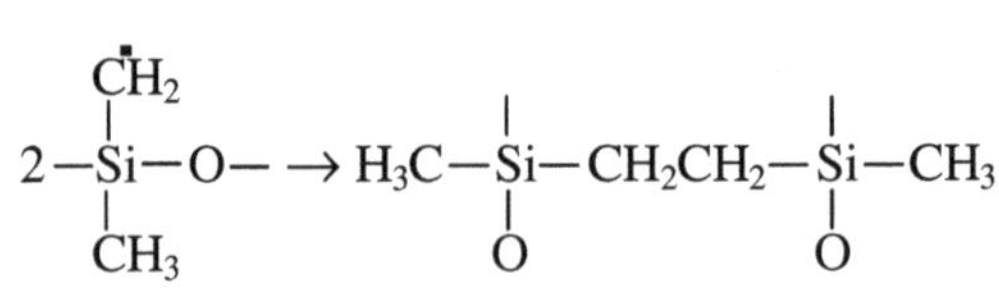

(8.11)

Very complex changes were detected during combined plasma and corona surface treatment, if oxygen was present. Unstable silyl compounds are formed in (peroxide and hydroperoxide) intermediate processes and decompose in decomposition products such as carboxylic acid and silanol groups. Si–O–Si compounds form if other condensation reactions occur [Eqs. (8.12)–(8.17)], [30].

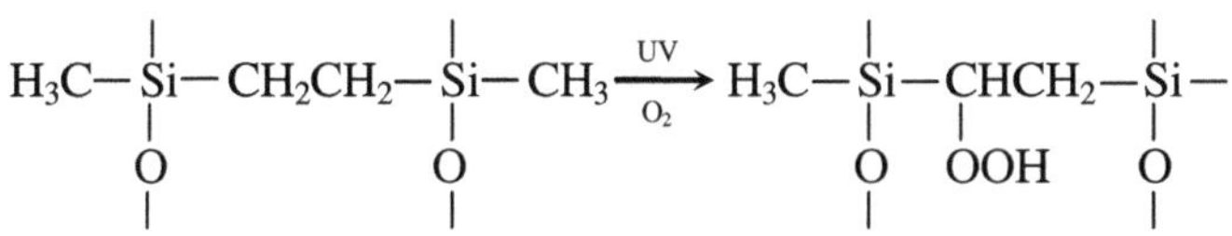

(8.12)

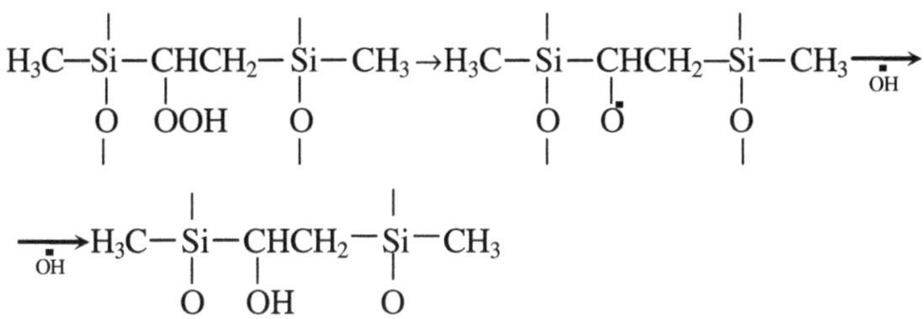

(8.13)

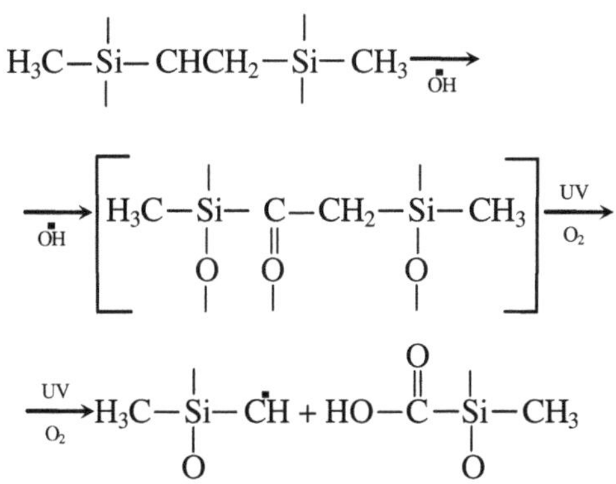

(8.14)

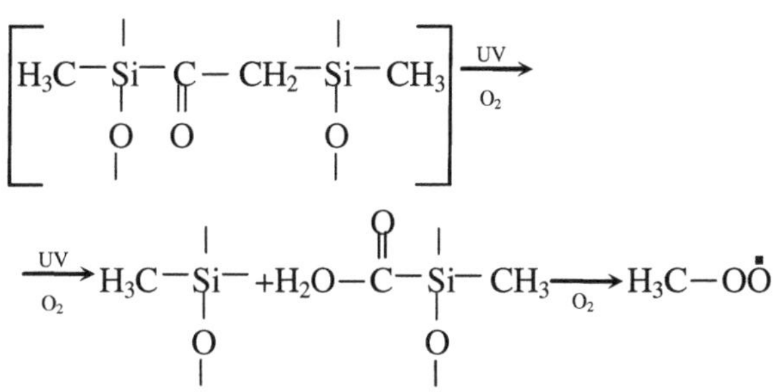

(8.15)

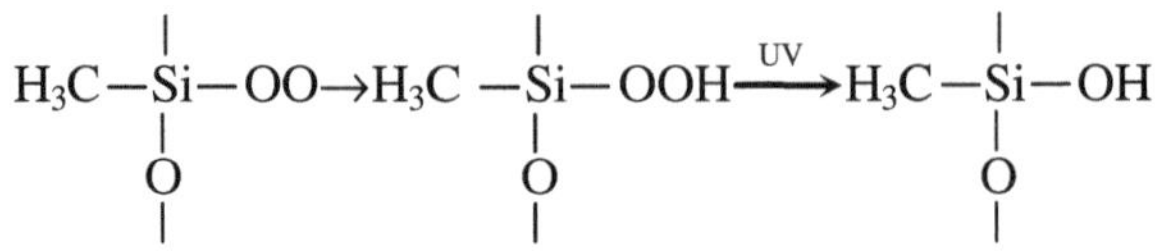

(8.16)

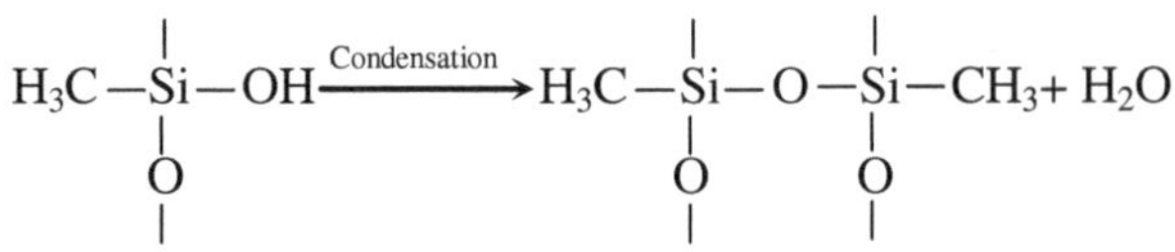

(8.17)

The equations show that a large number of decomposition products can be formed and reduce hydrophobicity as a result of their polar and/or radical nature. While the effect of the decomposition products may diminish (e.g. as a result of recombining or reorienting methyl groups, or under outdoor conditions during rainfall), the formation of SiO_x layers is an irreversible process that prevents the return of hydrophobicity (in particular, the diffusion of low molecular weight chains to the surface) if the layer thickness is adequate.

In addition to the effect of UV radiation, the thermal effect of corona discharges on various insulating material surfaces (each with >40 % filler by weight) was analysed whereby a distinction was made between the creation of corona from water droplet and corona from a metallic source [15, 31]. As a result of capacitive coupling, the currents associated with water droplet corona were typically within the 5–100 μA range. In the case of a corona that lasted for up to 60 s, only temperature rises up to 2.5 K at 5 μA and up to 25 K at 100 μA (when compared to the ambient temperature) were observed and, as a result, neither the thermal decomposition temperature of the insulating material nor the flashpoint of water was reached. Furthermore, it was established that there is a good correlation between the analytical considerations and the results of the measurements taken [15, 31].

The tests involving a metallic corona source then revealed that thermal destruction resulting from the local ignition of morning stars does not occur until currents reach the mA range. The term "morning star" refers to the shape of the intensive luminous effect that occurs if the thermal decomposition temperature of the insulating material is selectively exceeded. This is well in line with the measurements associated with a modified needle test (low current discharge IEC 61621 [32]), which ignites dry discharges on the surface of the insulating material tested and thermally stresses the material accordingly (Fig. 8.24 [33]). Various low viscosity silicone rubber grades to which no ATH was added were tested in the example; under these test conditions, the peak value of the morning star inception current was between 4 and 5 mA.

Therefore, the thermal effect of water droplet corona is considered to be low; however, an attack caused by acid formation is not. Irrespective of the insulating material, basic studies on the formation of air decomposition products that have an oxidative and hydrophilic effect or the formation of acids [34–38] have shown that

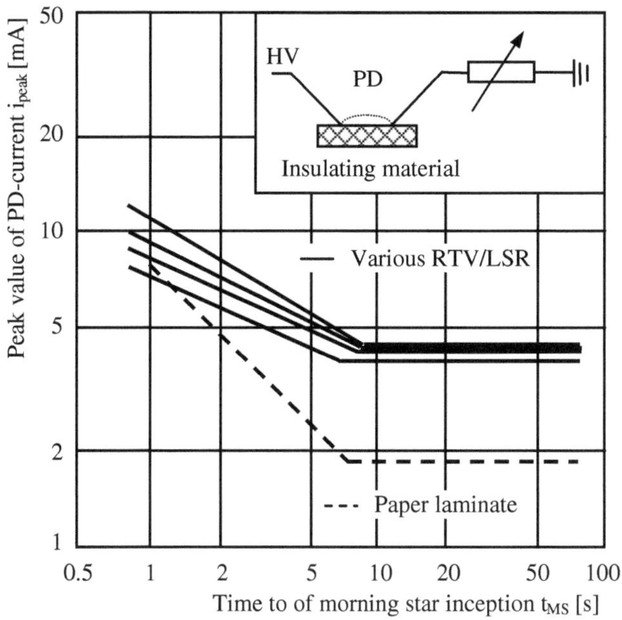

Fig. 8.24 Thermal damage to insulating material caused by morning star inception [33]

a range of chemical and physical processes that may destroy the insulating material occur.

8.5.2.2 Formation of Different Oxygen Species

Discharges with a low amperage (ozone formation)

$$e^- + O_2 \rightarrow 2\,O + e^- \tag{8.18}$$

$$2\,O + 2\,O_2 \rightarrow 2\,O_3 \tag{8.19}$$

Discharges with UV radiation (ozone decomposition and the formation of radicals)

$$e^- + O_2 \rightarrow \dot{O} + O + e^- \tag{8.20}$$

$$h/\lambda + O_3 \rightarrow \dot{O} + O_2 \tag{8.21}$$

$$\dot{O} + H_20 \rightarrow 2\,\dot{O}H \tag{8.22}$$

The OH-radicals have a high oxidation potential and can cause polymer compounds to break down. They also have a strong hydrophilic effect; a comparatively small quantity of $\dot{O}$H-groups (1.3–3.5) per nm^2 surface area increases wettability considerably.

8.5.2.3 Formation of Nitric Acid

$$N_2 + O_2 \rightarrow 2\ NO \tag{8.23}$$

$$2\ NO + O_2 \rightarrow 2\ NO_2 \tag{8.24}$$

$$2\ NO + O_3 \rightarrow 2\ NO_2 + 0.5\ O_2 \tag{8.25}$$

$$2\ NO_2 + H_2O \rightarrow HNO_3 + HNO_2 \tag{8.26}$$

$$HNO_2 + O_3 \rightarrow HNO_3 + O_2 \tag{8.27}$$

If this acid comes into contact with the rod as a result of sealing problems with the housing material or the seal itself, a brittle fracture (see also section "Choosing the Glass Fibre and Resin System") may occur due to an unfavourable choice of material (E-Glass). Depending on the resin matrix, pH values between 3 and 3.5 are sufficient here. In comparison, a pH value of as low as 0 was measured in ageing tests [39]!

A differentiated approach is required for the pH value of acid rain. pH values may be low at the beginning of low intensity rain but increase due to dilution if the rainfall occurs over a longer period of time.

Formed acids can also indicate a direct attack on the insulating material. This applies to silicone rubber with a low filler content [40], as well as materials filled with ATH or fillers that have an alkaline effect (e.g. $CaCO_3$). The use of $CaCO_3$ with or without the addition of ATH results in great sensitivity to acids [41, 42], while the use of untreated ATH must also be regarded as critical [43].

If one assumes a linear correlation, [15] proves that, when the pH value for a particular testing arrangement is reduced from 7 to 2.5–3, only total charges of 600–1,000 pC are required by the corona for a droplet size of 100 μl or, in the case of smaller droplets, only a brief corona exposure time of a few seconds is required. This scenario applies to dew (i.e. the absence of rain leads to a dilution of the acids formed).

8.5.2.4 Effects of Hydrophobicity Reduction by Corona

As the examples from service have shown, both the exposure time and the intensity of corona play an important role in terms of the damage that they cause. In [44], various silicone rubber materials of different thicknesses were exposed to corona discharges generated by means of ten 10 mm long wolfram needle electrodes. The exposure time was 12 min (Fig. 8.25). The receding contact angle measured (for an explanation, see section "Effect on the Recovery Behaviour") reacts more sensitively to corona treatment and indicates that a film layer is formed

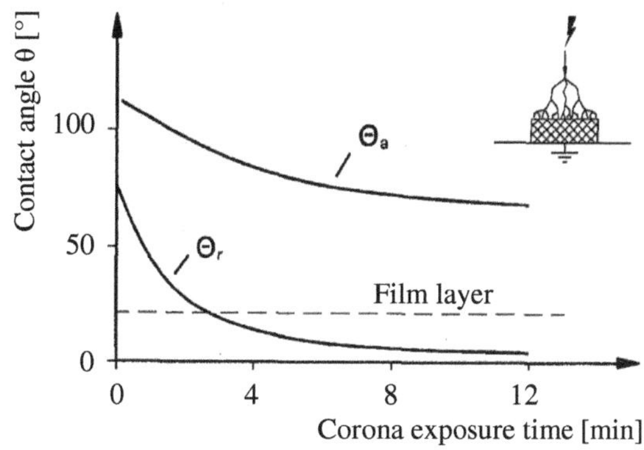

Fig. 8.25 Reduction in the dynamic contact angle in proportion to the corona exposure time [44]

after approx. 3 min. The material's recovery takes longer than its direct exposure time to the corona. This state is achieved after approx. 60 h (Fig. 8.26). The effect of the thickness of the test specimens was not determined in this series of tests. Both conclude that only minimal damage occurred (e.g. temporary enrichment with silanol and hydroxyl groups).

Another picture emerges if the intensity and associated exposure time exceed a material-specific value for a particular testing arrangement [45]. Test specimens made of HTV silicone rubber, with a diameter of 30 mm and a thickness of 1 mm, were exposed to corona discharges in a test chamber at room temperature with a defined relative humidity. The corona discharges were created using needle electrodes positioned 1 mm from the surface of the insulating material. The correlation between exposure time and intensity is shown in Table 8.4.

The static contact angle results shown below are mean values from each of the three individual measurements (Figs. 8.27, 8.28, 8.29, 8.30) with a droplet volume of 10 μl. As expected, a clear trend of cumulative damage emerges. While the initial state of hydrophobicity is achieved again for almost every intensity after an exposure time of 1 h, changes to the insulating material occur with E24, which result in only a low recovery. When both figures are compared (Fig. 8.28, 8.29), it is evident that corona discharges can have an accelerating or decelerating effect on the recovery of hydrophobicity. The formation of low molecular diffusion-mobile compounds competes with oxidative vulcanisation, which can produce "glass-like" layers that are difficult to penetrate. The extent to which the low sample size of 3 measurements influences the curve progressions is not analysed further. Once again, the trend in Fig. 8.30 is clear; the longer the exposure time and the greater the intensity, the longer the rise time.

The initial state is achieved again for A_{200} only, which represents a threshold value under these conditions. The relatively low thickness of the test specimens also influenced these results.

A further example from earlier analyses [46] can demonstrate the increase in layer conductivity (Fig. 8.31). The test set-up can be applied to situations where small drops form when dew occurs, the natural level of pollution is low and there is no cleaning by rain. This situation is not solely limited to indoor applications; it

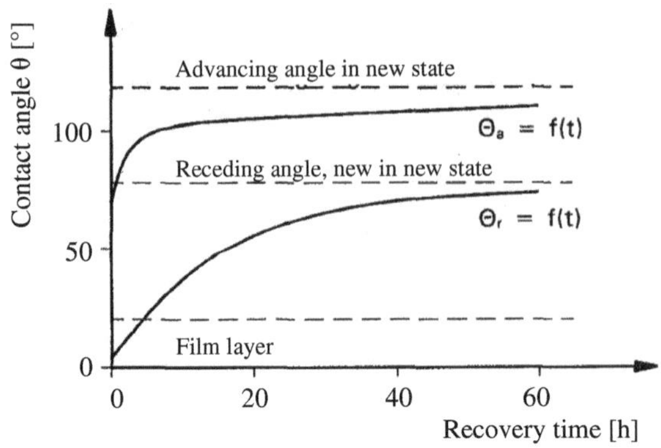

Fig. 8.26 Recovery of the dynamic contact angle after the removal of corona source [44]

Table 8.4 Test specimen code depending on the exposure time and stressing voltage [45]

Corona exposure time (h)	Voltage stress (kV)				
	$U_1 = 2$	$U_2 = 3$	$U_3 = 5$	$U_4 = 7$	$U_5 = 10$
1	A_1	B_1	C_1	D_1	E_1
24	A_{24}	B_{24}	C_{24}	D_{24}	E_{24}
200	A_{200}	B_{200}	C_{200}	D_{200}	E_{200}

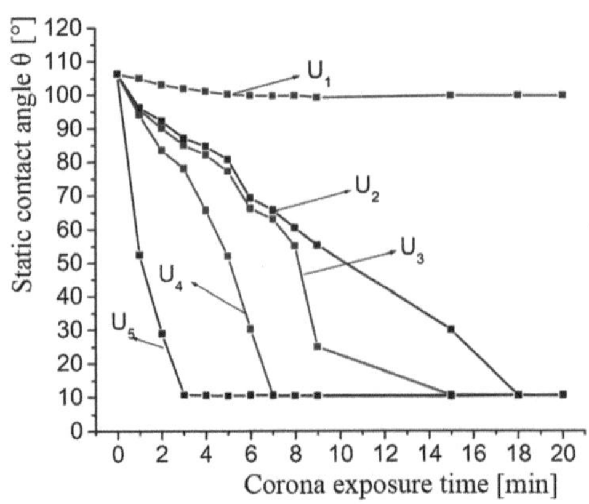

Fig. 8.27 Effect of intensity on a loss of hydrophobicity [45]

can also arise under outdoor conditions (see also the damage pattern in Fig. 8.9). Strong increases can be verified depending on the charge measured and the choice of insulating material. An absorbent strip probe was used to take the measurement. In other words, the effect of a loss of hydrophobicity by a hydrophilic species must be regarded as a further negative effect that has not been quantified in this test.

The effect of a strained insulating material is yet to be considered [31]. If a corona causes damage in the form of cracks on the surface of the insulating material, the mechanical stress in a strained material can intensify the formation of cracks, as in the case of a notch effect.

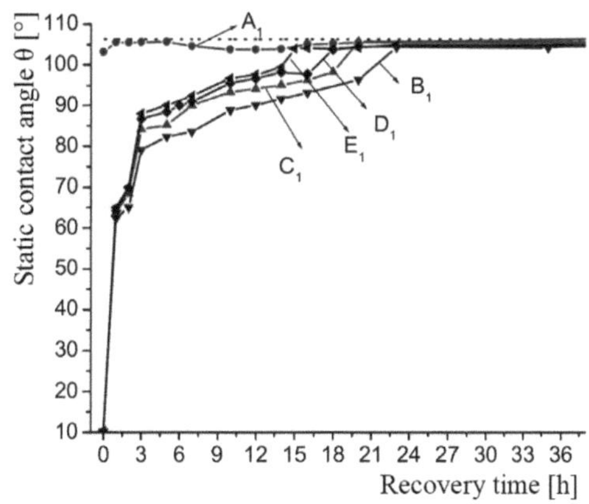

Fig. 8.28 Increase in hydrophobicity after an exposure time of 1 h [45]

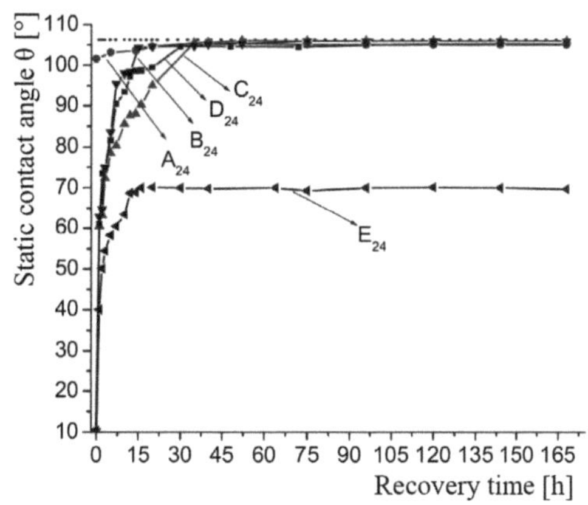

Fig. 8.29 Increase in hydrophobicity after an exposure time of 24 h [45]

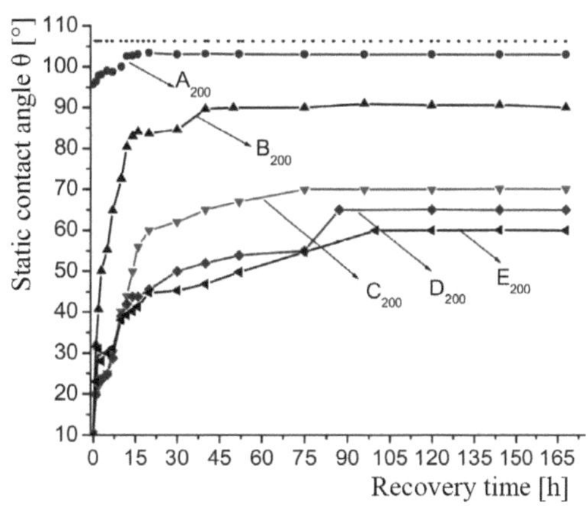

Fig. 8.30 Increase in hydrophobicity after an exposure time of 200 h [45]

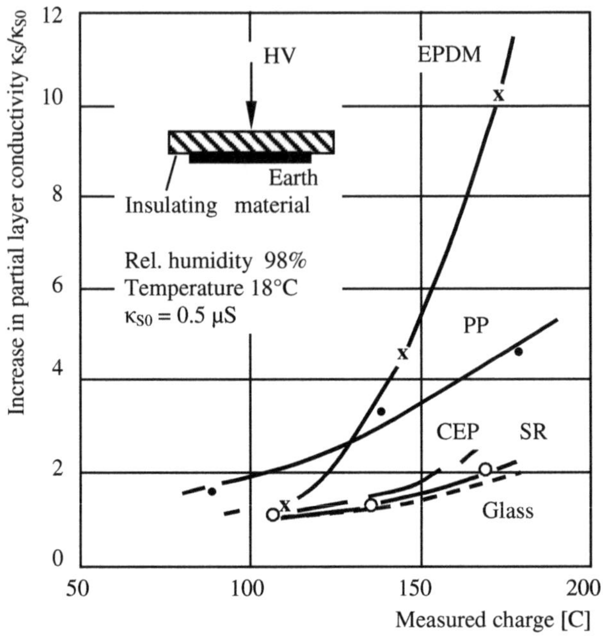

Fig. 8.31 Increase in partial layer conductivity as a result of decomposition products caused by corona [46]

8.5.3 Corona: Potential for Damage

The potential risk can be summarised and evaluated as follows:

- If a continuous corona occurs at the insulator fitting or at string elements close to the insulator, it can cause permanent damage to the insulator housing and seal.
- Major damage mechanisms include changes to the surface of the insulating material as a result of UV radiation and a chemical attack resulting from the formation of radicals or acid.
- A reduction in hydrophobicity as a result of water droplet corona can also occur while complying with the empirical threshold values for field stress and the high-grade hydrophobic properties of the insulator housing. Based on outdoor experiences, this reduction is deemed to be temporary and therefore not functionally relevant.

A general rule when using composite insulators is to check which additional design criteria can be defined to ensure an optimal set-up for an insulator or insulator set for the in situ conditions.

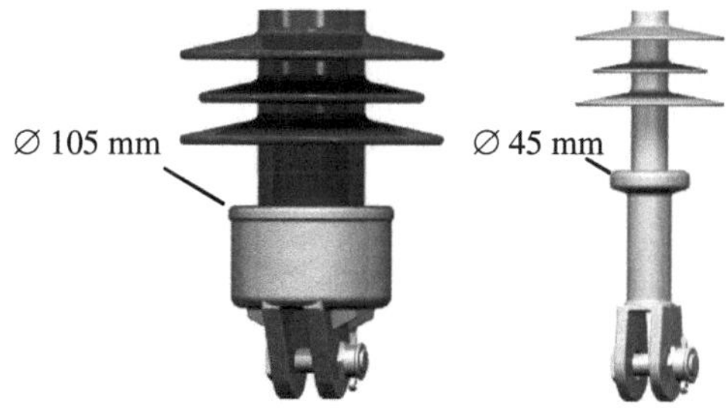

Fig. 8.32 Scale comparison between a porcelain long rod and a composite insulator

8.6 Additional Requirements for Composite Insulators or Composite Insulator Sets

8.6.1 Material and Geometry Differences

The aforementioned threshold values for the interference voltage and the corona extinction voltage (Sect. 8.4), which have proven their worth for many years in relation to largely inert insulating materials against environmental factors and in the case of the insulator end fitting dimensions required for a particular strength class, require further clarification and expansion for composite insulators. One reason for this is the different shank diameter, which is mentioned in the context of a long rod in Fig. 8.32. The porcelain long rod for a specified mechanical load of 120 kN has a shank diameter of approx. 65 mm, while the composite insulator has a rod diameter of only 16 mm, which corresponds to a shank diameter of approx. 23 mm. Even with a specified mechanical load of 1,500 kN, the composite long rod insulator would have a shank diameter of only 70 mm (In comparison, porcelain long rods can be manufactured up to a specified mechanical load of 500 kN). The smaller rod diameter in the composite insulator also results in longer insulator end fittings because the required friction area must be configured by crimping the fitting that has a small diameter in proportion to its length.

The main differences between a conventional insulator set and a composite insulator set can be contrasted as follows from the perspective of the effect on pollution behaviour and corona effects (Table 8.5).

A comparison of conservative insulator designs from the past with the lower-cost options available today also reveals a trend of using less material in an effort to reduce costs. The images shown in Fig. 8.33 are of various composite long rod insulator designs for a specified mechanical load (SML) of 160 kN.

The following are examples of some key reasons for the occurrence of an intermittent water droplet corona or continuous corona at the fitting:

- A reduction in the rim diameter at the fitting
- A lower field-homogenising effect of the conductor bundle in the case of an extended fitting, especially if there is no additional corona protection
- No type tests performed for the entire string; design errors not detected

Table 8.5 Comparison of technologies in terms of pollution behaviour and corona effects

Characteristics of a composite insulator set compared to an equivalent conventional insulator set	Effect on	
	Pollution performance	Corona
Smaller diameter of shank/housing	Improved	Negligible
Smaller diameter of end fitting	Negligible	The rim diameter of the insulator end fitting is to design carefully and corona rings must be installed at 200 kV onwards or even at lower voltages in the case of critical designs or environmental conditions
Longer distance between conductor bundle and end fitting rim	Minor effect	The entire insulator set must be coordinated and type-tested (medium to major effect in the case of extended fittings with no corona protection)
Hydrophobicity	Enhanced	Phenomenon of water droplet corona

A conservative design with a large rod diameter and rim diameter at the fitting

A reduced design with a smaller rod diameter

A cost-optimised design with a smaller rod diameter and rim diameter at the fitting

Fig. 8.33 Scale comparison of insulator design developments

- No overall responsibility because various suppliers are involved in the manufacture of the insulator and the hardware components of the insulator set; corona protection not adjusted to the components
- No correction for installations over 1,000 m in altitude
- A reduced fault tolerance for the insulator fitting in relation to production-related variations in surface roughness in the galvanisation (zinc layer) process.

The smaller diameters and longer insulator fittings (for technical reasons) compared to conventional insulators, and the optimisations performed today due to cost restraints, require a critical evaluation of the corona protection design for composite insulators. Previously, the following simplified rules from 1992 [47] were applied:

- Corona rings must be used on the high voltage side as of 220 kV; they must also be used at lower voltages if the shank diameter of the composite insulator falls below 20 mm.
- If uncertainty surrounds the behaviour (today's simulation processes and computing capacity were not available in 1992), corona tests that simulate the insulator arrangement (suspension configuration, tension configuration or insulated cross-arm) in accordance with the on-site conditions must be performed.

These first rules apply only to the radio interference voltage and corona extinction voltage; the phenomenon of water droplet coronas is not considered by.

8.6.2 Calculating the Electrical Field Stress

In order to prevent corona damage such as that described in the example in Sect. 8.2, including the consideration of the phenomenon of water droplet corona, it is necessary to comply with certain threshold values relating to the electrical field stress at the hardware components of the insulator set and on the surface of the insulating material. These threshold values must be defined in such a way that a balance is found between the commercial, increasingly aesthetic and technical requirements of the overhead transmission line for high-grade insulating materials such as silicone rubber.

In 2005, CIGRE Working Group B2.03 [48] published the Technical Brochure TB 284, which addressed the topic of using corona rings in transmission line composite insulators. Irrespective of the use of simulation software and the set-up of the insulator set (single string, double string, insulated cross-arm) or set position (suspension or tension configuration), the following constructive details are important factors for the electrical field stress (Fig. 8.34):

- Diameter and shape of the insulator end fitting
- Shed profiles
- Distance between the first shed and the insulator fitting
- Ring diameter and profile diameter of the protection ring
- Position of the ring.

The shank diameter of the insulator is indirectly taken into consideration through the diameters D_F and D_S.

Two main insulator designs were evaluated, one with and one without an overmoulded (subsequently sealed) fitting embedded into the housing material (Figs. 8.35, 8.36). The field simulation was performed using the 3D software Coulomb V6.1, the results of which were then compared against the practical measurements taken in a laboratory. Typical 145 kV insulators were used as test

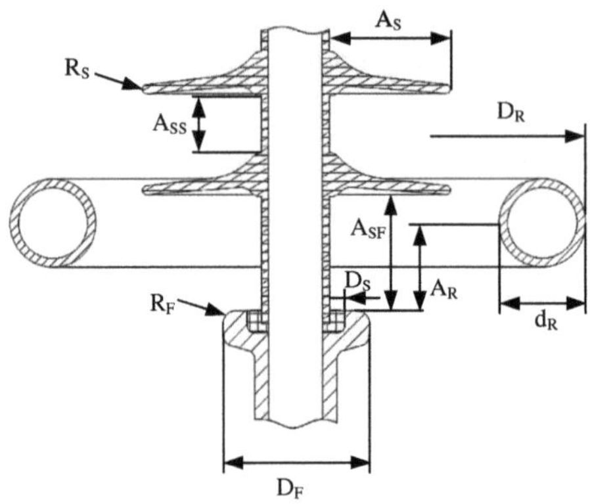

A_R = Distance between ring and end fitting
A_S = Shed overhang
A_{SF} = Distance between 1st shed and end fitting
A_{SS} = Distance between shed and shed
D_F = Diameter of end fitting
D_R = Outer diameter of ring
D_S = Diameter of seal
d_R = Diameter of ring profile
R_F = Radius of end fitting rim
R_S = Radius of shed edge

Fig. 8.34 Geometrical parameters for electrical field simulation

Fig. 8.35 Profile 1, overmoulded end fitting

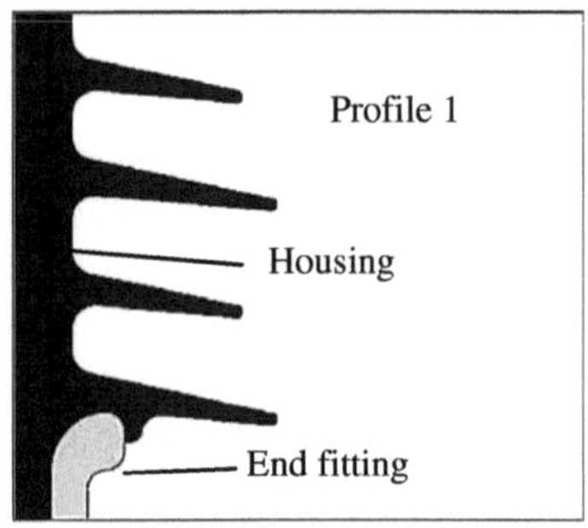

specimens (profile 1 with an insulator length of 1,384 mm and profile 2 with an insulator length of 1,270 mm). For the field simulation, it was assumed that a transgression of 22 kV/cm (RMS-value) at the insulator fitting on the high voltage side corresponds to the corona threshold value. There was a good correlation between the simulated and measured values (Table 8.6, RMS-values); the variances lie between −2 and +7 %. Under these conditions, the design of the insulator fitting and how it is embedded into the housing material result in better corona behaviour by profile 1.

For 420 kV insulators, the simulations continued to be performed using axis-symmetric 2D software (Electro V5.1) that standardises the simulated set-up with the results of the previous 3D simulation. Both profile types were calculated using an insulator length of 3,000 mm at an impressed voltage of 243 kV (RMS). It was confirmed that profile 1 results in lower field stresses (Fig. 8.37). The following dimensions were used in the simulations involving a corona ring: An outer diameter of 280 mm and a ring profile diameter of 60 mm. The distance x in mm is the varied distance between the top side of the corona ring and the top side of the insulator fitting (embedded for profile 1).

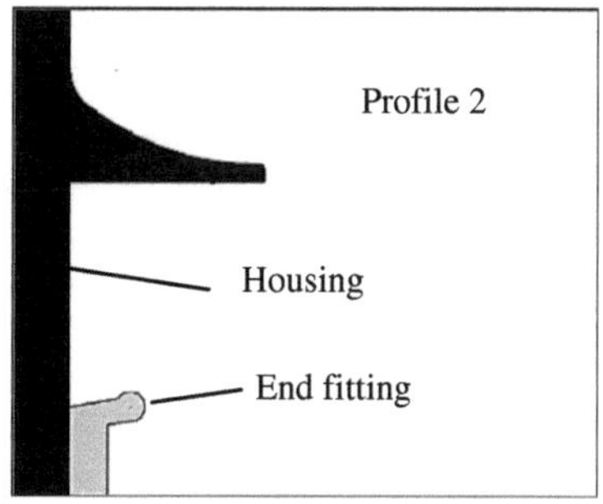

Fig. 8.36 Profile 2, cast end fitting

Table 8.6 Comparison of corona threshold values from a simulation and test [48]

Insulator design	Corona threshold values from a 3D-simulation (kV)	Corona threshold values from tests (kV)			
		Inception voltage[b]	Extinction voltage[b]	Inception voltage[c]	Extinction voltage[c]
Profile 1	142	160	147	171	147
Profile 2	79	87	85		
Profile 1 with a corona ring[a]	262	275	258	267	269
Profile 2 with a corona ring[a]	–	–	–	–	–

Explanatory notes
[a] Ring dimension 280 × 60 mm
[b] Laboratory I
[c] Laboratory II

The different field stresses for both profiles show that there is no generally applicable rule for different insulator designs (profiles). The analysis of highest field stresses on the silicone rubber surface is also interesting; apart from the simulation without a ring (simulated installation fault), these highest field stresses occur on the sheds or outer shed edges. The accelerating combined ageing tests mentioned in the context of 500 kV silicone insulators in Sect. 8.5.1 have empirically shown that, as long as there is compliance with certain field stress threshold values for the shank, the occurrence of droplet coronas on the sheds can virtually be ruled out and their occurrence at the shed edges can be tolerated.

It confirms that, under these conditions, the design of the insulator fitting and how it is embedded into the housing material result in better corona behaviour by profile 1.

The corona ring dimensions for profile 1 were varied (Fig. 8.38) and all variants for x = 25 mm in Fig. 8.37 were calculated. The specification (e.g. 20/280) describes a corona ring where $d_r = 20$ mm and $D_R = 280$ mm. As the outer diameter of the ring increases, the field stress on the corona ring decreases. However, the field-homogenising effect at the insulator end fitting (rim) might also fall as negative effect.

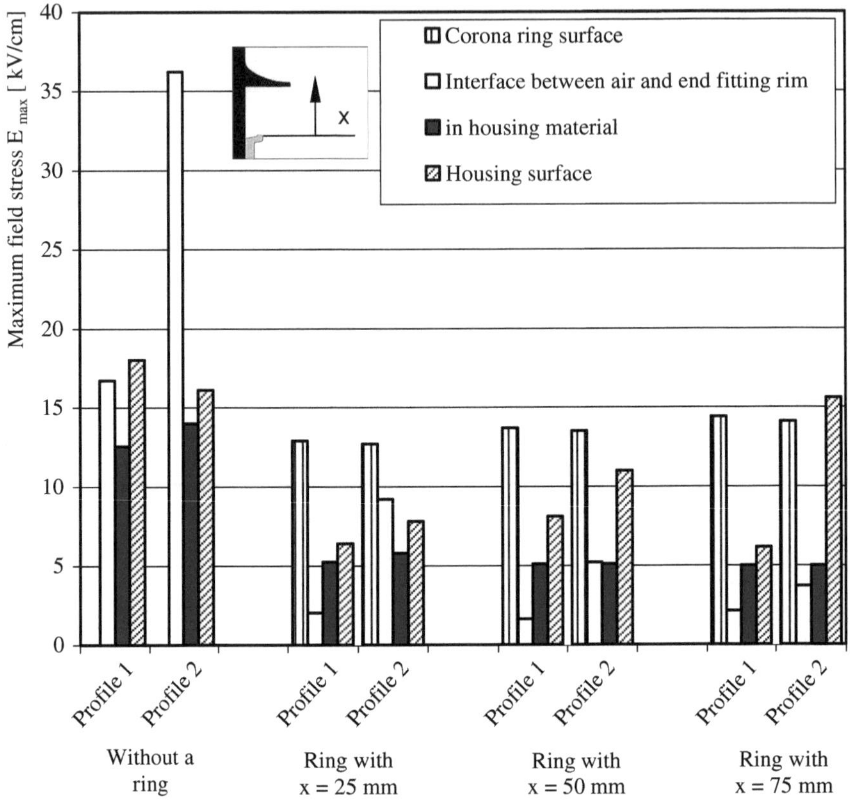

Fig. 8.37 Highest field stress at profiles 1 and 2 with/without a corona ring at various positions x [48]

The ring manufacturing process must be taken into account when varying the outer diameter and the ring profile diameter. In the case of bent aluminium tube profiles, the outer diameter to ring profile diameter ratio must be between 8 and 10. This limit does not apply when cast aluminium is used. However, it is still important to ensure that, when designing the tool, no mould lines occur at positions of high field stress. An optimised ring design for profile 2 was presented in [49], the content of which included the complex dependencies involving a parameter field and the implementation of a total field stress from the individual parameters. An optimal ring dimensions with an outer diameter of 390 mm and a profile diameter of 90 mm were found and confirmed that the ring diameter and its relative position to the insulator fitting are key factors here.

Thanks to increasingly powerful work stations, more complex simulations are possible in a smaller amount of time. In addition to the important factors relating to field stress (Fig. 8.34) the following can be considered:

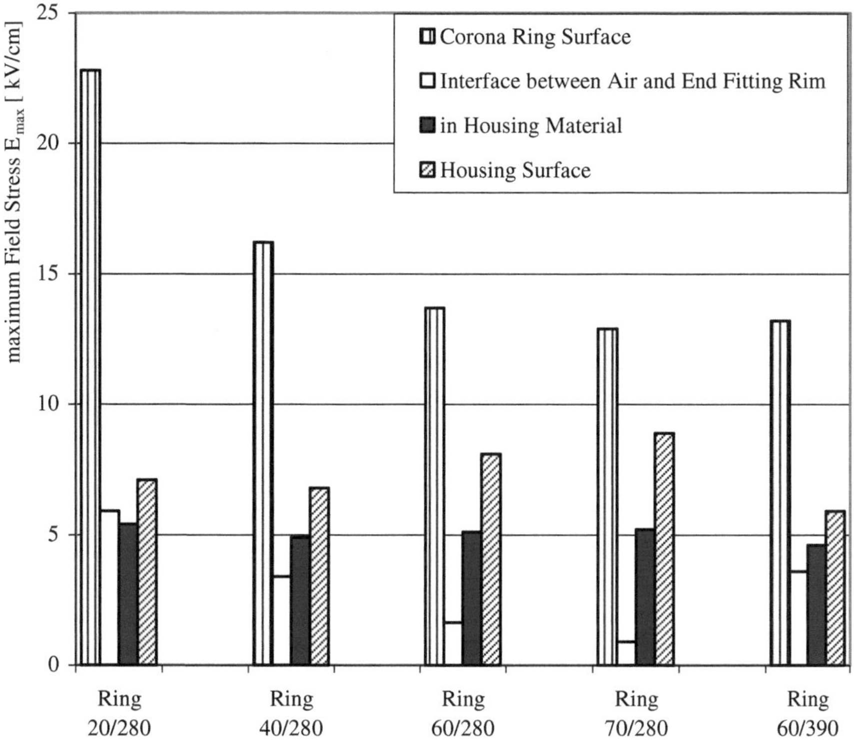

Simulation	Highest field stress on silicone surface			
	Profile 1		Profile 2	
	On the shed	At the shank	On the shed	At the shank
Without a ring		x		Seal
Ring with x = 25	Edge		Edge	
Ring with x = 50	Edge		Edge	
Ring with x = 75	Edge		Edge	

Fig. 8.38 Simulation results of corona ring variations of profile 1 [48]

- Geometry of hardware components such as connecting elements and suspension or tension clamps
- Entire set-up of the insulator set (single/multiple strings)
- Effect of the conductor bundle (single/multiple conductor)
- Positioning of the string in a suspension or tension configuration
- Simulation of adjacent phases
- Simulation of the earth or earth wire.

One example of a complex simulation is the field simulation for a project involving a 420 kV compact line (TenneT Holland [50]), which had the following ambitious (almost contradicting) requirements: an aesthetic design and for the combined corona and power arc protective fittings, the highest field stress at the shank could only exceed the value 2.5 kV/cm (2.5 criterion) at a length of 10 mm only. Despite an extensive computing capacity, the simulation needed to run for a period of 48 h.

For this simulation, four adjacent phases were used in accordance with the arrangement in Fig. 8.39 and the electrical field for the various relative phase relationships was calculated as an RMS value (Fig. 8.40). The phases have an asymmetrical arrangement (C–B, A–B), which is an established measure for reducing the electrical and magnetic field stress for towers having multiple voltage systems [51]. As a first step, the highest field stresses were determined at the string elements. The highest values (17.4 kV/cm) occurred on the power arc protective fitting of the bracing long rod insulator. The values at the C-shaped yoke are lower because the four-stranded conductor bundle comprising 32.4 mm conductors (type AMS 620) with a centre distance of 500 mm have an effective field-comparative effect. The number of conductors and their distance were chosen not only from an ampacity perspective but also from the perspective of interference voltage behaviour (acoustic, electrical, etc.) under dry conditions and during rainfall [52, 53].

The electrical field stress on the surface of the insulating material was also evaluated.

8.6.2.1 Bracing Long Rod Insulator

As can be seen in Fig. 8.41, two rings were provided on the high voltage side of the tensioning long rod insulator (at positions 5 and 15), namely an inner aluminium fitting for a "near" field reduction on the surface of the insulating material and an outer steel fitting with the dual function of "outer" field control and power arc protection. For this reason, the aluminium fitting is positioned below the steel fitting. In each case, only one power arc protective fitting (made from steel) is used on the earth side.

A corona shed was fitted to both ends of the long rod insulator fittings (see also Sect. 7.3.2.3). When the rear side (Fig. 8.42) and top side (Fig. 8.43) of the bracing long rod insulator were analysed, there was good compliance with the 2.5 criterion as a non-exceedable tangential component of the electrical field stress on the surface of the insulating material when the length exceeds 10 mm.

8.6.2.2 Post Insulator

An analysis of the simulation showed that, despite the larger diameter of the post insulator fitting, the sole use of a power arc protective fitting is not enough to ensure compliance with the 2.5 criterion. For this reason, the following two variants were evaluated: the use of an additional corona ring made from aluminium and the use of a corona shed made from silicon rubber.

Fig. 8.39 Test set-up with an emergency tower and an architectural model for TenneT's 420 kV-compact bipole line in Holland

From a technical perspective, the results favoured the use of the corona shed. Even though the aluminium corona ring achieved a field reduction, the range for the remaining highest field stress coincides directly with the seal (red line in Fig. 8.44). Consequently, a high electrical stress occurs on a functionally relevant position of the insulator. Since the solution involving the corona shed reduces the electrical stress on the seal considerably (Fig. 8.45), it was technically implemented.

While taking account of the 2.5 criterion, these simulations were also performed for other components in the compact line, namely the 420 kV double tension set of long rod insulators, the 420 kV horizontal posts of the jumper loops, and the insulated cross-arm and tension string for the 170 kV voltage level.

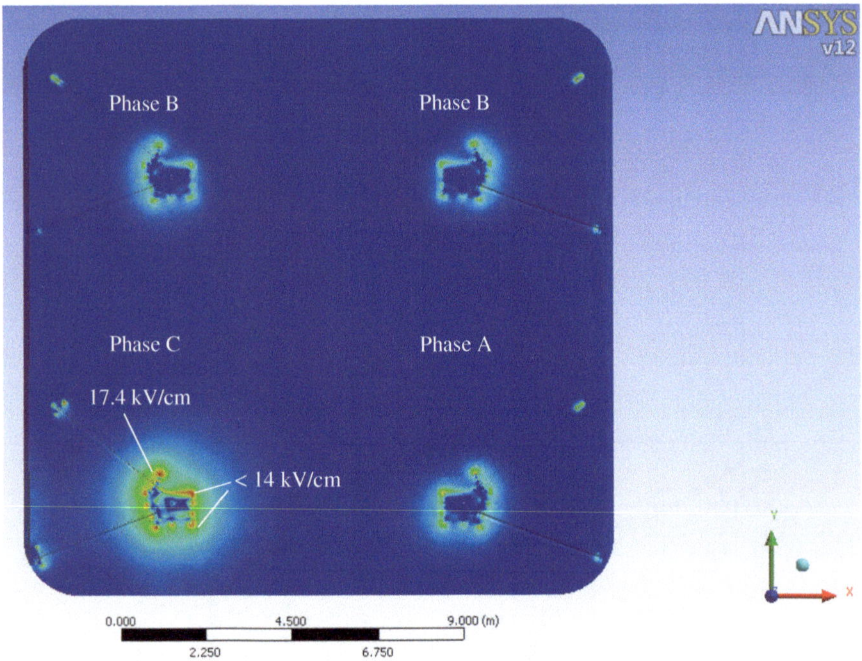

Fig. 8.40 Result of maximum field stress from simulation under consideration of neighbour phases

8.7 Empirical Threshold Values for Corona Prevention

Technical Brochure 284 explicitly does not provide any threshold values for permitted field stresses on the insulating material and hardware components. The permitted insulating surface field stress range for AC stressing, in order to prevent water droplet corona, is described in the literature on the basis of model testing and tests on silicone insulators up to 500 kV, and lies between 4 and 7.5 kV/cm [2, 9, 17, 54–56]. An IEEE publication from 2008 contains a summary of simulations, service experiences involving various manufacturers, model testing and materials testing [2].

The aforementioned permitted field stress range can be extended to include other parts of an insulator string/set (Table 8.7).

The value 17 kV/cm was obtained from local experiences in Canada where, after 20 years in service, the corona inception voltage can fall by 20 % due to corrosion of the surface of hardware components [1].

The results of studies on 420 kV composite insulators after 10 years in service will be presented in Sects. 10.5.1 and 10.5.2. The field stresses that occurred at these insulators and how these values compare to the empirical threshold values will be analysed as next.

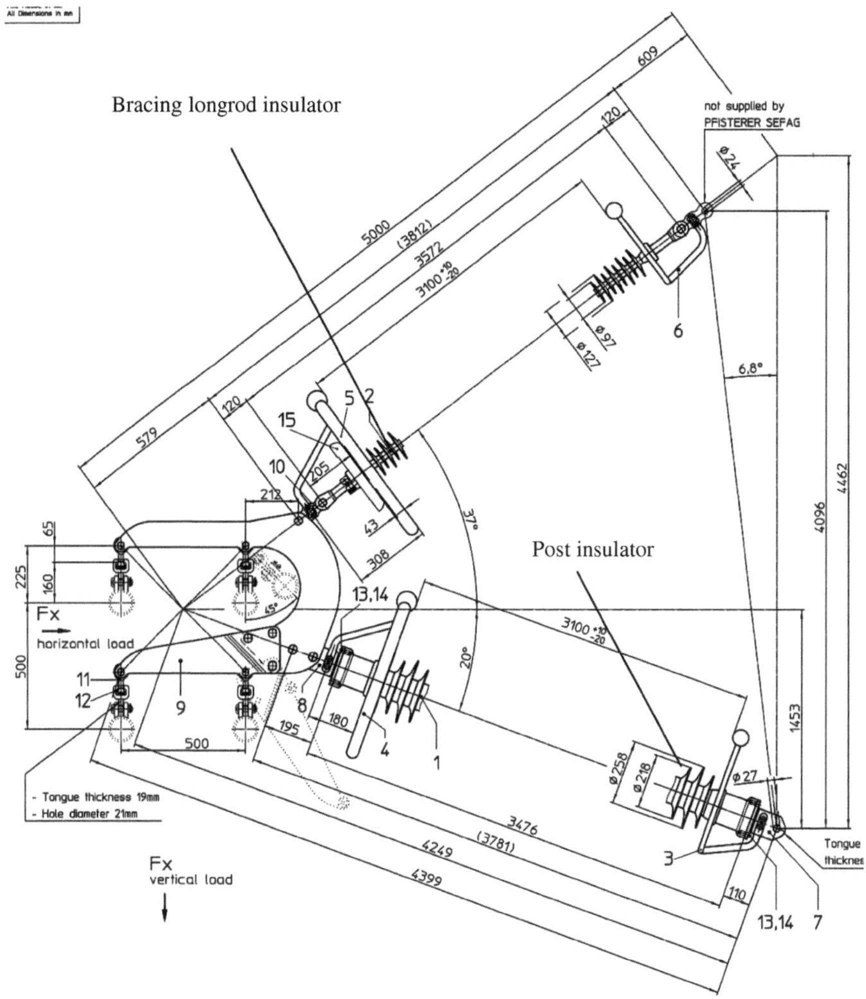

Fig. 8.41 Assembly drawing of TenneT's 420 kV-isolated cross-arm

8.8 420 kV Composite Insulator Sets After 10 Years in Service

The following evaluations of composite insulators with an HTV silicone rubber housing but manufactured using different production processes have the goal of correlating the changes found after 10 years in service with the electrical field stress values while the insulators were in service. The findings also facilitate an assessment of how to apply the threshold values in Table 8.7 when designing composite insulators.

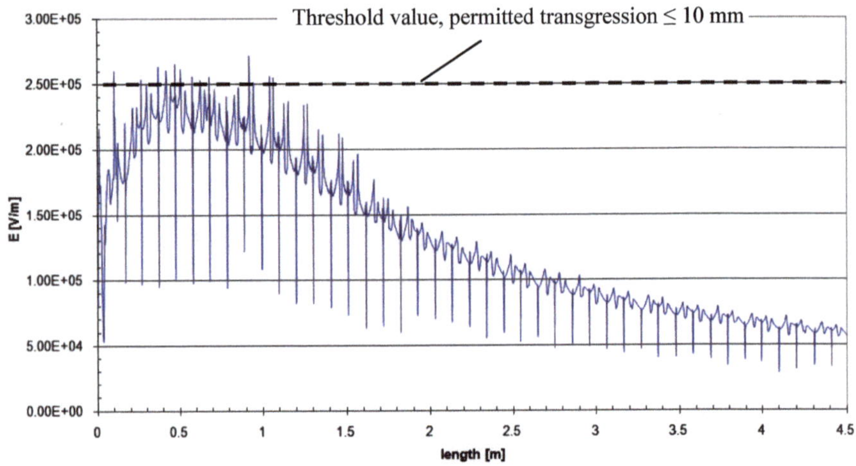

Fig. 8.42 Electrical field progression at long rod insulator (*rear*)

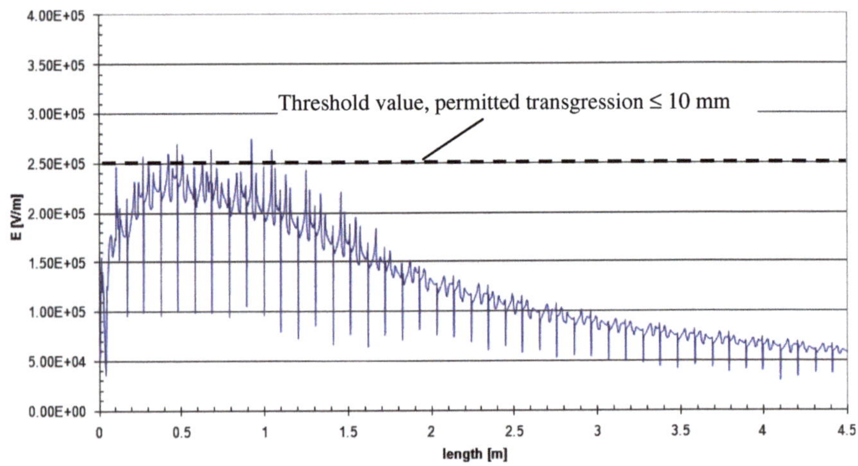

Fig. 8.43 Electrical field progression at long rod insulator (*top*)

8.8.1 Overmoulded Design

As it will be shown in Sect. 10.5.1, four composite insulators were withdrawn from a 420 kV network after 10 years in service and evaluated in terms of their state. Almost no changes were observed [57]. The withdrawn insulators had been installed in single suspension strings (Fig. 8.46). From the perspective of electrical field stress, the special feature of this insulator design was that the insulator fittings had been overmoulded. This set-up is comparable with profile 1 in Sect. 8.6.2.

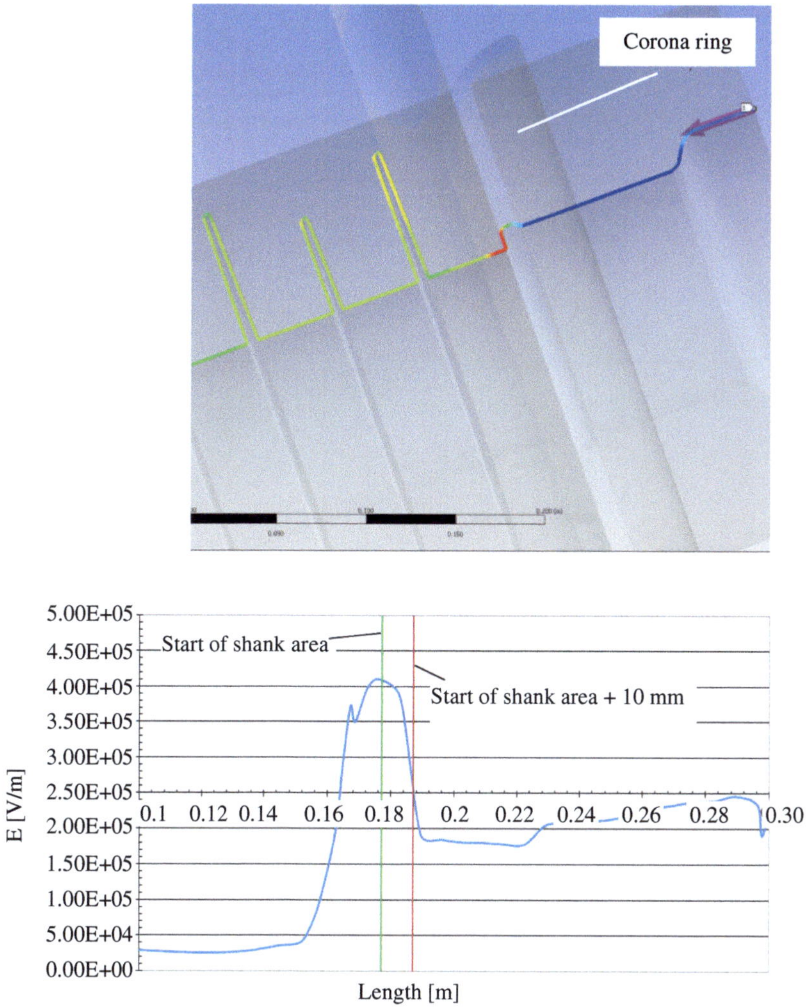

Fig. 8.44 Result of field simulation involving a corona ring

Only one twin bundle conductor configuration (type ACSR ø 38 mm) was used for 420 kV.

For this string, the electrical field stress was calculated in order to set the highest values in relation to the empirical threshold values. The calculation was performed using the Coulomb (Version 8) and ANSYS 13 programmes for a phase to earth voltage of 242 kV and the results were compared against each other.

Due to the eccentric ring position, there is a different stress on the surface of the insulating material (Fig. 8.47). The higher field stress is on the side with the greater distance to the ring.

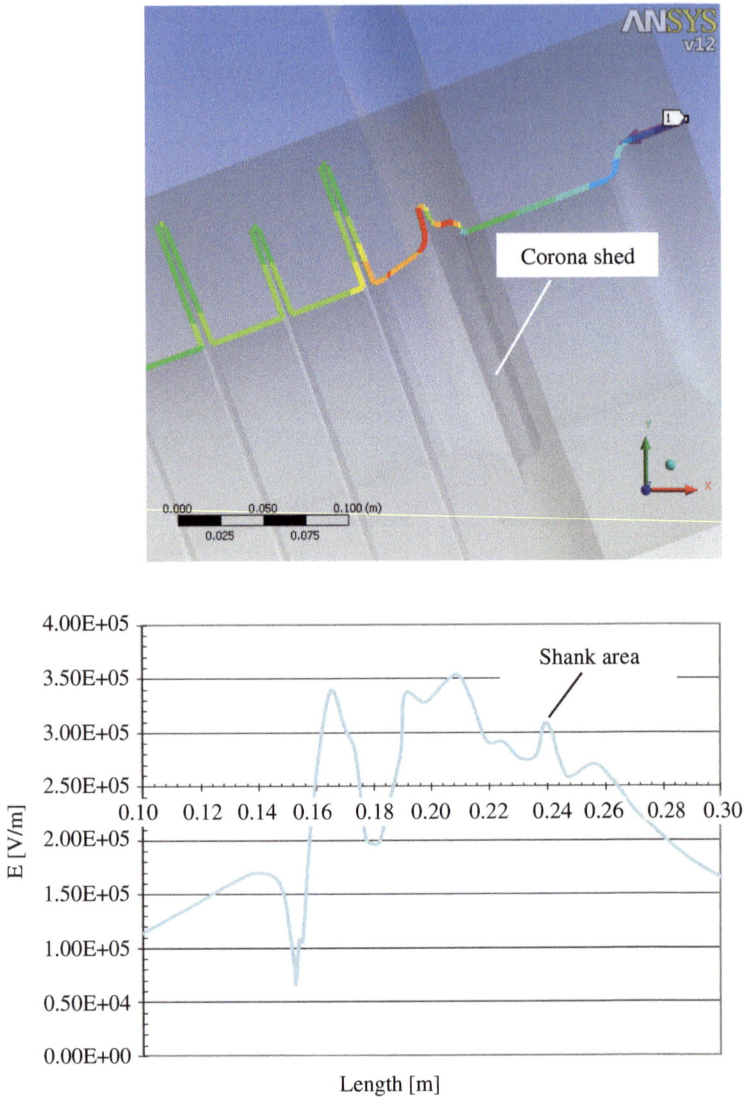

Fig. 8.45 Result of field simulation involving a corona shed

The results (Table 8.8) are to be interpreted as follows: Both programmes require careful modelling of curved contours, both in the case of metallic and non-metallic contours and, in particular, triple junctions (the points where air, metal and the insulating material converge) [58]. This is also the reason, why positions "F" (fitting) and "I" (insulating material) in Fig. 8.46 have almost the same values even though "I" has a greater diameter and is non-metallic. The high field stresses also have little effect in a salt fog test conducted as an ageing test. No dry zones

Table 8.7 Empirical threshold values for composite insulators (and composite insulator sets) in AC-applications

Occurrence of corona	Duration	Part of the insulator set	Threshold of highest field stress (kV/cm)	Verification by	
				Test under dry conditions	Simulation under dry conditions
Dry conditions	Continuous	Corona ring or combined corona/arc protection ring, insulator end fitting	$\leq 17^{b}$–21	x	x
Moist conditions	Intermittent	Insulator shank housing	$\leq$4.5 for a length $\leq$10 mm	–	x
Moist conditions	Intermittent	Seal[a]	3.5	–	x
Dry and moist conditions	Continuous/ intermittent	In the insulator (rod, housing and the interface between the two)	$\leq$30	–	x

Explanatory note

[a] For designs without an embedded fitting (similar to profile 2)

[b] Highly-corrosive environment

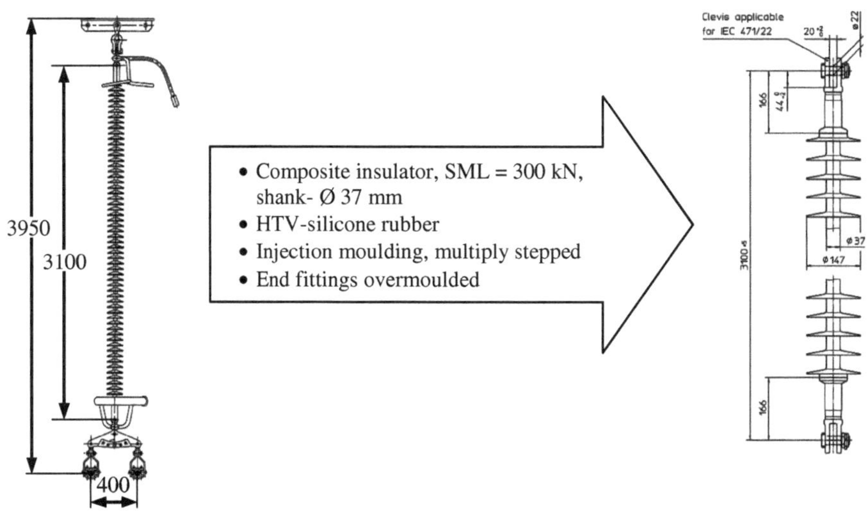

Fig. 8.46 420 kV single suspension set for field simulation and after-service evaluation (see also Sect. 10.5.1)

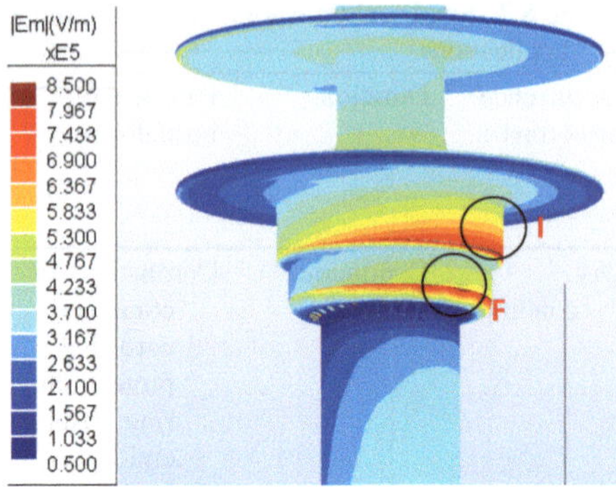

Fig. 8.47 Coulomb simulation of overmoulded end fitting

Table 8.8 Result comparison of BEM- und FEM-simulations

Position on the insulator	Highest field stress (kV/cm)	
	Coulomb	ANSYS
Overmoulded area below the first shed	7.9	7.7–8.75
On the surface of the first shed	5.3–5.7	5
At the first shank section	4.5–4.9	4.5
On the power arc protective fitting	15–27	15–31

can be detected in the region of the large diameter above the overmoulded fitting (see also section "Field Considerations"). The field stress on the first shed surface is particularly high in the region of the curvature where the field lines are concentrated at the junction between the air and the insulating material. The field stress at the shank is crucial from the perspective of water droplet corona. As a result of the overmoulded design, only values up to 4.9 kV/cm result even with the eccentric position of the power arc protective ring.

It is interesting to note the high values on the power arc protective fitting, which were determined at the radius of the inwardly facing upper ring opening. These values do not correlate with the corona measurements from the type test; corona inception occurred elsewhere at values higher than 242 kV. It can be assumed that, once again, the curvature issue associated with contours is responsible for this deviation from the test. A further refined network was not used to repeat the simulation at this point because the field stress on the silicone rubber surface was the focus of this evaluation.

8.8.2 *Modular Design*

As it will be shown in Sect. 10.5.2, nine composite insulators were withdrawn from a 420 kV network after 10 years in service. Their state was evaluated and virtually no functionally relevant changes were observed. The withdrawn insulators had been

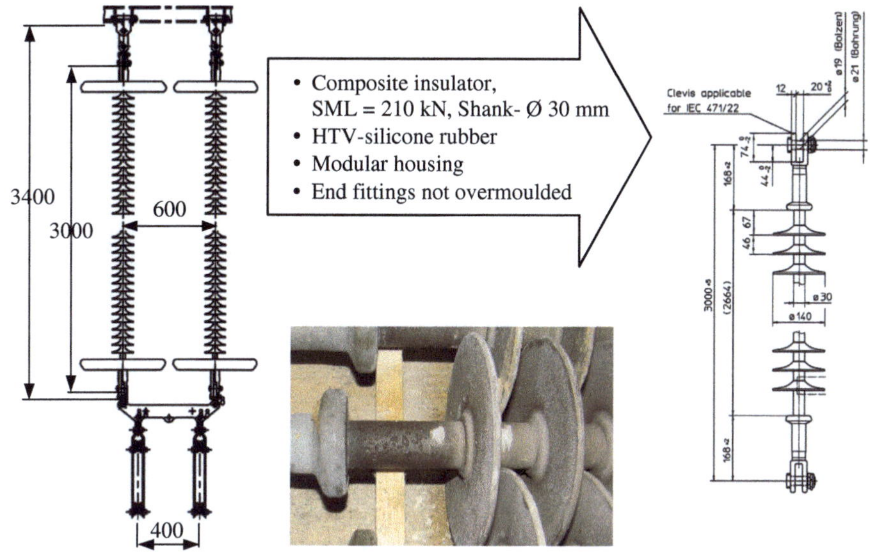

Fig. 8.48 420 kV double suspension set for field simulation and after-service evaluation (see also Sect. 10.5.2

installed in double suspension strings (Fig. 8.48). From an electrical field stress perspective, this insulator design corresponds to profile 2 in Sect. 8.6.2. A quad bundle conductor configuration was used for the transmission voltage of 420 kV.

In contrast to the insulators evaluated in Sect. 8.8.1, a dark pollution layer that had hydrophobic properties and was relatively strongly bonded to the surface was visible in the first shank area. A solvent was needed to remove the pollution layer. Under the pollution layer, there was no evident damage to the surface, which can be caused by corona discharges. The sealing surfaces were also covered with this pollution layer but were effectively restored to their new state after they were cleaned. The Coulomb programme was used to perform the field calculations for this string set-up. The results (Figs. 8.49, 8.50, Table 8.9) can be interpreted as follows:

There is also a certain asymmetry of the field stress values. The side facing the adjacent insulator has the smaller values. The highest field stresses occur at the insulator end fitting rim, a result that is confirmed by the calculations in Sect. 8.6.2.

However, the value is below the value for incepting continuous corona discharges. This was confirmed by the type tests. The values at the shank are higher than those with the overmoulded design: The value on the seal is approximately twice as high as the recommended threshold value for the droplet corona. The same applies to the shank section. Despite this transgression, the insulators did not experience any damage after 10 years in service. This is also due to their interaction with the environment (pollution class, humidification cycles, humidification levels, etc.) and the high quality of the insulating material used. For example, the

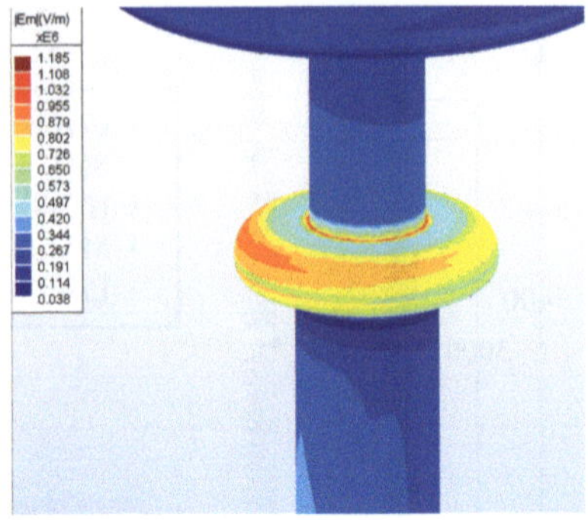

Fig. 8.49 Coulomb simulation, scale 1

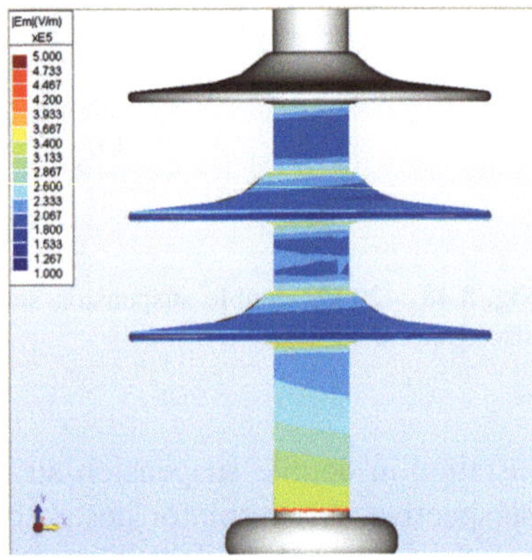

Fig. 8.50 Coulomb simulation, scale 2

Table 8.9 Results of the Coulomb simulation

Insulator position	Highest field stress (kV/cm) Coulomb
On the insulator end fitting	9.5
On the seal	7.2
On the shank	8
On the power arc protective fitting (ring surface)	13–19

HTV silicone rubber used does not contain any particularly acid-sensitive fillers (e.g. $CaCO_3$). Furthermore, the added ATH was silanised.

The dark pollution layer underwent a further evaluation, the goal of which was to verify SiO_x as an index for the effect of discharges caused by water droplet corona. The following processes were used here: Infrared spectroscopy and thermo-gravimetric analysis (TGA).

For these analyses, three different samples were taken from the same position on the shank:

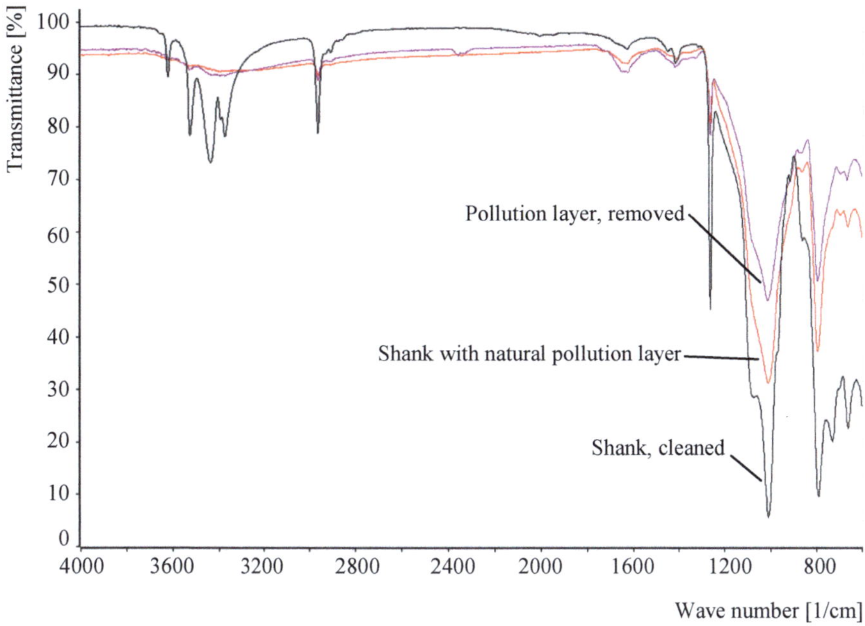

Fig. 8.51 Transmission spectra of different specimens taken from the 420 kV insulator after service

- From the shank with the original black layer (shank with natural pollution layer as provided)
- From the shank after it was cleaned using a solvent (shank, cleaned)
- From the black layer, which was mechanically removed from the shank (pollution layer, removed)

When evaluating the transmission spectrum (Fig. 8.51), the following wave numbers are of interest to an evaluation of the silicone rubber:

- 2,962 cm^{-1}: C–H symmetric stretching of CH_3
- 1,261 cm^{-1}: CH_3 asymmetric stretching and deformation of Si–CH_3
- 1,008 cm^{-1}: Si–O–Si symmetric stretching
- 792 cm^{-1}: Si–C symmetric stretching and CH_3 rocking

The following wave numbers are the range for function groups such as C = O, O–H, N–H, etc. (e.g. from peroxide): 4,000–1,300 cm^{-1}. Generally, these do not diffuse in contamination layers. Consequently, there are no verifiable bands, unlike in the case of a cleaned shank. The differences in transmittance are not a clear indicator of the presence of SiO_x.

For this reason, the same samples continued to be analysed by means of TGA (destructive testing). The measurement was taken within a temperature range of between 50 and 900 °C and an initial gradient of 10 K/min. Nitrogen with a flushing rate of 75 ml/min was used as an inert gas.

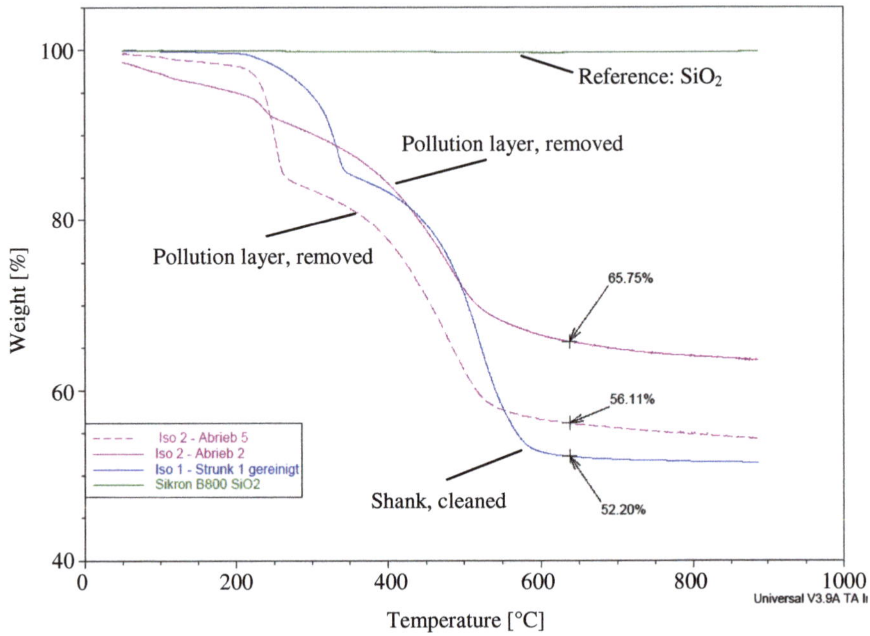

Fig. 8.52 Reduction in weight in proportion to temperature (TGA analysis)

For reference purposes, untreated SiO_2 (Sicron B800) was also tested. The result reveals a trend (Fig. 8.52), whereby the cleaned shank material contains more decomposable components (PDMS) and the contamination layer comprises more inert components (i.e. oxidated silicone rubber SiO_x) under these test conditions. However, it is difficult to interpret the result because the level of abrasion is relatively low. The use of other analytic processes will be tested.

8.9 Summary

By observing some simple rules, it is possible to prevent the occurrence of corona discharges on composite insulator string or sets. Current standards for conventional insulator sets are applicable insofar as the occurrence of a continuous corona at hardware components or at the insulator end fitting can be tested directly.

The phenomenon of water droplet corona, which depends on numerous factors, should be given special consideration in the case of hydrophobic composite insulators. The composite insulator design (i.e. the material chosen, the stability of the sealing system, etc.) and the stresses that occurs at the service location can play a significant role here. The threshold value 4.5 kV/cm can be regarded as a conservative recommendation, especially in view of the current trend whereby cost pressures are resulting in further developments with a lower fault tolerance.

As the example in Sect. 8.8.2 shows, high-grade insulating materials ensure safe operation even when the threshold value for water droplet corona is exceeded. However, this does not apply to insulator sets that are subjected to a continuous corona discharge. Even if acid-resistant (E-CR) glass is used in this scenario, there may be some cumulative damage, which will ultimately cause the composite insulator to fail.

References

1. Phillips A, Bologna F, Shaw T Application of corona rings at 115 kV and 138 kV. EPRI-Document 1015917 2008, CIGRE WG B2.21 IWD 063-2009
2. IEEE Taskforce on Electric Fields and Composite Insulators (2008) Electric fields on AC composite transmission line insulators. IEEE Trans Power Delivery 23(2):823–830 (Apr)
3. IEC 61284 Ed 2.0 (1997) Overhead lines—requirements and tests for fittings
4. IEEE 539 (2005) IEEE standard definitions of terms relating to corona and field effects of overhead power lines
5. IEEE 1829: Guide for conducting corona tests on hardware for overhead transmission lines and substations. Ongoing project
6. IEC 60437 Ed 2.0 (1997) Radio interference test on high-voltage insulators
7. BS 137 Part 2 (1973) Specification for insulators of ceramic material or glass for overhead lines with a nominal voltage greater than 1,000 V
8. Hofmann J (1995) Elektrische TE- und Ableitstrommessungen mit Impulsspannung an polymeren Isolierstoffoberflächen mit Tropfenbelägen zur Diagnose des Oberflächenzustandes – am Beispiel von Elastomeren. Dissertation TH Zittau, Germany
9. Phillips AJ, Childs DJ, Schneider HM (1999) Aging of non-ceramic insulators due to corona from water drops. IEEE Trans Power Delivery 14(3):1081–1089 (July)
10. Cheng ZX, Liang XD, Zhou YX, Wang SW, Guan ZC (2003) Observation of corona and flashover on the surface of composite insulators. In: IEEE Bologna PowerTech conference, Bologna, Italy, 23–26 June 2003
11. Keim S, Koenig D, Hinrichsen V (2003) Experimental investigations on electro hydrodynamic phenomena at single droplets on insulating surfaces. In: 2003 Annual report conference on electrical insulation and dielectric phenomena
12. Krivda A, Birtwhistle D (2001): Breakdown between water drops on wet polymer surfaces. In: 2001 Annual report conference on electrical insulation and dielectric phenomena
13. Dawson GA (1969) Pressure dependence of water-drop corona onset and its atmospheric importance. J Geophys Res 74(28):6859–6868
14. Phillips AJ, Jandrell IR, Reynders JR (1996) Consideration of corona onset from a water drop as a function of air pressure. IEE Proc: Sci Meas Technol 143(2):125-130 (Mar)
15. Moreno VM, Gorur RS, Kroese A (2003) Impact of corona on the long-term performance of nonceramic insulators. IEEE Trans Dielectr Electr Insul 10(1):80–95 (Feb)
16. Phillips AJ, Childs DJ, Schneider HM (1999) Water drop corona effects on full-scale 500 kV non-ceramic insulators. IEEE Trans Power Delivery 14(1):258–265 (Jan)
17. Braunsberger T, Dziubek A, Kurrat M (2004) Water drop corona on hydrophobic epoxy. In: International conference on solid dielectrics, Toulouse, France, July 2004
18. Bärsch R (2003) Bewertung der Hydrophobie sowie des Kriechstromverhaltens von Silikonelastomeren für Hochspannungs-Freiluftisolatoren. ETG-Fachbericht 93:S.97–S108
19. Cervinka R, Bärsch R, Exl F, Kindersberger J, Winter H-J (2008) Untersuchungen zur Beständigkeit der Hydrophobie von polymeren Isolierstoffoberflächen und ihrer Wiederkehr mit dem Dynamischen Tropfen-Prüfverfahren ETG-Tagung

20. Bärsch R, Cervinka R (2010) Oberflächenverhalten von Epoxidharz-Formstoffen unter Hochspannungsbeanspruchung und elektrolytischer Fremdschichtbelastung, Seminar "Epoxidharze in der Elektrotechnik", Technische Akademie Esslingen
21. CIGRE WG D1.14 (2010) Evaluation of dynamic hydrophobicity properties of polymeric materials for non-ceramic outdoor insulation—retention and transfer of hydrophobicity. Technical brochure 442, Dec 2010
22. Cervinka R, Bärsch R, Seifert J (2010) Einfluss der Prüfspannungsart auf das Oberflächenverhalten von polymeren Isolierstoffen unter elektrolytischen Fremdschichtbelastungen, ETG-Fachbericht 125, S. 69–75, VDE-Verlag Berlin Offenbach
23. Hillborg H, Gedde UW (1999) Hydrophobicity changes in silicone rubbers. IEEE Trans Dielectr Electr Insul 6(5):703–717 (Oct)
24. Yoshimura N, Kumagai S, Nishimura S (1999) Electrical and environmental ageing of silicone rubber used in outdoor insulation. IEEE Trans Dielectr Electr Insul 6(5):632–650 (October)
25. Reynders JP, Jandrell IR, Reynders SM (1999) Review of ageing and recovery of silicone rubber insulation for outdoor use. IEEE Trans Dielectr Electr Insul 6(5):620–631 (Oct)
26. Hackam R (1999) Outdoor HV composite polymeric insulators. IEEE Trans Dielectr Electr Insul 6(5):557–585 (Oct)
27. Ma B, Andersson J, Gubanski SM (2010) Evaluating resistance of polymeric materials for outdoor applications to corona and ozone. IEEE Trans Dielectr Electr Insul 17(2):555–565 (Apr)
28. Siegel S, Judeikis H (1965) EPR study of the biphotonic nature of the solute-sensitized photodecomposition of polydimethyl-siloxane. J Chem Phys 43(2):343
29. Delman AD, Landy M, Simms BB (1969) Photodecomposition of polydimethylsiloxane. J Polym Sci Part A-1 7:3375
30. Lacoste J, Israeli Y, Lemaire J (1996) Photoaging of substituted and unsubstituted silicones. Advances in chemistry series 249, Am Chem Soc
31. Moreno VM, Gorur RS (2001) Effect of long-term corona on non-ceramic outdoor insulator housing materials. IEEE Trans Dielectr Electr Insul 8(1):117–128 (Feb)
32. IEC 61621 Ed 1.0 (1997) Dry, solid insulating materials—resistance test to high-voltage, low-current arc discharges
33. Bärsch R, Jahn H, Lambrecht J, Schmuck F (1999) Test methods for polymeric insulating materials for outdoor HV insulation. IEEE Trans Dielectr Electr Insul 6(5):668–675 (Oct)
34. Goldman M, Sigmond RS (1982) Corona and insulation. IEEE Trans Electr Insul EI-17(2): 90–105 (Apr)
35. Goldman M, Goldman A, Sigmond RS, Sigmond T (1991) On the role of water in the aging of polymers in air-insulated electrical systems. IEEE Trans Electr Insul 26(4):770–775 (Aug)
36. Goldman M, Goldman A, Sigmond RS (1989) Analysis of air corona products by means of their reactions in water. In: Proceedings 9th international symposium of plasma chemistry, Pugnachinoso, Italy
37. Brisset JL, Leievre J, Doubla A, Amouroux J (1990) Interactions with aqueous solutions of the air corona products. Rev Phys Appl 25:535–543
38. Chughtai AR, Smith DM, Kumosa LS, Kumosa M (2004) FTIR analysis of non-ceramic composite insulators. IEEE Trans Dielectr Electr Insul 11(4):585–596 (Aug)
39. Al-Hamoudi IY (2008) Field test results of composite silicone rubber insulators at shoaiba of Saudi Arabia. In: GCC power 4th CIGRE conference 2008, Bahrain
40. Kopetjkova D, Sklenicka V (2006) Investigation of HV equipment composite insulators ageing phenomenon. EURODOBLE 2006
41. Koshino Y, Umeda I, Ishiwari M (1998) Deterioration of silicone rubber for polymer insulators by corona discharge and effect of fillers. In: Conference on electrical insulation and dielectric phenomena, 1998. Annual report
42. Koshino Y, Nakajima I, Umeda I (1998) Effect on the electrical properties of fillers in silicone rubber for outdoor insulation. In: Proceedings of 1998 international symposium on electrical insulating materials, in conjunction with 1998 Asian international conference on

dielectrics and electrical insulation and the 30th symposium on electrical insulating materials, Toyohashi, Japan, 27–30 Sept 1998
43. Ansorge S, Schmuck F, Aitken S, Papailiou KO (2010) Improved performance of silicone rubbers for the use in composite insulators. Paper D1_105_2010, CIGRE
44. Bärsch R, Lambrecht J, Jahn H (1996) On the evaluation of the hydrophobicity of composite insulator surfaces. In: IEEE annual report—conference on electrical insulation and dielectric phenomena, San Francisco, Oct 1996
45. Liang Y, Li CR, Ding L (2009) Study on the hydrophobicity of HTV SIR treated by different corona intensity. In: Power and energy engineering conference, APPEEC 2009. Asia-Pacific
46. Bärsch R, Pilling J, Schmuck F (1989) Zum Problem der Prüfung von Kunststoffisolatoren. 34. Internationales Wissenschaftliches Kolloquium, TH Ilmenau
47. CIGRE WG 22.03 (1992) Use of stress control rings on composite insulators. ELECTRA 143 (Aug)
48. CIGRE WG B2.03 (2005) Use of corona rings to control the electrical field along transmission line composite insulators. Technical brochure 284, Dec 2005
49. Schümann U, Barcikowski F, Schreiber M, Kärner HC, Seifert JM (2002) FEM calculation and measurement of the electrical field distribution of HV composite insulator arrangements. CIGRE session 2002, paper 33–404
50. Hoekstra HE, van Wolven JF, van der Wekken AJP (2010) Development of a compact bipole 380 kV overhead line. CIGRE session 2010, paper B2_112_2010
51. Papailiou KO, Schmuck F (2011) Past experience and future trends with compact lines to solve the right of way issues. CIGRE symposium in Bologna
52. Trinh NG, Maruvada PS (1977) A method of predicting the corona performance of conductor bundles based on cage test results. IEEE Trans Power Apparatus Syst PAS-96(1):312–325 (Jan/Feb)
53. Chartier VL, Blair DE, Stearns RD, Lamb DJ (1994) Effect of bundle orientation on transmission line audible and radio noise. IEEE Trans Power Delivery 9(3):1538–1544 (July)
54. Nixon K, Reynders JP, Hill RJ (1998) Corona and sustained arcing on conventional and novel insulator designs. South Africans universities power engineering conference 1998 Stellenbosch, Jan 1998
55. Swift DA (1993) AC flashover mechanism for water droplets on an hydrophobic insulator. In: 8th ISH, Yokohama, August 1993, paper 44.09
56. Windmar D (1994) Water drop initiated discharges in air. PhD Thesis, Uppsala University
57. Ansorge S, Camendzind A, Pratsinis SE, Ammann M, Schmuck F, Papailiou KO (2008) Evaluation of silicone rubber housing interfaces after service exposure and performance improvements by nanofillers enriched silicone rubbers. Paper B2-208, CIGRE 2008
58. Craigen D (2011) Benchmark problems for simulating electric fields near triple junctions. http://www.integratedsoft.com/papers/Benchmark/TripleJunctions.pdf Feb 2011

Chapter 9
Power Arc Protective Fittings for Composite Long Rod Insulators

Symbols and Abbreviations

A_E	Cross-section of rod electrode
ATH	Aluminium trihydrate, Aluminium hydroxide
B_{PA}	Magnetic field of arc
B_R	Magnetic field of current-carrying rod
CIGRE	Conseil international des grands reseaux electriques (international council for large electric systems)
F_M	Migration force by self field (on the rod electrode)
F_{AR}	Force by regeneration of optimal arc root conditions
F_A	Force by aerodynamic resistance
F_L	Force by outer current loop
F_T	Thermal ascending force
F_{LT}	Force by longitudinal pulling
I	Current
I_T	Cumulative current
I_{T1}	Cumulative current for feed 1
I_{T2}	Cumulative current for feed 2
I_{SC}	Short-circuit current
I_{PA}	Power arc current
I_{P1}	Partial current 1
I_{P2}	Partial current 2
IEC	International Electrotechnical Commission
P_{MD}	Required power for material decomposition
TC	Technical committee
PD	Partial discharge
TG	Glass transition temperature
U_{PA}	Power arc voltage
c	Thermal capacity
d	Diameter of rod electrode

K. O. Papailiou and F. Schmuck, *Silicone Composite Insulators*, Power Systems,
DOI: 10.1007/978-3-642-15320-4_9, © Springer-Verlag Berlin Heidelberg 2013

k_L	Factor between 0…1, which determines the part of the arc power to heat the end burning point
t	Time
t_{SC}	Time of short circuit
t_{PA}	Time of power arc
I_{PA}	Arc current
v	Arc movement rate
x	Distance/exponent
λ	Thermal conductivity
ϑ_0	Temperature on the rod front end

9.1 Introduction

Various factors determine whether or not power arc protective fittings need to be used. Network parameters such as the current, duration and frequency of a short circuit are very influential. However, the basic set-up of the insulator string is also significant: Here, it is necessary to differentiate between the cap and pin design and the long rod design, which behave differently when there is a power arc and therefore influence the philosophy of power arc protection. Composite insulators have a long rod design. Unlike porcelain long rod insulators, however, they have the distinctive feature of being able to be manufactured in one piece without any intermediate fittings, even for the highest voltage levels. As a result of this classification, many "power arc" experiences and rules associated with porcelain long rod insulators also apply to composite long rods. There is also one notable difference: The slenderness of the composite long rods results in smaller fitting diameters, which makes it necessary to protect the insulator against corona at lower transmission voltages. Consequently, aluminium ring constructions, which are to be used solely for the purpose of corona protection, have also found their way into string designs. This chapter will present the relevant design criteria for power arc protective fittings while considering the current practice of porcelain long rods as well as the many years of experience with string components and silicone composite insulators, all of which shaped the content of the CIGRE Technical Brochure TB 365 [1].

9.2 Power Arcs as a Physical Phenomenon

9.2.1 Causes of Flashover on a Composite Insulator String/Set

A power arc can occur over an insulator or insulator string/set or between conductors for the following reasons:

a. Inner or outer over voltage (i.e. as a result of a surge caused by switching operations or lightning)
b. Pollution flashover
c. Adequate bridging of the striking distance by animals (either by direct contact or indirect contact through "bird streamers")
d. Electrical insulator failure (flashunder, dielectric breakdown, etc.)
e. Conductor approach as a result of wind or galloping after ice shedding
f. Reduced electrical strength of air if fires occur below an overhead transmission line

Flashovers caused by (a), (c), (e) and (f) are determined by air clearances and depend either marginally or not at all on the insulator technology used. In contrast, the pollution flashover is heavily influenced by the insulator technology: Composite insulators have a smaller mean diameter and, in the case of housing materials with a lasting hydrophobic effect, they have a higher insulation strength than conventional insulators when pollution occurs. An electrical insulator failure can occur with cap and pin insulators (especially porcelain cap and pin insulators). It can also occur as an interface flashover or after prolonged tracking in the case of low-grade composite insulators.

9.2.2 The Power Arc

Comprehensive studies have been conducted on power arcs and adequate protection for porcelain long rod insulators [2–6], which are widely used in Europe. A detailed history and description of the technology is provided in [7].

The power arc over an insulator set burns in air, which is a self-regenerating insulating medium. In other words, the overhead transmission line can operate again as soon as the power arc has been extinguished and the atmosphere has regained its electrical strength. Appropriate measures are taken to protect the insulator and hardware components of the insulator set against the thermal effect of a power arc. The structure of a power arc can be described as shown in Fig. 9.1: It comprises five parts, namely the arc roots, stems (often also known as plasma or plasma jet) and bow. The highest temperatures occur in the root and can be up to 18,000 K. The temperature of the arc stem is approx. 11,000 K. Both values are determined by the shape of the electrode, the electrode material and the arc current [3]. Due to its dimensions, the arc bow has a larger surface and is better able to release energy into the atmosphere. Studies have also confirmed that the arc bow is less critical in terms of its thermal effect. Tests involving a 24 kA power arc have shown that the arc bow touching the porcelain insulator did not cause any damage until a burning time of 3 s had elapsed [3].

A power arc is subject to the effect of different types of forces, which can also be utilised to guide it (Fig. 9.2, among others [7]). The thermal ascending force F_T always takes effect in an upward direction, in particular, to the arc bow. This must be taken into consideration for tower super-structures with V-strings and large

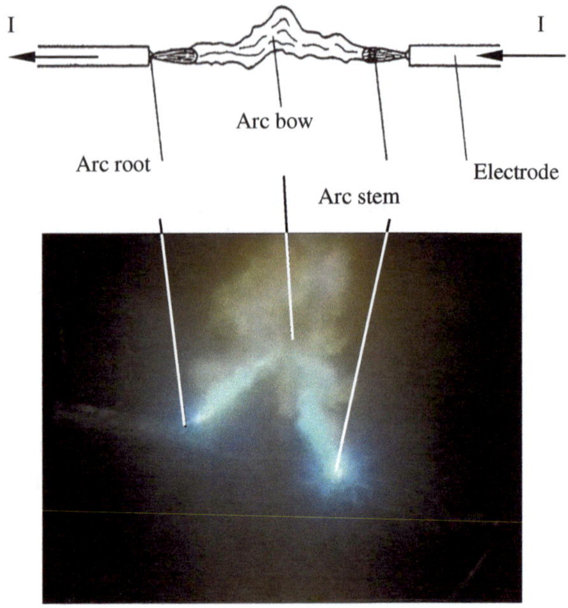

Fig. 9.1 Structure of a power arc

angles of Vee (>90°), for example. The force by longitudinal pulling F_{LT} takes effect if the power arc burns on a curved trajectory and "tries" to stretch. The power arc aims to migrate along the power arc protective fitting until it reaches the front ends of the round electrodes, which is its end burning point. The force by aerodynamic resistance F_A and the delay F_{AR} due to "thermal preparation" (charge carriers) of the new root position reduce the migration speed.

The force by outer current loop F_L has an important effect on the overall orientation of the power arc or the arc bow; in addition to the string set-up, it is important to consider how the current is fed in and out (e.g., the proximity of a transmission line to a station or "centre of impedance", or the string position at the outer crossarm or in the centre of the tower).

If one performs a qualitative comparison of various forms of discharge at an insulator string, their period of effect and the range of values for the current are very different (Fig. 9.3). As shown in Sect. 8.2.3, a stable corona discharge that is "solely" characterised by the μA range but has a continuous effect in the worst-case scenario can result in considerable damage to the insulator housing and trigger an insulator failure (e.g., as a result of a brittle fracture). Partial discharges (PDs) stabilised by a pollution layer occur only as long as the supplying pollution layer is sufficiently conductive (i.e., ions and solvent of the pollution layer). The insulator housing can be damaged by erosion or prolonged tracking. Pre-arcs occur as thermally ionised PDs having already the U-I-characteristics of a power arc. They typically ignite in the shed separations above the insulator shank. If these partial discharges become instable (i.e., they can unite to form one overall power arc), flashover occurs. The following empirical rule applies: Approx. 2/3 of the creepage distance must be bridged for this instability to occur.

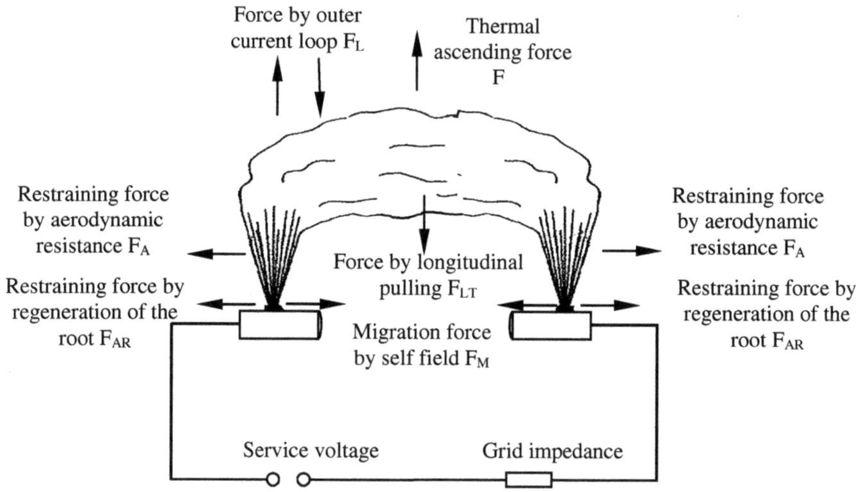

Fig. 9.2 Duration and amperage of different forms of discharge

9.2.3 Principles of Power Arc Protection

While the effect of the outer current loop (Fig. 9.3) must be considered at the planning stages of an overhead transmission line project, certain rules are applied to power arc protective fitting designs to ensure that the power arc is guided safely. The interaction between the magnetic field of the supplying rod steel fitting B_R and the magnetic field of the power arc B_{PA} (in particular, the arc stem (Fig. 9.4)) is a significant factor. It causes the power arc to migrate from the supply point. This principle is applied when designing power arc protective fittings. Round rod profiles typically made from galvanised steel are used here. It is worth noting that the migration speed decreases when the diameter increases [4].

If the power arc is guided to an end burning point, it must be intensified in such a way that thermally induced material consumption (through melting, vaporisation, etc.) does not affect functionality after the power arc has been extinguished. Sharp-edged melting points, for example, may cause corona discharges. In general, spheres or, in the case of higher transmission voltages, hollow spheres are used; their diameters are determined by the arc current and corona behaviour of the entire string. When porcelain long rod insulators were used, it was also important to consider the distance between the insulator and the end burning point as well as the associated decrease in thermal radiation on the surface of the insulator.

The effectiveness of a power arc protective fitting will be discussed below (Figs. 9.5 and 9.6). Once ignited, the protective fitting should quickly take over the power arc and guides it to the end burning point. The use of ring-shaped power arc protective fittings has proven itself for a quick take over because a pollution flashover can occur at any position on the insulator periphery. When arcing horns are used on their own without a ring to "catch" the pollution flashover, the process

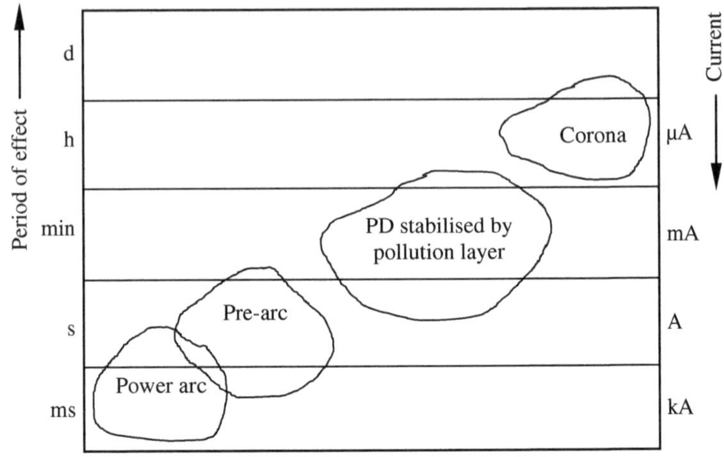

Fig. 9.3 Effect of forces on a power arc

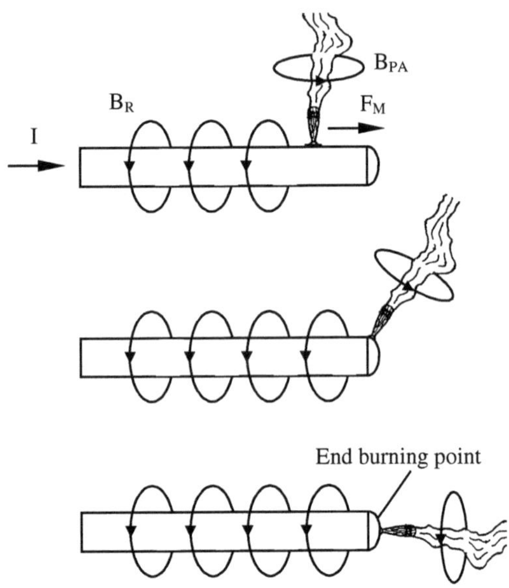

Fig. 9.4 Guiding the power arc through the magnetic self field at rod electrodes

of taking over the power arc can be delayed in such a way that the insulator is already damaged. The ring takes over the power arc by means of direct contact, thus a ring that has been placed over the insulator end fitting (in axial direction) accelerates the take over of the power arc.

The distance x (Fig. 9.6) between the fitting rim of the insulator and the edge of the ring must be defined accordingly and, in the case of higher voltages or the use of composite insulators, it must be coordinated with corona behaviour. The uni-directionally supplied power arc in Fig. 9.5 "slowly" migrates to the side opposite

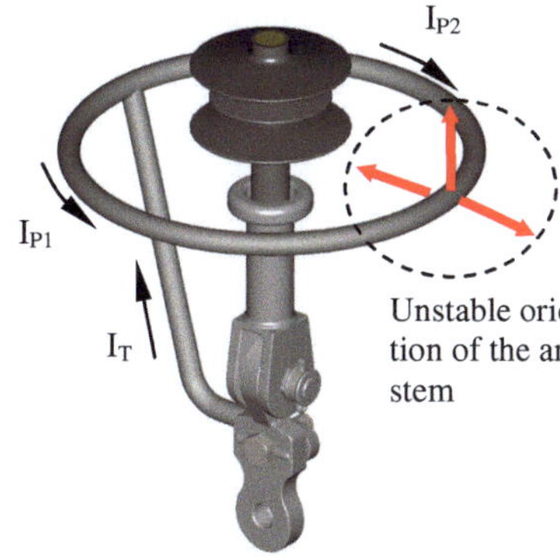

Fig. 9.5 Guiding the power arc in the case of a unidirectional supply (but two partial currents), a closed ring and no end burning point

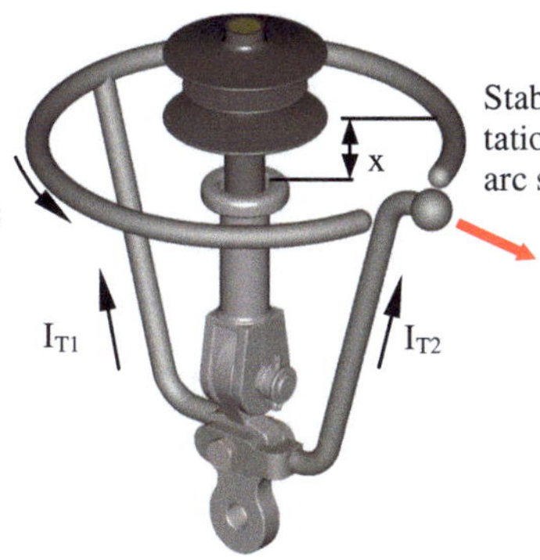

Fig. 9.6 Guiding the power arc in the case of a unidirectional supply, an open ring and an end burning point

to the supply side, either accelerated or delayed by the two partial currents I_{P1} and I_{P2}. Due to the absence of an end burning point, however, its blow direction is undefined and the insulating material is endangered. Optimal guidance is achieved in accordance with the set-up in Fig. 9.6. Once the power arc has been taken over by the ring, it migrates to the front surfaces of the ring electrode (migration along the lower half of the ring is shown) and is taken over by the arcing horn with the end burning point when the arc comes into contact with. The power arc supply is always unidirectional, thus forcing it to be stable in one direction (stable orientation).

Various solutions have been developed for the combined requirements for corona and power arc protection. These include the C-profile ring [5] and the tandem ring (also known as a bundle ring [6]).

With regard to the design for the end burning point, [7] makes reference to a German imperial patent from 1941 [8] in which a "crater-shaped" recess has the following advantages for power arc guidance and the secondary thermal effect on the end burning point:

- Prevention of an exposed formation of welding beads (which are a source of corona)
- Preferential burning of the power arc from the recess itself or from the edge of the recess
- Rotation of the power arc on the edge of the "crater"

Fig. 9.7 Insulator end fitting with a corona shed after a power arc test

This idea was developed further and the shape and depth of the recess and its plasma pressure relief in the end burning point were optimised [9–12]. Tests on power arcs have demonstrated the effect of this recess: A single suspension string for 420 kV was tested in accordance with IEC 61467 [13] (Figs. 9.7, 9.8, 9.9).

The testing parameters were 50 kA for 0.5 s. The introduction of a copper wire short circuit for variable axial positions directly at the composite insulator simulates a pollution flashover triggering a short circuit; the power arc must cross over from the insulator fitting to the ring on the power arc protective fitting.

After the test, a thin layer of vapour-deposited copper remains on the surface of the insulating material (Fig. 9.7). The seal is not subjected to a thermal attack because it is protected by a corona shed made from highly ageing-resistant HTV silicone rubber (see also Sect. 7.3.2.3).

The insulator set with power arc protective fitting passed the test (Fig. 9.8). The melting loss at the end burning point and lower ring are of interest (Fig. 9.9): It is clear that the migrating power arc remains on the lower ring until it switches to the end burning point. The end burning point acts like a surface electrode. Suppression of the current (skin effect) at 50 Hz, the use of steel and a reduction in the number of "migration-barrier" inverting flow lines through the interrupted surface causes the power arc to migrate to the recess (desired behaviour). As a result of the low magnetic field in the recess, the arc root is retained at this position of the end burning point. If one assumes that the current supply at the edge of the recess is irregular, the power arc may rotate around the edge of the recess. The fact that this effect occurred in this test is most likely due to the absence of any significant welding beads.

Fig. 9.8 Power arc protective fitting after a power arc test

Fig. 9.9 A closer look at the melting loss at the recess point

In terms of mechanics (i.e., the accessibility of power arc protective fittings), designs with a unidirectional supply are less resilient than power arc protective fittings with a symmetrical design.

Because its unidirectional mechanical (free) side faces the supply point, the tandem ring, in particular, can become so deformed that its corona protection effect can no longer be ensured. It is recommended that the installation team receives appropriate training. If composite insulators are used, the team should also receive training on their proper handling [14].

The following is a summary of key criteria for power arc protective fittings irrespective of the insulator technology used:

- Use of galvanised steel
- Use of the smallest possible diameter for the ring electrode
- An open ring at the end burning point
- Unidirectional power arc supply for both the ring and end burning point
- Guidance of the power arc through recesses at the end burning point in the case of high short-circuit currents
- Dimensioning of the end burning point from the perspective of corona behaviour and the expected power arc energy
- Use of open rings as a replacement without opening the entire string

9.3 The Power Arc Effects

9.3.1 Porcelain Long Rod and Porcelain as well as Glass Cap and Pin Insulators

In addition to the melting effect on unprotected insulator end fittings, which applies to conventional and composite insulators, the insulating material is directly endangered by the heat transferred via the arc root or stem. Insulating materials are generally poor heat conductors and cannot dissipate the heat fast enough.

The local temperature rises cause mechanical stresses in the material; due to the brittleness of glass and porcelain, the fracture, which can appear as a partial shed fracture or, in the case of porcelain long rods, as a core fracture, occurs when the limit of elasticity is exceeded. In the case of unprotected porcelain long rods, power arc resistance diagrams were determined experimentally in the past (Fig. 9.10). Here, a distinction was made between two damage modes, namely glaze abrasion and a reduction in strength, when a certain power arc volume was exceeded [7]. This type of test does not exist for composite insulators. Furthermore, conventional insulators are now usually tested in accordance with IEC 61467, which was first published as a Technical Report in 1997 [15]. Depending on the burning conditions of the power arc and the level of insulator protection, the effects shown in Figs. 9.11, 9.12 and 9.13 may occur in porcelain insulators.

The defects that power arcs cause in glass cap and pin insulators are similar to those that occur in porcelain cap and pin insulators (Figs. 9.14, 9.15, 9.16).

In the case of power arc tests conducted in the laboratory, damage resulting from melting loss or a shed fracture is permitted if the residual strength is at least 70 or 80 % of the specified mechanical load (depending on the version of the standard). In order to satisfy the minimum requirement for cap and pin insulators, the first three, the last three and the middle three elements of a string must undergo

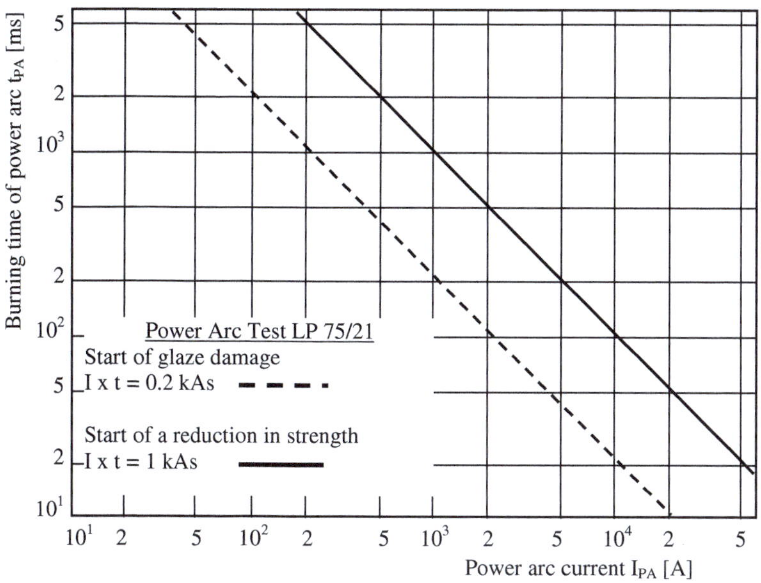

Fig. 9.10 Power arc resistance diagram [7]

Fig. 9.11 Molten glaze [16]

mechanical testing. For (cap and pin) insulators that may be prone to dielectric breakdowns, an electrical test is also performed to diagnose potential dielectric breakdowns.

9.3.2 Composite Insulators

The reaction of an unprotected composite insulator to a power arc (in comparison to conventional insulators) must be assessed under equivalent conditions as outlined below. The result after a power arc of 50 kA and a burning time of 0.3 s in

Fig. 9.12 Cracked porcelain cap [17]

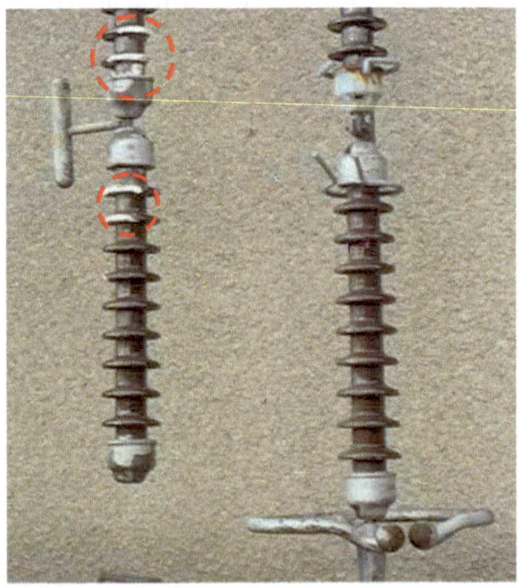

Fig. 9.13 Damage to a porcelain long rod

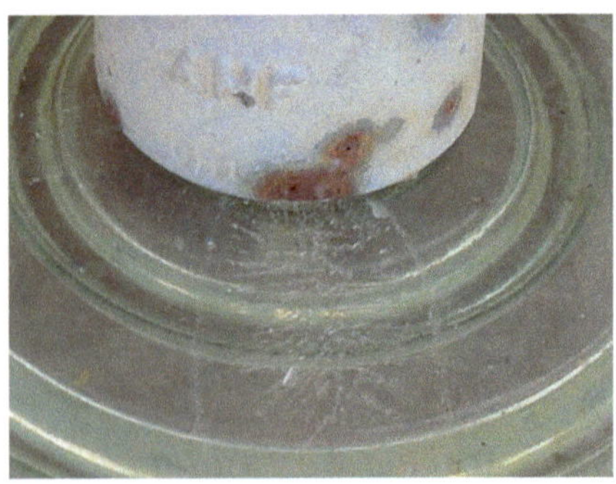

Fig. 9.14 Molten glass [16]

Fig. 9.17, 9.18 and 9.19 is considered to be representative for non-flammable housing materials. In accordance with the relevant standard, a short-circuiting wire was used to initiate the power arc. Once the short circuit has been activated, it

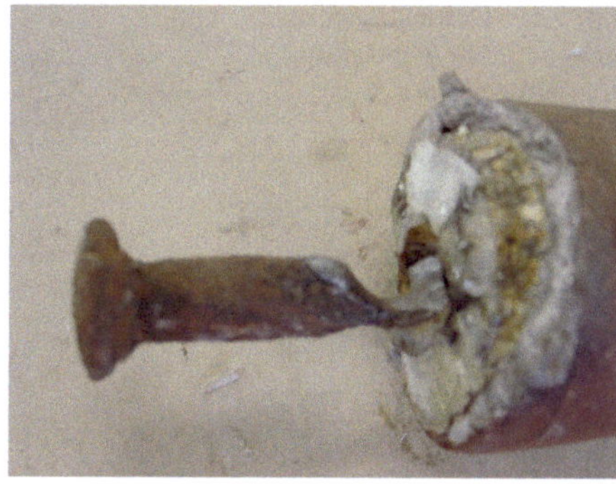

Fig. 9.15 Cracked cap and molten ball [16]

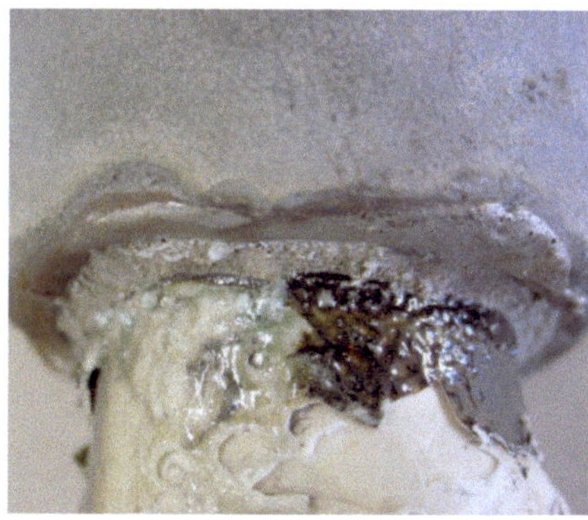

Fig. 9.16 Molten cement [16]

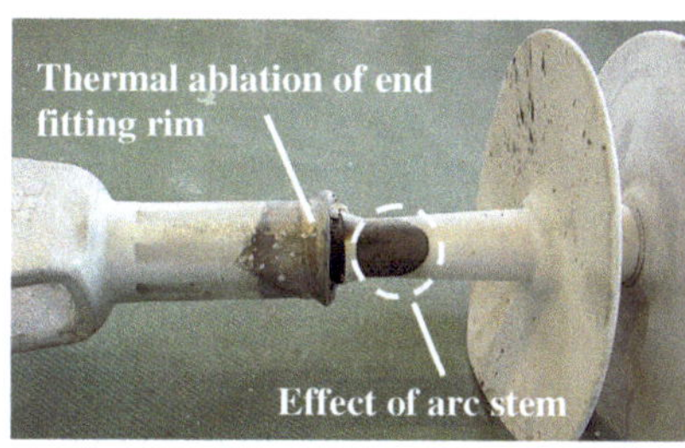

Fig. 9.17 Defunctionalised seal and molten end fitting rim on the HV side

Fig. 9.18 Housing appearance of the insulator centre

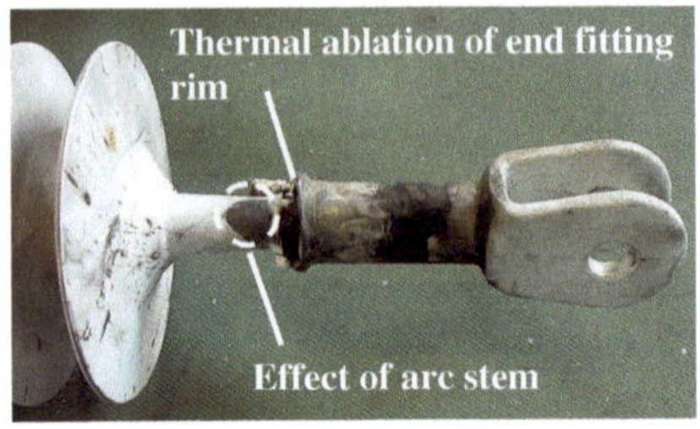

Fig. 9.19 Defunctionalised seal and molten end fitting rim on the earth side

Fig. 9.20 Earth side of a 25 kV-catenary insulator after a flashover caused by a bird

vaporises the wire and triggers the power arc whose roots can burn directly on the rim of the steel fitting and melt it (Figs. 9.17 and 9.19). The arc stems resulted in minor ablation of the insulator housing over the fittings on the rod; no material fractures occurred at these points. In contrast to Fig. 9.7, this insulator was sealed by casting the end fitting. If the power arc burns in this area and produces the damage pattern shown, the sealing system is also affected. The vapour-deposited remains of the ignition wire are visible at the centre of the insulator (Fig. 9.18); the insulator housing was not damaged (dielectric breakdowns, rod exposure, etc.). The mechanical residual strength was 102 kN, which means that the 80 % requirement was fulfilled (SML = 120 kN) [13].

The first conclusion to be drawn is as follows: If, as in the case of conventional insulators, a power arc damages load-transmitting steel components, which usually have the same connecting dimensions as composite insulators (e.g., in accordance with IEC 60120 [18] for ball and socket), so badly that the specified mechanical load is reduced, this applies irrespective of the insulator technology used. In contrast to conventional insulators, however, the insulator housing of composite insulators is not damaged by thermally induced mechanical stresses.

This is also confirmed by a case involving the 25 kV railway network of a high-speed route (Figs. 9.20, 9.21). The power arc was initiated by a parrot-like bird whose bite marks were evident on the sheds. The power arc burnt between the tongue end fitting on the earth side and the tubular end fitting on the high voltage

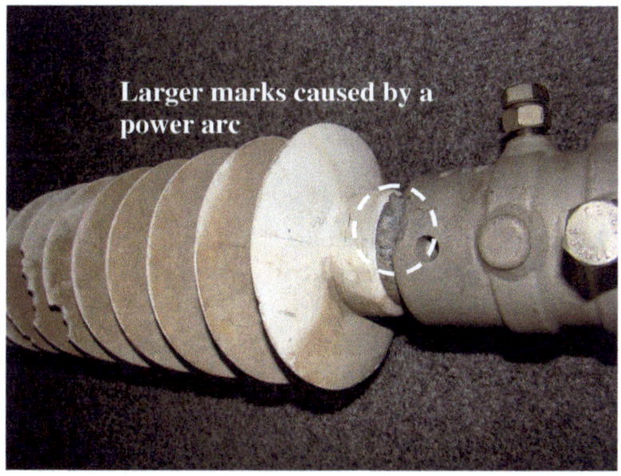

Fig. 9.21 HV side of a 25 kV-catenary insulator after a flashover caused by a bird

side. The tongue end fitting has very few power arc marks because the clevis fitting surrounding it took over the power arc. As it will be shown later, aluminium's reaction is more critical when it is in contact with the root. The low melting loss is due to the fact that network protection was triggered quickly. In contrast to Figs. 9.17 and 9.19, the sealing interface remains intact because it was sealed by means of overmoulding and the mechanical strength is not reduced. The insulator was removed so that a root cause analysis could be performed.

9.3.3 Summary Comparison of Effects

The examples provided here, as well as the many years of service experiences, show that both conventional and composite insulators may be damaged by the "power arc" event if they are not protected in accordance with the relevant operating conditions. This particularly applies to the thermal effect on load-transmitting metal fittings (apart from the rail sector or in the case of device insulators generally made from hot-dip galvanised steel). The main difference lies in the behaviour of the insulating material; polymeric materials do not suffer from fractures or oxidative decomposition if the composition is correct (flame-retardant fillers or additions). From an operational behaviour perspective, and on the basis of many years of experience, other differences have emerged, which are presented in the comparison below (Table 9.1).

9.3.4 Peripheral Protection

When protecting an insulator string/set, one must consider having a safe burning zone in which the power arc can burn until it is extinguished, without damaging

Table 9.1 Probability of occurrence and effects of power arcs

	Porcelain long rod Glass or porcelain cap and pin insulator	Composite insulator (long rod)
Probability of occurrence of a pollution flashover[a]	High	Low, for a hydrophobic surface effect[b] and due to a smaller diameter
Probability of occurrence of an overvoltage-caused flashover	In the first approach: same for an identical striking distance, power arc in the shortest air clearance or the sum total	
Thermal damage to the insulator if the power arc is poorly guided[c]	Insulator may tear or split	No tears or splits, no significant abrasion of materials that have a high thermal and erosion resistance[d]
Effect on the mechanical insulator function when the power arc is poorly controlled	Loss of mechanical strength if the attachment to the fitting (cement) is thermally degraded and experiences an irreversible reduction in strength or if the (long rod) insulator fractures	If the attachment to the power arc protective fitting is unsuitable, the fault current flows over the insulator fitting or, if there is no power arc protection, the crimp connection may lose some of its strength when the rod's TG is exceeded[e]
Effect on the electrical insulator function when the power arc is poorly guided	A reduction in the withstand values for the flashover voltage if the striking distance or creepage distance is shortened as a result of sheds that have been destroyed	Damage to the seal and housing critical[f], inappropriate housing materials can form conductive residues

Explanatory notes

[a] Assuming that pollution flashovers may occur

[b] If there is a hydrophobic effect (recovery, transfer, etc.)

[c] The duration, stability and intensity of the power arc are damaging factors

[d] Elastomers that are filled with ATH and have successfully obtained the flame classification V0 IEC 60695-10-11 [19]

[e] Glass transition temperature [20]

[f] The following are critical: damage to the seal at the triple junction or exposure of the rod to environmental conditions

adjacent parts. This includes preventing molten steel from migrating to or dripping onto the conductor, which has outer aluminium wires (Fig. 9.22). Old conductors with reduced conductivity between the adjacent outer aluminium wires are at particular risk if the fault current is managed by only a small number of wires, which are then thermally overloaded.

The following is another example from 25 kV rail operations: For various different reasons, cumulative flashovers occurred on railcars and caused peripheral damage (Fig. 9.23). The challenge in this application scenario was that, due to space constraints, there was effectively no safe burning zone. This problem was solved by implementing a combined package of measures to eliminate the causes

Fig. 9.22 Damage to the arcing horn and conductor in a 123 kV network

Fig. 9.23 Virtually no safe burning zone due to the compact design of the railcars

of flashovers (extending the striking distance or creepage distance and defining a cleaning regime).

In the next section, the design of power arc protective fittings will be discussed from the perspective of material selection and the permitted current density.

9.4 Designing Power Arc Protective Fittings

9.4.1 Material Selection

When evaluating materials, a distinction is made between steel and aluminium because both are used in insulator strings/sets: Steel is specifically for power arc protection, which can be combined with corona protection, while aluminium is for

Table 9.2 Material comparison between aluminium and steel from the perspective of power arc protection [1]

Material	Density [g/cm^3]	Melting point [°C]	Combustion heat [kJ/g]	Melting heat [Ws/cm^3]	Thermal conductivity [W/(K cm^3)]	Tensile strength [N/mm^2]	Specific resistance [Ω cm]
Steel	7.85	1500	5.6	2104	0.54	400	15×10^{-6}
Cast aluminium	2.7	620	30	1025	1.88	200	4×10^{-6}

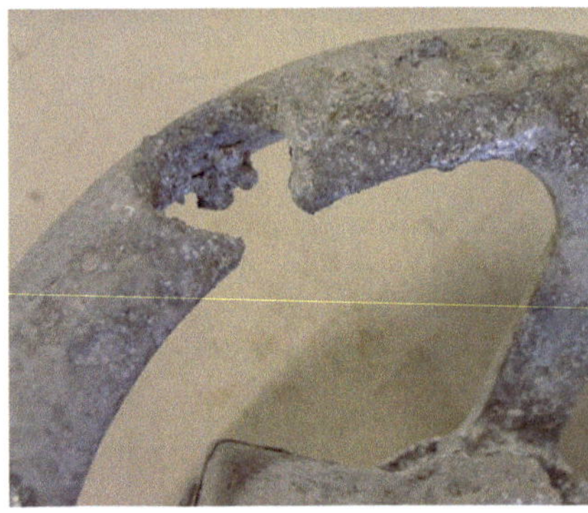

Fig. 9.24 Power arc damage to an aluminium corona ring [16]

Fig. 9.25 Power arc damage to an aluminium corona collar [16]

corona protection only. If a power arc burns over a string that only has corona protection rings made from aluminium, the rings are exposed to the power arc in accordance with the arc roots. The material properties of both are shown in Table 9.2. Note that aluminium has a low density and could have the edge over steel in terms of weight savings for a comparable geometry. However, the other properties for a material exposed to a power arc are very unfavourable: The arc movement rate on a steel fitting is 50 % higher than on an aluminium fitting. In other words, the damage-inducing combination of a slow movement rate and a low melting point occurs with aluminium and can cause holes to form in the profile (Fig. 9.24). If aluminium is ignited, the heat of combustion is five times higher

Fig. 9.26 Intensive thermal deterioration of the polymeric housing due to the aluminium combustion [16]

Fig. 9.27 Use of a steel protective fitting to protect an aluminium end fitting (31 kA for 0.5 s)

than when steel is ignited. The thermal radiation is also more intensive. In the case of solutions that involve mounting aluminium rings close to the insulator fitting (Fig. 9.25), considerable damage to the insulating material may occur if the direction of the power arc is unstable (Fig. 9.26).

The better thermal conductivity of aluminium is quite detrimental when combined with a low melting point because the critical temperature is quickly transferred to the inside of the profile and deformities may occur.

The negative effect of this combination of properties was confirmed in a power arc test where the melting loss for aluminium electrodes was four to five times higher under the same conditions [3]. It was also observed that material abrasion was not only oxidative but also occurred as metal vapour, which can deposit itself on insulators of adjacent strings as a permanently conductive layer.

The literature contains references to various values associated with a melting loss resulting from the power arc effect: As expected, steel has a lower value (0.4 $cm^3/(kA\ s)$) compared to aluminium (1.4 $cm^3/(kA\ s)$). This difference by a factor of 3.5 is smaller than the factor of 4–5 mentioned above and is due to the fact that different tests were performed. If one calculates the value that causes complete abrasion when there is large spherical end burning point with a diameter of 80 mm, which is often used in 420 kV networks in order to comply with the interference voltage values (Fig. 9.27), the result is 670 kA s.

With such a high level of redundancy, however, it should be noted that the formation of a sharp-edged melting loss geometry can cause corona discharges. The melting loss shown without the formation of welding beads occurred during a power arc test of 30 kA for 0.5 s. The aluminium fitting on the post insulator was securely protected by a power arc protective fitting made from steel.

In addition to the amperage or current density, the time required to bring the metal at the end burning point up to the decomposition temperature is a key factor for the thermal effect of the power arc, which can cause material abrasion. In [3] and [21] reference is made to the threshold value $S^2 \times t$. Virtually no melting loss is experienced under this threshold value. This value is $5 \times 10^4\ (\mathrm{A/cm^2})^2\mathrm{s}$ for iron/steel. It has been analytically proven that virtually no melting loss is experienced when the iron/steel is exposed to the power arc for a very short time. Here, the power available for material decomposition was calculated as follows:

$$P_{MD} = U_{PA} \times I_{PA} \times k_H - k_P / \sqrt{t} \tag{9.1}$$

and

$$k_P = \sqrt{\lambda \times c/\pi} \times A_E \times \vartheta_0 \tag{9.2}$$

P_{MD}	Required power for material decomposition
U_{PA}	Power arc voltage in V
I_{PA}	Power arc current in A
k_H	Factor between 0…1, which determines the part of the arc power to heat the end burning point
t	Time in s
λ	Thermal conductivity in W/(cm K)
c	Thermal capacity in Ws/(cm^3 K)
ϑ_0	Temperature on the rod front end
A_E	Cross-section of rod electrode

9.4.2 Density of Fault Current in the Elements of an Insulator String

In addition to the actual thermal effect of the power arc, the heat-generating effect of the fault current in the string elements is another important design criterion. As already mentioned in Sect. 9.2.3, small ring electrode diameters are beneficial to the migration speed. Tests involving steel electrodes and arc currents of 1 kA for an electrode distance of 25 mm have led to the empirical relationship $v \sim d^{-x}$ (with $x = 0.66$ for the current of 1 kA) [4]. The larger the diameter d, the more likely the migration speed v is hindered by inverting current flow lines, which supply the power arc in the opposite direction of its migration.

In this context, it is worth mentioning that corona rings manufactured from aluminium have a comparatively large diameter for a good effect of electrical field control. The aforementioned material "shortcomings" in the case of a power arc are further intensified as a result of the reduced migration speed.

Out of the desire for a small diameter, however, physical limits are set from the perspective of combined corona protection and the permitted ampacity of the steel cross-section. The current flow through a steel cross-section resulted in a temperature rise to which the zinc layer on the surface, which is an effective coating for galvanisation protection, is exposed. Zinc has a melting point of 440 °C, which restricts the permitted temperature rise. If one assumes a standardised current density of 1 s and an ambient temperature of 35 °C, this produces a value of 70 A/mm^2, which increases the temperature of the fitting to 400 °C, if the fault current flows through a cross-section of the fitting.

For this reason, the following two current densities for string components are defined in the German standards, in particular [22], and have proven their worth for many years:

- String/set components that participate directly in transmitting the load should not exceed a fault current density of 70 A/mm^2 over a period of 1 s. These include shackles, eyes, and so on.
- String/set components that do not contribute directly to the load transmission should not exceed a fault current density of 80 A/mm^2 over a period of 1 s. These include power arc protective fittings, for example.

The current density is defined without taking the skin effect into consideration. As next, tests relating to the power arc resistance of composite insulators will be discussed below.

9.5 Tests Related to Power Arc Protection

9.5.1 Materials Testing of Housing Materials (Design Test)

In 1982, CIGRE Working Group WG 22.10, which was founded in 1978, issued a paper entitled the "Technical basis for minimal requirement for composite insulators" to Technical Committee TC 36 (Insulators) of the IEC, which became a landmark foundation document for IEC (6)1109 [23], published in 1992. This paper was published in the ELECTRA journal in 1983 [24]. Even at this time, a distinction was made between the flame resistance of the polymeric housing material and its behaviour in the event of a power arc. As shown in Fig. 9.3, the

Table 9.3 Criteria for obtaining the V0 or HB40 classification [19]

Criterion	Vertical burning method—required to successfully obtain the V0 classification	Horizontal burning method—required to successfully obtain the HB40 classification
After flame time of individual specimens	$\leq$10 s	Minutes
Progress of the after flame or afterglow up to the specimen holding clamp	Not permitted	Visible afterburning not permitted, *but if it occurs:* The burning distance should not exceed 100 mm, *but if it occurs:* The burning rate of $\leq$40 mm/minute should not be exceeded.
Ignition of cotton indicator pad by dripping flames or particles	Not permitted	Permitted, must be logged.

housing material can be stressed by various forms of discharge with varying amperages and exposure times. Performing and passing material tests is an important part of choosing suitable materials. However, it does not replace power arc tests, which evaluate the behaviour of the entire string/set.

When the product standard IEC 61109 was first supplemented in 1995 [25], a successful flame test in accordance with IEC 60707 [26] was implemented as a design test. However, not all polymeric insulating materials (especially for the distribution level) successfully obtained the required classification V0. For this reason, and to evaluate the testing accuracy, CIGRE Working Group D1.14 performed comprehensive material tests including a test to assess the fire risk in accordance with IEC 60695-10-11 [19], which effectively took over from the standard IEC 60707. Here, the required classification V0 was compared against HB40, which was proposed as an alternative classification (Table 9.3). It is evident that both procedures are not equivalent and the requirements associated with a tested material obtaining the HB classification are lower. Systematic analyses confirmed that materials that obtain a V0 classification also safely obtain the HB40 classification. However, the reverse is not true.

The standard IEC 60695-11-10 permits different selectable thicknesses of the material to be tested. The tested thickness (up to 13 mm is permitted) must be logged. The greater the material thickness, the easier it is to obtain a particular classification. The results of tests involving different thicknesses and materials are not directly comparable. For this reason, the standard IEC 62217 [28] specifies a material thickness of 3 mm for the flame test. From a technical perspective, 3 mm corresponds, for example, to the minimum required thickness of the insulator housing on the rod at the insulator shank.

The following is a summary of the results of various materials tested for outdoor insulators in distribution and transmission networks (Table 9.4).

In addition to these material analyses, power arc tests were performed and current service experiences were analysed from a fire perspective.

Table 9.4 Results of systematic flame tests [27]

Material	VB method		HB40 method	
	Without thermal pre-stressing	With thermal pre-stressing	Classification in accordance with IEC 60695-11-10 (test specimen thickness: 3 mm)	More stringent classification with a permitted burning distance of < 25 mm (test specimen thickness: 3 mm)[7)]
Epoxy resin[1)]	Failed	Failed	Passed	Passed
EVA (ethylen-vinyl acetat)[2)]	Failed	Failed	Passed	Passed
EPDM (ethylen propylen dien monomer)[3)]	Passed V0	Passed V0	Passed	Passed
Silicone rubber[4)]	Passed V0	Passed V0	Passed	Passed
Silicone rubber[5)]	Passed V0	Passed V1	In the case of flexible test specimens, can only be performed using supporting resources	
Silicone rubber[6)]	Passed V0	V0… failed[8)]	In the case of flexible test specimens, can only be performed using supporting resources	
Ertacetal (POM-polyoxy-methylen)	Failed	Failed	Passed, but with a burning distance of 75 mm	Failed
Ertalon (PA6-polyamid)	Passed, but V2 only	Failed	Passed	Passed

Explanatory notes:
Grey = service experience in the distribution network, white = service experience in the distribution and transmission networks, dark grey = reference material for the purpose of comparison

1) - Commercially available materials based on epoxy resin
2) - Special formulation by a manufacturer
3) - Commercially used material
4) - HTV-silicone rubber with an ATH filler
5) - RTV-silicone rubber without ATH
6) - LSR-silicone rubber without ATH
7) - The 25mm limit was specified due to the maximum burning distance for HB40-materials that have had a positive long-term service experience in the distribution network.
8) - The results revealed a strong dependency on the material formulation and thickness tolerance.

Fig. 9.28 Operational insulator housing damaged by a power arc during the use of a highly flammable housing material

Table 9.5 Re-Specifying the flammability requirements for polymeric insulating materials

Application	Satisfied the V0 criterion[a]	V1 criterio[a]	HB40-25 mm criterion[b]	Inclined Plane Test with a constant 3.5 kV[c]
Line insulators ≤72.5 kV			x	x
Line insulators >72.5 kV	x			
Other insulators incl. device insulators ≤145 kV			x	x
Other insulators incl. device insulators >145 kV		x		

Explanatory notes
[a] IEC 60695-11-10 [19]
[b] CIGRE D1.14 [27]
[c] IEC 60587 [29]

It has been demonstrated that unsuitable materials (Fig. 9.28) can be identified in a combined flame and erosion resistance test.

The criteria for materials testing were specified for use in the distribution or transmission network [27] and proposed to IEC Working Group WG 12 for a revision to the standard IEC 62217. These re-specifications can be summarised as shown in Table 9.5.

9.5.2 String/Set Test from a Current Density Perspective: Short Circuit Test

The first standard in relation to conducting power arc tests on insulator strings was published in 1997 [15]. Up until that time, a short circuit test was frequently specified as a way to qualify insulator strings that short-circuited the flashover path at a low resistance.

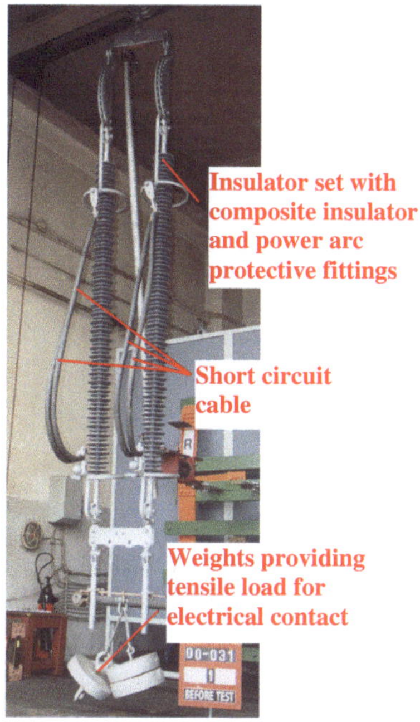

Fig. 9.29 Set-up for the short circuit test

Short circuiting occurred by means of a cable with a sufficient cross-section to ensure that no power arc occurred. Examples of this test are shown in Figs. 9.29 and 9.30. The string components were tested solely from the perspective of their resistance to the resulting current density, not from the perspective of the more complex behaviour in relation to a power arc. For this reason, only a minor colour change was diagnosed. The zinc layer did not peel away and, as expected, there was no thermal material abrasion (Fig. 9.30).

9.5.3 String/Set Test from a Power Arc Effect Perspective: Power Arc Test

The behaviour of an insulator string is evaluated in the more event-relevant power arc test for which IEC 61467 provides good guidance. Figure 9.31 shows the moment when a burning 30 kA power arc occurs over the post insulator of an insulated cross-arm. As a supplement to the short circuit test, the following key information is available in the power arc test:

- Forces and effects of the current density on the entire insulator string
- Formation of the power arc
- Movement of arc stems
- Thermal effect of the power arc on the insulator and metallic string components

Fig. 9.30 Result after the short circuit test

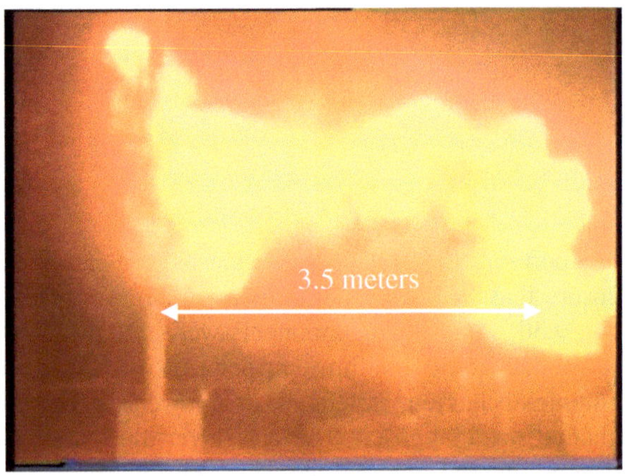

Fig. 9.31 Power arc over a post insulator of a 420 kV insulated cross-arm (30 kA)

The standard IEC 61467 uses various test scenarios to consider the different installation positions in an overhead transmission line (Fig. 9.32). Both Figs. 9.32a and b are critical cases because the "outer current loop" (see also Fig. 9.2) has a stabilising effect on the power arc, and the blow-off direction (safe burning zone) of the string is determined solely by how the power arc protective fitting guides the power arc. For this reason, the tests, which are described as test "Series X" [13] and are simulated by the set-up in Figs. 9.32a and b are often implemented as being representative of the scenarios in Figs. 9.32c and d.

Recommendations have been made for different combinations of the short-circuit current I_{SC}, time of short circuit t_{SC} and installation location in order to simulate typical network conditions (e.g., response time of network protection). Other than these, real-life network conditions can be simulated using modified values. Experience has shown that smaller short-circuit currents cause less mobile

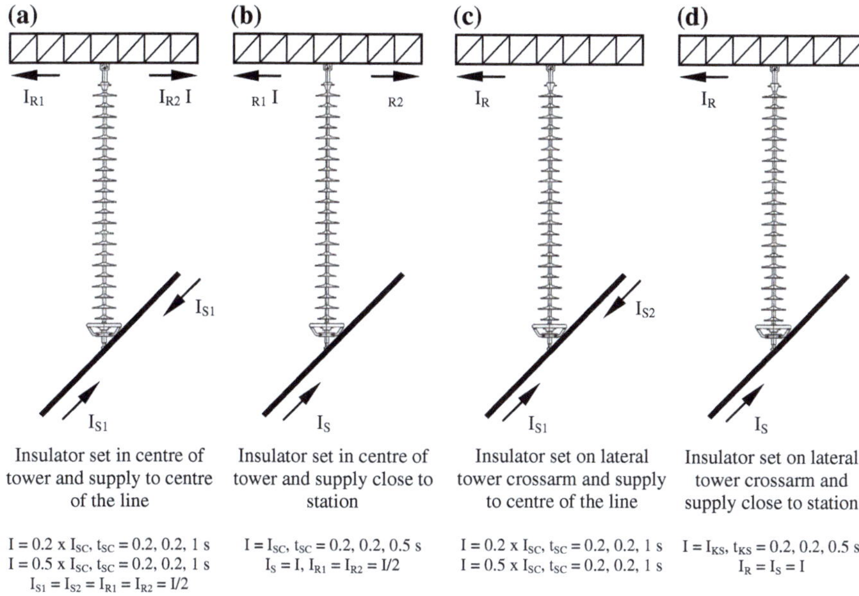

Fig. 9.32 Test scenarios in accordance with IEC 61467—from *left to right*: **a**, **b**, **c**, **d**

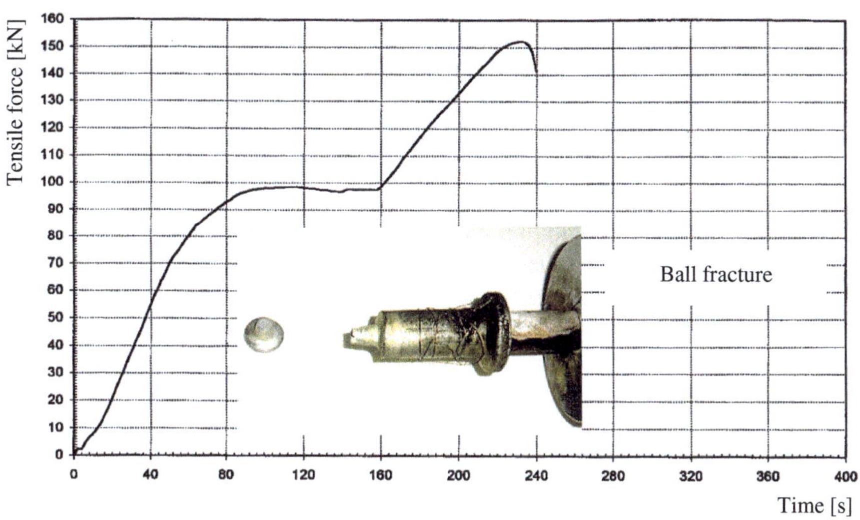

Fig. 9.33 Result of tensile test after 25 kA/0.5 s power arc test

power arcs and when typically tested with a longer short-circuit time, they can have a more damaging effect.

After the power arc test, diagnostic tests are performed, which involve a visual inspection in accordance with certain criteria, a mechanical test and, in the case of cap and pin insulators, an electrical test. Figure 9.33 shows the example of a tensile test after a power arc test; the ignition wire is still identifiable. The composite insulator was designed for a specified mechanical load of 120 kN and, due to the effective power arc protective fitting, its mechanical strength is not affected by the power arc.

9.6 A Selection of Project Examples

9.6.1 Misuse of Power Arc Protective Fittings for Cap and Pin Insulator Strings in Composite Insulator Strings

If one compares an insulator string that comprises cap and pin insulators with a one-piece composite long rod insulator, the different design for the composite long rod insulator results in a shorter striking distance for a given installation length. The striking distance is shortened further as a result of the mandatory installation of corona protection rings (HV side, at least as of 245 kV Fig. 9.34).

This known difference is irrelevant when erecting new overhead transmission lines because the requisite striking distance for the composite insulator can be set over a longer installation length. The situation is different if an existing line of cap and pin insulators is retrofitted with composite insulators. Then, it is only feasible to extend the composite insulators if the minimum distance between the conductor and the ground can be maintained within the spans. In most cases, the installation length of the insulators is retained and the insulation level is reduced. In the example shown, which refers to a 245 kV network, the level of the lightning impulse withstand voltage falls from 1,050 to 950 kV [30]. This compromise is widely accepted because the line performance is enhanced due to the significantly reduced probability of a pollution flashover compared to cap and pin insulators.

This substitution becomes technically critical if composite insulator strings are not procured as "one unit" (packet solution) and the components from individual suppliers are not evaluated in terms of their compatibility with one another. Alternatively, composite insulators are procured and hardware stocks available for cap and pin insulator strings are installed. Such an example is shown in the following diagrams of a 245 kV single suspension string (Figs. 9.35, 9.36, 9.37).

The conventional design for cap and pin insulators does not require any corona protection at 245 kV. Here, rackets are used for power arc protection (Fig. 9.35). If flashover occurs, the power arc burns between the protective fitting on the earth side, which is designed as a horn, and the racket on the high voltage side. This type of string design for cap and pin insulators has proven its worth for many years and is considered to be the state-of-the-art design.

As a result of a high level of pollution, the only way to increase the reliability of the overhead transmission line was through the use of composite insulators. Their

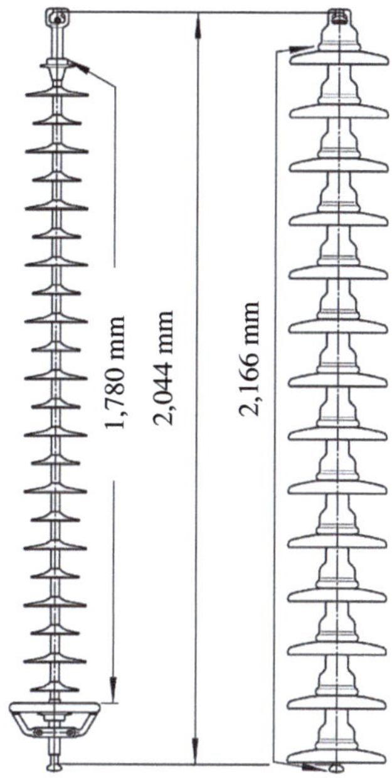

Fig. 9.34 Comparison between the striking distance for composite long rod and cap and pin insulators

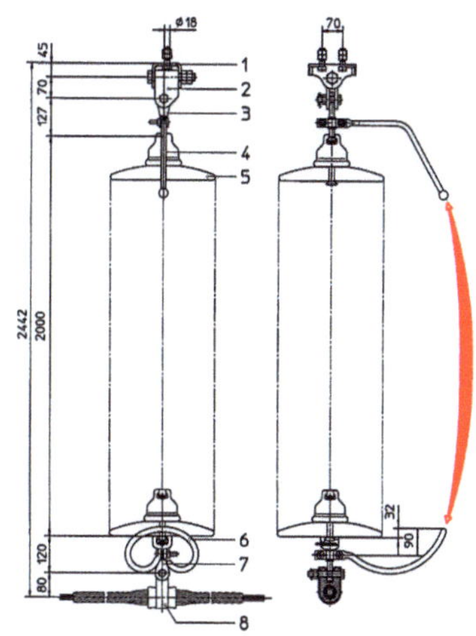

Fig. 9.35 Conventional insulator set *1* U-bolt; *2* Hinge; *3* Ball-tongue fitting; *4* Power arc horn; *5* Cap and pin insulator; *6* Racket protective fitting; *7* Socket-tongue fitting; *8* Suspension clamp

installation length had to correspond to the installation (Fig. 9.36—2,000 mm). At the project planning phase, it was fortunately known that the user intended to replace only the insulator and retain the peripheral string elements.

Consequently, the racket on the high voltage side would not have been able to provide protection in the event of a flashover (Fig. 9.36). Furthermore, the level of corona protection is inadequate. The solution was to use a combined corona and

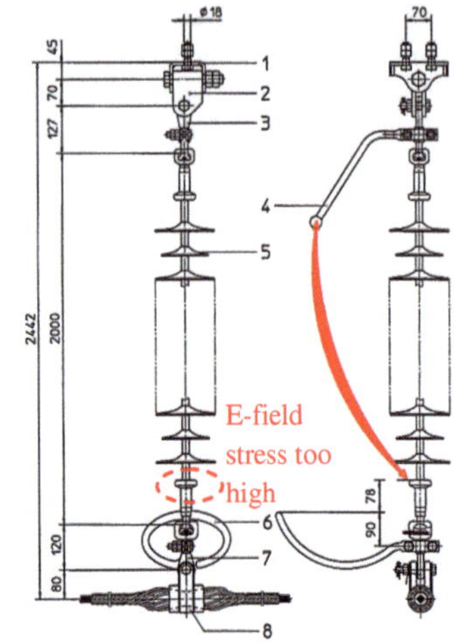

Fig. 9.36 Composite insulator set with racket *1* U-bolt; *2* Hinge; *3* Ball-tongue fitting; *4* Power arc horn; *5* Composite insulator; *6* Racket protective fitting; *7* Socket-tongue fitting; *8* Suspension clamp

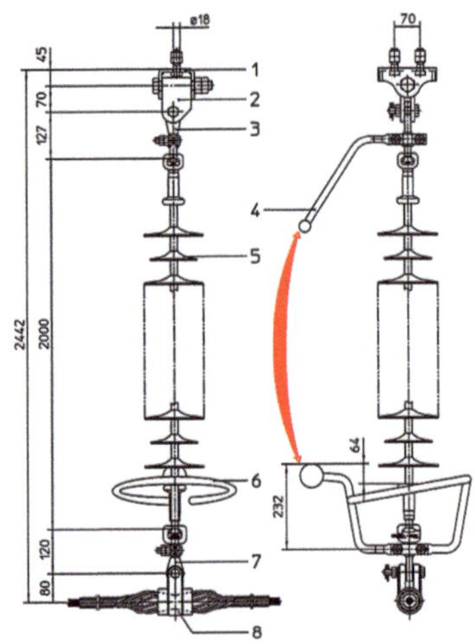

Fig. 9.37 Composite insulator set with adapted correct ring *1* U-bolt; *2* Hinge; *3* Ball-tongue fitting; *4* Power arc horn; *5* Composite insulator; *6* Ring for combined corona and power arc protection; *7* Socket-tongue fitting; *8* Suspension clamp

power arc protective fitting made from galvanised steel whose overall height was adapted to the length of the composite insulator fitting (Fig. 9.37). The diameter of the spherical end burning point was determined by the radio interference voltage and corona values specified. If a power arc occurs, now it burns between the horn and the protection ring. In the case of higher transmission voltages, a ring must also be provided with an end burning point on the earth side in order to ensure adequate field control.

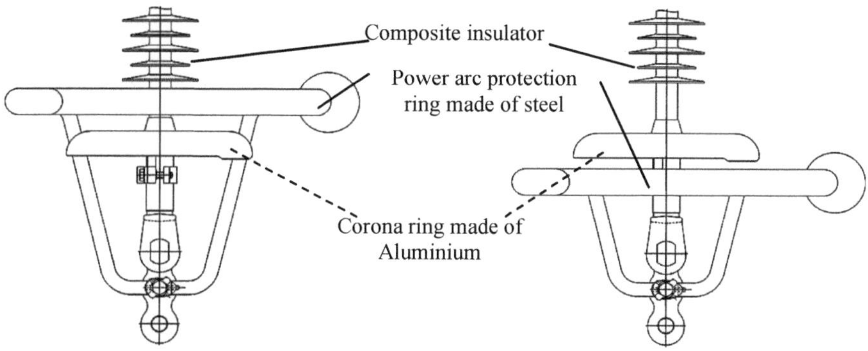

Fig. 9.38 **a** Correct coordination. **b** Incorrect coordination

9.6.2 Coordination Between a Corona Ring and Power Arc Protective Fitting

Because of the wide range of production processes and solutions for the sealing system, etc., the corona protection ring design is not standardised for composite insulators. For this reason, it has been established that the composite insulator must be delivered with corona protection that has been tailored to its needs (one or two rings). If components are procured separately, it is important to coordinate corona and power arc protection. An example of correct coordination is shown in Fig. 9.38a whereby the power arc protective fitting is pushed forward to shorten the striking distance and takes over a power arc relatively quickly. Since the aluminium corona ring does not have to guide the power arc until it is extinguished, it can be mounted directly onto the composite insulator fitting. An example of incorrect coordination, which can be critical to operations, is shown in Fig. 9.38b. Such a situation can occur, for example, if a relatively small rod diameter is chosen in order to save costs and the crimp interface has to be extended.

This results in a longer composite insulator fitting and causes the corona ring on the insulator fitting to be moved. If this is positioned over the power arc protective fitting (as shown in the example), it is unlikely that it will take over the power arc. The aluminium corona ring will have to carry the power arc until it is extinguished and, depending on the power arc parameters, it may be thermally destroyed as a result of unsuitable material properties (see also Sect. 9.4.1). This, in turn, may cause the insulator to experience some (critical) secondary effects (e.g., metal vapour condensation).

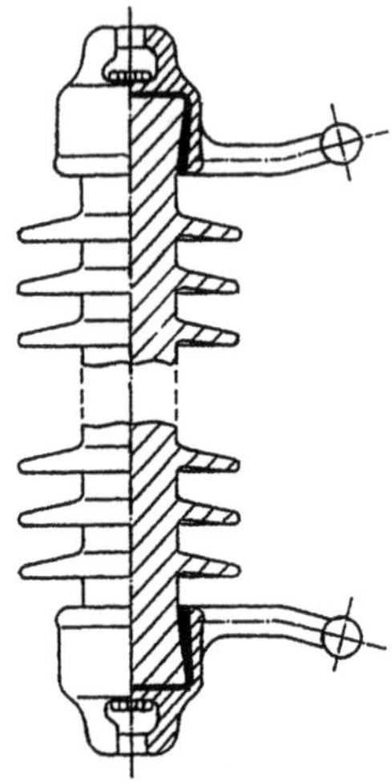

Fig. 9.39 Porcelain long rod with cast-on arcing horns

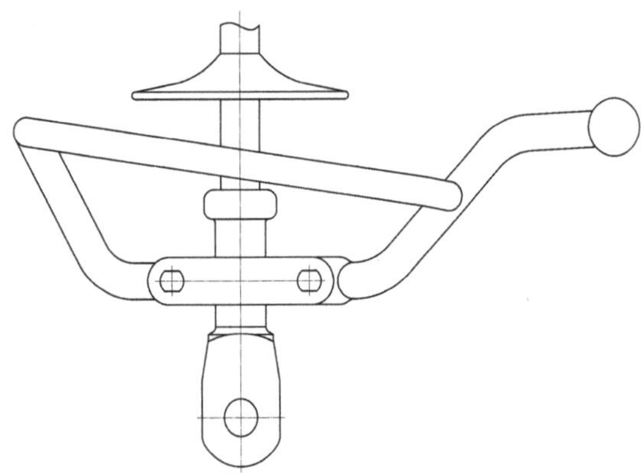

Fig. 9.40 Composite insulator with a directly mounted power arc protective fitting

9.6.3 Direct Mounting of Power Arc Protective Fittings onto Composite Insulator End Fittings

In individual cases, the difference in striking distance shown in Sect. 9.6.1 will attempt to compensate for extending the composite insulator. If, however, the string length is not to be increased in comparison to for cap and pin insulators, the power arc protective-fittings are frequently not mounted onto the intermediate fittings but directly onto the composite insulator end fittings. This solution has already been used in porcelain long rod insulators (Fig. 9.39 [7]). However, service experience has shown that protection against a power arc due to pollution flashover is low if the horns do not take over the power arc immediately and the maximum limit for routing the fault current over the ball fittings with a required current density of 70 A/mm^2 is 14 kA over a period of 1 s. This solution was used

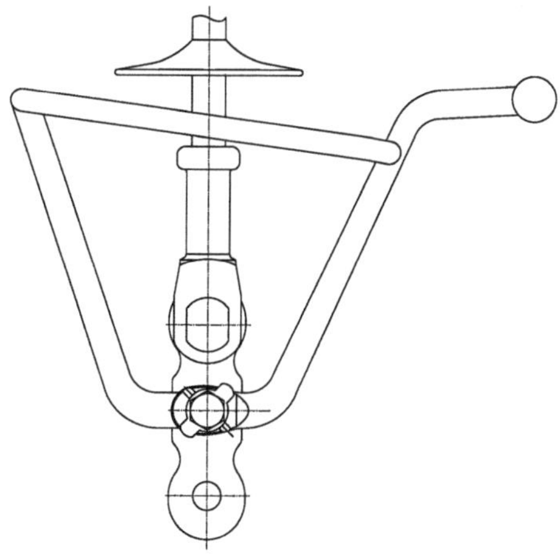

Fig. 9.41 Composite insulator with an intermediate fitting for mounting the power arc protective fitting

solely for small fault currents in the distribution network. This also applies to composite long rod insulators (Fig. 9.40). If, as in the example shown, a ring is used to take over a power arc more quickly and to facilitate corona protection, an adequate effect is ensured from the perspective of power arc protection.

The fault current flows directly over the composite insulator end fitting. However, this type of direct mounting is associated with the following disadvantages:

- Depending on the ball diameter used (in accordance with IEC 60120), the 1 s residual current is restricted to the following values: 16 mm for 14 kA and 20 mm for 22 kA in order to adhere to the criterion of 70 A/mm^2.
- The fault current may cause the temperature of the insulator fitting to rise, which, in turn, can cause the glass transition temperature of the rod to be exceeded. In this case, compression between the rod and fitting diminishes and the tensile strength is also reduced after adjusting the cold conditions. Tests involving a specific type of glass fibre rod, impregnated with epoxy resin and a glass transition temperature (TG) of 110 °C, have shown that a temperature rise to 120 °C can result in a 20 % reduction.
- The position of the ring is not clearly defined and requires precise instructions from the installation team.
- If conductor vibrations occur, the ring may loosen due to the gravitational force and move downwards if the tightening torque of the screws was not defined and was therefore too low.

If the fault currents are high, and one wishes to avoid incorrect installations, the most reliable solution is to use intermediate fittings to explicitly mount the power arc protective fitting (Fig. 9.41). Consequently, the fault current does not flow directly over the composite insulator fitting. Tensile tests conducted after power arc tests on these designs have confirmed that there is no critical heat transfer from

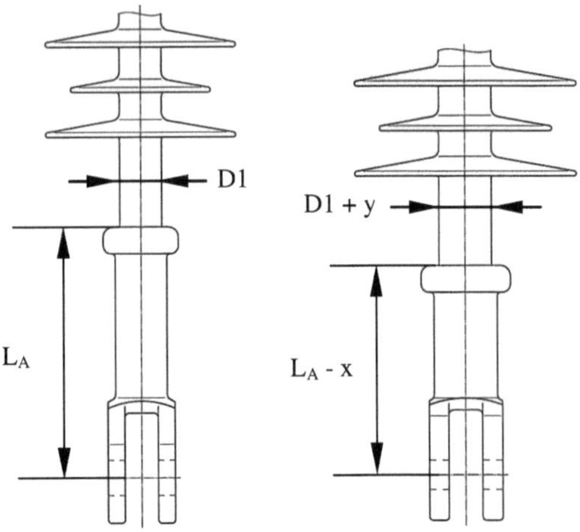

Fig. 9.42 Shortening the length of the end fitting by increasing the rod diameter of composite insulators

the intermediate fitting to the composite insulator fitting. This is also in line with tests conducted by CIGRE Working Group B2.03 on the effect of high temperature conductors on the temperature of fittings on conventional and composite insulators [31]. These tests were able to demonstrate that, at a conductor temperature of 200 °C and higher, the temperature of the composite insulator fittings rises by only 30 K when compared with the ambient temperature. The smaller shed diameter of composite insulators in comparison to cap and pin insulators, and the longer length of the crimped end fitting, counteract large increases in temperature.

Since the wide range of string designs did not have to be simulated in this test, a corresponding check is recommended when using high temperature conductors. This recommendation applies to conventional and composite insulators.

One optimised, reliable solution for reducing the increase in length and ensuring that intermediate fittings are used to mount power arc protective fittings is to shorten the composite insulator fitting (Fig. 9.42). Since the crimp area, which comprises the rod periphery and crimp length, and transmits the load between the rod and fitting, is important from a geometry perspective, the overall length of the insulator fitting can be reduced accordingly when using a larger rod diameter and, as a result, the striking distance can be increased. This is successful, for example, if one wants to achieve direct interchangeability for porcelain long rod insulators.

9.6.4 Effect of the String/Set Design on the Tower

The end burning point in the string on the tower is preferably arranged in such a way that the outer current loop (Fig. 9.2) can provide support in guiding the power arc. In the case of double suspension strings that are transverse to the direction of the line, the opposite orientation of the end burning points (left or right) is often practiced for composite insulators (Fig. 9.43). If the left string experiences a

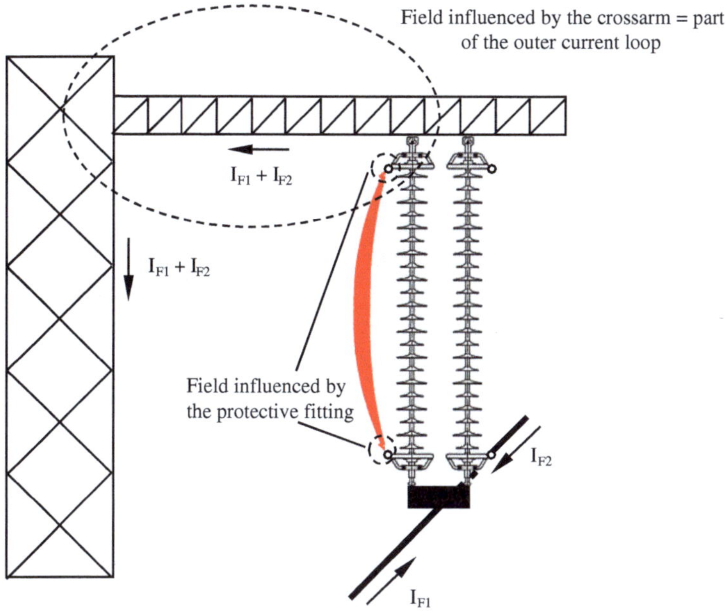

Fig. 9.43 Basic diagram of the magnetic fields influenced by the cross-arm and protective fitting

Fig. 9.44 Orientation of the end burning points when double suspension strings of composite insulators are installed transversely

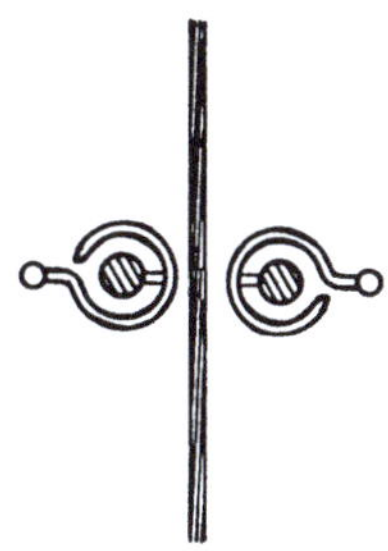

Fig. 9.45 Orientation of the end burning points when double suspension strings of composite insulators are installed longitudinally

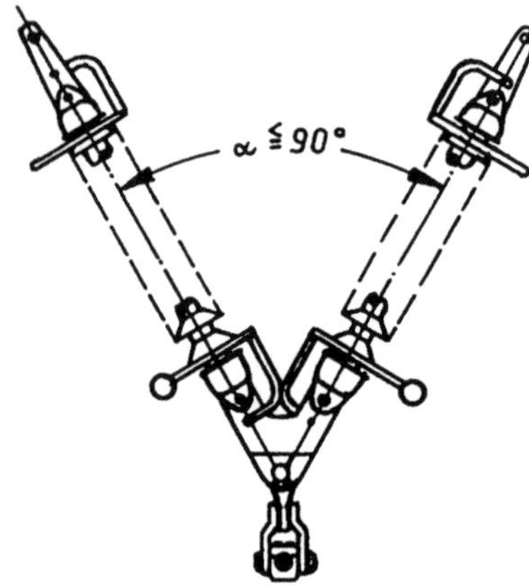

Fig. 9.46 Orientation of end burning points in a V-string with a small angle of Vee

flashover, the field influenced by the cross-arm will attempt to depress the power arc at the insulator if the fault current is supplied from the left-hand side. If the power arc on the protective fitting has a unidirectional supply, the field influenced by the fitting is allowed to dominate and align the arc stems to the left (against the field influenced by the cross-arm). Due to the thermal insensitivity of a composite insulator with a silicone rubber housing, there is no risk of material damage as in the case of glass or porcelain, and the end burning point orientations shown in Figs. 9.43 and 9.44 are often applied in the case of composite insulators. If the double suspension string is longitudinal to the direction of the line, the end burning points are orientated in the same way (Fig. 9.45). The orientation of the end burning points in the direction of the line is applied to situations characterised by a fault current with a unidirectional supply e.g., proximity to a station (see also Fig. 9.32b and d).

The angle of Vee is important in V-strings: For angles of Vee $\alpha \leq 90°$, the end burning points are oriented in accordance with Fig. 9.46. When composite insulators are installed, larger angles of Vee are wanted in order to prevent the compression-loaded insulator from buckling when mechanical horizontal forces act on it. In this case, the end burning points are oriented inwards and the power arc is guided into the V-aperture and can navigate to the cross-arm as a result of the thermal ascending force. When additional arcing horns are used directly on the cross-arm, this ensures safe burning of the arc without any damage to the cross-arm.

In the case of anchor towers or dead-end towers, the end burning points are directed upwards in order to benefit from the effect of the thermal ascending force (Fig. 9.47). When measuring radio interference voltages and corona, this usage situation is often more critical because the exposed end burning points are barely protected. Consequently, the diameter of the end burning point can be larger for this installation situation than for a suspension string with the same fault current.

Fig. 9.47 50 Hz arc across a tension set, low power high voltage test

9.7 Summary

In addition to the criteria in Sect. 9.2.3, please note the following in relation to power arc protection for composite insulators:

- Depending on the design and installation situation concerning the composite insulator, corona protection must be provided as of 145–245 kV. This generally involves ring-shaped aluminium profiles. Since corona protection is not resistant to power arcs, it must be coordinated with power arc protection for the steel fittings used.
- "Packet solutions" in the form of entire strings/set are recommended whereby corona and power arc protection are combined with a steel fitting.
- Generally, it is not possible to fit composite insulators with power arc protective fittings that are used for conventional insulators; this is due to the different fitting lengths and the requisite corona protection.
- Due to their limited ampacity, power arc protective fittings can only be mounted directly onto the composite insulator fitting if the fault currents are small.
- The proven values for the permitted current density in the event of a fault (namely 70 and 80 A/mm^2) are valid irrespective of the insulator technology used and therefore are also valid for composite insulators.

All in all, the more than thirty years of service experience with composite insulators that have a lasting hydrophobic effect has shown that the probability of a pollution flashover occurring has been reduced significantly and the power arc protective fitting rarely has to provide any protection. However, the string design must continue to be closely scrutinised in order to ensure that the string is safe.

References

1. de Tourreil C, Schmuck F on behalf of CIGRE working group B2.21 (2008) On the use of power arc protection devices for composite insulators on transmission lines. CIGRE TB 365, Dec
2. Klengel H (1990) Über Möglichkeiten der Stabilisierung des langen freien Hochleistungs-Lichtbogens in Luft. Dissertation TU Dresden
3. Möcks, L (1982) Power arc protection devices for striking distances in overhead lines and substations. ETZ Report 16, VDE Verlag
4. Gönenc I (1960) Power arc movement along round rods. ETZ Archive 81
5. Strnad A, Schuster M (1980) Protection of insulator strings against high-power arcs, EVS
6. Strnad A (1979) Lichtbogenschutzarmaturen für das 380 kV Netz, Elektrizitätswirtschaft 1979 Nr. 3099
7. Klengel H (2009) Isolatoren und Armaturen für Isolatorketten in Starkstrom-Freileitungen, Online Verlag new-ebooks.de
8. Patent DRP 916312 (1941) An den Enden von Lichtbogenschutzhörnern anzubringender, zweckmäßig aufzuschraubender Elektrodenkörper kugeliger, zylindrischer oder birnenförmiger Gestalt
9. Patent DE 2610348 (1976) Lichtbogenschutzarmatur für Isolatoren von Hochspannungsfreile itungen
10. Patent DE 2659101 (1976) Lichtbogenschutzarmatur für Isolatoren von Hochspannungs freileitungen
11. Patent DE 2814994 (1978) Brennelektrode für einen Lichtbogenschutzring
12. Lichtbogenschutz und Feldsteuerarmaturen. RIBE-Katalog, 10/2011
13. IEC 61467 Ed 1 (2008) Insulators for overhead lines—Insulator strings and sets for lines with a nominal voltage greater than 1 000 V—AC power arc tests
14. CIGRE WG 22.03 (2001) Composite insulator handling guide. TB 184
15. IEC TR 61467 Ed. 1 (1997) Insulators for overhead lines with a nominal voltage above 1000 V—AC power arc tests
16. Vosloo WL (2007) Power arc damage on insulator end fittings. Study of Stellenbosch University on the ESKOM network
17. Gorur RS, Cherney EA., Burnham JT (1999) Outdoor insulators. ISBN 0967761107
18. IEC 60120 Ed. 3: (1984) Dimensions of ball and socket couplings of string insulator units
19. IEC 60695-11-10 1st Ed. 1999 + Amendment 2003 Fire hazard testing: test flames—50 W horizontal and vertical flame test methods
20. P-IEC 61006 Ed 2 (2004) Electrical insulating materials—methods of test for the determination of the glass transition temperature
21. Kesselring F (1968) Theoretische Grundlagen zur Berechnung der Schaltgeräte. Sammlung Göschen, Band 71, Verlag de Gruyter
22. EN 50341-3-4 (2001) Overhead electrical Lines exceeding 45 kV
23. P-IEC 61109 Ed1 (1992) Composite insulators for a.c. overhead lines with a nominal voltage greater than 1000 V—Definitions, test methods and acceptance criteria
24. CIGRE WG 22.10 (1988) Technical basis for minimal requirement for composite insulators, ELT_088_3
25. IEC 61109 Amendment 1 (1995) Composite insulators for a.c. overhead lines with a nominal voltage greater than 1000 V—Definitions, test methods and acceptance criteria

26. P-IEC 60707 Ed 2 (1999) Flammability of solid non-metallic materials when exposed to flame sources—List of test methods
27. CIGRE D1.14, Project Group Flammability: Requirements on Testing Flammability of polymeric Materials for Outdoor Insulation. TB 489 2012
28. IEC 62217 Ed 1 (2005) Polymeric insulators for indoor and outdoor use with a nominal voltage >1 000 V—General definitions, test methods and acceptance criteria
29. IEC 60587 Ed. 3 (2007) Electrical insulating materials used under severe ambient conditions—Test methods for evaluating resistance to tracking and erosion
30. IEC 60071-1 Ed. 8.1 + Amendment (2011) Insulation co-ordination—Part 1: Definitions, principles and rules
31. Stephen R, Seppa TO, Douglass D, Pirovano G, Hill R, Troppauer W, St-Louis M, Macey RE, Yoshida S, Schmuck F, Sunkle DC, Dulhunty P, Asselin J-M, Hearnshaw D (2007) Considerations relating to the use of high temperature conductors. CIGRE TB 331 Oct

Chapter 10
In-Lab Evaluation of Composite Insulators following their Withdrawal from the Network

Symbols and Abbreviations

ANSI	American National Standards Institute
ASDD	Apparent salt deposit density
CEPS	Czech power utility
CIGRE	Conseil International Des Grands Reseaux Electriques (International Council For Large Electric Systems)
DC	Direct current
DSC	Differential scanning calorimetry
E-CR or ECR-Fibres	Corrosion-resistant glass fibre for electrotechnical applications
EDS	Every day stress
E-Glass	Glass fibre for electrotechnical applications
ELECTRA	Professional journal of CIGRE
EPDM	Ethylene propylene diene monomer (insulant)
EPRI	Electric Power Research Institute
ESDD	Equivalent salt deposit density
ESKOM	South African power utility
EVA	Ethylene vinyl acetate
EVU	Power utility
FMEA	Failure mode and effects analysis
FTIR	Fourier transform infrared spectroscopy
GFK	Glass-fibre reinforced plastic
HT	Hydrophobicity transfer
HTM	Hydrophobicity transfer mechanism (transfer procedure)
HTV	High temperature vulcanising
IEC	International Electrotechnical Commission
IR	Infrared
k	Index of service experience
MTBF	Mean time between failure
MTTR	Mean time to repair
NSDD	Non-soluble deposit density

K. O. Papailiou and F. Schmuck, *Silicone Composite Insulators*, Power Systems,
DOI: 10.1007/978-3-642-15320-4_10, © Springer-Verlag Berlin Heidelberg 2013

SEM	Scanning electron microscope
RTE	French power utility
RTV	Room temperature vulcanising
SML	Specified mechanical load (nominal force of insulator)
SR	Silicone rubber
T	Time of use
TE	Partial discharge
TGA	Thermo-gravimetric analysis
Triton-X	Wetting agent
UV	Ultra-violet
WC 1..7	Wettability class
XRF	X-ray fluorescence
XRD	X-ray diffraction
n_a	Number of failures
n_b	Total number of insulators in service
n_E	Number of insulators still in service
λ	Failure rate
$\hat{\lambda}$	Empirical failure rate

10.1 Introduction

This chapter is dedicated to the in-lab evaluation of aged, "old" or failed line composite insulators. The term *in-lab evaluation* is used, because the state of the insulators or the insulator strings/sets is evaluated as *off-line diagnosis* [1], which is different to the inspection method of an *on-line diagnosis* [1] (not considered in this chapter, more information for example in [2]). It describes various methods, philosophies and resources that make it possible to evaluate their technical state and thus estimate the remaining time of use of composite insulators that are the same type as those being tested. The composite insulators to be evaluated are specifically withdrawn from the line in order to obtain representative results for this generation of insulators. Frequently, analyses are also conducted if old composite insulators are available as part of scheduled replacement schemes. If a certain behaviour has already been identified as being critical to operations, it is recommended to evaluate at least ten insulators. If the line route includes various pollution zones or loads that are otherwise different, a sufficient number of each insulator must be inspected representative for the corresponding in situ conditions. When withdrawing insulators from the network, it is important to ensure that the installation conditions are adequately documented and can be clearly assigned. In many cases, the manufacturer of the composite insulators can provide additional data to identify the product (technical state of a particular generation of insulators, production date, material batches, etc.). In critical cases, it is recommended to perform the evaluation in conjunction with the manufacturer and a specialist independent laboratory. The identification of intensive ageing phenomena that effectively signify the end of service life for an insulator can prompt the withdrawal of other insulators of the same type/generation.

This chapter is primarily concerned with the evaluation of composite long rod insulators. However, many aspects of the evaluation philosophy can be applied to other insulator applications such as post insulators or hollow core insulators if reference is made to the relevant product standard in each case.

10.1.1 Is it Necessary to Evaluate Composite Insulators?

The composite insulator has undergone the various development phases and standardization stages typically associated with technical products. The composite insulators used today, often also known as "third-generation composite insulators", are characterised by high-grade components with a long-term history of use. Given the different components available and their structure, there is a range of interface areas whose functionality must be verified in design tests on brand-new insulators. Such verifications then apply to a design, irrespective of the number of finished units and the insulator lengths (types). The corresponding product standards provide well-developed guidelines for these tests.

There are still no mandatory tests for evaluating existing composite insulator installations. Developing such testing recommendations was part of the work undertaken by CIGRE Working Group B2.21 "Insulators" [3]; which was published as Technical Brochure 481 by end of 2011 and whose recommendations are considered in this chapter. The need to test composite insulators after a certain period of time in service was initiated by user of composite insulators. Thematically, it is comparable to the work undertaken by CIGRE Working Group B2.03 in relation to evaluating conventional insulators [4], which was published in the Technical Brochure 306 in 2006. The purpose of these tests is to evaluate the current life expectancy of a batch or generation of insulators that have been supplied.

In addition to evaluating existing composite insulator installations, the market situation whereby composite insulators are frequently procured as commodity products, without any project specifications tailored to their intended purpose, gives rise to the following reasons for a possible need for an evaluation:

- Logging the current state of a particular batch or earlier generation of insulators
- Detailed analysis of the quality of new supplied insulators from the perspective of pricing pressure and international competition
- The wide variety of materials and production processes used

10.1.2 Interface Areas as a Key Difference for Composite Insulators

One characteristic of composite insulators is their many interface areas, which can be divided into macroscopic and microscopic interface areas [5, 6]. Macroscopic interface areas are formed, for example, between the rod and insulator housing or between the rod and fitting, while microscopic interface areas are formed (for example, during the production of individual composite insulator components) between the glass fibres, their size and the resin matrix when manufacturing rods or between the axial and radial vulcanising zones when manufacturing the insulator housing. Since all of the interface areas are key factors in ensuring that a composite insulator works reliably, many testing programmes aim to evaluate these areas.

10.2 Reasons for Evaluating Composite Insulators Following Their Withdrawal from the Network

If withdrawal of insulators from the network is scheduled, the testing goals to be derived must be coordinated with the reasons for the withdrawal. For example, three reasons are provided as possible scenarios.

10.2.1 Evaluation of an Insulator Installation in Terms of its (Ageing) State

The evaluation of an insulator installation that has not experienced any failures is a preventive measure for one or a combination of the following reasons:

- A need to obtain information about the state of a particular insulator (e.g. due to negative experiences in other areas of application, discontinuation by suppliers, etc.)
- Confirmation or redefinition of inspection intervals
- Revision of previously valid design philosophies, specifications, new technologies or environmental changes (e.g. new assignment of pollution levels within a pollution map, new threshold values for permitted field stresses, need for power arc protective fittings, etc.).

Certain defined steps have proven their worth in this procedure (Fig. 10.1).

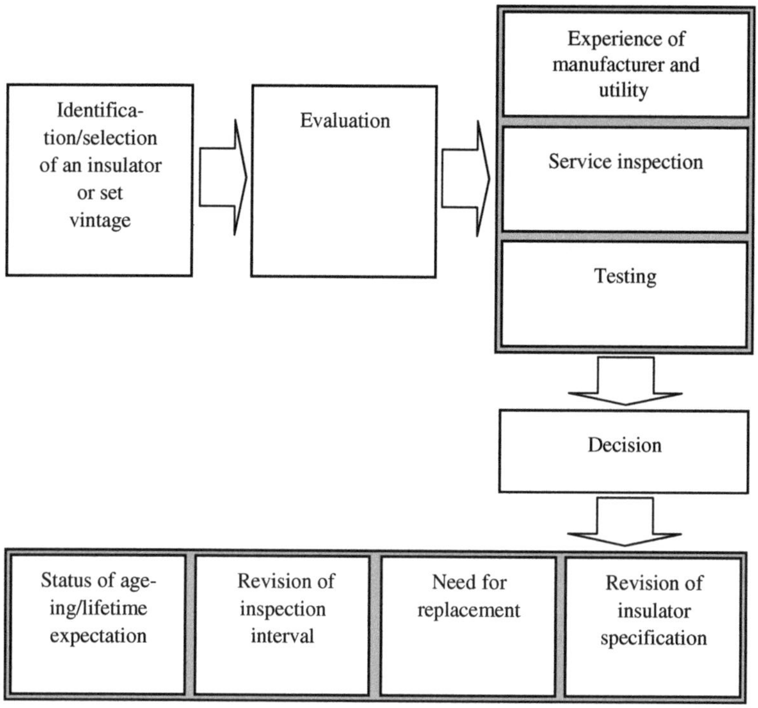

Fig. 10.1 Chart outlining the procedure for evaluating the ageing state

10.2.2 Evaluation of an Insulator Failure or Insulator Type with a High Risk of Failure

One important reason for evaluating an insulator is the occurrence of a failure or the detection of functionally relevant damage during an inspection. In these situations, the user needs to know if it is an isolated problem (e.g. caused by rough handling) or a far-reaching problem (e.g. widespread (systematic) failure). Before selecting and determining the number of insulators to be withdrawn from the network for testing purposes, it is fundamentally important to understand the fault, damage and cause. Otherwise, there is a danger of withdrawing insulators indiscriminately and therefore running the risk of excluding the types concerned from the tests.

A simplified example of a systematic procedure is shown in Fig. 10.2. It is important to gather all necessary information:

- Unique identification of the date of manufacture (for traceability purposes)
- Service time, which may differ from the actual age if the insulator was in storage for a period of time
- Effect of operating conditions (pollution, lightning, periods of humidification, network behaviour, human or animal influences, etc.)

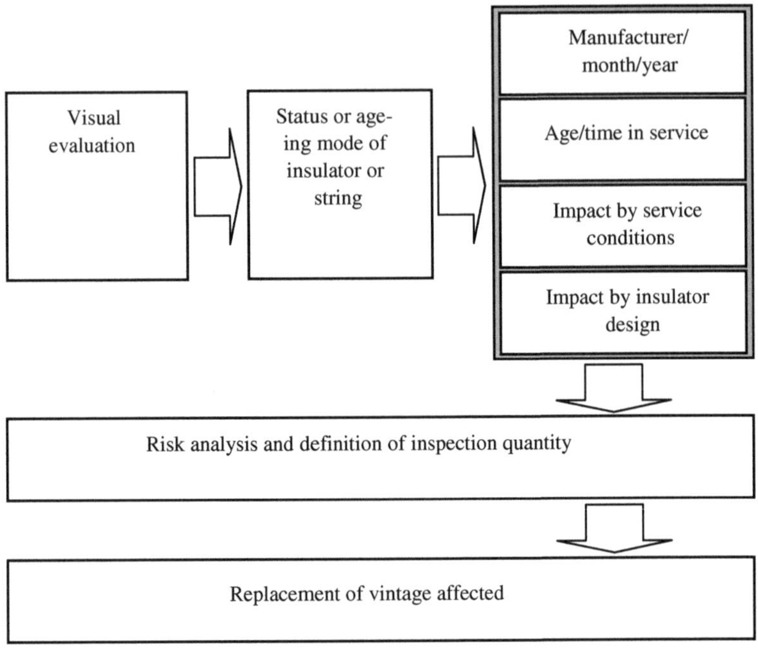

Fig. 10.2 Sample procedure for systematic fault diagnostics

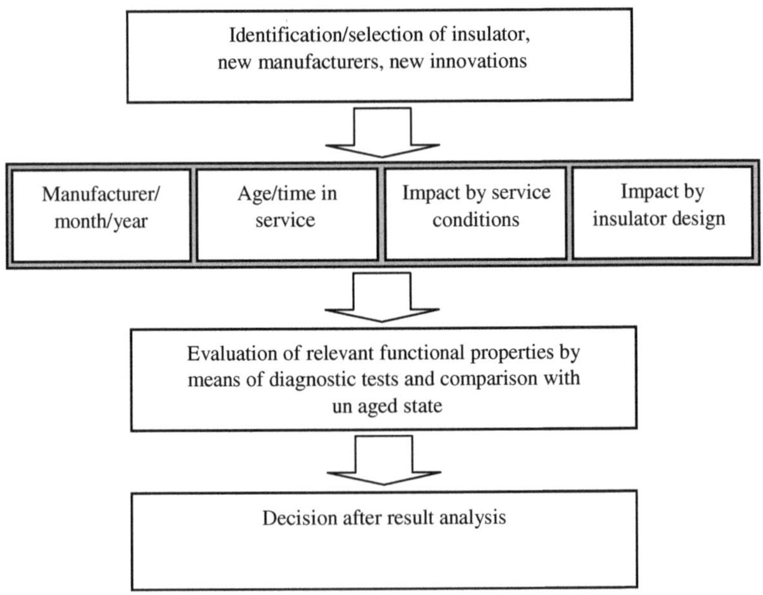

Fig. 10.3 Sample procedure for research programmes

- Insulator design (material selection, dimensioning, string design, etc.) in relation to operating conditions.

This analysis also requires the following: Knowledge of how differently designed composite insulators behave and how the application itself behaves. For this reason, it is recommended to involve the manufacturer of the relevant composite insulator and, if necessary, an independent expert or a specialist laboratory in the analysis.

10.2.3 Evaluation of Composite Insulators for Research Purposes

Today's well-developed standards and the knowledge of how composite insulators behave are also based on the scientific evaluation of materials and insulators that were not stressed and aged under accelerated conditions in the laboratory but under real outdoor conditions. Therefore, the sample procedure outlined in (Fig. 10.3) is very similar to that for systematic fault diagnostics.

Some EVUs set up and operate test stations or test towers for this purpose; these are typically set up in areas exposed to harsh environmental conditions to accelerate ageing [7]. Examples include the ESKOM test station in Koeberg, South Africa, the RTE test station in Martiques, France or, as shown in Sect. 10.5.5 the CEPS test tower in the Czech Republic. One benefit of test stations is the simultaneous recording of environmental data (pollution levels, weather, etc.), which has an immediate impact on the insulators, as well as easy access to test specimens.

10.3 Composite Insulator Failure Rate and Failures

This section deals with the information available in relation to composite insulator failures. This includes the surveys contained in the EPRI database in the USA, which is updated annually. A selection of results from a CIGRE survey and the latest experiences from China are also presented here.

10.3.1 USA/EPRI

The various causes of insulator failures in the USA and the rate at which they occur are shown in Fig. 10.4. By 2011, a total of 315 failures had occurred among an estimated 3 million composite insulator installations, resulting in a cumulative failure rate of 0.0105 %. Brittle fracture or interface failures dominate the statistics, but other reasons may be associated with a direct cause of failure (e.g. damage to the insulator housing, a string design that is susceptible to coronas, rod

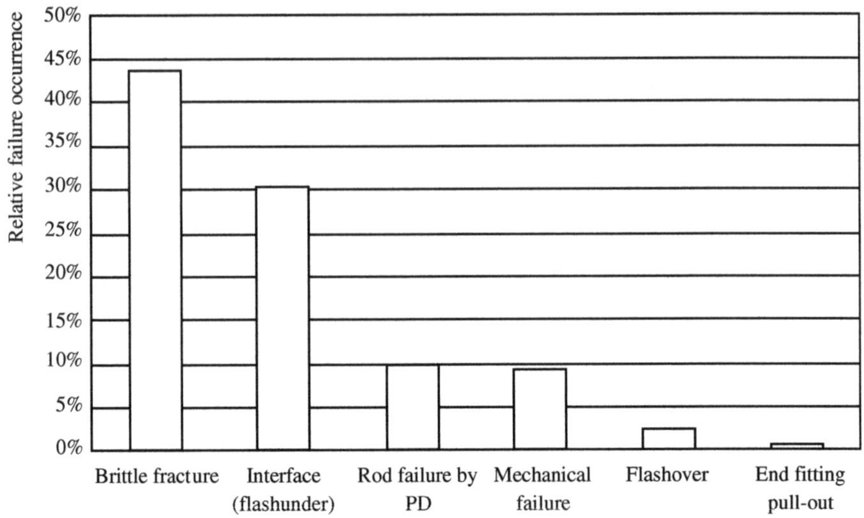

Fig. 10.4 Causes of failure for composite insulators in the USA [8]

exposure, etc.). As a result of a cumulative analysis, the survey also contains information about first-generation insulators, which are particularly prone to a brittle fracture failure when non-acid-resistant glass fibres are used (E-Glass) and do not have any tolerance for faults, if the damage is such that it permits the rod to be attacked by acid.

This comparatively frequent occurrence of brittle fractures has resulted in extensive analyses [9–17] (see also Sect. 7.2.2.1). The resulting current state-of-the-art technology has led to the availability of glass fibres and impregnation methods that facilitate the production of electrically high-grade, acid-resistant rods. Today, the use of E-CR-Fibres is a recognised standard for composite insulators that are subjected to a permanent tensile load (long rod insulators).

IEC 62662 [18] can serve as a useful guide when assessing the risk of insulators that do not yet correspond to this state-of-the-art technology. In this technical report of IEC, failure mode and effects analysis (FMEA) is used to assess those factors that make the entire insulator design or string design prone to brittle fractures.

10.3.2 CIGRE Survey, Published in 2000

CIGRE Working Group B2.03 had previously conducted two surveys, the results of which were published in the professional journal ELECTRA in 1990 [19] and 2000 [20]. CIGRE Working Group B2.21 is currently working on drafting a new and updated questionnaire. The survey that produced the results published in 2000 considers many aspects of use, including insulator applications (suspension/tension insulator, interphase spacers, posts, etc.), reasons for use (instead of using

Table 10.1 Inspection frequency of EVUs that inspect composite insulators in use (partly = >1 year, periodically = monthly – ≤1 year) [20]

Inspection frequency	Africa	Asia	Australia	Canada	Europe	South America	Middle East	Total
Non	1/3	2/7	3/8	0/10	1/33	0/10	1/3	8/74
Partly	1/3	0/7	3/8	5/10	14/33	3/10	1/3	27/74
Periodical	1/3	3/7	2/8	2/10	14/33	4/10	1/3	27/74

Table 10.2 Distribution of causes of faults in accordance with the CIGRE survey conducted in 2000 [20]

Cause of failure	Service voltage (kV)				Total
	<200	200 … < 300	300 … < 500	500	
Insulator housing	25	8	0	0	33
Interface housing-rod	51	10	6	2	69
Rod	23	8	101	1	133
End fitting pull-out	4	2	0	0	6
End fitting	2	0	0	0	2
Total	105	28	107	3	243

Table 10.3 Presentation of comparative figures [20]

Analysis	Service voltage (kV)				Total
	<200	200 … <300	300 … <500	500	
No. of insulators recorded in the survey	456,835	174,482	59,446	1,413	692,176
Failure rate in %	0.023	0.016	0.18	0.21	0.035
Index of service experience k (1,000 insulators/year)	2,991 (160)	1,102 (42)	558 (81)	28 (9)	4,679
No. of failed insulators per 10,000 units/year	0.35	0.25	1.9	1.1	0.52

conventional insulators), housing material, pollution level, etc.). The inspection frequency also revealed an interesting result (Table 10.1). It is assumed that this had already implemented in countries that use conventional insulators and perform an inspection as a fixed component of operating overhead transmission lines and that this continued after such countries changed to using composite insulators. The inspections are not limited to insulators and insulator strings. Rather, they include an evaluation of the conductors and dampers (Stockbridge dampers or field spacers).

A direct comparison between the EPRI database and the list of individual causes of faults and the frequency with which they occur is possible (Table 10.2). The "failure" criterion was defined in such a way that an insulator is no longer able to assume its mechanical and/or electrical function. It is obvious that failures also dominate in the context of the rod component. This includes faults due to inadequate handling or a poor installation.

The survey also enquired about the approximate time in service of the insulators under consideration and derived corresponding comparative figures (Table 10.3). The index of service experience is calculated in such a way that the quantity of a particular type of insulator is multiplied by its service time and the values obtained are standardised to 1,000 insulators/year. The lower values from the first survey [19] are provided in brackets. The statistics available in relation to glass cap and pin insulators indicate failure rates of between 0.5 and 1.5 failed insulators per 10,000 units; a similar order of magnitude can be assigned to porcelain cap and pin insulators. However, these values cannot be compared directly with the failure rates for composite insulators because the type of failure associated with glass cap and pin and porcelain cap and pin insulators is different to that associated with composite insulators [21–23]. This also applies as a result of the way in which the failure criterion is defined in this survey: Often, glass cap and pin insulators with a broken shed remain in the insulator string if a certain quantity is not exceeded. On the other hand, a composite insulator damaged in accordance with the defined criterion must be replaced quickly.

10.3.3 Service Experiences in China

The following analysis is taken from the article [24] presented in 2011. An impressive number of composite insulators is deployed in electrification programmes within China (Fig. 10.5). The total number was approx. 2.2 million at the end of 2006. If one conservatively assumes that 200,000 new units were installed per year (which reflects growth in the period 2005–2006), this produces a total number of 3 million units up to and including 2011. For high service voltages (750–1,000 kV and 500 kV DC), in particular, this reveals a clear trend of using composite insulators with a silicone rubber housing. The percentage share for each service voltage level reflects this trend:

(kV)	Percentage
U = 66–110	57.15
U = 220	33.21
U = 330–500	9.15
U = 750	0.02
U = ± 500	0.47

In the service statistics, a distinction is made between flashovers (as interferences) and failures. Over voltages, primarily due to lightning, top the statistics at approx. 45 % (Fig. 10.6). If an insulator string/set has been designed appropriately (i.e. with power arc protective fittings), this is a controllable scenario for silicone composite insulators. The biological cause of a bird streamer is a phenomenon that is less associated with the insulator technology than with the striking distance. In these cases, it is recommended to use a "bird protector" at the tower cross-arm [25]. Thanks to hydrophobicity, pollution plays only a minor role here.

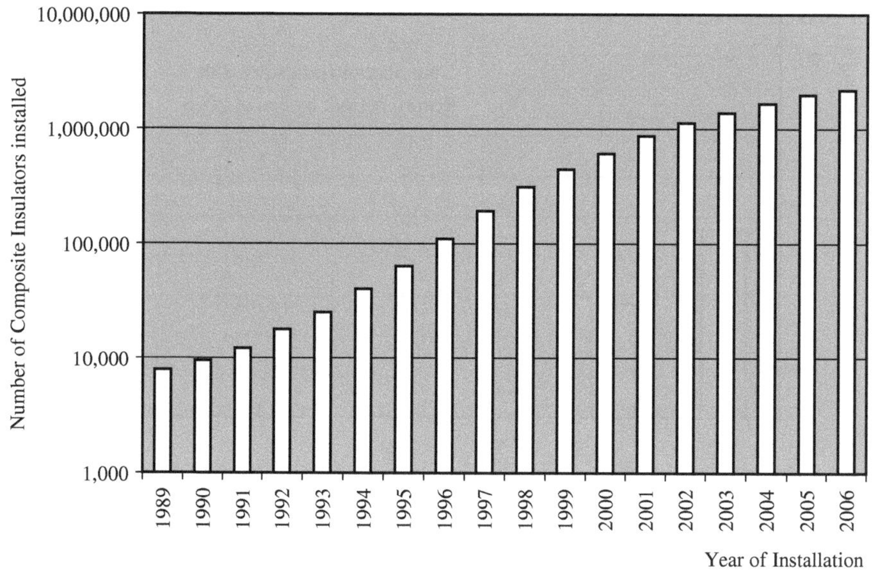

Fig. 10.5 Composite insulators installed in China [24]

When compared with the root cause analysis performed by the EPRI (Fig. 10.4), the percentage of failures caused by ageing or a brittle fracture is almost the same (Fig. 10.7). This may be due to the fact that the EPRI analysis makes another differentiation and a brittle fracture caused by ageing is only recorded cumulatively.

Furthermore, more stringent environmental rules apply in the USA. Therefore, the effect of industrial pollution during ageing will become more significant in China.

The percentage of faults resulting from a product defect is a critical indicator. In particular, the low percentage of failures caused by installation faults can be attributed to the provision of appropriate training for the installation team.

Overall, the cumulative failure rate is 0.0026 %, which is lower in comparison to the EPRI and CIGRE studies. It should be noted that these two studies also include numerous first-generation composite insulators, which may have a higher potential for failure.

10.4 Commonly Used Test Strategies

10.4.1 Identification of Test Specimens

Determining/identifying insulator test specimens or complete insulator strings is an important prerequisite when devising a testing programme, interpreting the results and, if required, assessing risk in relation to the remaining service life.

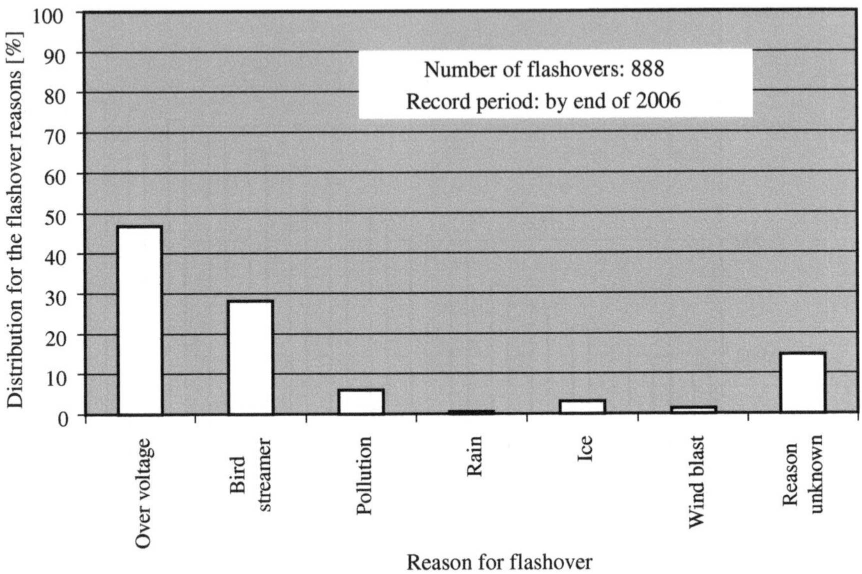

Fig. 10.6 Reasons (in %) for flashovers in China [24]

Photographic documentation is not only helpful but, thanks to today's digital technology, easy to integrate into any test documentation. Clearly assigned information is preferable when identifying test specimens. Such information can include the following:

- The manufacturer and drawing number
- The date of manufacture (year, month and week)
- The generation of insulator manufactured
- The material used and its batches (housing material, rod, fittings, etc.)
- The installation details: line name/designation, nominal voltage, string orientation, environmental and/or pollution conditions, time in service, EVU's fault report (if applicable).

10.4.2 Test Sequence

The number of tests or test strategies to be applied is very varied. This decision should be based on the number of insulators available, their design, possible damage patterns, etc. in order to obtain the maximum amount of information. It is always recommended that a comprehensive visual inspection precedes the planning stage of a testing programme (Fig. 10.8). The state of externally visible damage patterns then defines the selection of suitable tests, which are relevant for quantifying reduced properties. Consideration must be given to the root causes (Sect. 10.2) and, if necessary, the involvement of individual specialists. A more

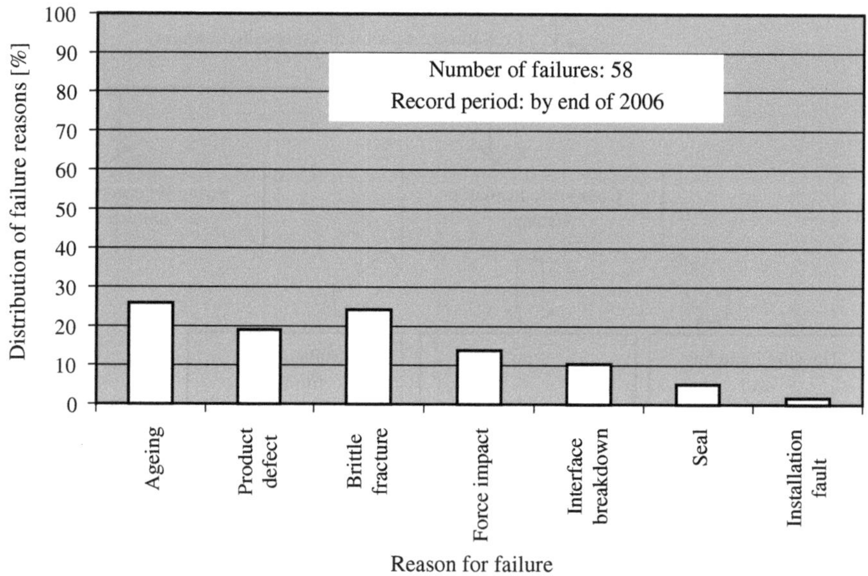

Fig. 10.7 Reasons (in %) for composite insulator failures in China [24]

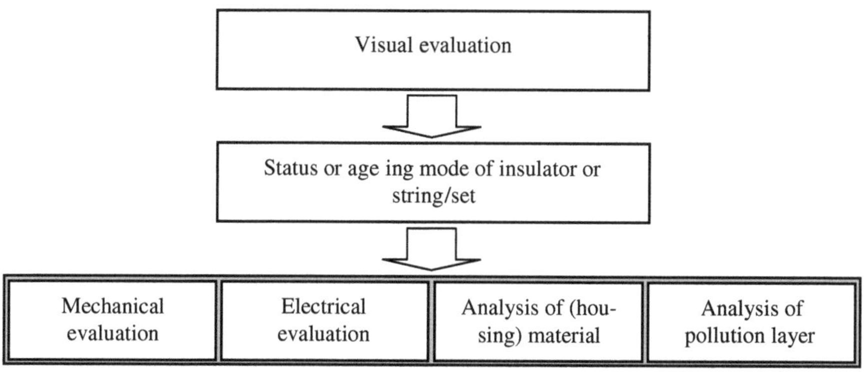

Fig. 10.8 Recommended sequence of evaluation

scientifically based evaluation will also require more time than available in the event of a fault, which, in the worst-case scenario manifests itself in the form of a conductor being "dropped" and requires immediate counter-measures.

10.4.2.1 Visual Evaluation

The visual evaluation is performed on the insulator or, if supplied, the entire insulator string/set. It has proven worthwhile to structure the evaluation according to the components of a composite insulator (Fig. 10.9).

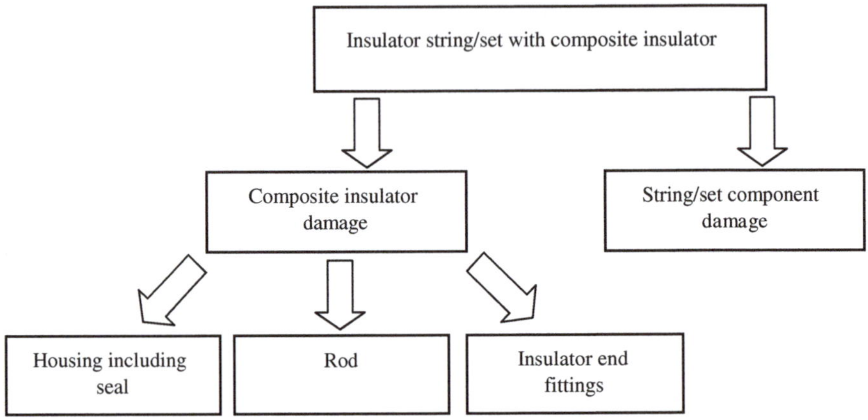

Fig. 10.9 Evaluation structure

Fig. 10.10 Shed cracking [26]

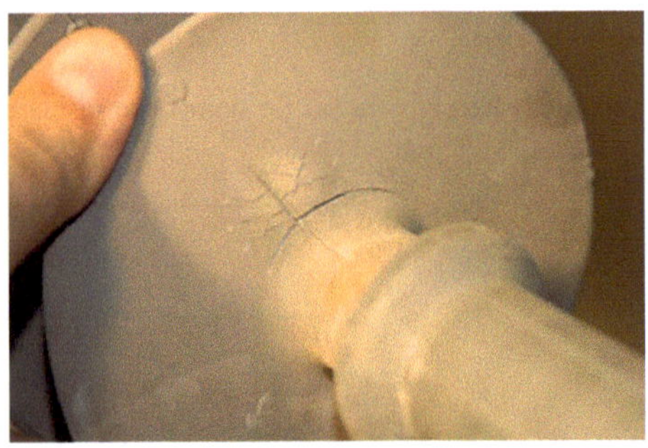

Every piece of damage or abnormality must be recorded. The photographic documentation provided in Figs. 10.10, 10.11, 10.12, 10.13, 10.14, 10.15, 10.16, 10.17, 10.18, 10.19, 10.20, 10.21, 10.22, 10.23, 10.24, 10.25, 10.26, 10.27, 10.28 provide a visual basis for cataloging damage. The examples shown [7, 25, 26] have varying degrees of severity and represent the damage patterns of composite insulators from different generations. If insulators or insulator strings/sets have already failed, other examples are listed in the EPRI Field Guide [26] for comparative purposes. The photographic documentation shown follows the structure of Fig. 10.9.

Damage to the insulator housing, including the seal

Figure 10.10: This damage is more commonly associated with early insulator materials, which have a lower mechanical strength or UV resistance. If this damage is caused by a permanent corona discharge, the insulator housing must be thoroughly inspected.

Figure 10.11: This may be caused by a material fault or it may occur during the manufacturing process. This is critical because the rod is exposed.

Fig. 10.11 Shed cracking [26]

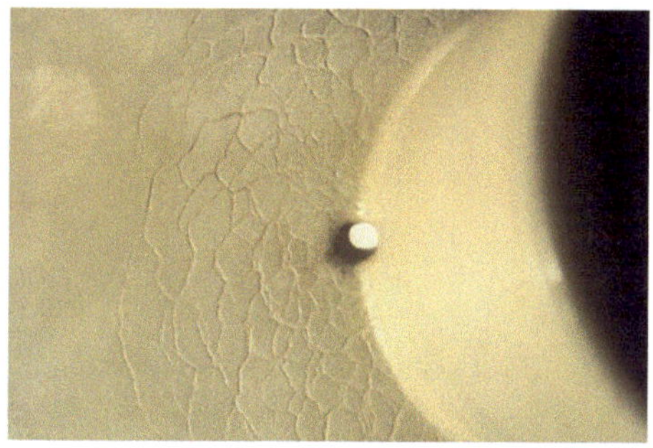

Fig. 10.12 Shed puncture [7, 25]

Fig. 10.13 Shed puncture with prolonged tracking [26]

Figure 10.12: The shed's creepage distance has effectively short-circuited.

Figure 10.13: Since the rod is not protected, further damage to the interface is possible; short-circuiting of striking and creepage distances having tracking.

Figure 10.14: This occurred because the material was unsuitable or the creepage distance was undersized. The insulator is effectively bridged by the conductive track.

Figure 10.15: Bulk erosion is a non-conductive housing damage. Effect here: The creepage distance of the first shed is no longer effective and the sealing system may be damaged.

Figure 10.16: This type of damage may be observed if materials are susceptible to UV or acid. Continued alligatoring may result in the rod being exposed. Reduction of the flashover performance in comparison to hydrophobic state very likely.

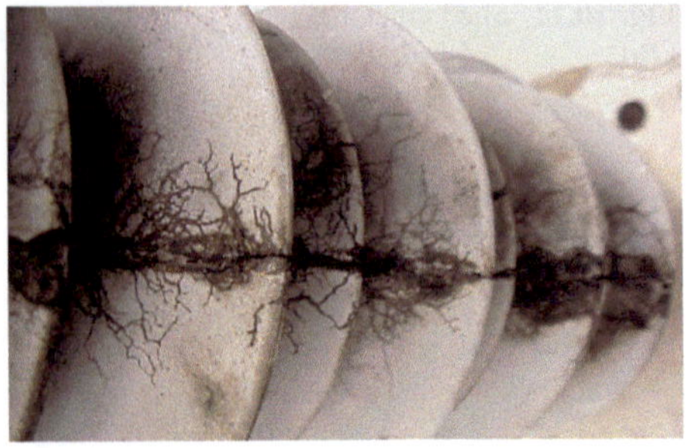

Fig. 10.14 Extended (full) tracking [7, 25]

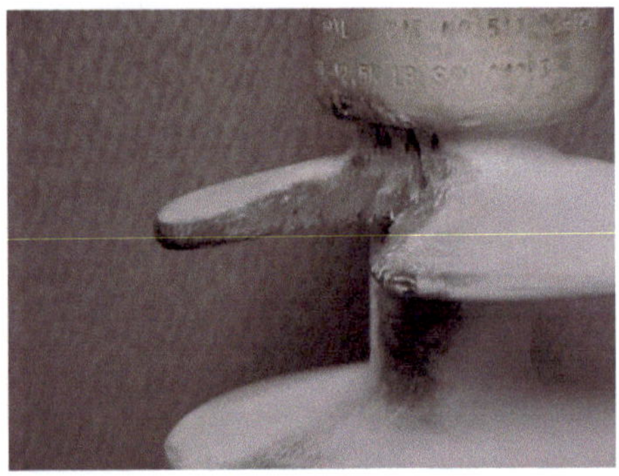

Fig. 10.15 Bulk erosion [26]

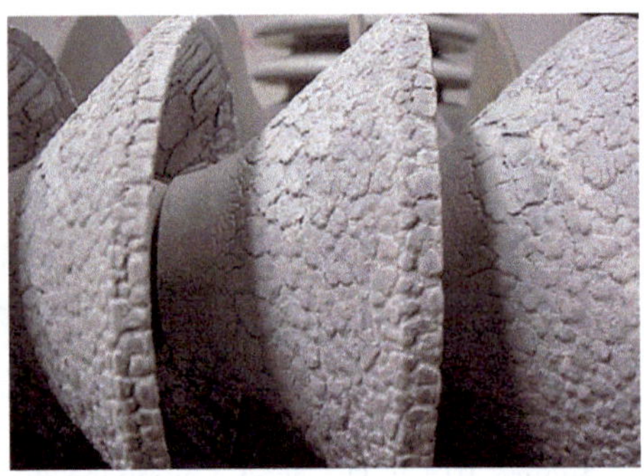

Fig. 10.16 Heavy embrittlement (alligatoring) [26]

Fig. 10.17 Handling-related damage to shed/housing [26]

Fig. 10.18 Damaged seal [26]

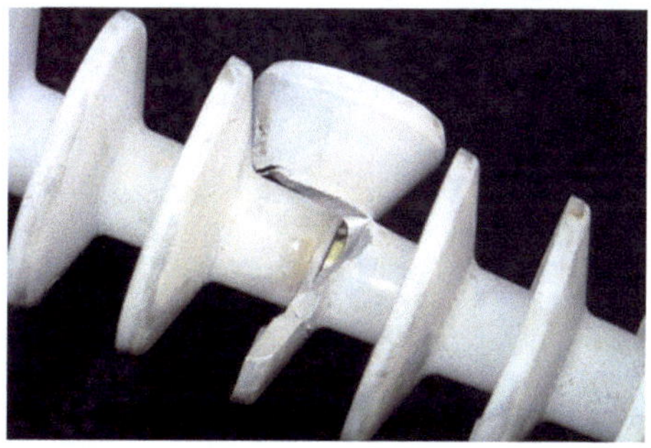

Fig. 10.19 Rod exposure [26]

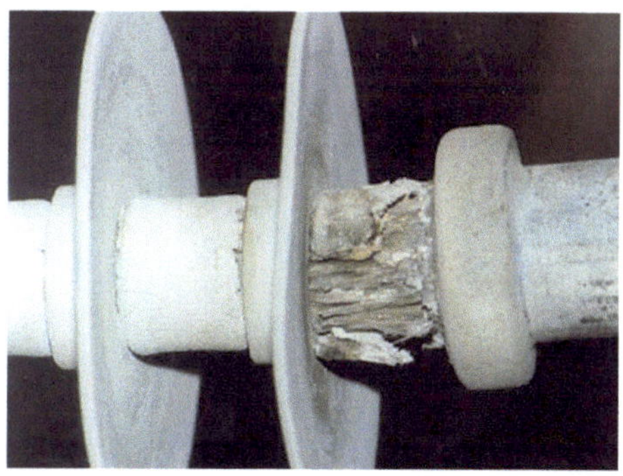

Fig. 10.20 Rod deterioration [26]

Fig. 10.21 Damage caused by an animal [26]

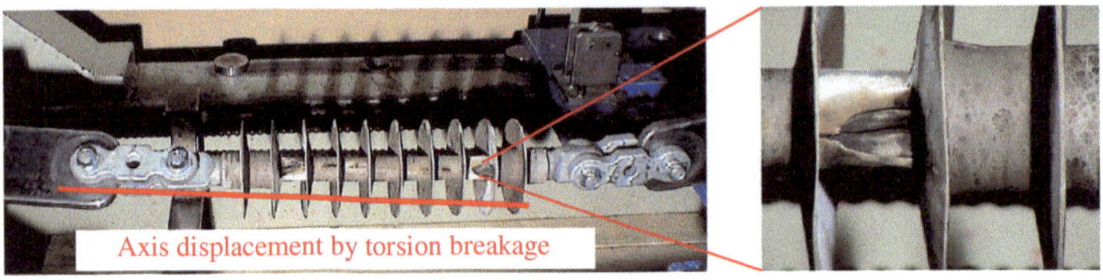

Fig. 10.22 Interface deterioration

Fig. 10.23 Housing separation [26]

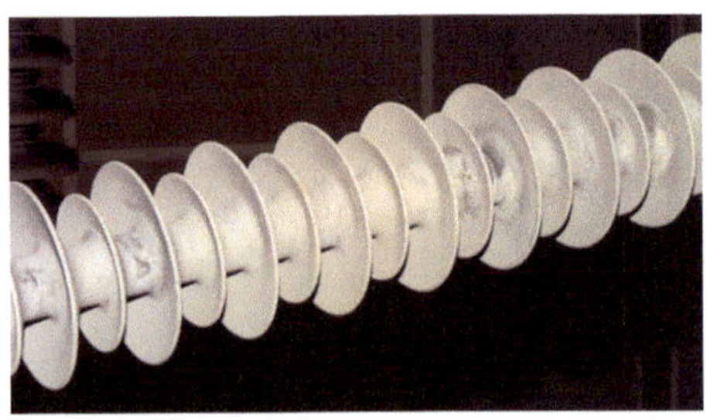

Fig. 10.24 Gun shot damage [26]

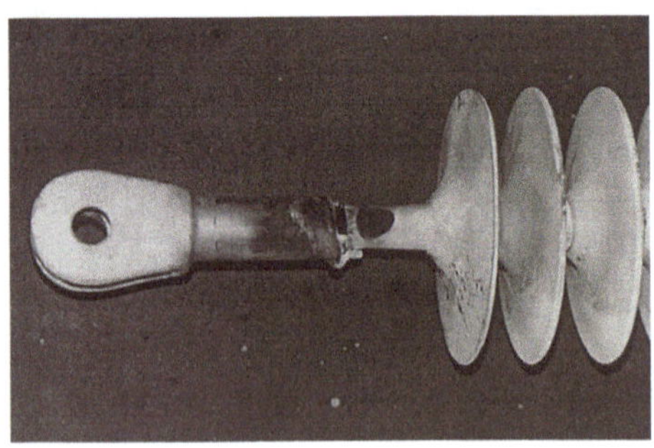

Fig. 10.25 Molten end fitting

Fig. 10.26 Heavy corrosion and damage to the seal

Fig. 10.27 Damaged corona collar [25]

Fig. 10.28 Power arc marks on a protective steel fitting

Figure 10.17: This reduces the effective creepage distance of the shed affected. This is critical if the insulator housing at the shank has been damaged.

Figure 10.18: Moisture (acid) may penetrate. This example was most likely caused by corona discharges (high voltage fitting).

Rod damage

Figure 10.19: This type of damage can often be identified in earlier housing materials (weak mechanical strength) as a result of rough handling. This is critical if the rod is exposed.

Figure 10.20: The exposed rod has most likely suffered thermal damage. This may be caused by permanent corona discharges followed by stable partial discharges.

Figure 10.21: This type of damage frequently occurs while the insulator is in storage or before a line is switched on. Sheds may also be removed (bitten off by a bird) while the insulators are in service.

Figure 10.22: During the installation, the insulator rod was overloaded by torsion, which resulted in an axial rod breakage. Due to the water vapour permeability of the housing material, a conductive track may form along the fracture and cause interface degradation or a flashover/dielectric breakdown.

Figure 10.23: This damage can occur on insulators manufactured using swaging technology. The pre-fabricated shed sections are fitted to the rod without the use of a chemical compound providing a bond.

Figure 10.24: Experience has shown that composite insulators are less attractive to be target of vandalism due to their non-shattering characteristic unlike their conventional counterparts and the slimmer shape.

Damage to the fitting or to string elements mounted directly onto the fitting

Figure 10.25: The insulator was not protected and exposed to a power arc of 50 kA/0.3 s. Due to melting loss, the seal is destroyed and, when the glass transition temperature of the resin is exceeded, the crimp connection may exhibit signs of reduced crimping.

Figure 10.26: Combined (seal and galvanizing) damage after continuous corona activity.

Figure 10.27: Typical damage pattern for aluminium; it can now act as a source of corona.

Figure 10.28: The fittings can remain installed after a power arc of 30 kA/0.5 s, provided that the melting points do not generate any corona.

The danger associated with a particular damage pattern and the resulting need for action can be classified according to severity. The definition of such a severity classification is unique to each individual set of circumstances and depends, among other things, on the strategic importance of the overhead transmission line affected, the statutory requirements, the safety concept of the operator and the route the line takes through inhabited territory.

The following list of effects for the severity level 0–5 serve as an example:

0: None/Cosmetic (technical functions not compromised)

1: Very low (barely detectable effect on technical function)

2: Low (detectable effect, but not critical to the technical function)

3: Significant (presence of a defect that may have a significant effect on the technical function)

4: Critical (remaining service life shortened as a result of the defect)

5: Very critical (final state reached, end of service life (EOL), technical function virtually non-existent)

The use of severity levels has also been proven to be a useful tool, when various insulator designs were comparatively evaluated in accelerating laboratory or outdoor tests [27–29]. CIGRE Working Group B2.21 used its expertise to evaluate the documented damage patterns (Table 10.4), which can serve as a guide in cases that are difficult to interpret. The term "early generation insulators" stands for first and second generation composite insulators, while the term "current generation insulators" stands for insulators and insulator strings that have successfully undergone a design test and type test in accordance with the latest standards (IEC 61109 [30] or ANSI 29.11 [31]). It should be importantly noted, for example, that an incorrectly designed insulator string that experiences permanent corona discharges as a result of an unverified type test must be assigned to early generation insulators from a risk perspective.

It is worth evaluating "global" hydrophobicity during the visual evaluation of insulators. This is true if the housing material exhibits the property of hydrophobicity transfer. This test is of little significance for materials such as EVA, EPDM or epoxy resin because there is no hydrophobicity transfer or no hydrophobicity transfer that is effective and relevant for operations. The spraying method (IEC 62073 [32], Method C) has proven its worth as a simple test; reference photos are used to determine the wettability class (WC 1–7) in this standard. For this purpose, the entire surface of the insulator (lower shed surface, upper shed surface, shank, etc.) are sprayed. The standard recommends a spraying time of between 20 and 30 s for each section to be evaluated. The following empirical rule applies: The point of "water saturation" characterised by downwards running droplets should be reached in order to achieve the formation of receding contact angles. If significant differences in hydrophobicity come to light after this global evaluation, the material sections affected can be inspected in greater detail.

In the spraying method, reference photos (Fig. 10.29) are used and the surface wetting qualitatively compared. The reference photos being based on a hydrophobicity classification guide, which was introduced by the Swedish Transmission Research Institute (STRI) in 1992 [33].

10.4.2.2 Mechanical Evaluation

The mechanical function of a composite insulator is defined by its Specified Mechanical Load (SML). This value must be interpreted as a withstand value. In other words, the actual threshold values (failure or ultimate value) are greater and statistically. The difference between the withstand value and the threshold value is not defined in the standard; today, a value of 20 % or more is common between the smallest threshold value and the withstand value. Depending on the application and the statutory requirements, a factor of 2.5–3.75 can occur between the every day stress (EDS) and the specified mechanical load (SML). Experience has shown that composite insulators whose load did not exceed the damage limit load [30] while in service also reach the specified mechanical load after 30 years. The following two tests were introduced to qualify (new) composite insulators, but they also apply to the evaluation of old composite insulators:

Table 10.4 Severity evaluation of examples shown in the photographic documentation Figs. 10.10–10.28

Description of damage	Reference to figure	Severity classification 0…5[a]	Probability of damage to early generation insulators	Probability of damage to the current generation of insulators (design and type test) in accordance with IEC 61109
Shed cracking	Fig. 10.10	1…2…5[b]	High	Low
Shank splitting	Fig. 10.11	5	Medium	Low
Shed puncture	Fig. 10.12	2	Low	Low
Shed puncture with prolonged tracking	Fig. 10.13	5	Low	Low
Extended (full) tracking	Fig. 10.14	5	Medium[c]–low[d]	Medium[d]–low[d]
Bulk erosion	Fig. 10.15	5[e]	Medium[c]–low[d]	Medium[c]–low[d]
Heavy embrittlement (alligatoring)	Fig. 10.16	3	Medium	Low
Handling-related shed/housing damage	Fig. 10.17	2	High[f]	Medium[g]
Damaged seal	Fig. 10.18	5	High	Medium[h]
Rod exposure	Fig. 10.19	5	High[g]	Low[g]
Rod deterioration	Fig. 10.20	5	High	Low
Damage caused by an animal	Fig. 10.21	5[i]	Medium[i]	Low[i, j]
Interface deterioration	Fig. 10.22	5	High	Low[g]
Housing separation	Fig. 10.23	5	High	Low[k]
Gun shot damage	Fig. 10.24	5[l]	Low	Low
Molten end fitting	Fig. 10.25	5	Depends on the insulator/set design and fault parameters of the grid	
Heavy corrosion and damage to the seal	Fig. 10.26	5	High	Low
Damaged corona collar	Fig. 10.27	3…4[m]	Depends on the insulator/set design and fault parameters of the grid	

(continued)

Table 10.4 (continued)

Description of damage	Reference to figure	Severity classification 0…5[a]	Probability of damage to early generation insulators	Probability of damage to the current generation of insulators (design and type test) in accordance with IEC 61109
Power arc marks on a protective steel fitting	Fig. 10.28	2…3	Depends on the insulator/set design and fault parameters of the grid	

Explanatory notes:
[a] Applied to the number of insulator or insulator sets available
[b] Applicable if caused by a permanent corona discharge that is stable locally
[c] For housing materials with a low filler content (generally RTV and LSR-silicone rubber) and a high level of pollution
[d] For housing materials with a high filler content (HTV silicone rubber or EPDM)
[e] If the rod is exposed or the seal is damaged
[f] For materials with a low mechanical strength (e.g. the first RTV)
[g] Handling and transportation rules are now fully defined and, in general, contained in the relevant EVU's documentation; the mechanical strength of elastomers (polymers) improved significantly compared with the first generations
[h] If the current design criteria for corona protection (droplet corona) are taken into account and the sealing system cannot be damaged by normal handling
[i] As shown, critical as a result of rod exposure; if individual sheds are affected, the "damaged sheds" description of damage (Fig. 10.17) applies
[j] A local problem whereby special installation rules, for example, are defined for the solution
[k] Today, most insulators are manufactured with a chemical bond between the housing material and rod
[l] If the projectile has damaged the rod
[m] It must be noted that the fault current flowed over the crimped insulator fitting (possible crimp relaxation)

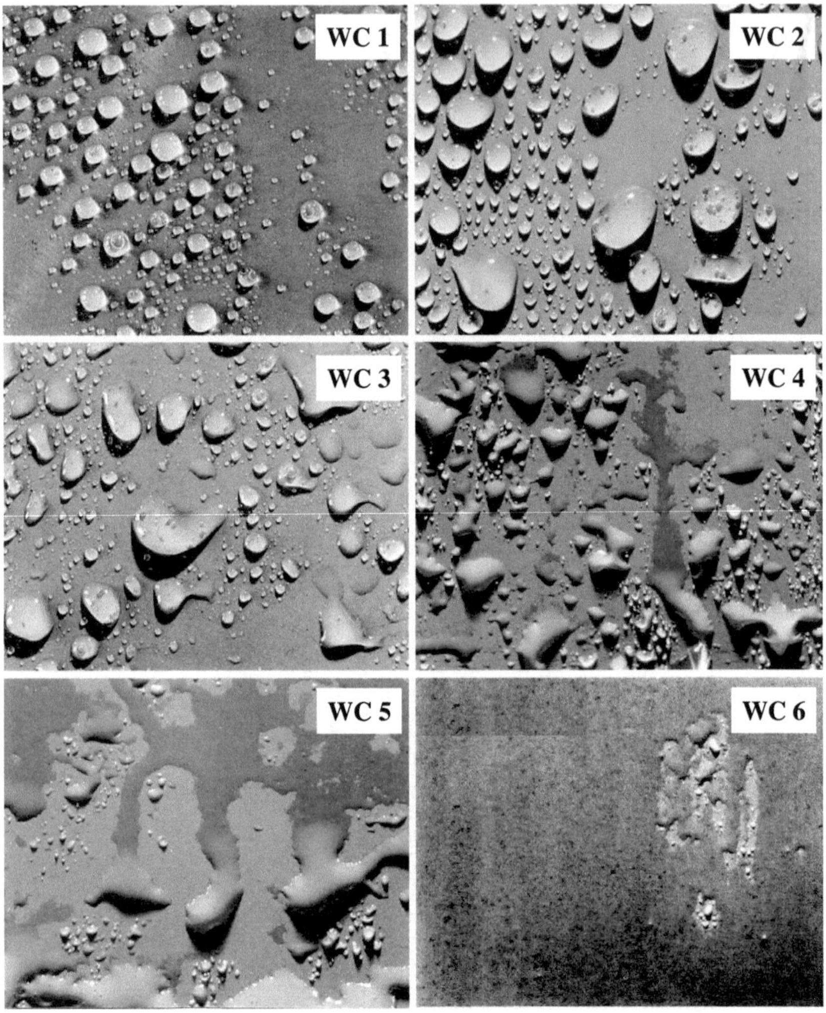

Fig. 10.29 Reference photos of the wettability classes [32]

- Verification of the current specified mechanical load (SML) in accordance with [30]
- A 96 h load-time test; this test method is contained in [30] as a design test (load = 60 % of the mean ultimate load from three tensile tests) and type test (load = 70 % of the SML). Due to the frequently low number of units, this procedure (as in the case of the type test) is more practical because only three (as opposed to six) insulators are required.

Old composite insulators should also pass the test chosen. The relevant graphs of the test must be drawn.

10.4.2.3 Electrical Evaluation

Electrical tests are usually performed to quantify the properties of outer insulation. Tests for composite insulators are described in IEC 61109 and IEC 60383 [34] and, analogous to the type tests for conventional insulators or insulator strings, have proven their worth for many years now. When evaluating old insulators, it has been shown that the use of standard tests in dry or rainy conditions is unsuitable for diagnosing hidden defects or defects that are visible on the surface. Under dry test conditions, in particular, hardly any differences are measurable. If a conductive defect that has moved forward axially is of a sufficient length, a certain reduction in the flashover voltage can be detected under artificial rain. Better diagnostic results are achieved in the case of interface defects, in particular, if a steep front voltage test is performed (with or without prior boiling) on the entire insulator or on sections of the insulator (Fig. 10.30). This test often forms part of evaluations performed after a certain period of time in service [35]. Even though the use of modern UV and IR cameras is a relatively new test method, it is being used increasingly as a result of the commercial availability of different product suppliers [36]. The tests are frequently performed when the voltage increases. Faults tend to be easier to detect when using a UV camera.

A number of short-term tests have been developed to evaluate the pollution layer on the insulator and the current state of hydrophobicity. Often, the pollution layer on the insulator is not a representative quantity but rather constitutes a snapshot because the following factors, among others, may effect insulation strength: Heavy rain before withdrawal from the network, changes (wiping the layer) during disassembly and transportation, and hydrophobicity transfer acceleration when the insulator is stored at an indoor temperature. The short-term tests are described below:

Quick Salt Fog Test [38]

This test is based on the standardised salt fog test in accordance with IEC 60507 [39]. However, due to time and cost restraints, it has been modified to include variable stressing. Over a conditioning period of 20 min, a voltage value that corresponds to approximately 90 % of the flashover voltage value is applied when assessing the specified fog conductivity. After the conditioning period, the voltage is increased in increments of 5 % and, in each case, this value is retained for one minute until flashover occurs. Immediately after flashover, the original starting voltage (90 %) is applied to the insulator and the procedure for increasing the voltage is repeated five times until flashover occurs. This concludes both the conditioning period and baseline recording of values. In the second part, 90 % of the mean value recorded for the five preceding flashovers is used as the starting voltage. The voltage is then increased every five minutes in increments of 2.5–3.5 % until flashover occurs. After flashover, the test continues with 90 % of the last flashover voltage value as the starting voltage. When the flashover voltage values have stabilised, the mean value of the individual implementations is determined as the characteristic value.

Fig. 10.30 Steep front voltage test on an intact (*left*) and defective composite insulator, with partial interface breakdown (*2nd of left*) due to conductive tracks inside the interface rod-housing (*both photos to the right*) [37]

Rapid Clean Fog Test [38]

If several insulators are available and the goal is to evaluate the electrolytic conductivity generated by the natural pollution layer, the rapid clean fog test can be used [36, 40, 41]. The insulator is exposed to a fog stress and voltage stress simultaneously. In this case, the voltage is increased incrementally until flashover occurs. If there are several insulators of the same type and they have comparable pollution layers, this test must be performed once for each insulator. This test has also proven its worth in terms of substantiating the hydrophobicity transfer. The test specimen may have a natural or artificial pollution layer on its surface. The test is performed as described above and repeated after 48 h [36, 41]. Between these two tests, the insulators are stored under normal indoor conditions. If a hydrophobicity transfer has occurred, higher flashover values are determined in the second test.

Modified Clean Fog Test

As it is the case with the previous two procedures, this test also uses the testing facilities in accordance with IEC 60507. Since relative comparisons should always be feasible, it is also possible to use other means of producing fog.

The testing voltage is 1.0–2.0 times the phase to earth voltage. The leakage current is logged continuously. The test duration is determined by the type of pollution layer. Empirically, it has been established that the test can be concluded if the leakage current reduces by 30 % of the maximum value measured. The fog also results in the removal of the pollution layer. Furthermore, due to a lack of salt content, it no longer increases conductivity. Test times of up to two hours are quite common.

Depending on the type of pollution layer and the behaviour of the housing material (hydrophilic or hydrophobic), etc., the aforementioned procedures permit a relative comparison of different insulators within a single procedure. However, it is not possible to directly compare the results of the three procedures or to translate these results into direct operational behaviour. For better evaluation of the results, conventional insulators are often also tested within these procedures for reference purposes.

CIGRE Working Group B2.21 has summarised the experiences gained as a result of testing composite insulators in accordance with the three procedures (Table 10.5) to provide assistance on the choice of the appropriate procedure, "fit for the corresponding purpose".

From the perspective of a "classic" pollution layer test, work is currently underway with a view to creating reproducible pollution layers on hydrophobic housing materials [45].

10.4.2.4 Material Analysis

Materials testing can be used successfully to evaluate the various types of materials used and their ageing state. If available, the results of such tests can be compared against storage samples (e.g. an insulator from warehouse stock). Major deviations may be suggestive of ageing. The overview described in Fig. 10.31 will have test procedures in the following chapters that have proven successful for many years and, in some cases, form part of the latest product standards.

Housing Materials

- High voltage test after a water diffusion test (IEC 61109)
- Power arc resistance test involving small currents (IEC 61621 [46])
- Inclined Plane Test (IEC 60587 [47])
- Thermo-gravimetric analysis (TGA) and differential scanning calorimetry (DSC) to determine the filler content and as characteristic values of the "fingerprint"
- Fourier transform infrared spectroscopy (FTIR) to determine certain chemical groups and as a characteristic value of the "finger-print"
- Various X-ray technologies for material analysis such as X-ray fluorescence (XRF) to detect the elemental and oxide content and X-ray diffraction (XRD) to detect crystalline phases or compounds

Table 10.5 Summary of experiences associated with acquiring information from the aforementioned procedures [36, 38, 40–44}

	Evaluation of the current level of pollution	Evaluation of current hydrophobicity	Effect of intensive ageing (e.g. erosion or prolonged tracking)
Quick salt fog test	No	Yes	Yes
Rapid clean fog test	Only limited information possible	Only limited information possible	Only limited information possible
Modified clean fog test	No	Only limited information possible	Only limited information possible

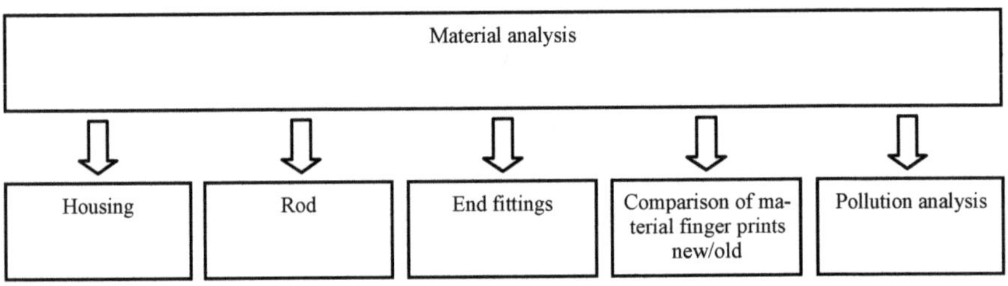

Fig. 10.31 Material analysis of composite insulator components after service

- Flammability test (IEC 61109)
- Hardness test, which is performed as a relative comparison before and after a boiling test (IEC 62217 [48])
- Surface roughness (ISO 4287 [49])
- Interface test (boiling and steep front voltage—IEC 61109)
- Tightness test (IEC 61109)
- Ageing test (IEC 62217)

The Dynamic Drop Test [50–52] can be used to quantify the hydrophobicity of insulator housing sections. Systematic evaluations e. g. by a Round-Robin-Test are not available yet. For the investigation of complete (full-scale) insulators, the test procedures of Table 10.5 are applicable.

The costs and time associated with a repeated test, especially in the case of design tests, need to be weighed up against the expected test results. The state of the respective insulator type determines which tests are conducted. For example, the erosion resistance of an insulator design that passes the ageing test in its new state will also not change after several years of outdoor use if no intensive ageing occurred while in service. It would not make sense to perform a corresponding 1,000 or 5,000 h test, then.

Differences may be measured in terms of its hydrophobicity, which can be lower or higher than when the insulator was in its new state.

Another problem is choosing suitable test sections from the entire insulator because comparatively small insulators, which are intended to be representative of real-life insulators, are tested in the design tests. The parts of the insulator that experienced the highest electrical stress in service should be used. This is typically the area around the fitting on the high voltage side.

Rod Material

- High voltage test after a water diffusion test (IEC 61109)
- Thermo-gravimetric analysis (TGA) and differential scanning calorimetry (DSC) to determine the filler content and as characteristic values of the "finger-print"
- Fourier transform infrared spectroscopy (FTIR) to determine certain chemical groups and as a characteristic value of the "finger-print"
- Dye penetration test

End Fittings

- Zinc layer thickness (IEC 60383-1)
- Structural analysis using a scanning electron microscope

Housing Material Finger-Print

CIGRE Working Group WG D1.27 evaluates analytical methods currently used to uniquely identify polymeric materials with a view to choosing the most suitable procedures from the many available.

10.4.2.5 Pollution Layer Analysis

While the housing materials and any changes to these materials are evaluated in Sect. 10.4.2.4, the next section concerns the interaction between the housing material and the pollution layer.

Adsorption of the Pollution Layer

The pollution layer can be evaluated using the methods in IEC 60815-1 [53]. In addition to measuring the equivalent salt deposit density (ESDD), which concerns equivalent conductive pollution layer components, and the non-soluble deposit density (NSDD), which concerns non-conductive pollution layer components that nevertheless contribute to the thickness of the pollution layer), surface conductivity can be used to characterise pollution. Examples are provided in [42, 43, 54]. Experience has shown that composite insulators (especially when silicone rubber is used as the housing material) can adsorb a greater pollution layer than equivalent conventional insulators under the same conditions [55]. If an insulator's hydrophobicity is transferred into the pollution layer, its overall conductivity is not effective during humidification as a result of the known pollution layer hydrophobising treatment; the insulation strength is higher than the ESDD and NSDD measurements would expect [38]. At this point, it must be mentioned again that a pollution layer evaluation is a snapshot of purely relative information.

Surface Changes

Surface changes in the sense of increased roughness can result in more intensified adsorption of the pollution layer. The procedure of ISO 4287 can be used to perform an evaluation here.

Wettability Class

Ideally, this simple spraying test is performed during the visual evaluation phase [32].

Efficacy of the Hydrophobicity Transfer Mechanism

In addition to measuring hydrophobicity, the remaining capacity for a hydrophobicity transfer is an important property, especially after a long-time in service [56]. CIGRE Working Group D1.14 has created a guide for such tests [52], which is based on the results of a series of round-robin tests. A defined *artificial* pollution layer comprising quartz powder, isopropanol and distilled water is applied to the material test specimens, which may also be resected parts from an insulator housing (e.g. sheds). The pollution layer is dried under controlled conditions, preferably in a desiccator. The static contact angle is measured at certain times. The measurement times after a droplet has been applied have proven to be important: The contact angle of the droplet is measured after 30 and 180 s. The first measurement is taken to verify that hydrophobicity transfer has taken place, while the second measurement is the criterion for its sustainability. If the contact angle returns to 0° after 180 s, the measurement for this test time is also set to the value zero, even if a higher value was detectable after 30 s. An example of this relatively new procedure is shown in section "Hydrophobicity".

Hydrophobicity Transfer

This additional procedure for measuring hydrophobicity transfer serves more to evaluate housing material with a *natural* pollution layer [57]. In this test, the scope of the hydrophobicity transfer is determined as follows: HT = (ESDD-ASDD)/ESDD. ESDD is the conventional measurement in accordance with IEC 60815-1, while the apparent salt deposit density (ASDD) is measured using a conductivity cell, which is applied directly to the surface. In contrast to the ESDD measurement, comparatively less water is available and not all of the hydrophobised pollution layer components are dissolved. If HT values close to 1 are measured, this means that an effective hydrophobicity transfer has taken place.

Some of the procedures listed in Sect. 10.4.2.3 can also be used to evaluate pollution layers.

10.4.3 Summary of Test Methods and Failure Criteria

Given the diverse range of appearances and damage patterns, it is not possible to specify a testing programme that is universally valid. However, the previous sections can be used as a basis for compiling an individually structured programme. All of the tests named here have already been used to evaluate composite insulators. The ground-breaking role that visual evaluation plays in compiling the testing programme is a crucial step at the start of evaluations and should therefore be performed with great care and skill.

In the interest of clarity, those damage patterns that indicate the end of service life for a composite insulator are, once again, listed below:

- Any sign of a brittle fracture [58, 59] or potential triggers for a brittle fracture (IEC 62662)
- Heavily corroded fittings, exposed steel without a zinc layer and/or the occurrence of arc roots
- Bulk erosion, which exposes the rod
- Defective seals
- Insulator housing that has been damaged by mishandling and has an exposed rod
- Shed punctures, which reduce the creepage distance. Depending on the safety margin built into the creepage distance configuration and the total number of insulators of this type affected in the network, a 10–20 % reduction in the creepage distance may be permitted as an indication that is not specific to any particular scenario. This also applies to other types of shed damage.
- Any sign of a shank puncture
- A loss of adhesion between the rod and insulator housing in the case of insulators that have a chemical composite between the rod and insulator housing.

10.5 Examples of Testing Programmes and Their Results

The sections below will provide examples of how to perform an evaluation. The various approaches to testing will be discussed in conjunction with their results.

10.5.1 *Example 1—a 420 kV Composite Insulator (Stepped Injection Moulding Variant) After 10 Years in Service*

10.5.1.1 Test Specimens and Test Strategy

Four composite insulators were withdrawn from a European 420 kV network after 10 years in service [60]. The insulators had been installed in single suspension strings. In Sect. 8.8.1, an analysis of the field stress of this insulator string was compared against today's usual threshold values and the state in which the insulators were found.

The insulators have the technical characteristics shown in Fig. 10.32.

The visual evaluation of insulators did not produce any findings, which is most certainly also due to the low level of pollution (pollution class II). For this reason, the programme focused on the evaluation of hydrophobicity and the interface areas. A test matrix was created in tabular form (Table 10.6). The insulator string elements used (rings, horns, etc.) had no signs of rust or flashover. As a result of this diagnosis, the elements were retained in the string and only the insulators were replaced. Consequently, the electrical values for a type test (including the interference voltage test and corona test) were not measured.

10.5.1.2 Results and Discussion

Hydrophobicity

After "global" hydrophobicity was measured during the visual evaluation, the parts of the insulator housing that had a lower hydrophobicity in comparison to adjacent sections were inspected more closely. Their hydrophobicity in the as-received state with a pollution layer, after cleaning, and as a new insulator was determined using an optical bench that compared the static contact angles.

The simplicity of this measurement makes it possible to generate good statistics. Twenty measurements were taken in each case. The sheds and shanks were covered with a grey layer that was difficult to remove, even with the use of a solvent. This indicates that hydrophobicity transfer took place, which is further confirmed by the measurements (Fig. 10.33). The contact angles of the polluted surface are, on average, somewhat higher than the contact angles when the surface is in its new state or has been cleaned. This known phenomenon [25] is confirmed by the operational behaviour of this overhead transmission line, which has not experienced any flashovers since composite insulators were used. It also reflects the common practice of not cleaning composite insulators that have a silicone rubber housing, if a hydrophobicity transfer can take place.

In addition to evaluating hydrophobicity, the effect of the hydrophobicity transfer mechanism was also analysed. The tests were performed in accordance with the recommendations of CIGRE Working Group D1.14 [52], namely one

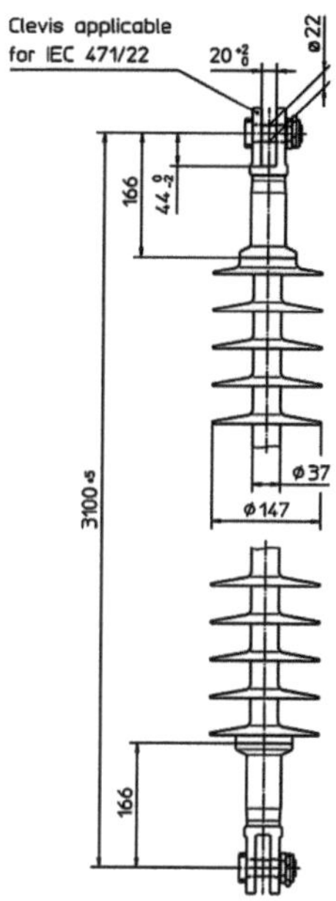

- Creepage distance:..........................8600 mm
- Striking distance:..........................2768 mm
- SML: ....300 kN
- Lightning withstand voltage*:..............1543 kV
- Switching withstand voltage, wet:..........1192 kV
- 1-minute PF withstand voltage, wet:..........782 kV

Design details:

- Rod diameter 31 mm (E-Glas)
- Housing made of HTV-Silicone Rubber
- Housing made by injection moulding, multiply stepped
- End fittings overmoulded

* measured with power arc protection devices

Fig. 10.32 Summary of technical details

defined artificial pollution layer and two measurement times for the contact angle. Shed sections were used as test specimens (Fig. 10.34). The measurements were able to clearly demonstrate that the pollution layer was sufficiently hydrophobised after approx. 8 h (Fig. 10.35). The time after which hydrophobising occurs and the values achieved correspond to the values from the round-robin test of CIGRE Working Group D1.14 involving materials in mint condition [52], which are classified as being transferable. Due to the newness of this procedure, there are no threshold values according to which a material classification can be made.

However, it has been shown that setting a defined relative humidity and the relevant device for measuring the contact angle can have a significant effect on the measurement.

Specified Mechanical Load and Ultimate Load Test

The test was conducted in accordance with the requirements of the applicable standard IEC 61109. To further intensify the test, the specified mechanical load was held for 30 min and the tightness test was applied during this time in accordance with ISO 3452 (the standard value is 70 % of the SML).

Table 10.6 Test matrix for the four insulators taken off the line

	Insulator 1	Insulator 2	Insulator 3	Insulator 4
Hydrophobicity				
Insulators as received (IEC 62073)	Measurements taken on upper and lower shed surfaces	Measurements taken on upper and lower surface of sheds		
Hydrophobicity transfer (CIGRE D1.14)				
SML/Ultimate tensile test (IEC 61109)	Tightness test at SML-level	Tightness test at SML-level		
Interface SR/SR	SEM analysis, tear test under a microscope	SEM analysis, tear test under a microscope	Boiling followed by steep front voltage test	Boiling followed by steep front voltage test
Interface Rod/SR				
Interface SR/End fitting	Tightness test at SML-level	Tightness test at SML-level		
Dye penetration test	after ultimate tensile test, 5 specimens	after ultimate tensile test, 5 specimens	5 specimens, in each case with and without prior boiling	5 specimens, in each case with and without prior boiling
Inclined Plane Test (IEC 60587)	5 specimens	5 specimens		

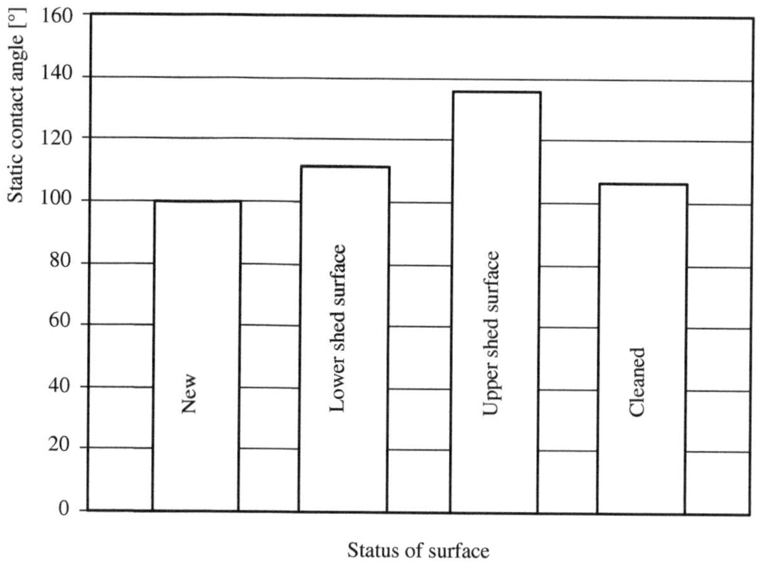

Fig. 10.33 Measuring the static contact angle of various surface states (mean values from 20 measurements)

There was no evidence of cracks or the insulator housing detaching from the over moulded end fitting (Fig. 10.36). The photograph also shows that there was no corrosion of the end fitting.

After the 30 min-long preliminary load with a 100 % SML, 378 or 362 kN was the maximum force achieved (Fig. 10.37). Both values represent the range of values that were measured 10 years earlier as part of a sample test on the new insulators and correspond to the design rule whereby the lowest damage limit load value for an insulator in its new state should be at least 20 % above the specified mechanical load. The failure occurred at both measurements when the rod slipped out from the fitting. The rod did not break or split.

Interface Tests

Since 1975, composite insulators have been manufactured by gradually stepwise casting silicone rubber sheds and shank sections. The property of silicone rubber whereby complete homogenous vulcanisation occurs again between a vulcanised section of the insulator housing and a section of the insulator housing that was subsequently casted for the corresponding process parameters is deployed here and in the modular process. In the 1990s, the experiences associated with "RTV stepping" were assigned to "HTV stepping". The HTV multi-injection procedure successfully applied since this time is characterised by axial interfaces and more than one radial interface. The type of insulator to be tested was

Fig. 10.34 Shed sections with a pollution layer in a desiccator

manufactured using HTV stepping and the fittings were embedded by means of over moulding. Despite the reliable chemical mechanisms of silicone rubber vulcanisation are also known, these different interface areas were of particular interest after 10 years in service.

Microscopic Evaluation and SEM Analysis

For the visual evaluation, parts of the radial silicone rubber/silicone rubber interface vulcanised in two successive vulcanisation steps (transition area) were removed from the insulator housing. This area can only be found due to a small radial mould line (and not in the material); the silicone rubber in both vulcanisation steps is completely vulcanised. This is also confirmed under the microscope (Fig. 10.38). The test specimen with the transition area was strained by 20 % without the occurrence of a stress crack.

A scanning electron microscope (SEM) can also be used to verify that the radial interface of the transition area has been fully vulcanised (Fig. 10.39). SEM exposures were taken of the transition area and of a bulk section without an interface. Both exposures are effectively identical, thus confirming the quality of the stepped transition interface. It is well known that the comparatively high vulcanising temperature and the pressure that occurs in the HTV injection moulding tool have a positive effect on both parts of the insulator housing.

Boiling Test and Steep Front Voltage Test

This procedure was used to evaluate the following interface areas (Figs. 10.40 and 10.41):

- Radial silicone rubber/silicone rubber interface
- Axial rod/silicone rubber interface
- Steel/silicone rubber interface of the over moulded end fitting

The test specimens were trimmed to the standard test size for design tests (approx. 800 mm). Since the resulting cut surface at the open end was not sealed by a cover, this further intensified the test. Pre-stressing by means of boiling took

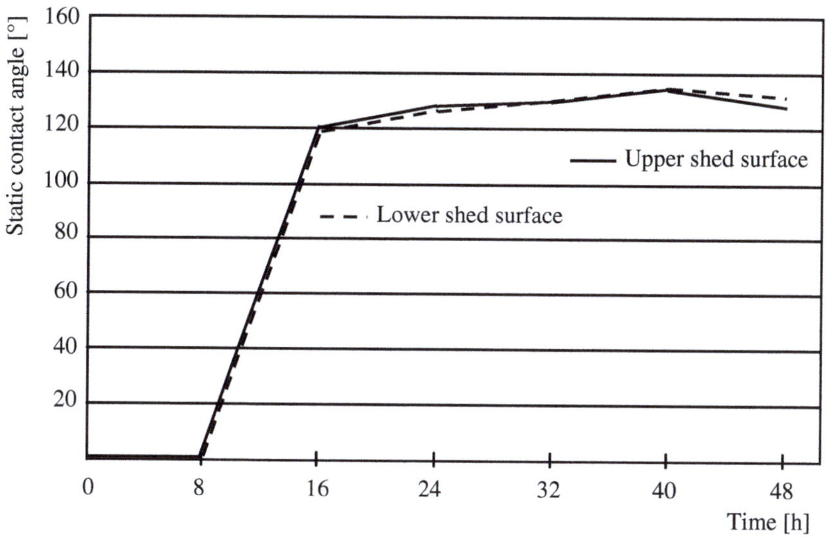

Fig. 10.35 Increase in the contact angle as a result of a hydrophobicity transfer

Fig. 10.36 Dye penetration test during SML test

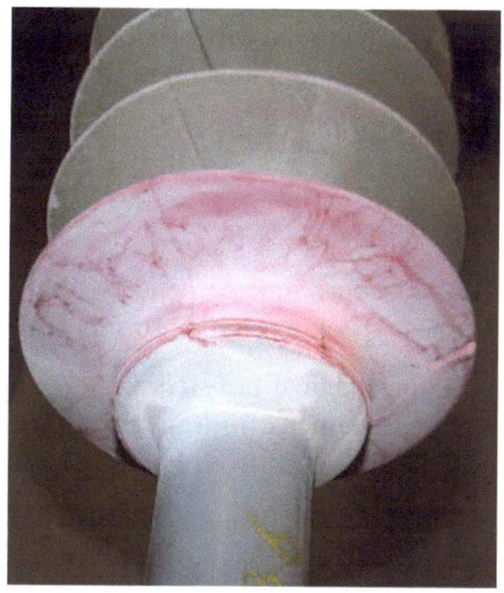

place over a period of 42 h, followed by the steep front voltage test involving 25 positive and negative pulses in each case. The electrodes were positioned differently in order to individually test each interface area of interest.

Each steep front voltage application resulted in an external flashover without a dielectric breakdown. In other words, the quality of the interface areas was fully ensured after 10 years in service.

The question of whether 96 hours of boiling is an adequate stress for a prestressing test is often asked. When answering this question, reference can be made to the experiences of IEC 61109, which introduced this procedure at its

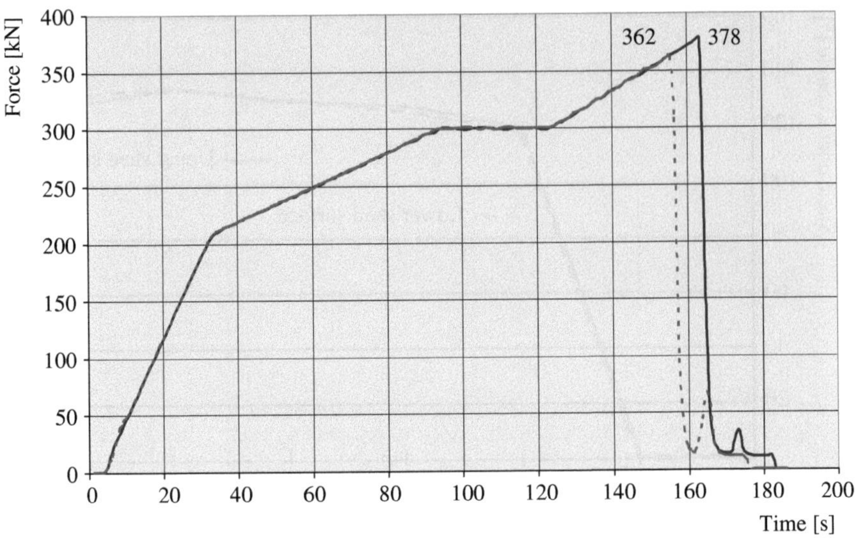

Fig. 10.37 Force-time progression of the ultimate limit load measurements

introduction in the early 1990s, because of its simplicity rather than a desire to use the Arrhenius relationship to establish a correlation. Service experience has shown beyond doubt that designs that pass this test have a significantly smaller probability of failure as a result of interface problems.

Dye Penetration Test

The procedure in accordance with the standard IEC 61109 was used for this test. The test specimens were 10 mm in length and were prepared as follows:

- No mechanical pre-stressing
- Test specimens from insulators used in the tensile test as described before
- Test specimens from both groups, boiled for a period of 42 h

There were 20 test specimens in each group. In other words, a total of 80 specimens were tested. None of the test specimens showed evidence of penetration after the standard test time of 15 min. When the test was extended to 1 h, the results remained the same.

Inclined Plane Test

The Inclined Plane Test has been an established material test method for evaluating erosion and tracking resistance since 1977. In order to accelerate and level the test conditions, a wetting agent is added to the test solution; hydrophobicity is not part of these evaluations.

Due to its simplicity, this test is often used for screening purposes within quality assurance or if new or modified material formulations are to be tested. Plate-shaped

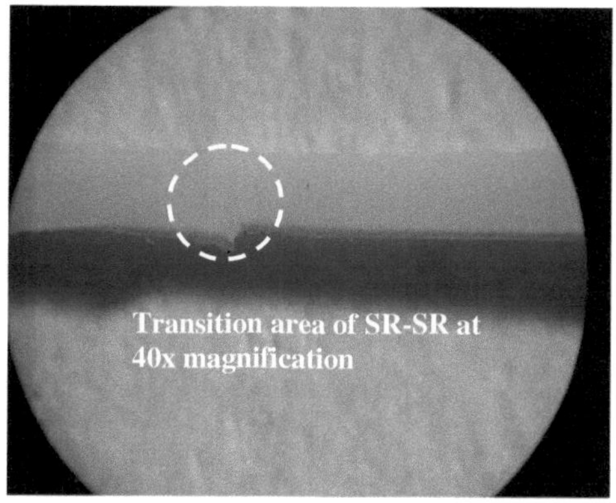

Fig. 10.38 Strained interface 20 %

test specimens are mandatory for this conformity test from the perspective of standardisation.

However, this test also has the potential to test insulator housing parts of composite insulators that have already been manufactured. This can be done within the framework of a sample test on delivered composite insulators and thus verify that the insulator material processed conforms with the specification of supply [61]. "Old" composite insulators can also be tested.

Test specimens are prepared in such a way that the insulator housing is carefully detached from the rod and any residual primer is removed (Fig. 10.42). The length of the test specimens is set in accordance with the mounting frames used for the standardised test (Fig. 10.43). The test is then performed in accordance with the standard (Fig. 10.44).

In general, it is always possible to test the interior area of the insulator housing; this is recommended for sample tests (Fig. 10.45). Due to the curvature that occurs, it is often more difficult but nevertheless possible to test the exterior of the insulator housing (Fig. 10.46).

In comparison to the standardised test, the test specimens are more heavily stressed because of the geometry of the insulator. The smaller insulating material width can lead to a concentration of partial discharges, while the lower insulating material thickness may cause (early) hole erosion. However, various measurement campaigns have shown that insulator housing sections made from highly resistant silicone rubber (HTV) will pass the test with low erosion similar to that associated with the standardised plate-shaped test specimens. One such example is the test result for insulator housing sections extracted from the four composite insulators in this testing programme example (Fig. 10.45). The test was successful after 6 h, with very little erosion at 4.5 kV. The brown discoloration at the edges is the residue of decomposed electrolytes (Triton X, in particular).

Figure 10.46 shows the result of a test performed on the exterior of an HTV insulator housing after 10 years of outdoor use. The test was also successful after 6 h, with very little erosion at 4.5 kV.

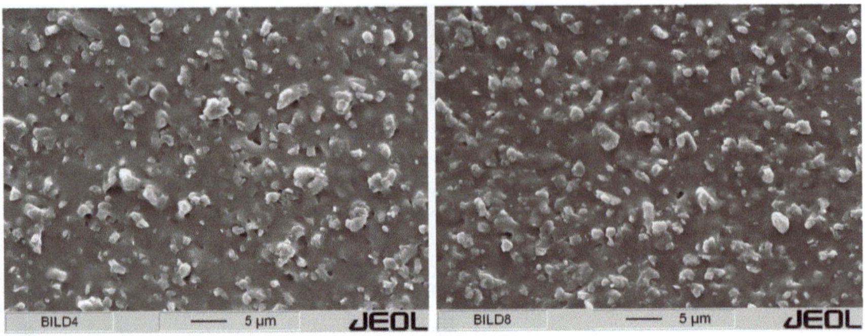

Fig. 10.39 SEM exposures with the transition (*left*) and without the transition (*right*) when magnified by a factor of 2,000

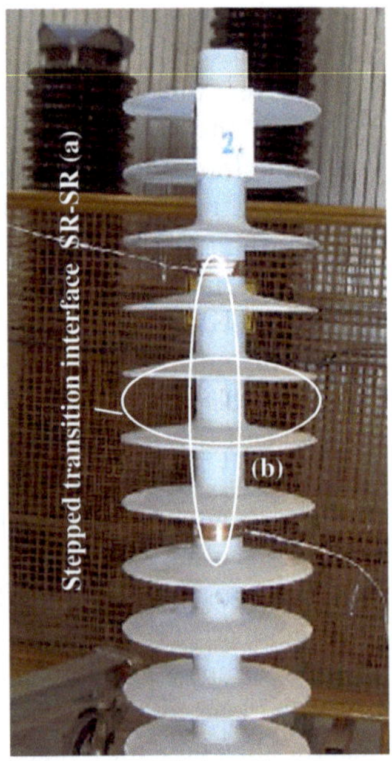

Fig. 10.40 Interface tests a + b

10.5.1.3 Summary of Results

The composite insulators evaluated were in very good condition:

- Hydrophobicity and the hydrophobicity transfer could be verified as being effective.
- The ultimate load values measured were within the range of values measured 10 years earlier when the new insulators were involved in a sample test.
- The interface tests usually performed within the context of design tests on brand new composite insulators do not reveal any evidence of a reduction.

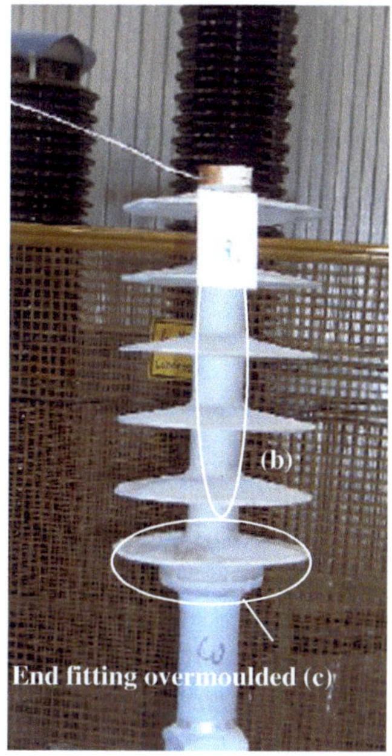

Fig. 10.41 Interface tests b + c

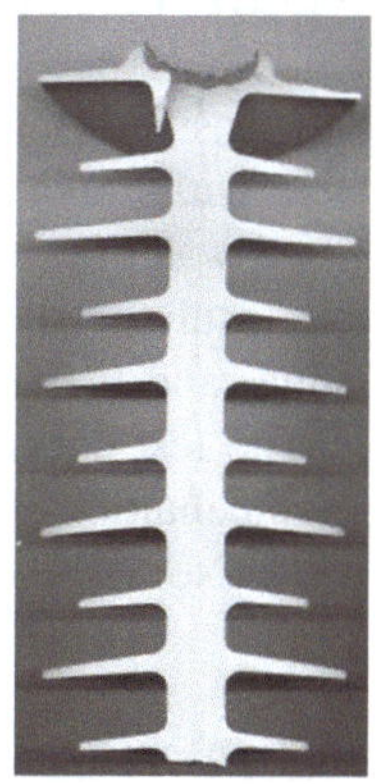

Fig. 10.42 Detached insulator housing

Fig. 10.43 Section of insulator housing in the standard frame

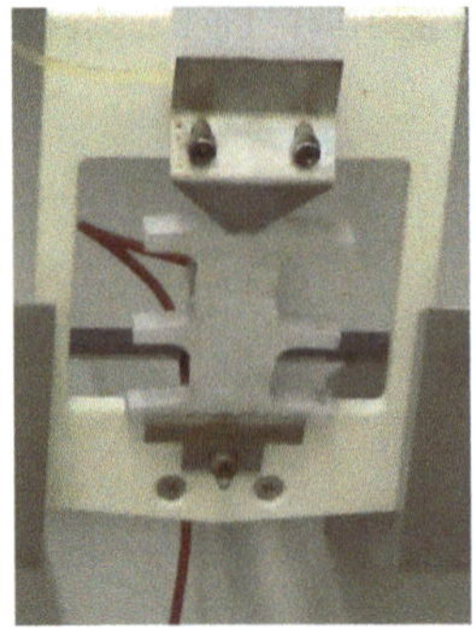

Fig. 10.44 "Standard" test

- The Inclined Plane Test was successfully performed on sections of the insulator housing.
- Nothing was found to indicate that the end of service life was imminent.

10.5.2 Example 2—a 420 kV Composite Insulator (Modular Variant) After 10 Years in Service

10.5.2.1 Test Specimens

Nine insulators (Fig. 10.47) were tested after 10 years in service. The insulators were withdrawn at random from a 420 kV line and subjected to various tests.

As was the case in Example 1 (Sect. 10.5.1), an analysis of the field stress for this insulator string was compared in Sect. 8.8.2 against today's usual threshold values and the state in which the insulators were found.

The test specimens were assigned a customer number and the accompanying documentation was extremely detailed in terms of each installation location. They were also assigned a reference number for testing purposes (Table 10.7).

10.5.2.2 Visual Evaluation

The insulators did not have any identifiable defects and were deemed to be in good general condition. The sheds had several visible paint splatters (Fig. 10.48) as a result of applying tower corrosion protection. There was also evidence of various forms of pollution as a result of storing the insulator on the ground during disassembly (Fig. 10.49). Insulator S9 was heavily polluted with bird excrement on one side, decreasing axially from the earth-side end.

10.5.2.3 Testing Programme

Following the detailed visual evaluation of all nine composite insulators, the following testing programme was defined (Table 10.8).

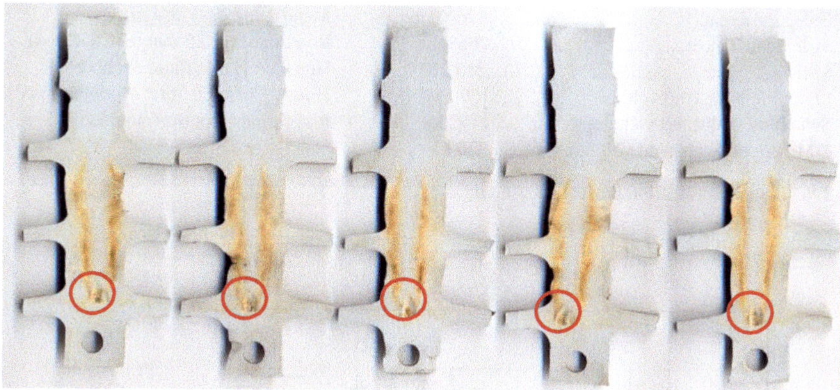

Fig. 10.45 Passed test, performed on specimens made of HTV silicone rubber, interior housing side tested

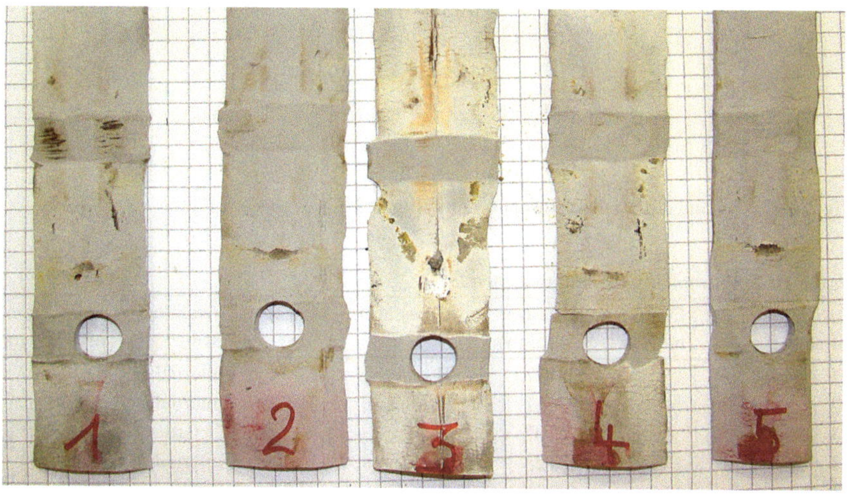

Fig. 10.46 Passed test, performed on specimens made of HTV silicone rubber, exterior housing side tested

10.5.2.4 Results and Discussion

Evaluation of Hydrophobicity

It is relatively easy to evaluate hydrophobicity. It has proven successful to horizontally position of the insulator supported on the fittings and to spray in opposite directions from both sides. When documenting the visual evaluation, the insulator housing can be displayed in a graphic and the number of the respective wettability class can be inserted into this graphic (Fig. 10.50).

Creepage distance: ....8200 mm
Striking distance:.....................................2682 mm
SML:...210 kN
Lightning withstand voltage*:.................1599 kV
Switching withstand voltage, wet*:.........1192 kV
1-Minute PF-withstand voltage, wet*:.....770 kV

* Without arc protection devices measured

Material/process design details
Rod diameter: 25 mm (E-CR-Glass)
Housing: HTV silicone rubber
Housing manufacture: Modular process
End fittings: not overmoulded

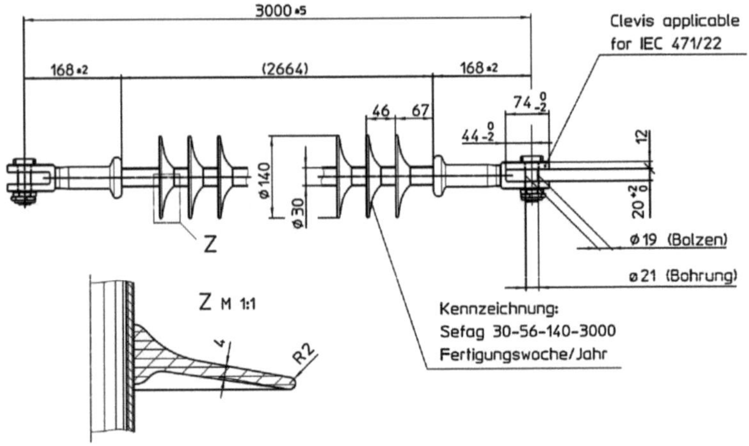

Fig. 10.47 Technical details of insulators investigated

Table 10.7 Information on test specimens

Test reference number	Customer number	Tower number	Cross-arm	Phase	Insulator string	Manufactured in week/year
S1	6	107	II external	U	Internal	47/99
S2	4	107	II internal	W	Internal	47/99
S3	1	107	I	V	External	47/99
S4	11	125	II external	U	Internal	49/99
S5	14	137	I	V	External	49/99
S6	10	125	II internal	W	External	49/99
S7	7	125	I	V	Internal	49/99
S8	16	137	II internal	W	external	49/99
S9	17	137	II external	U	internal	49/99

For practical reasons, and to subjectively attain a degree of accuracy, a summary of wettability classes is recommended. In the case of negligible levels of pollution, the evaluation can be reduced to three positions along the axis of the insulator; the high voltage side should always be evaluated. In the case of special types of pollution, a complete evaluation may be necessary if, for example, axial pollution is evident as a result of bird excrement or biological settlements (fungi, moss, etc.).

A loss or reduction of hydrophobicity (Table 10.9) mainly occurs in the case of pollution associated with coloration, traces of bird droppings and soil deposits.

Fig. 10.48 Paint splatters on insulator S1

Fig. 10.49 Soil particles on insulator S6

This type of soil pollution suggests that the composite insulators were stored without any protection after they were removed. As already discussed in Sect. 10.4, it is necessary to take into consideration any changes to pollution layers when withdrawing, storing and transporting the insulator, and when interpreting the results.

Evaluation of Seal

Given the importance of seals in ensuring that the insulator functions correctly, all 18 seals were evaluated in detail. The seals consist of two layers, silicone gel with a permanently low viscosity, and an outward-facing covering layer.

The primary function of the silicone gel is to seal; it is protected by a covering layer made from silicone rubber (see also Sect. 7.3.2.3). Here, all seals are coated with a highly adhesive pollution layer (Fig. 10.54). The pollution layer on the earth side is less intensive than the pollution layer on the high voltage side. The seals were cleaned to facilitate a better assessment (Fig. 10.55). Since the pollution layer was highly adhesive and had good hydrophobic properties, a solvent was used. The adhesion of the covering layers at the steel fittings and shank sections was fully intact. There was no evidence of detachment or penetration. After the covering layers on selected insulators were removed, the silicone gel was stripped

Table 10.8 Test matrix for the nine insulators taken off the line

	SEFAG-number								
Tests conducted	S1	S2	S3	S4	S5	S6	S7	S8	S9
Visual evaluation	x	x	x	x	x	x	x	x	x
Evaluation of hydrophobicity	x	x	x	x	x	x	x	x	x
Evaluation of seal	x	x	x	x	x	x	x	x	x
Adhesion test						x	x	x	
96-hour-tensile test with 75 % of SML followed by ultimate tensile test			x	x	x				
Test of galvanizing	x	x	x	x	x	x	x	x	x
Dye penetration test	x	x	x						
Ultimate tensile test without pre-stress	x	x	x			x	x	x	

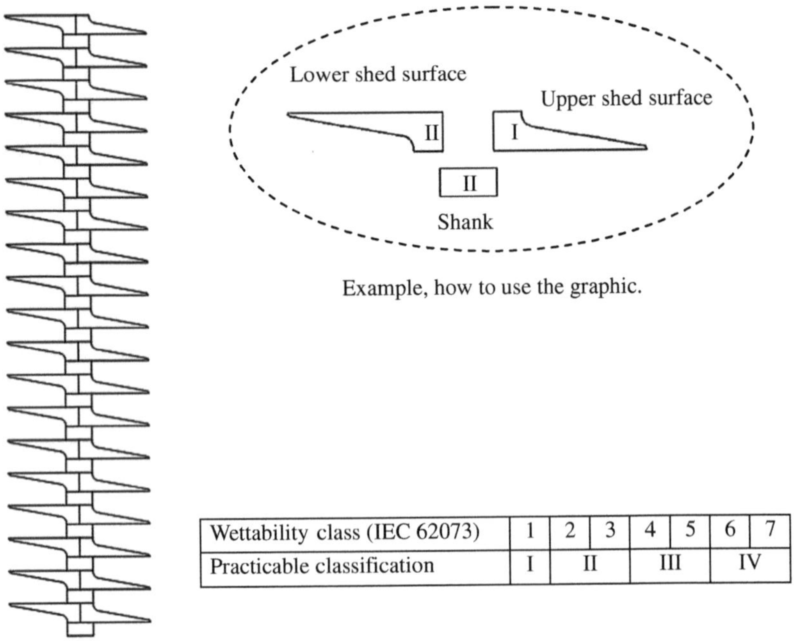

Fig. 10.50 Visualising the insulator housing in order to document hydrophobicity

(Fig. 10.56) in order to compare its adhesion on the interface against its adhesion with a new material. There was virtually no discernible difference.

Adhesive Test

Six samples were taken from the insulators, each at a 90° angle to the longitudinal axis of the insulator. In accordance with the standard IEC 61109, the test specimens were stored in boiling de-ionised water that contained 0.1 % NaCl by weight for a period of 100 h. A stainless steel container was used for this purpose.

Table 10.9 Result of hydrophobicity evaluation

Test specimen	Test section					
	Lower shed surface		Upper shed surface		Shank and seal	
	Findings	Class	Findings	Class	Findings	Class
S1	Well hydrophobic (Fig. 10.51)	II	Partially hydrophilic (Fig. 10.52)	III–IV[a]	Well hydrophobic (Fig. 10.53)	II
S2	Well hydrophobic	II	Well hydrophobic	II	Well hydrophobic	II
S3	Well hydrophobic	II	Well hydrophobic	II	Well hydrophobic	II
S4	Well hydrophobic	II	Partially hydrophilic	III–IV[a]	Well hydrophobic	II
S5	Well hydrophobic	II	Well hydrophobic	II	Well hydrophobic	II
S6	Well hydrophobic	II	Well hydrophobic	II	Well hydrophobic	II
S7	Well hydrophobic	II	Partially hydrophilic	III–IV[a]	Well hydrophobic	II
S8	Well hydrophobic	II	Well hydrophobic	II	Well hydrophobic	II
S9	Well hydrophobic	II	Well hydrophobic	II	Well hydrophobic	II

Remarks [a] In areas with paint splatters, soil and/or bird excrement

Fig. 10.51 Hydrophobic effect on lower shed surfaces

Fig. 10.52 Partially hydrophilic effect on upper shed surfaces

Fig. 10.53 Hydrophobic effect on the HV side

Fig. 10.54 Pollution on HV side

This test also evaluates the adhesive strength of the covering layer. It does this by applying a selective compression load, which has a shearing effect on the interface close to the fitting or shank (Fig. 10.57). The adhesive strength at the shank is tested by exposing and cutting into small areas of the shank and testing their ability to detach (Fig. 10.58). Once again, there was no detectable loss of adhesion.

96-Hour Tensile Load with 75 % of the SML Followed by an Ultimate Tensile Test

This test corresponds to the type test in accordance with IEC 61109, but with a 5 % higher tensile load value. After this preliminary test, which is conducted over a period of 96 h, the tensile load steadily increases from 0 to 75 % of the specified mechanical load and then increases incrementally during a period of 30–90 s until it reaches the specified mechanical load. After this test sequence, the tensile load is increased until the insulator fails/breaks.

The results shown in Table 10.10 were obtained, thus determining that the preliminary load did not have any effect on the limit load.

Test of Galvanizing

The zinc layer test was conducted on both end fittings on all insulators in accordance with IEC 60383-1. The values contained in Table 10.11 were measured. These values are above the value specified in the standard (85 μm). Since there are no measurements for fittings in their new state, it is not possible to quantitatively estimate the level of corrosion.

Dye Penetration Test (IEC 61109)

Nine samples were taken from an insulator, each at a 90° angle to the longitudinal axis of the insulator. For this test, the samples were 10 ± 0.5 mm in length and consisted of a GFK core and peripherally vulcanised housing material. Both the rod and the integrity of the interface between the rod and silicone rubber housing were tested.

There was no evidence of penetration by the test liquid.

Ultimate Load Test Without a Prior Applied Load (IEC 61109)

This test corresponds to a 1-min. durability test on the specified mechanical load. After this test sequence, the tensile load is increased until the insulator fails/breaks. The values (Table 10.12) show a greater variation than those after the 96-hour preliminary test (Table 10.10).

Fig. 10.55 After cleaning

Fig. 10.56 Peel test to prove adhesion

Fig. 10.57 Adhesion test on covering layer

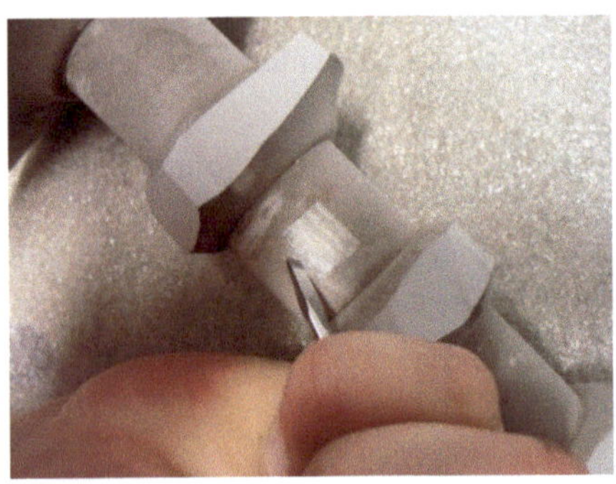

Fig. 10.58 Shear test at the shank housing (*sheds removed*)

Table 10.10 Results of ultimate tensile test after pre-stressing with 75 % of SML for 96 h

Test specimen	Failure mode	Ultimate failing value (kN)
S4	End fitting pull-out	289.2
S5	End fitting pull-out	289.2
S6	End fitting pull-out	289.0

Table 10.11 Results of the galvanizing test

Test specimen	Values (μm)							
	HV end fitting				Earth end fitting			
S1	156	109	112	94	115	100	99	116
S2	207	209	226	191	93	97	102	113
S3	154	165	94	134	128	164	92	134
S4	112	105	187	128	117	101	124	98
S5	115	121	177	154	111	94	101	87
S6	164	158	134	157	120	162	154	115
S7	135	94	154	112	108	92	157	187
S8	154	105	164	98	96	104	179	136
S9	126	134	94	108	135	99	152	98

The values are also within the range of values that were measured 10 years earlier as part of a sample test on the new insulators. The design rule generally applied to new composite insulators, whereby the lowest damage limit load value should be at least 20 % above the specified mechanical load, is also adhered to after 10 years in service.

10.5.2.5 Summary of Results

The summary of results is shown in Table 10.13.

10.5.3 Example 3—15 kV Composite Insulators After 15 Years in Service in a Railway Tunnel System

10.5.3.1 Test Specimens and Testing Programme

Since 1979, critical sections of Swiss railway lines have been provided with silicone composite insulators [43, 44, 62]. After 15 years in service, the railway operator wanted to conduct a comparative and exemplary test involving the porcelain and epoxy resin posts used in the tunnel and the silicone composite post insulators. It must be remembered that, historically, porcelain insulators with a specific creepage distance of approx. 50 mm/kV (Unified Creepage Distance—IEC 60815-1) had been installed since 1913. With an increase in the number of trains passing through the

Table 10.12 Results of ultimate tensile test without pre-stressing

Test Specimen	Failure mode	Ultimate failing value (kN)
S1	Rod breakage	255.4
S2	End fitting pull-out	285.6
S3	Could not be loaded up to the point of failure because the bolts were sheared off prematurely and therefore the clevis splayed.	
S7	End fitting pull-out	295.4
S8	End fitting pull-out	276.7
S9	End fitting pull-out	282.4

tunnel, it was necessary to perform at least two cleaning cycles per year. Silicone composite post insulators could be fitted at the same overall height with a creepage distance of approx. 80 mm/kV as a result of the smaller shed thickness and the proportionate overcasting of the fittings. In their as-received state, the insulators showed evidence of heavy pollution (Fig. 10.59). Due to the low number of units, it was only possible to conduct exploratory tests. The epoxy resin post had to be replaced after 2 years in service because discharges were observed and there were some evidence of erosion after the pollution layer was removed.

The following tests were conducted:

- Pollution layer measurements
- Modified clean fog test
- Ultimate cantilever bending force test

10.5.3.2 Results and Discussion

Pollution Layer Measurements

The ESDD and NSDD measurements were taken in accordance with IEC 60815-1. The silicone composite post insulator had the highest values, a phenomenon already confirmed as a result of hydrophobising the pollution layer (Table 10.14). This effect is also reflected in the wettability class. If the values for the porcelain post are comparatively low, it must be taken into consideration that there can be a maximum of 6 months between the last cleaning cycle and this evaluation. The silicone composite and epoxy resin post insulators were not cleaned while they were in service.

Modified Clean Fog Test

The posts with the original pollution layer were simultaneously humidified in a fog chamber containing clean fog for a period of 2 hours.

Then, the service voltage of 15 kV (but with 50 Hz) was applied to all posts and the leakage current was measured over a period of 30 min. The maximum values are provided in Fig. 10.60. Apart from the porcelain post, these results were as

Table 10.13 Summary of results

Tests conducted	Results
Evaluation of hydrophobicity	All insulators were in good conditions without any functional limitation.
Evaluation of seal	Good hydrophobicity on the lower shed surface and at the shank. There is a partial loss of hydrophobicity on the upper shed surfaces.
Adhesive test	There is no evidence of damage to the seals. The seals are also clean under the covering layer and still very much intact.
Adhesion test	Insulators S7–S9 were boiled. The adhesive tests were successful in accordance with IEC 62217.
96-hour-tensile test with 75 % of SML followed by ultimate tensile test	This test was conducted in accordance with IEC 61109; no failure occurred during the 96-hour test (fracture or tearing of the core or fracture of the metal fitting).
Test of galvanizing	All of the zinc layer thicknesses were above the minimum value of 85 μm.
Dye penetration test	This test was successfully conducted in accordance with IEC 62217.
Ultimate tensile test without pre-stress	The individual values were between 255 and 296 kN and therefore well above the SML value of 210 kN.

expected. It can be assumed that the porcelain surface underwent intensive cleaning as a result of the humidification step that took place beforehand. Because of its hydrophilic behaviour, and as a result of having the largest diameter, the epoxy resin post has the highest values.

The silicone composite post, currently the only technology used, was cleaned after this test and the test was repeated with a new silicone post insulator of a very similar design (Fig. 10.61). The result was as expected. In other words, the maximum current values for the cleaned and new post are, once again, a decade lower than with the natural but hydrophobic pollution layer. When one looks at the pollution layer characteristic curves for porcelain insulators that are geometrically similar, it is possible to estimate the required leakage current for flashover as 80 mm/kV, which produces a value of approx. 10 A [63, 64]. The difference between the measured leakage current with hydrophobic pollution and the value of 10 A is the safety margin of the chosen silicone composite post design without cleaning, which is confirmed by the fact that the insulators have been in service for more than 30 years without experiencing flashover.

Ultimate Cantilever Bending Force Test

This mechanical test is a destructive test. The posts were individually loaded with a bending force until they fractured (Fig. 10.62). There are significant differences in terms of the values and the failure modes. The silicone composite post achieves

Fig. 10.59 As-received state of insulators after 15 years (*silicone composite post, left*), 2 years (*epoxy resin post, centre*) and 13 years (*porcelain post, right*) in service

Table 10.14 Results of pollution layer measurements

	Silicone composite post insulator	Epoxy resin post insulator	Porcelain post insulator
ESDD (mg/cm^2)	0.56	0.6	0.18
NSDD (mg/cm^2)	11.4	7.4	6.1
Wettability class (WC) (IEC 62073)	1…3	7	7

a value that is twice that of the porcelain or epoxy resin post. In addition to its great strength, this can be attributed to another criterion for this application, namely the modulus of elasticity for the (GFK) rod, which is smaller than the modulus of elasticity for the porcelain post and means that a larger diameter than would have been required from the perspective of the damage limit load was used.

The failure mode demonstrates the advantage of using composite technology; while porcelain and epoxy resin fracture as brittle materials, the silicone composite post fails in "safe failure mode" (Figs. 10.63, 10.64, 10.65). The remaining residual strength of a (silicone) composite post suffices in most cases and the conductor does not drop.

10.5.3.3 Summary

The results of these tests correlate with the service experience whereby, in the case of approx. 4,000 silicone composite post insulators of this type, which are installed in tunnel operations under difficult conditions (no ability to self-clean), the

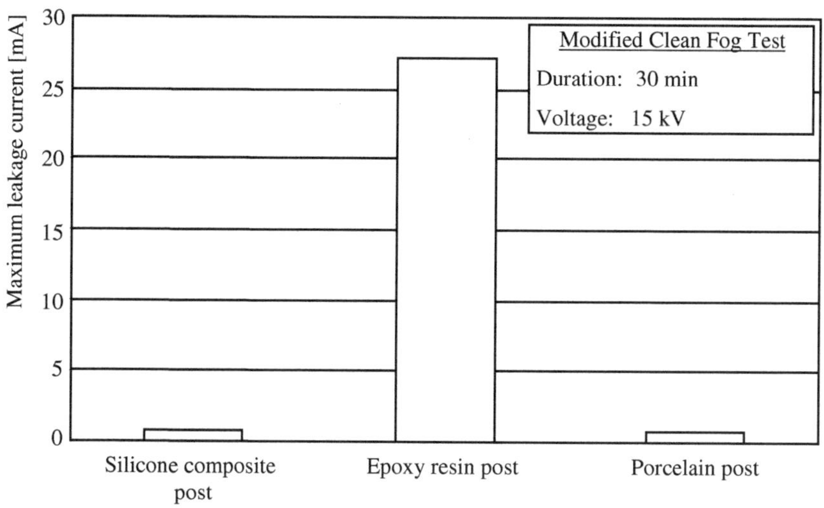

Fig. 10.60 Maximum leakage currents in the modified clean fog test

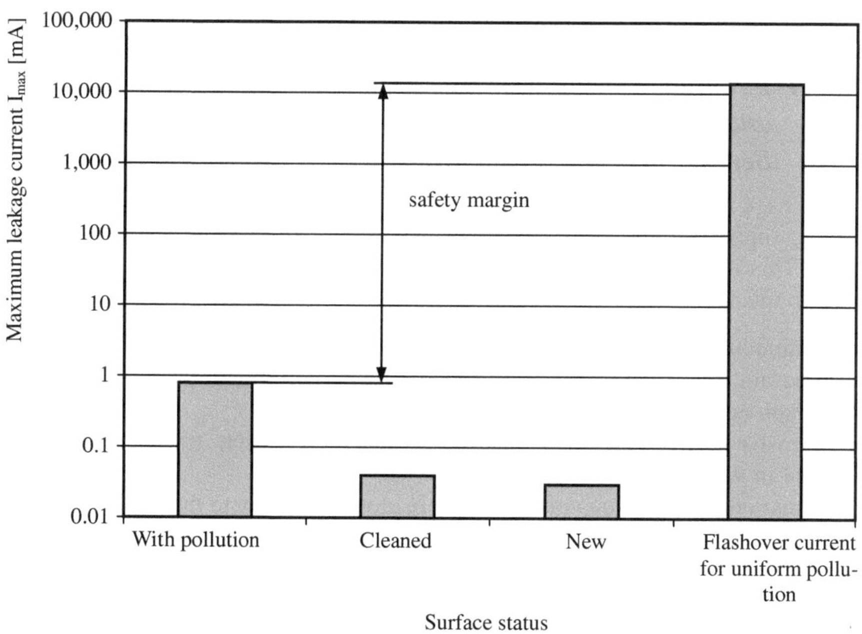

Fig. 10.61 Maximum leakage current for the various states associated with the silicone composite post insulator

reliability associated with setting cleaning measures simultaneously could increase significantly with a change in technology from the porcelain post to the silicone composite post.

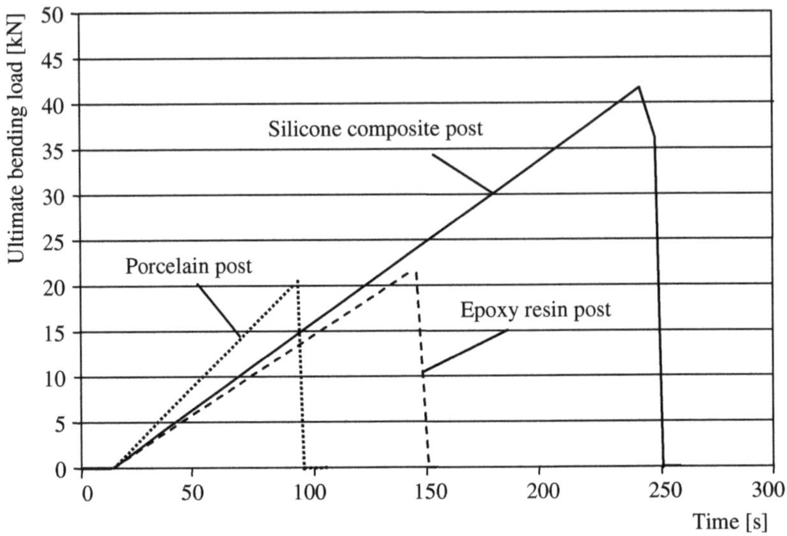

Fig. 10.62 Ultimate bending load of the various post insulators

10.5.4 Example 4—Analysis of Zinc Layer Thicknesses and the Hydrophobic Effect After 30 Years in a 15 kV Service in a Railway Tunnel

Tunnel supporting structures, including insulators, are exposed to special conditions. This is due to the operating conditions in a tunnel, which lead to the following special stresses:

- Continuous accumulation of pollution due to the inability to self-clean by rain
- A certain intrinsic conductivity associated with pollution (abrasion of the pantograph and conductor)
- Corrosive salt deposits and their accumulation if vehicle transportation takes place in winter
- Formation of critical quantities of dew in close proximity to the portals when the dew point is exceeded.

In the case of new railway tunnel projects, it was necessary to clarify, from the perspective of the specification, which corrosion protection is to be applied to steel parts. The following options were taken into consideration:

- Hot-dip galvanised steel with a standard layer thickness of 85 μm
- Hot-dip galvanised steel with a standard layer thickness of 85 μm and coated with an insulating layer (duplexing)
- Hot-dip galvanised steel with a greater layer thickness of at least 120 μm
- Stainless steel

Fig. 10.63 Fracture pattern of a porcelain post

Fig. 10.64 Fracture pattern of a epoxy resin post

10.5.4.1 Procedure

The silicone composite posts tested in Sect. 10.5.3 have been deployed in the Swiss Lötschberg-Scheitel-Tunnel for more than 30 years. It is virtually impossible to find a more ideal evaluation of long-term behaviour under practice-oriented conditions. For this reason, the tests were conducted in the tunnel with the aim of recording the pollution levels and susceptibility to corrosion of exposed components and composite insulators and confirming or, if necessary, modifying the corrosion protection design rules applied 30 years ago. The measurements were taken using a Deltascope MP3, which is calibrated on an annual basis. Before taking any measurements, the surface of the king posts (which are representative of exposed components) and the surface of the composite insulators were cleaned using distilled water and the pollution layer removed from the surface was stored for analysis in technically clean, sealable containers.

A scanning electron microscope was used to determine the composition of the pollution layers. For this purpose, the pollution layers removed in order to measure zinc layer thicknesses (pre-cleaning measurement) were transferred in powder form by means of vaporisation.

The tests were conducted directly in the portal area as well as inside the tunnel at a distance of +400 and +800 m from the portal area.

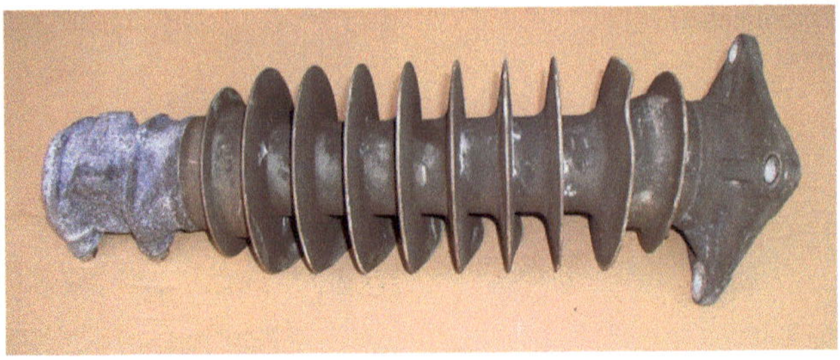

Fig. 10.65 “Safe-Failure-Mode”of a silicone composite post

10.5.4.2 Visual Observations

The king posts are square hollow sections and were soiled on both the inside and outside of the posts (Fig. 10.66). The pollution layer was easily removed using distilled water. Visually, there was no evidence of a complete loss of the zinc layer. The greater the distance from the portal, the greater the pollution layer.

All of the insulators exhibited signs of continuous, adhesive pollution (Figs. 10.67, 10.68). In part, there was also no rust on the zinc-coated steel fittings, which would mean a total loss of the zinc layer. A pollution layer sample was taken from the installed insulator shown in Fig. 10.68.

The pollution layer was easily removed using distilled water and a paint-brush. The pollution layer was hydrophobic.

Other installed insulators underwent a visual evaluation. The object was a measuring loop with cable terminations, which was installed for reference and test purposes with the first insulators in 1979. In comparison to the insulator shown in Fig. 10.68 the pollution layer was difficult to remove; the hydrophobicity of the pollution layer was weaker (Fig. 10.69).

None of the insulators exhibited any signs of partial discharge activity.

10.5.4.3 Results and Discussion

Zinc Layer Thickness

A comparative presentation of the mean values (arithmetic mean) of the individual measurements for zinc layer thicknesses is provided in Fig. 10.70. Since there are no reference values for portals in their new state, only a certain trend can be analysed. If one assumes that the king posts of the various measurement positions (both inner and outer) had a comparable zinc layer thickness when they were in their new state, a decrease in the corrosive effect of the portal inside the tunnel must be recorded. The lowest value for zinc layer thickness is associated with the inner surface of the king post of the portal. It can be assumed that, when a train enters the tunnel, the moisture is increasingly pushed into the king post, which is open from below, and due to the absence of an exposed surface, evaporation of the precipitated

Fig. 10.66 King post in the portal area

Fig. 10.67 Example of a pollution layer manifestation

moisture or electrolyte is delayed. The other measurements are above 150 μm, which is an indicator of a sharp deterioration in the moisture content in the tunnel and the lack of an attack by salts with a hygroscopic and corrosive effect.

If the king posts were installed when the silicone composite insulators were first used, this implies a minimum service time of 30 years.

The state of the insulator fittings after 30 years in service also corresponds to today's usual demands for standards for new insulators; almost all individual values measured were above the required value (for new insulators) of 85 μm (Tables 10.15 and 10.16).

Pollution Layer Analysis

The pollution layer analyses in Figs. 10.71 and 10.72 are representative of two types of specimen surface locations, namely king posts and silicone composite post insulators. The different pollution layers are very similar; they contain approx. 30 % carbon and 70 % rust components. The higher silicone content in Fig. 10.72

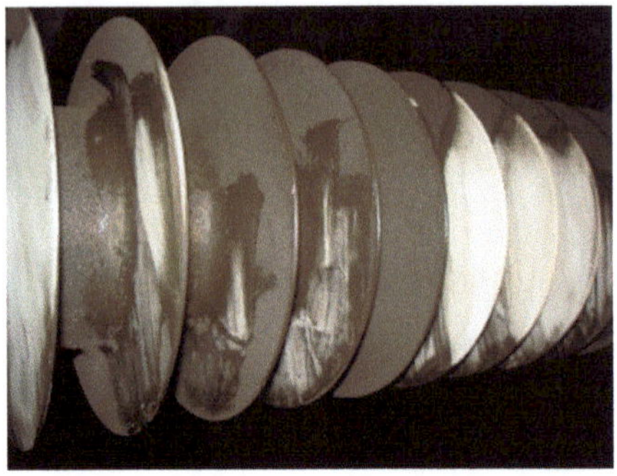

Fig. 10.68 Silicone composite post insulator following removal of the pollution layer

Fig. 10.69 Cable termination of the measuring loop (installed since 1979)

is due to the fact that these pollution layers were taken from the silicone surface of the insulators and, as a result of the hydrophobicity transfer, the pollution layer has been enriched, to a certain extent, with low molecular silicone compounds.

10.5.4.4 Summary

The zinc layer thickness measurements taken at the portal and in the tunnel after 30 years in service demonstrated that the use of hot-dip galvanised steel ensures sufficient corrosion resistance under these operating conditions. The greater zinc layer thicknesses that have since become available to ensure that fittings are securely crimped without any delaminations have proven themselves in highly corrosive environments such deserts. For this reason, an increased layer thickness of at least 120 μm was recommended. The results of crimping tests involving an additional duplex layer have tended to be promising, but there are no long-time experiences involving insulators. In particular, this applies to the situation where partial dielectric breakdowns of the layer may occur. For cost reasons, the use of stainless steel was not pursued further.

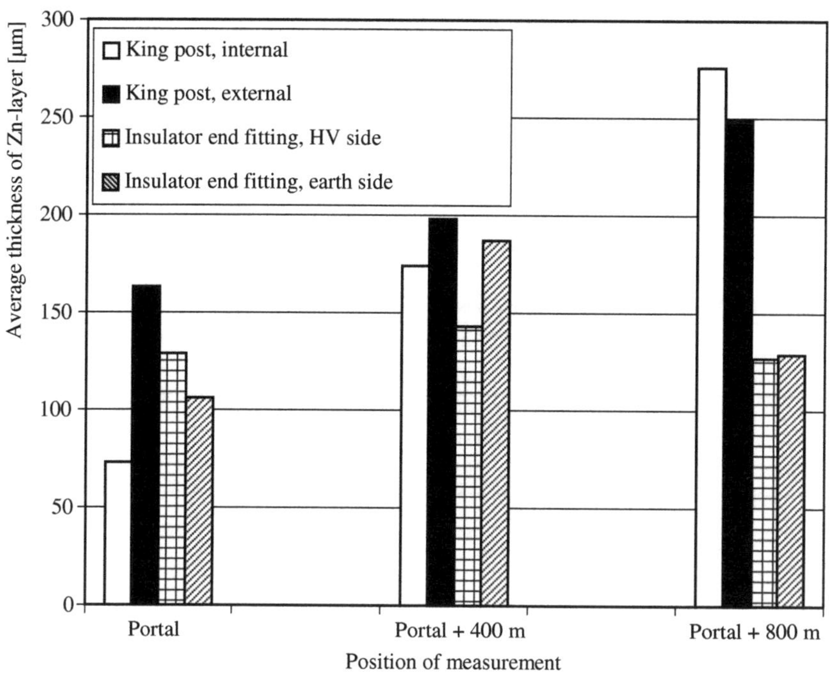

Fig. 10.70 Mean zinc layer thicknesses of king posts and insulator fittings

10.5.5 Example 5—Evaluation of 123 kV Insulators for the Purpose of Product Qualification

10.5.5.1 Insulator Specimens

Six types of composite insulators from various manufacturers were subjected to accelerated laboratory tests as well as being installed in the test tower of a 123 kV overhead transmission line under difficult outdoor conditions [65]. The technical details of the composite insulators are as follows (Table 10.17):

10.5.5.2 Testing Programme

Inspection of Composite Insulators at a Test Tower

The composite insulators were inspected annually and the following measurements were taken:

Table 10.15 Minimum values for insulator end fittings

	Minimum thickness for Zn-layer (μm)	
Insulator position	HV-side	Earth side
Portal	87	73
+400 m	93	166
	81	108
+800 m	109	90

Table 10.16 Minimum values of kink posts

	Minimum thickness for Zn-layer (μm)	
Position of kink post	Inner surface	Outer surface
Portal	69	119
+400 m	159	173
+800 m	223	251

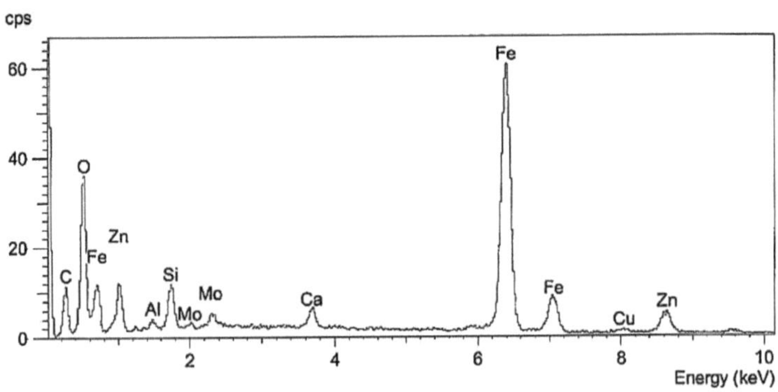

Fig. 10.71 Composition of a pollution layer on a king post

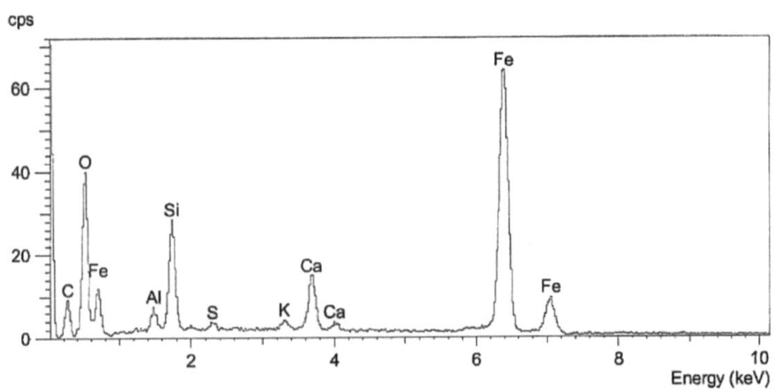

Fig. 10.72 Composition of a pollution layer on a silicone composite post insulator

Table 10.17 Technical details of the insulator specimens [65]

Insulator	Housing material	Length (mm)	Creepage distance (mm)	Shed diameter (mm)	Shed number	Installed since
A	Silicone rubber	1 200	3 160	148	22	1995
B	Silicone rubber	1 440	3 280	130/95	15/14	1995
C	Silicone rubber	1 397	3 510	128	24	1995
D	EPDM	1 275	3 230	105/90	19/18	1995
E	ESP	1 500	3 100	93	32	1995
F	Silicone rubber	1 295	2 673	110/70	15/14	2000

- Visual evaluation
- Measurement of surface conductivity (probe measurement in accordance with IEC 60507)
- Evaluation of hydrophobicity (spray test in accordance with IEC 62073)

In order to describe the ageing state, four categories were defined for the purpose of evaluating states:

Excellent:

No functional changes in comparison to a new insulator of the same type

Good:

An insulator with negligible ageing phenomena that are not functionally relevant (cosmetic, superficial such as chalking, colour change, surface erosion at the sheds, reduction in the hydrophobic effect, etc.)

Acceptable:

An insulator with greater ageing phenomena that are not yet functionally relevant (heavy chalking or alligatoring, bulk erosion without rod exposure, onset of tracking, shed puncture, etc.). In this case, the interval between inspections was shortened and the composite insulator was replaced if there was any further damage.

Failed:

The insulator has suffered considerable damage to the extent that its functioning is no longer firmly guaranteed (rod exposure, pronounced tracking, a loss of adhesion at the rod/insulator housing interface). The insulator must be replaced immediately.

Quick Salt Fog Test

This test was conducted with a salt content of 80 g/l; the fog was produced in a test chamber in accordance with IEC 60507. The test corresponded to the description in Sect. 10.4.2.3. After each flashover cycle, the insulators were cleaned with water. Each insulator was tested at the following times:

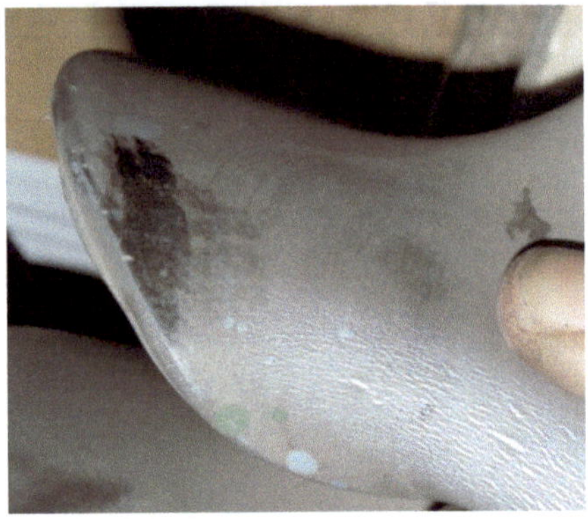

Fig. 10.73 Surface cracks on insulator A [65]

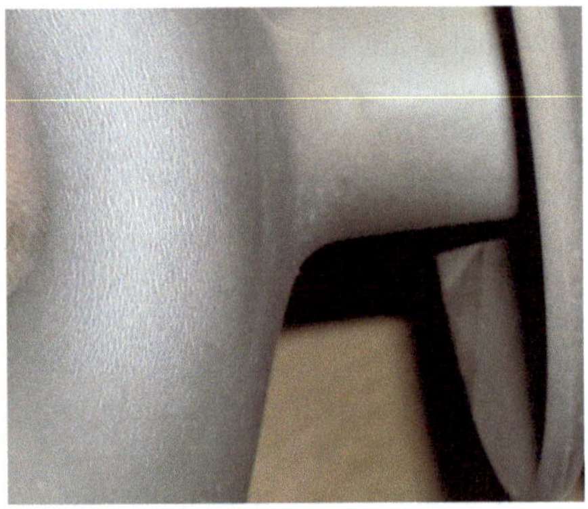

Fig. 10.74 Surface cracks on insulator B [65]

- New state
- After a 1,000-hour salt fog test (2.5 g/l)
- After 6–11 years of outdoor use at a test tower.

Evaluation of the Insulation Strength of the Interfaces

The steep front voltage test was applied in accordance with IEC 62217, but without any prior boiling. Each insulator was stressed with 10 positive and negative voltage pulses.

10.5.5.3 Results and Discussion

Insulator Specimen Inspection on the Test Tower

When interpreting the results, the following details were highlighted. Two types of insulators had surface cracks if the sheds were bent (Figs. 10.73, 10.74), while one insulator type experienced delaminations in the seal (Fig. 10.75).

Fig. 10.75 Seal delamination on insulator C [65]

The results of the tower inspections are summarised in (Table 10.18). Please note that only the years in which changes were detected are listed.

After 11 years of outdoor use, no critical changes were detected in 5 of the 6 insulators. Only Type C had a damage pattern that may be an indicator for the end of service life if further seal delaminations occur. However, this has not yet been verified in the steep front voltage test.

Quick Salt Fog Test

A comparison of the flashover voltages is shown in Table 10.19, 80 g/l was the chosen salt content because it represents a typically higher pollution layer value in Europe. Empirically, this value was chosen so that an unduly high salt concentration does not impact on any differences in behaviour by the various insulators.

The results reveal a trend whereby outdoor use for a period up to 11 years results in slower ageing than the salt fog test conducted for 1,000 h. This does not apply to insulator type C.

Insulation Strength of the Interfaces

All test specimens (including insulator type C) successfully passed the steep front voltage test. Afterwards, all insulators were loaded with 50 % of the SML value for a period of 10 s. They also passed this test.

10.5.5.4 Conclusion

Composite insulators A, B, D, E, F are deemed to be in good condition after 11 years in service. Composite insulator C shows some evidence of seal

Table 10.18 Summary of results relating to visual in situ tower inspections [65]

Insulator	Years	Change	State
A	1995	None	Excellent
	2000	Change of colour	Good
	2006	Surface cracks[a], increase of hardness	Good
B	1995	None	Excellent
	2000	Change of colour	Good
	2006	Surface cracks[a]	Good
C	1995	None	Excellent
	2000	Change of colour, seal delamination[b]	Good[c]
	2006	Change of colour, seal delamination[b]	Good[c]
D	1995	None	Excellent
	2000	Change of colour	Excellent/good
	2006	Change of colour	Excellent/good
E	1995	None	Excellent
	2000	Chalking, loss of hydrophobicity	Good
	2006	Chalking, loss of hydrophobicity	Good
F	2000	None	Excellent
	2006	Change of colour	Excellent/good

Explanatory note
[a] As shown in Figs. 10.73 and 10.74
[b] As shown in Fig. 10.75
[c] An indicator for end of life

delaminations, which will need to be monitored more closely. After being used outdoors, C also displays a greater reduction in the flashover voltage than after the accelerated ageing test. All of the insulators were re-installed on the test tower after the in-lab evaluation.

10.5.6 Example for Deriving Reliability Figures

For the Gotthard Base Tunnel in Switzerland, it was necessary to perform a reliability analysis of the behaviour of silicone composite insulators. The fact that there were virtually no failures during more than 30 years in service complicated matters in a positive way. For this reason, the experiences of porcelain insulators and projects implemented worldwide were used as reference.

10.5.6.1 Empirical Values

Service Experience of Composite Insulators

It was possible to research the following service experiences of some proprietary products.

Table 10.19 Summary of results from the quick salt fog test [65]

Insulator	Specific flashover voltage (kV/cm)			Reduction in comparison to the new state (%)	
	New state	After the ageing test	After 11 years of service	After the ageing test	After 11 years of service
A	0.41	0.31	0.34	24.4	17.1
B	0.41	0.33	0.37	19.5	9.8
C	0.36	0.34	0.32	5.6	11.1
D	0.39	0.29	0.33	25.6	15.4
E	0.28	0.27	0.28	3.6	0.0
F	0.34	–	0.32	–	5.9[a]

Explanatory note [a] After 6 years in service

Lötschberg-Scheitel-Tunnel

Service time in years[a]	Up to 30
Number of insulators in service	4,000
Number of failures (electrical or mechanical)	Not known

Remark: [a] Retrofitting from porcelain posts to silicone composite posts was continuously

England, Spain, Malaysia, Holland, Rumania, Switzerland, India, Turkey

Service time in years	Up to 10
Number of insulators in service	160,000
Number of failures (electrical or mechanical)	1 × mechanical failure after 10 years (spontaneous fracture after a conductor broke when it came into contact with a pantograph), 1 × electrical failure after 2 years (bridging of the insulator by an animal), 1 × electrical failure after 1 year as a result of an installation fault (torsion fracture)

Service Experience of Porcelain Insulators

Mechanical Failure

According to the Ref. [66], the German and Russian rail networks experienced 0.98 and 1.46 failures respectively in 1995 for a track length of 1,000 km.

Electrical Failure

There is no record available on porcelain insulator flashovers.

10.5.6.2 Assumptions

If one were to consider the 2 × 57 km length of track in the Gotthard-Basis-Tunnel, this would produce values of 0.11 or 0.17 failures per year. By using modern technology with composite insulators, the risk of fracture falls considerably in comparison to porcelain insulators and the value cited in the literature serves as a conservative comparative reference.

Due to the fact that the silicone composite insulators in the Lötschberg-Scheitel-Tunnel did not experience any flashovers over a 30-year period and that the creepage distance in the Gotthard-Basis-Tunnel was increased by at least 8 %, the mean time between failure (MTBF) value for an electrical failure is estimated to be twice that of a mechanical failure.

10.5.6.3 Statistical and Mathematical Calculation

The following legend and formulas relate to the MTBF value and the empirical failure rate [67];

MTBF

$$MTBF = \frac{1}{\lambda}$$

$$\lambda = \frac{n_a}{n_b \times T}$$

Empirical failure rate

$$\hat{\lambda} = \frac{n_a}{n_E \times T}$$

n_a	Number of failures	
n_b	Number in service	
T	Service time	[h]
n_E	Number still in service	

The following values can be used with the information available in relation to service experiences: $n_a = 3$, $n_b = 1{,}60{,}000$, $n_E = 1{,}59{,}997$, T = 5 years (The time was halved because the projects and therefore the number of units were implemented over 10 a.).

The MTBF values are very high (Table 10.20). The mean time to repair (MTTR) value, which is often used as a planning parameter, is sufficiently measured over a period of 2 h because, thanks to product standardisation, optimal storage of replacement parts is possible. As a result of this service experience, the empirical failure rate is also very low (Table 10.21).

Empirical failure rate

The values determined here usually continue to be used in failure mode and effects analysis (FMEA) and failure mode, effects and criticality analysis (FMECA).

Table 10.20 Results of MTBF and MTTR analysis

Place of service	Failure mode	Part of failure (%)	MTBF (h)	MTTR (h)
Tunnel	Breakage	67	3.49×10^9	2
Tunnel	Flashover	33	7.08×10^9	2
Outdoor	Breakage	50	4.67×10^9	2
Outdoor	Flashover	50	4.67×10^9	2

Table 10.21 Deduction of empirical failure rate

Place of service	Failure	Part of failure (%)	Empirical failure rate (1/h)
Tunnel	Breakage	67	2.87×10^{-10}
Tunnel	Flashover	33	1.41×10^{-10}
Outdoor	Breakage	50	2.14×10^{-10}
Outdoor	Flashover	50	2.14×10^{-10}

10.6 Summary

When evaluating a composite insulator, it is also necessary to consider various topics from the worlds of electrical engineering, mechanics, physics and chemistry. Two facts need to be established: Firstly, standardisation, while interactively taking into consideration the outdoor experiences of composite insulators, has achieved a high level of maturity, which is the basis for defining the minimum requirements for designing and selecting those materials for use in composite insulators. Secondly, the development of polymeric materials has given rise to a variety of independent product developments over the past 30 plus years. Both facts make it possible to conduct standardised tests that evaluate the condition of an insulator and are useful in terms of the relevant product design. In this chapter, this was demonstrated by means of examples involving various housing materials, which can be used as a reference structure for deriving a testing programme for a special insulator design.

References

1. State of the Art Report On behalf of Study Committees 15 and 33 presented by Working Group 33/15.08 (1990) Dielectric diagnosis of electrical equipment for AC applications and its effects on insulation coordination. CIGRE Technical Brochure 059
2. Schmuck F, Seifert J, Gutman I, Pigini A (2012) Assessment of the condition of overhead line composite insulators. CIGRE Session 2012, Paris, Paper D2-214
3. CIGRE Working Group B2.21 (2011) Guide for the assessment of composite insulators in the laboratory after their removal from service. Technical Brochure 481
4. CIGRE Working Group B2.03 (2006) Guide for the assessment of old cap and pin and long rod transmission line insulators made of porcelain or glass: what to do and when to replace. Technical Brochure 306
5. Papailiou KO (1999) Grenzflächen bei Silikon-Verbundisolatoren. (Bulletin SEV/VSE 21)

6. Gubanski SM, Derfalk A, Andersson J, Hillborg H (2007) Diagnostic methods for outdoor polymeric insulators. IEEE Trans Dielectr Electr Insul 14(5), October
7. CIGRE Working Group B2.03 (2007) Guidance for the establishment of naturally polluted insulator testing stations. Technical Brochure 333
8. EPRI Database, presented at the WG-Meeting of CIGRE B2.21 in Winterbach 2008, update in 2011
9. Kuhl M (2001) FRP rods for brittle fracture resistant composite insulators. IEEE Trans Dielectr Electr Insul 8(2), April
10. Montesinos J, Gorur RS, Mobasher B, Kingsbury D (2002) Mechanism of brittle fracture in nonceramic insulators. IEEE Trans Dielectr Electr Insul 9(2), April
11. Montesinos J, Gorur RS, Mobasher B, Kingsbury D (2002) Brittle fracture in nonceramic insulators: electrical aspects of microscopic flaws in glass reinforced plastic (GRP) rods. IEEE Trans Dielectr Electr Insul 9(2), April
12. de Tourreil C, Schmuck F on behalf of CIGRE Working Group B2.03 (2004) Brittle fractures of composite insulators—field experience, occurrence and risk assessment. ELECTRA 214, June
13. de Tourreil C, Schmuck F on behalf of CIGRE Working Group B2.03 (2004) Brittle fractures of composite insulators—failure mode chemistry, influence of resin variations and search for a simple insulator core evaluation test method. ELECTRA 215, August
14. Chughtai AR, Smith DM, Kumosa LS, Kumosa M (2004) FTIR analysis of non-ceramic composite insulators. IEEE Trans Dielectr Electr Insul 11(4), August
15. Kumosa M, Kumosa L, Armentrout D (2005) Failure analyses of nonceramic insulators part 1: brittle fracture characteristics. IEEE Electr Insul Mag 21(3), May/June
16. Kumosa M, Kumosa L, Armentrout D (2005) Failure analyses of nonceramic insulators: part 2: the brittle fracture model and failure prevention. IEEE Electrical Insulation Magazine 21(4), July/August
17. Kumosa LS, Kumosa MS, Armentrout DL (2005) Resistance to brittle fracture of glass reinforced polymer composites used in composite (nonceramic) insulators. IEEE Trans Power Delivery 20(4), October
18. IEC 62662 Ed1 (2008) Guidance for production, testing and diagnostics of polymer insulators with respect to brittle fracture of core materials
19. CIGRE SC 22 WG 03.01 (1990) Worldwide experience with HV composite insulators. ELECTRA 130, December
20. CIGRE WG 22.03 (2000) Worldwide service experience with composite insulators. ELECTRA 191, August
21. Bonzano R, Ricca M, Garbagnati E, Marrone G, Pigini A (1991) Experimental research on the behaviour of HV cap and pin insulator strings with failed units. Eur Trans Electr Power 1(1), January/February
22. Mishra AP, Gorur RS, Venkataraman S, Kingsbury D (2006) Condition assessment of porcelain and toughened glass insulators from residual strength tests. Annual Report Conference on Electrical Insulation and Dielectric Phenomena
23. Rawat A, Gorur RS (2008) Electrical strength reduction of porcelain suspension insulators on AC transmission lines. Annual Report Conference on Electrical Insulation Dielectric Phenomena
24. Zhicheng G, Xiaoxing W, Zhidong J, Ruobin Z (2011) Application and research of composite insulator in China. INMR Symposium 2011 in Seoul, South Korea
25. Vosloo WL, Macey RE, De Tourreil C (2004) Outdoor high voltage insulators. Crown Publications, Johannesburg
26. Guide Field (2009) Visual inspection of polymer insulators. EPRI, Palo Alto, p 1020289
27. Seifert J, Bärsch R (2006) Bewertung des Designs von Silikon-Verbundisolatoren unter dem Aspekt der Isolierstoffoberflächen und Fremdschichtverhaltens. RCC-Fachtagung "Werkstoffe für Isolatoren, Überspannungsableiter, Kabelgarnituren, Schaltgeräte, Berlin April, Tagungsband S. 113–124

28. Seifert J, Bärsch R (2007) Design evaluation of silicone rubber composite lnsulators under the aspect of surface pollution stress, 15. International symposium on high voltage engineering, Ljubljana, Slovenia, August 27–31, proceedings
29. Seifert JM, Bärsch R, Vosloo WL (2008) Dimensioning of the housing profile of silicone rubber composite lnsulators for harsh marine pollution conditions, CIGRE Session 2008, Paris, paper D1—303
30. IEC 61109 Ed. 2 (2008) Composite suspension and tension insulators for a.c. overhead lines with a nominal voltage greater than 1 000 V—Definitions, test methods and acceptance criteria
31. ANSI 29.11 1989 (R 1996) (1989) For composite insulators - test methods
32. IEC TS 62073 Ed. 1 (2003) Guidance on the measurement of wettability of insulator surfaces
33. STRI-Guide 1, 92/1 (1992) Hydrophobicity classification guide
34. IEC 60383-1 Ed 4 (1993) Insulators for overhead lines with a nominal voltage above 1000 V—part 1: ceramic or glass insulator units for a.c. systems—definitions, test methods and acceptance criteria
35. Gutman I, Solomonik EA, Solomatov VN, Yashin YuN (1993) Operation and field tests of overhead line composite insulators with silicone rubber cover. 8th International symposium on high voltage engineering, Yokohama, Japan, 23–27 August, 47.13
36. Berlijn S, Halsan K, Gutman I, Dernfalk A (2009) Assessing ten years of service experience with composite line insulators at HVDC. World Congress and Exhibition on Insulators, Arresters and Bushings, Crete, 11–13 May
37. Gayvoronski A (2008) The damages of composite insulators and their on-line diagnostics in service. IWD 047-2008. Presented at CIGRE WG Meeting
38. CIGRE Task Force 33.04.01 (2000) Polluted insulators: a review of current knowledge. Technical Brochure 158, June
39. IEC 60507 Ed 2 (1991) Artificial pollution tests on high-voltage insulators to be used on a.c. systems
40. Gutman I, Dernfalk A (2010) Pollution tests for polymeric insulators made of hydrophobicity transfer materials. IEEE Trans Dielectr Electr Insul 17(2):384–393
41. Gutman I, Halsan K, Seifert J, Vosloo W (2011) Service experience on pollution and ageing of DC line composite insulators: service/field analysis and laboratory testing. World Congress and Exhibition on Insulators, Arresters and Bushings, Seoul, Korea, 17–20 April
42. Bärsch R, Kuhl M (1999) Betriebserfahrungen und Untersuchungen an Kunststoffisolatoren in einer 20 kV-Leitung auf der Insel Nordstrand. ETG-Fachbericht 76:249–257
43. Huiber W, Papailiou KO, Peter M, Schmuck F (2001) Increased installation performance and application solutions using composite insulators—a manufacturerer's philosophy. INMR World Insulator Congress 2001 Shanghai
44. Papailiou KO, Peter M, Fluri W, Schmuck F (2003) A review of material development, recent 420 kV braced line post designs and long-term evaluation of composite insulators in Silicone rubber technology. INMR Insulator Symposium 2003 Marbella
45. CIGRE WG C4.03.03 (2009) Proposal for the round robin pollution test for polymeric insulators, October
46. IEC 61621 Ed. 1 (1997) Dry, solid insulating materials—Resistance test to high-voltage, low-current arc discharges
47. IEC 60587 Ed. 3 (2007) Electrical insulating materials used under severe ambient conditions—test methods for evaluating resistance to tracking and erosion
48. IEC 62217 Ed 1 (2005) Polymeric insulators for indoor and outdoor use with a nominal voltage >1 000 V—general definitions, test methods and acceptance criteria
49. ISO 4287 (1997) Geometrical product specifications (GPS)—surface texture: profile method—terms, definitions and surface texture parameters
50. Bärsch R (2003) Bewertung der Hydrophobie sowie des Kriechstromverhaltens von Silikonelastomeren für Hochspannungs-Freiluftisolatoren. ETG-Fachbericht 93:97–108
51. Cervinka R, Bärsch R, Exl F, Kindersberger J, Winter H-J (2008) Untersuchungen zur Beständigkeit der Hydrophobie von polymeren Isolierstoffoberflächen und ihrer Wiederkehr mit dem dynamischen Tropfen-Prüfverfahren ETG-Tagung

52. CIGRE WG D1.14 (2010) Evaluation of dynamic hydrophobicity properties of polymeric materials for non-ceramic outdoor insulation—retention and transfer of hydrophobicity. Technical Brochure 442, December
53. IEC/TS 60815-1 Ed. 1 (2008) Selection and dimensioning of high-voltage insulators intended for use in polluted conditions—part 1: definitions, information and general principles
54. Seifert J, Zembsch T (2007) Long-term in-service pollution performance of composite insulators in arid and coastal areas. 2007 GCC CIGRE Conference, Dubai, November, U.A.E
55. Hayashi T (2008) Pollution accumulation and live-line washing withstand voltage performance of silicone rubber composite insulators. B3. Preferential Subject PS1, Question 9, CIGRE
56. Kindersberger J, Kuhl M (1989) Effect of hydrophobicity on insulator performance, 6th ISH New Orleans
57. Gutman I, Wieck H, Windmar D, Stenström L, Gustavsson D (2007) Pollution measurements to access the performance of naturally exposed silicone rubber composite insulators. IEEE Trans Fundam Mater 127(9), pp 513–518
58. CIGRE SC 22 WG 22.03 (1992) Guide for the identification of brittle fracture of composite insulators FRP rod. ELECTRA 143
59. IEEE Task Force Report (2002) Brittle facture in nonceramic insulators. IEEE Trans Power Delivery, 17(3), July
60. Ansorge S, Camendzind A, Pratsinis SE, Ammann M, Schmuck F, Papailiou KO (2008) Evaluation of silicone rubber housing interfaces after service exposure and performance improvements by nanofillers enriched silicone rubbers. Paper B2-208, CIGRE
61. Schmuck F, Aitken S, Papailiou KO (2010) A proposal for intensified inspection and acceptance tests of composite insulators as an addition to the guidelines of IEC 61109 Ed. 2, 2008 and IEC 61952 Ed. 2, 2008. IEEE Trans Dielectr Electr Insul 17(2), April
62. Kocher M (1993) Experience with silicone composite insulators in the tunnels of BLS Lötschberg. Railway Technology
63. Djafri S (1995) Zur getrennten Bewertung der Einflussgrössen auf den Fremdschichtüberschlag von Isolatoren bei Wechselspannung. Dissertation TH Zittau
64. Pilling J, Berndt L (1991) Flashover voltage and flashover current of polluted insulators. 7th ISH Dresden, August 1991, Paper 43.12
65. Sklenicka V, Zeman I (2001) Utilization and service experience with composite insulators in Czech power system. In: Proceedings of world congress and exhibition on insulators, arresters and bushings—Shanghai, November
66. Kissling F, Puschmann R, Schmieder A, Schmidt PA (1998) Fahrleitungen elektrischer Bahnen: Planung, Berechnung, Ausführung. (2.Auflage). Vieweg + Teubner, Wiesbaden
67. Meyna A, Pauli B (2003) Taschenbuch der Zuverlässigkeits- und Sicherheitstechnik—quantitative Bewertungsverfahren. Hanser Verlag, Praxisreihe Qualitätswissen

Chapter 11
Overview of Standards and Tests Concerning Composite Insulators

Symbols and Abbreviations

A	Constant
AC	Alternating current
ATH	Aluminium trihydrate, Aluminium hydroxide
CIGRE	Conseil International des Grands Reseaux Électriques (International Council for Large Electric Systems)
DC	Direct current
DSC	Differential scanning calorimetry
EAP	Early aging phase
EPDM	Ethylene propylene diene monomer
FTIR	Fourier transform infrared spectroscopy
HTV	High temperature vulcanising silicone rubber
HV	High voltage
HVDC	High voltage direct current
i	Current
i_{max}	Maximum leakage current
i_{Peak}	Peak value of leakage current
IEC	International electrotechnical commission
L	Internal inductive reactance of voltage source
LAP	Late ageing phase
LMW	Diffusible low molecular weight silicone components
LSR	Liquid silicone rubber
PD	Partial discharge
R	Ohmic resistance of pollution layer in the switch-on moment
RTV	Room temperature vulcanising
TB	Technical brochure
TGA	Thermo-gravimetric analysis
t_{BT}	Burn time (of PD)
t_D	Test duration
t_F	Time to failure

K. O. Papailiou and F. Schmuck, *Silicone Composite Insulators*, Power Systems,
DOI: 10.1007/978-3-642-15320-4_11, © Springer-Verlag Berlin Heidelberg 2013

t_{IE}	Time to erosion inception
t_{LH}	Time to loss of hydrophobicity
u_L	Voltage across internal inductive reactance
u_{PA}	Arc voltage
u_R	Voltage across pollution layer in the switch-on moment
u_S	Switch-on voltage
U_F	Flashover voltage
U_{FF}	Flashover voltage at film layer state
U_{F0}	Flashover voltage at new state
d_c	Creepage distance
x	Length of pre-arc
Θ	Contact angle
Θ_R	Dynamic receding contact angle
Θ_A	Dynamic advancing contact angle
κ	Electrolyte conductivity

11.1 Introduction

This chapter deals with various aspects of standards and tests for composite insulators. The first part lists important IEC standards according to which composite insulators are tested and designed. Since composite insulator strings share many similarities with conventional—especially long rod—insulators, the relevant standards are also listed here for comparative purposes. The second part deals with some testing principles. By using the example of the Inclined Plane Test (IEC 60587 [1]), which was initially introduced in 1977 as a material test method for evaluating erosion and tracking resistance, it is possible to show that other properties of polymeric insulating materials can be evaluated using this test principle. The third part discusses analytical methods that can be used to identify the polymeric materials used in a composite insulator. This topic, known as "fingerprinting", is also part of the work currently being undertaken by CIGRE Working Group D1.27.

Figure 11.1 provides examples of tests that need to be conducted. Material evaluation tests are specifically used to quantify certain properties (e.g. erosion resistance, hydrophobic behaviour, etc.). Often, simple test specimen geometries are used to prevent the material properties from impacting on the test specimen design. These tests are used to develop new materials and, thanks to their simplicity and the high level of automation that can be achieved, they are also used to develop in-process quality assurance procedures. In the case of product qualification tests, a distinction can be made between design tests (also known as prototype tests) or type tests in accordance with current standards. Sample tests are also a form of product qualification. Here, however, it is necessary to verify the conformity of a batch; design tests and type tests are generally concluded at this

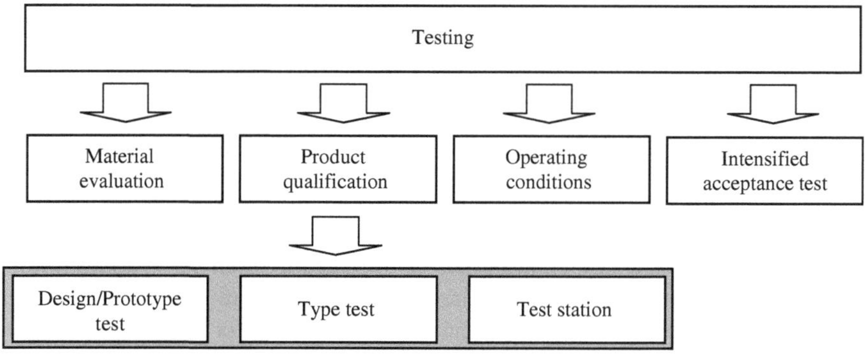

Fig. 11.1 Example of tests conducted on composite insulators

time. When compared against a sole material evaluation, the design testing is more complex because both material properties and geometry influence are evaluated on short composite insulator specimens, which are functionally complete.

The type test is usually conducted on composite insulators in their original size; certain tests must already be performed involving string elements (e. g. corona rings or power arc protective devices). The current version of IEC 61109 [2] permits an interpolation of the lightning withstand voltage, the switching withstand voltage and the power frequency withstand voltage values if the striking distances of the insulators describing the limits of the interpolation range does not exceed a factor of 1.5.

Some power utilities use test stations [3] or test towers [4] to correlate the results from accelerated ageing tests in the laboratory with the results of natural weathering and simultaneous stressing. Often, these special installations are at locations that are subjected to accelerated ageing as a result of high pollution layer stresses.

As described in Chap. 10 (both methodically and by means of examples), it may be necessary to perform a detailed evaluation of composite insulators in service. Such an evaluation goes beyond a visual evaluation.

Intensified acceptance tests are used if some uncertainty surrounds a batch's conformity to the order specification [5]. Tests that go beyond the sample test normally undertaken and are targeted at the additional evaluation of specific material properties are performed.

The next section provides an overview of IEC standards. Now for a brief word on the difference between the terms "norm" and "standard" because the English term "standard" can apply to norms and standards: A norm is usually the outcome of a regulated standardisation process and requires the consensus of all parties involved (e.g. agreement among the members involved in creating IEC norms). It reflects the current state of science and technology. A norm can be used voluntarily. Often, however, such use is binding (e.g. when it is cited in contracts). A standard, on the other hand, "merely" describes a unifying recommendation. Often, it can be interpreted as being a precursor to a norm.

11.2 Current IEC Standardisation for Composite Insulators and Similarities in Comparison to Conventional Insulators and Insulator Strings/Sets

With the increasing use of composite insulators, there was a need for standards to ensure product quality and, in particular, substitutability in comparison to known conventional technologies (cap and pin or long rod insulator, glass or porcelain, etc.). CIGRE made a key contribution in this regard when its Working Group 22.10 formulated the minimum requirements for composite insulators in a comprehensive study in 1982 [6]. This document was the basis for the product standard IEC (6)1109 [7], which was published in 1992 and whose test philosophies were pursued in subsequent standards. In the meantime, many years of service experience with composite insulators have been qualified in accordance with this first standard and their validity has been successfully verified. As is the case with many standards, however, the minimum requirements to be followed apply, but the on-site conditions may involve more stringent rules.

The comparison between the standards for composite insulators and the standards for conventional insulators begins with the material selection procedure (Fig. 11.2). IEC 62039, which was published in 2007, is based on the contributions of the CIGRE Working Group D1.14, which published the Technical Brochure TB 255 [8].

Since composite insulators also share similarities despite various end applications such as line, apparatus or station post insulators, two "common clause" standards that provide a full description of certain tests to be verified are available, namely IEC 62217 and IEC 62730.

The respective product standards then give way to general rules for designing insulators. Particularly noteworthy are those parts of IEC 60815 that consider the different behaviour of conventional insulators and composite insulators under pollution conditions when planning the creepage distance.

When testing insulator strings, a distinction is no longer made between conventional insulators and composite insulators. The task here is to evaluate the behaviour of the entire system, which comprises one or more insulators and string/set elements.

Overall, it should be noted that composite insulator standards have reached a level that is deemed to be equivalent to the technology of conventional insulators and considers the differences between both technologies (e.g. the interface areas of a composite insulator) in detail.

Relevant and valid insulator or product standards are listed in Fig. 11.2. The following standards can also supplement the material test methods: In addition to the Inclined Plane Test (IEC 60587), there are other test methods for evaluating erosion resistance. These include a further wheel test (IEC 61302 [9]) and the resistance test to high voltage, low-current arc discharges (IEC 61621 [10]).

The method for the determination of the proof and the comparative tracking indices of solid insulating materials in accordance with IEC 60112 [11] is still

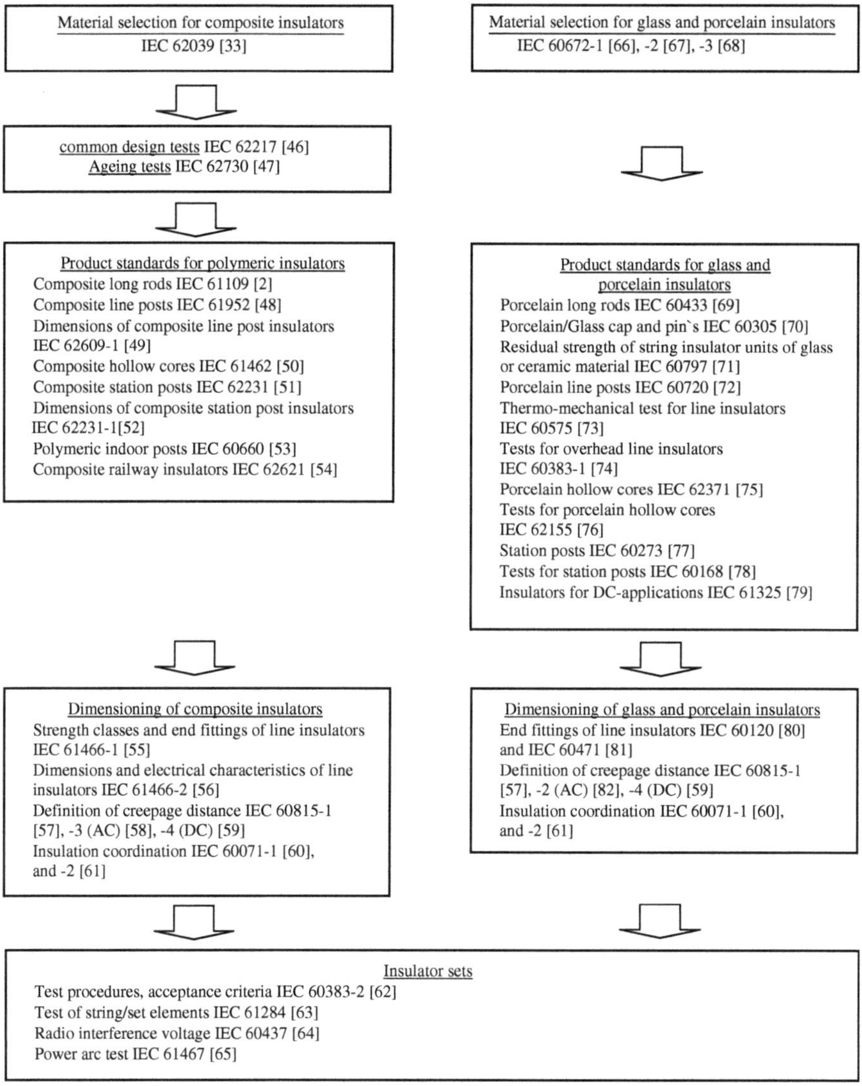

Fig. 11.2 Overview of IEC tests conducted on composite insulators, conventional insulators and insulator strings/sets

often mentioned in specifications. In most cases, this procedure cannot sufficiently differentiate between the ageing resistance of the insulating materials used today [12]. A test that determines the resistance of polymeric surfaces to corona discharges is available with IEC 60343 [13]. However, it is preferable to evaluate other influencing factors (e.g. simultaneous tensile stress) during this test [14].

With respect to pollution tests, IEC 60507 [15] and IEC 61245 [16] are used for AC and DC tests on conventional insulators respectively. To some extent, both are also used to test composite insulators. Due to the hydrophobic effect, however, the results may differ greatly. For this reason, CIGRE Working Group C4.03.03 is in the process of developing a test method that can also be used for composite insulators [17].

11.3 Special Flame Resistance Tests

The need to verify that outdoor insulators are flame resistant was already discussed in Chap. 9. Special applications, however, require further proof. These include railway applications (e.g. fixed installations and vehicles). Since it does not make sense to use the cable standards (which are well developed in regard to flame testing) associated with composite insulators, the standard pR CEN/TS 45545-2 [18] for example, distinguished between indoor and outdoor insulators. The testing accuracy is higher for indoor insulators. In both cases, the following tests need to comply with established threshold values:

- Determination of burning behaviour by oxygen index (EN ISO 4589-2 [19])
- Smoke generation: Determination of optical density by a single-chamber test (EN ISO 5659-2 [20])
- Fire tests, analysis of gaseous effluents (NF-X 70-100-1 [21] und -2 [22]).

11.4 Test Methods for Evaluating Certain Properties of Polymeric Housing Materials

The behavior of polymeric insulator materials with a hydrophobic surface effect can be shown as a model of the relationship between the flashover voltage and the conductivity of the electrolytic pollution layer (Fig. 11.3). In the laboratory, measuring the flashover voltage is a practical method for quantifying the state of the surface [23, 24]. The flashover processes differ in terms of the state of the droplet or film layer: It can be assumed that, in an ideal scenario, a droplet layer comprises droplets insulated by air clearances. If the voltage level is sufficient, the droplets become deformed and initial discharges occur between them in order to bridge the remaining air clearances (see also Fig. 8.18). For this reason, the flashover voltage values are higher and the dependency on electrolyte conductivity is rather low (slope of the curve of approx. 0.01–0.1). In the case of the film layer, the flashover occurs as a "classic" pollution flashover (i.e. via the stages of dry zone formation, pre-arc ignition, prolongation and aggregation).

The flashover voltage values are therefore lower for a given electrolyte conductivity and the dependency on electrolyte conductivity (slope of the curve of

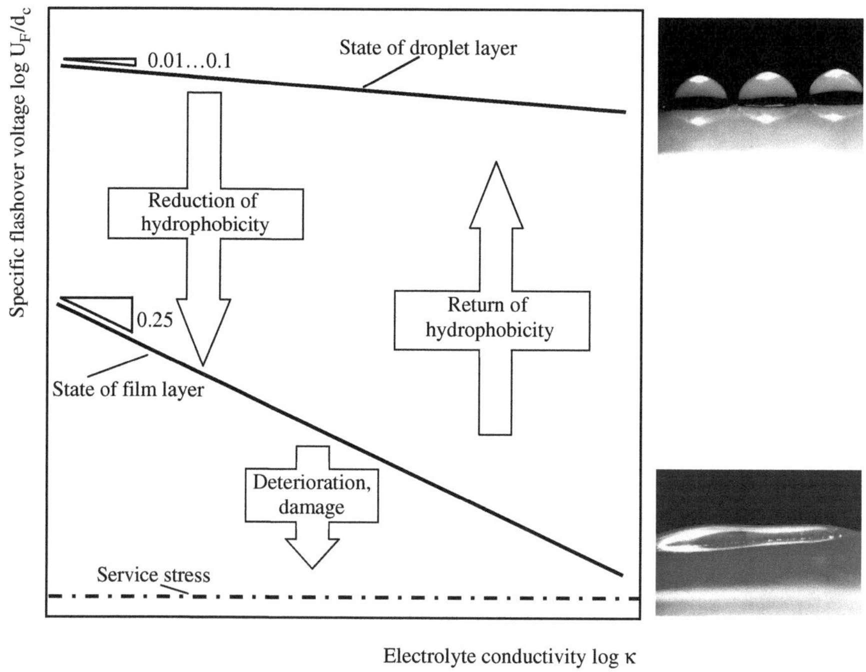

Fig. 11.3 Flashover voltage as diagnostic criterion of different surface states

approx. 0.25) is correspondingly higher than in the case of a droplet layer. The schematically represented reduction in hydrophobicity can occur by many ways:

- Humidity caused by rain or fog
- Pollution layers caused by the environment
- UV radiation
- Corona discharges
- Biogenous formation of layers

It has been known for many years that the hydrophobicity of silicone rubber can return even without implementing any cleaning measures. Here, a distinction can be made between the recovery of hydrophobicity process (reorientation of hydrophobic groups in silicone rubber) and the hydrophobicity transfer process (diffusion of low molecular hydrophobic groups (known as *LMW chains*) into the pollution layer).

Deterioration or damage are irreversible and can manifest themselves in the form of bulk erosion (see also Fig. 7.25) or in the formation of conductive tracking (Fig. 10.14). It usually begins with a brightly illuminated, point-shaped phenomenon, known as a morning star, at a point of high current density.

Since this damage will generally progress, the film layer values will fall below a certain threshold and the component may fail if there is further cumulative damage during periods of operating stress.

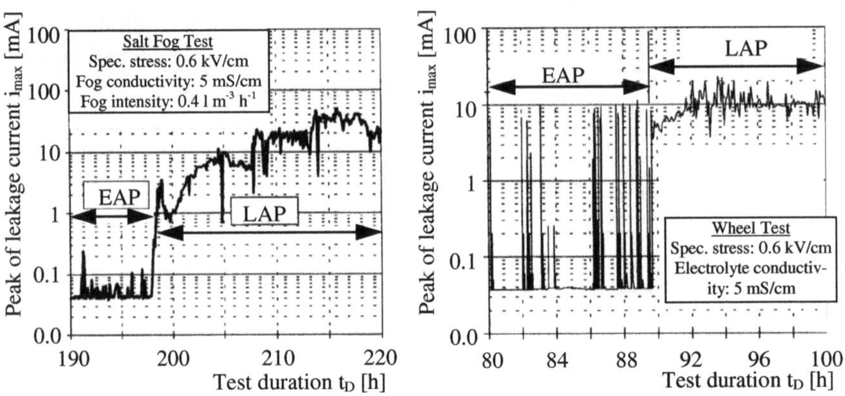

Fig. 11.4 Characteristics of AFP and ASP in the salt fog test and wheel test [25, 26]

While past test methods were solely aimed at evaluating erosion and tracking resistance, today's testing parameters are additionally chosen in such a way as to differentiate between *Early Ageing Phase* (EAP) and the *Late Ageing Phase* (LAP), especially in development tests. The comparison in Fig. 11.4 [25, 26] is one such example. At the start of the salt fog or wheel test, the mean leakage current measured is below 0.1 mA under the chosen test conditions and the silicone rubber surface is hydrophobic. When a loss of hydrophobicity is experienced in a sufficient axial length, this current value rises to over 10 mA and the silicone rubber surface is also evaluated in terms of its erosion and tracking resistance in the ASP state. In comparison to the salt fog test, it is clear that a greater acceleration occurs in the wheel test. Despite the greater conductivity of the fog generated (5 mS/cm), the transition between AFP and ASP already occurs in the wheel test with 1.5 mS/cm in 90 h.

This greater acceleration can also be demonstrated using the flashover voltage (Fig. 11.5 [24]). In this diagram, the flashover voltages U_F are based on the value for the new state U_{F0}. In both tests, identical testing parameters were configured with a specific stress of 0.4 kV/cm and an electrolyte conductivity of 16 mS/cm. The film layer state, which is characterised by U_{FF}/U_{F0}, is already achieved in the wheel test after approx. 20 h, but only after approx. 70 h in the salt fog test. The flashover voltages U_F/U_{F0} measured after cleaning the test specimens no longer achieve the values for the new state because a factor $U_F/U_{F0} < 1$ comes into play. This reduction is due to an increase in surface roughness as a result of surface erosion. After a longer rest phase, however, as was used in this test, the hydrophobicity transfer processes can achieve the original values. The low molecular weight chains can achieve a certain level of smoothing and hydrophobising in terms of roughness.

In [26], the dynamic contact angle measurement described in "Effect on the Hydrophobicity Behaviour" was performed on test specimens earmarked for the wheel test and compared against the increase in the leakage current (Fig. 11.6). It

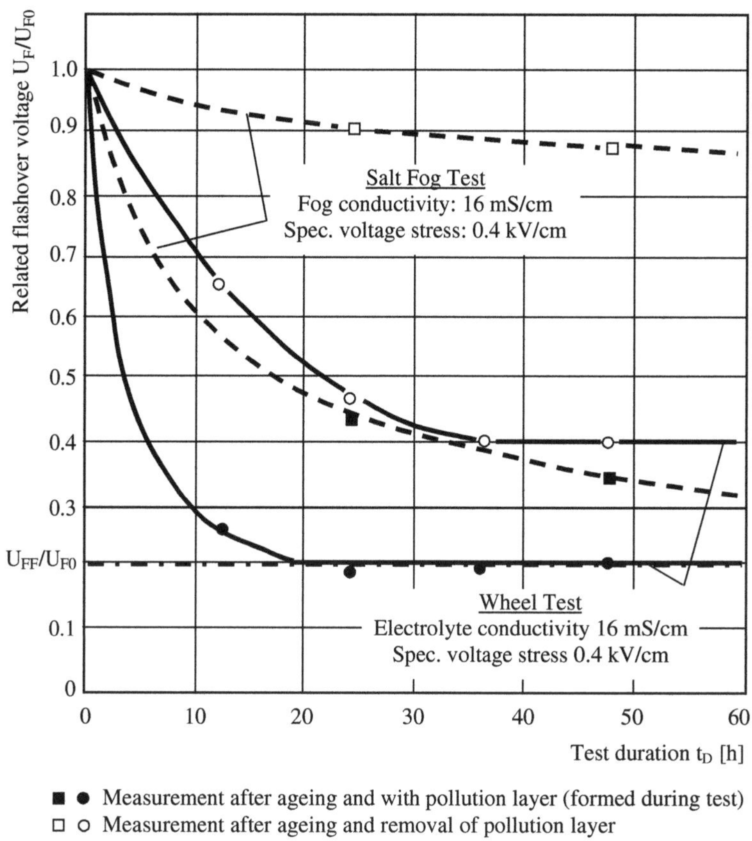

Fig. 11.5 A reduction in the flashover voltage obtained in the wheel test and salt fog test

shows that an increase in the leakage current by a couple of μA in the mA range, which is indicative of a loss of hydrophobicity, is accompanied by a large reduction in the size of the dynamic receding contact angle. The dynamic advancing contact angle, on the other hand, experiences only a minor reduction in size.

The next section discusses the Inclined Plane Test and provides detailed examples in relation to the multi-functional use of this test principle.

11.5 Inclined Plane Test

11.5.1 Erosion and Tracking Resistance

The main purpose of this test is to evaluate the ageing resistance to erosion or prolonged tracking under AC stressing. Here, hydrophobicity is removed by

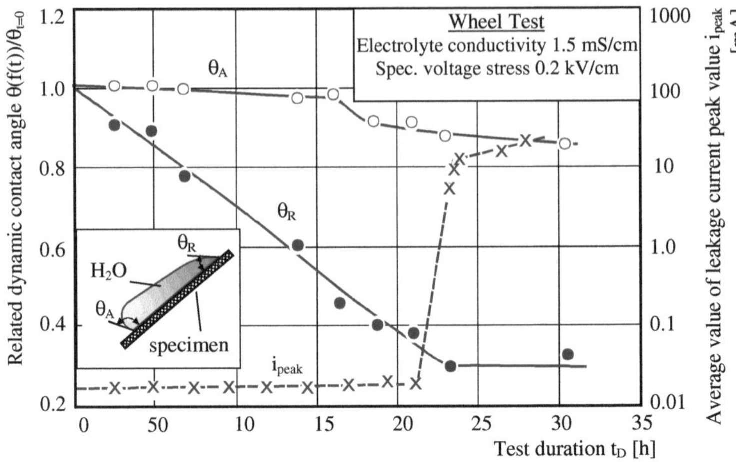

Fig. 11.6 Using the example of the wheel test to chart the progression of a leakage current, dynamic advancing contact angle and dynamic receding contact angle [26]

adding a wetting agent to the test solution. The level of damage to the insulating material depends on numerous factors [27]:

- Leakage current and root temperature of the PDs
- Mobility of the discharge (e.g. as a result of electrode roughness or an electrode contour)
- Voltage source (impedance and voltage stability)
- Physical structure of erosion or tracking (e.g. cavity in the insulating material)

Tests conducted in different laboratories on test specimens from one material batch and an identical manufacturing process have repeatedly produced different results. For this reason, the next section discusses those factors that may influence the test result.

11.5.1.1 Factors that Influence the Test Result

Effect of the Test Specimen Holder

The test specimen holder described in the standard must be designed in accordance with Fig. 11.7 (left). The purpose of this continuous opening is to prevent the formation of electrolyte traces on the rear side of the test specimen. The absence of a supporting effect can cause flexible materials to become deformed (Fig. 11.7 right). In the case of badly deformed materials, it was observed that the electrolyte may drain away without reaching the counter electrode [28, 29]. Consequently, the test specimen is stressed for a shorter time, can pass the test easily and the ability to reproduce the test is restricted.

Fig. 11.7 Test specimen holder in accordance with the standard IEC 60587 [1]

To improve the situation, the suggestion was made to use small bars to prevent deformities [28, 29] (Fig. 11.8). Practical experience has shown that, if the test set-up is prepared carefully, no electrolytes will form on the rear side between the bar(s) and the test specimen. This modification to the test specimen holder is recommended for tests involving flexible materials (such as RTV or LSR specimens).

Effect of Electrode Roughness

Electrode roughness can be differentiated as follows:

- Differences in the new state as a result of different manufacturing processes
- Melting loss after a test (Fig. 11.9).

Different types of roughness can have different effects that influence electron work flow and the local electrical field stress. This can result in the formation of a preferential burning point for the PDs, which then damages the insulating material more intensively.

Since there may be significant differences between the roughness of new electrodes and the roughness after a melting loss, both trends were identified: New electrodes or electrodes that have experienced a melting loss can result in more intensive damage. For this reason, the following recommendations were made: Use sand blasting to "condition" new electrodes, use new electrodes for each test and, in the case of comparative tests in different laboratories, use electrodes from only one manufacturer [28, 29].

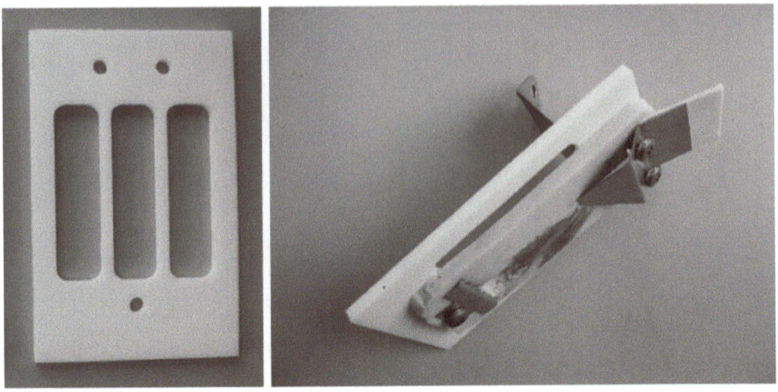

Fig. 11.8 Modified test specimen holder for flexible insulating materials [28, 29]

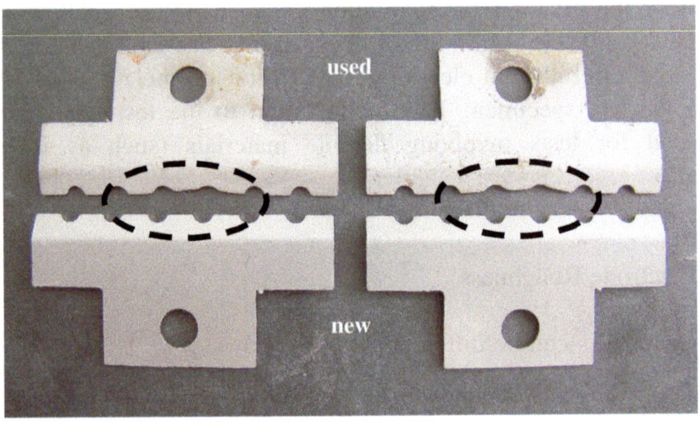

Fig. 11.9 Electrode melting loss (upper electrodes) after 6 h at a voltage level of 6 kV

Effect of the Voltage Source

The properties of the voltage source can affect ageing behaviour in many different ways. The pollution flashover model can be used for this purpose [30]. If the internal impedance of a weak voltage source is too high or if this restricts the requisite increase in the current di/dt or the current level, flashover voltage values that are too high in comparison to the real network are produced. This effect can already start in the dry zone formation phase if the applied voltage collapses over the ohmic pollution layer and the necessary vaporisation energy cannot be provided. If a voltage is applied to a pollution layer that is assumed to be full layer and this ignites a pre-arc, the electrical circuit may be reproduced as shown in Fig. 11.10.

Kirchhoff's voltage law applies:

$$u_S = u_L + u_R + u_{PA} \tag{11.1}$$

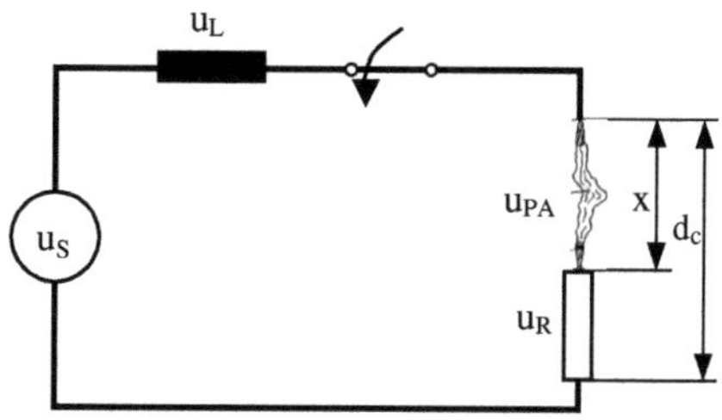

Fig. 11.10 Simplified equivalent circuit diagram for the ignition of a pre-arc over a dry band zone [30]

$$u_S = L\,\frac{di}{dt} + i\,R\left(1 - \frac{x}{d_c}\right) + \frac{A d_c}{i}\,\frac{x}{d_c} \tag{11.2}$$

A	Constant
i	Current i(t)
L	Internal inductive reactance of voltage source
u_S	Switch-on voltage
u_L	Voltage across internal inductive reactance
u_R	Voltage across pollution layer in the switch-on moment
u_{PA}	Arc voltage
d_c	Creepage distance
x	Length of pre-arc
R	Ohmic resistance of the pollution layer in the switch-on moment

In Fig. 11.11, it is shown that the voltage drop u_L has a maximum and, as a result, the current i has achieved its largest increase. If differentiated, this results in:

$$\frac{di}{dt} = \frac{u_S}{L} - \frac{2RAki}{L\left(Ri^2 + Ak\right)} \tag{11.3}$$

The time constant can be specified as follows:

$$\tau = \frac{L}{R} \tag{11.4}$$

If, for example, the time is calculated for an assumed time constant of 0.85 ms for a test transformer and an applied voltage of $u_s = 1.1u_F$ (Fig. 11.12), the result is 17 ms. When compared to a powerful network, the process duration is longer than a half-wave and a higher switch-on voltage is required for flashover, e.g. $u_s = 2u_F$. However, this value is false with respect to a comparable pollution layer situation in a powerful network where flashover would occur at u_F. Quickly switching on a charged shunt capacitance is one way of supplying the prolonging pre-arc with the required energy and reducing the delaying effect of inductive reactance.

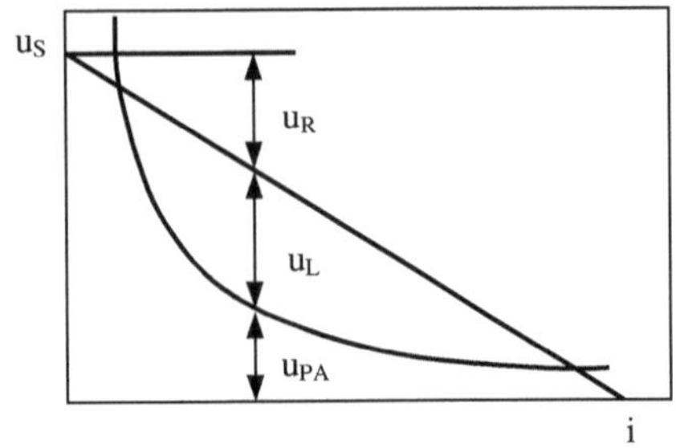

Fig. 11.11 Partial voltages for an equivalent circuit diagram [30]

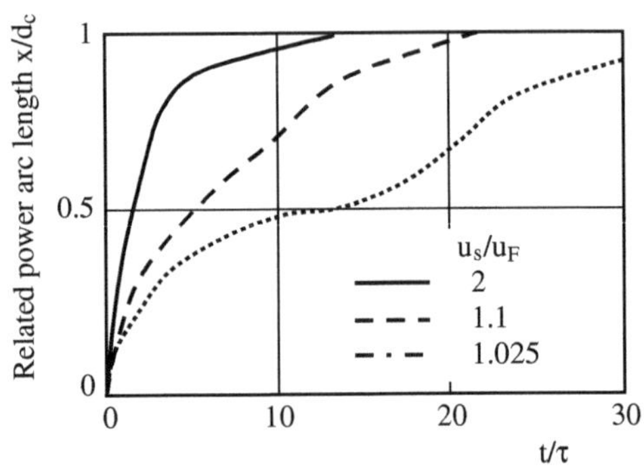

Fig. 11.12 Arc length obtained depending on the time, which is based on the time constant [30]

How does the voltage source affect ageing caused by thermally ionised discharges?

Their dynamic processes for dry zone formation, pre-arc ignition and pre-arc prolongation are comparable to those for pollution flashover, but without inducing the flashover event. To verify the effect, the Inclined Plane Test was modified and the electrolyte conductivity was varied (Fig. 11.13 [23, 24]).

The time to failure has a minimum at 1.5 mS/cm under these test conditions and the material evaluated (HTV without ATH). However, the intensity of damage was different (Fig. 11.14). The lowest time to failure at 1.5 mS/cm was accompanied by a fast but very localized growth of the erosion path. The higher the electrolyte conductivities, the more damage to the material occurred.

The effect of the voltage source was examined at an electrolyte conductivity of 32 mS/cm because relatively high currents occur at this value. It shows that the voltage source operated at 4 kV for a nominal voltage of 35 kV results in a significantly lower failure time for the same time to erosion inception in comparison to the voltage source for a nominal voltage of 10 kV.

A video set-up was used to determine the length of stay (burn time) for discharges. For localised fixations, a needle electrode was used at the point at which the discharges always ignited. It showed that the shorter failure time in Fig. 11.13 is due to a longer burning time for the discharges (Fig. 11.15 [24]). It is clear that the unsuitable voltage source prevents the formation of a sufficient number of dry

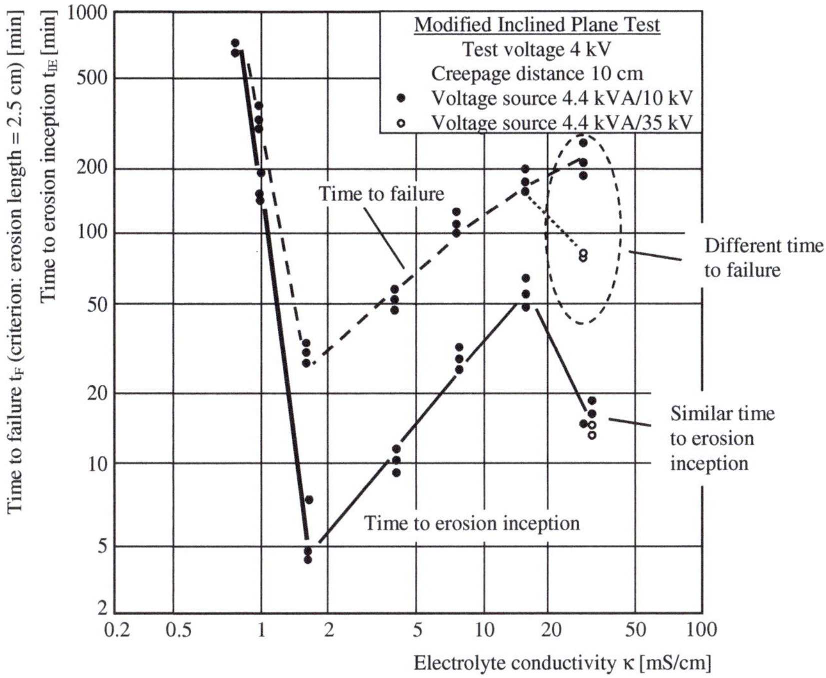

Fig. 11.13 Influence of the voltage source to failure and inception time

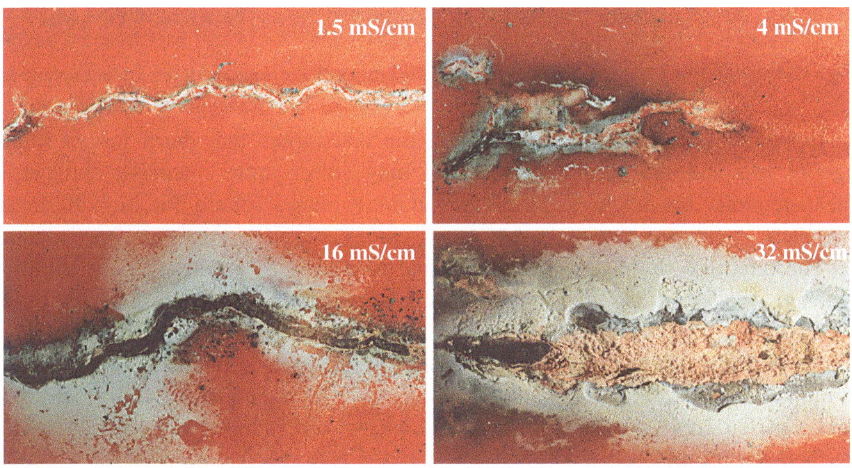

Fig. 11.14 Influence of electrolyte conductivity to the damage pattern [24]

zones for quenching purposes or it prevents the pre-arc from reaching its quenching length.

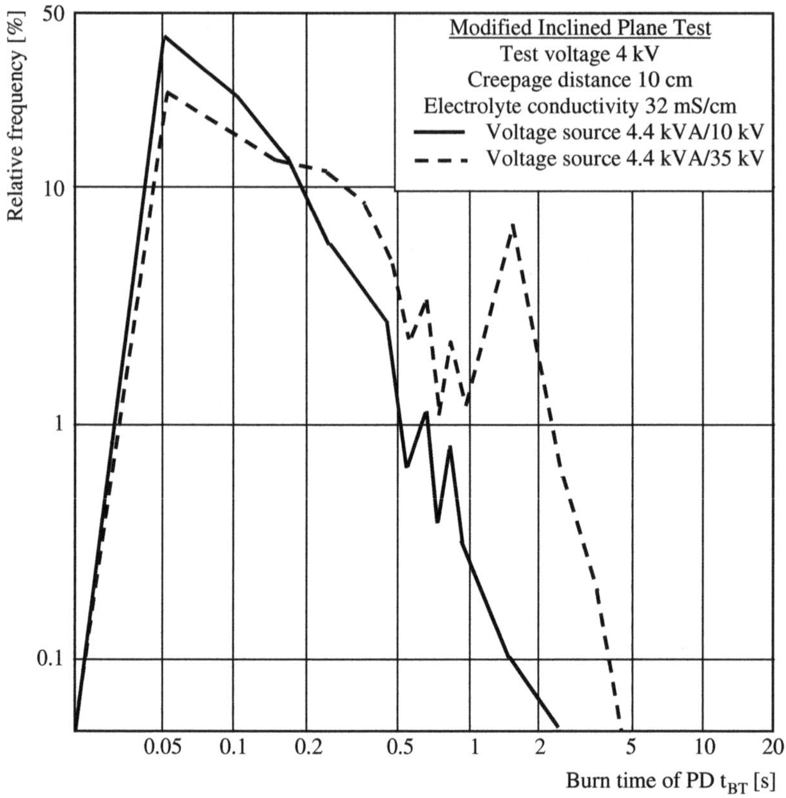

Fig. 11.15 Relative frequency of burn time for different voltage sources in the inclined plane test

In contrast to a pollution flashover, the discharges may burn longer in the ageing test and the insulating material may suffer more intensive damage, if measurements are taken using an unsuitable voltage source for high flashover voltages.

The effect of shunt capacitance on the time until a certain tracking length is reached was examined in [31]. A 5 kVA transformer was used and the testing voltage was 4 kV. The result was determined in accordance with (Table 11.1).

In a certain analogy to the pollution flashover, shunt capacitance also helps to simulate stresses comparable to real network in the ageing tests. A similar effect was verified in [31] when stray capacitances were simulated using additional earthed electrodes.

If one assumes that different results from two Inclined Plane Test devices are due to the different characteristics of the test transformers or test set-up (e.g. proximity of earthed device parts), a video image of the discharge burning time on an inert sheet of glass can be used as a calibration measurement.

Table 11.1 Effect of a shunt capacitor on the mean failure time and its standard deviation [31]

Capacity of the shunt capacitor [μF]	Time until the failure criterion is reached	
	Mean value [min]	Standard deviation [min]
0	94	38
0.001	231	115
0.005	257	79
0.025	277	98
0.050	745	167

Effect of the Sample Size

The standard specifies that a minimum of five test specimens must be tested. Due to the variation of the ageing process and of the material properties, it is beneficial to increase this number to between 15 [29] and 20 [27] in development tests or if a test specimen barely passes a chosen voltage level. While testing different types of silicone rubber, it was established that, when there is a greater variation among measurements (e.g. as a result of reducing the filler), the empirical distribution function adapts well to a 2-parameter Weibull distribution function.

11.5.1.2 Testing Silicone Rubber Coatings

The use of silicone rubber coatings is an established mitigation measure for reducing pollution layer problems in substations with porcelain apparatus insulators, for example. The effectiveness of the coating lies in the fact that it bestows hydrophobic properties on the hydrophilic porcelain surface. Depending on the care with which the coating is applied and the level of pollution layer stress, the coating may need to be re-applied after a certain period of time in service. Different systems are available on the market. If the coating experiences a loss of hydrophobicity, leakage currents occur within the mA range and dry zones form, there is a risk of erosive abrasion, which is also the case with polymeric solids. If the coating abrasion is so severe that the porcelain surface is exposed again, the device's insulation strength is comparable to that before the coating was applied. If, however, conductive tracking forms, this reduces the insulation strength further (see also Fig. 11.3). Concerning this matter, CIGRE Working Group D1.14 conducted analyses in the form of an international round-robin test using the Inclined Plane Test on commercially available silicone rubber coatings [32]. The testing voltage was 3.5 kV, which was a minimum requirement [33], while the mean material thickness applied was approx. 2 mm. The test results, which contributed to this round-robin test, can be summarised as follows:

Material 1 is a coating with a low filler content and pronounced hydrophobicity. It took up to 30 min for a wetted track to develop. All test specimens passed the test (Fig. 11.16). Only in the case of one test specimen was the layer up to the

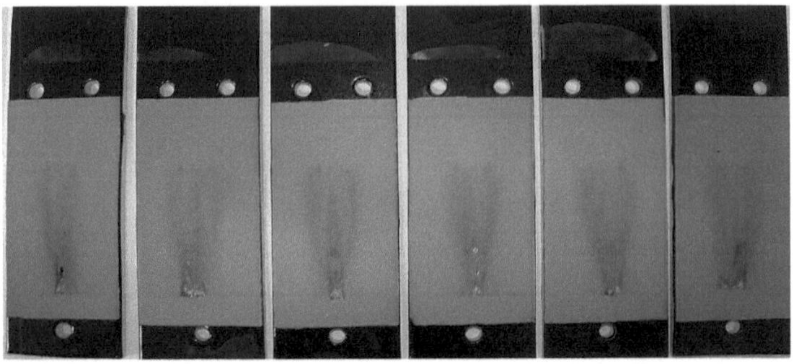

Test time [min]	360	360	360	360	360	360
Weight loss [mg]	5	11	5	45	45	42
Comment	Erosion up to the support material					

Fig. 11.16 Results of an Inclined Plane Test on RTV coating 1

porcelain eroded. Nevertheless, this test specimen, which experienced the lowest mass loss, did not show any signs of cumulative damage.

Material 2 is a coating with a higher filler content. It also passed the test (Fig. 11.17). However, four of the six test specimens exhibited signs of erosion up to the porcelain. As expected, the greater the size of the defect, the greater the mass loss.

In the course of these analyses, Material 3 was a negative example that demonstrated that unsuitable compositions of silicone rubber coatings can also cause failures (Fig. 11.18). None of the test specimens reached a test time of 60 min. Given the rapid growth of the tracking and the accompanying loss of the material, it can be assumed that this concerned conductive tracking.

When operating a device that has this coating, this scenario must be classified as critical because the insulation strength significantly falls below the level of an untreated porcelain surface and a flashover can already occur in the case of a service voltage.

11.5.1.3 Testing Within the Context of more Intensive Acceptance Tests

As already shown in Sect. 10.5.2.1, the Inclined Plane Test can also be conducted on test specimens taken from the insulator housing of an insulator. As is already known from materials testing for standardised dimensions, results can be extremely varied.

For various types of silicone rubber filled with ATH, it was established that a voltage level of 4.5 kV will definitely pass the test despite an increased test accuracy (material thickness, material width, etc.) (Fig. 11.19).

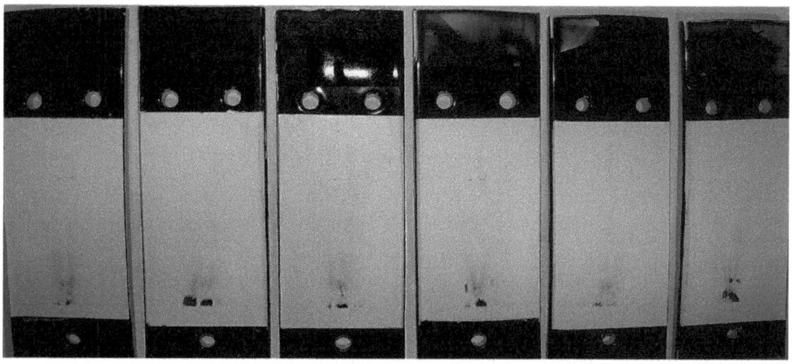

Test time [min]	360	360	360	360	360	360
Weight loss [mg]	13	38	9	6	18	35
Comment		Erosion up to the support material	Erosion up to the support material	Erosion up to the support material		Erosion up to the support material

Fig. 11.17 Results of an Inclined Plane Test on RTV coating 2

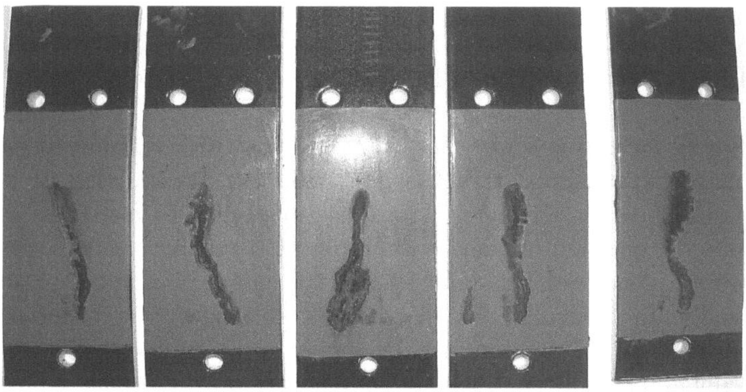

Test time [min]	47	40	63	51	42
Weight loss [mg]	147	170	328	148	249
Comment	All test specimens quickly failed the test. A rapid growth in erosion or tracking was observed; the progression of the damage was characterised by an intensive morning star inception. The porcelain surface was exposed in all cases.				

Fig. 11.18 Results of an Inclined Plane Test on RTV coating 3

Test specimens taken from other insulators with a silicone insulator housing were completely destroyed. The test was concluded after a test time of only 42 or 72 min. As it was the case with the sample test specimens shown in Fig. 11.20 there was a noticeable tendency for the test specimens to ignite and, to some extent, suffer complete oxidation [34].

A critical scenario arose while testing the shed sections of a 145 kV EPDM post insulator (Fig. 11.21). Despite a comparatively large material thickness (approx.

Fig. 11.19 Results of the Inclined Plane Test at 4.5 kV on housing specimens after 6 h

8 mm), the test specimens caught fire within a few minutes of the test commencing under strong smoke emissions and damaged the test device. Since the housing material EPDM loses its initial hydrophobicity, it is very likely that partial discharges will occur in the same way in which they are simulated in the Inclined Plane Test. A material formulation that ignites itself in this situation is deemed to be critical from the perspective of using such a formulation in a power utility.

11.5.1.4 Tests Involving a DC Voltage

With the increasing use of energy transportation over very long distances with high voltage direct current (HVDC) up to 800 kV, new milestones in energy transmission have been reached. From the perspective of materials testing, this also poses the question of transferability of AC stressing results acquired during the inclined plane test. This topic is currently undergoing intensive analysis internationally. In particular, the question as to whether the material classifications measured for AC stressing can be applied to DC stressing is being examined.

Initial exploratory tests do not appear to confirm an equivalent AC-DC-behaviour for every family of materials [35]. These independent tests were performed with three families of materials (Fig. 11.22). Material A is a structural material typically used mostly for distribution applications. Material B is a castable Silicone Rubber with no added filler to improve its erosion resistance. Material C is a HTV Silicone Rubber with ATH filler. For material A, practically all tests under DC-stress resulted in failure with a conductive path causing an excess of the 60 mA current criterion.

If materials B and C failed then by a different failure mode as material A did. B and C failed only by erosion and not by excess of the current criterion. And the test time required for this erosion failure to occur was always much longer than for material A. The lower failure susceptibly of material C is attributed to the ATH content, which improves the tracking and erosion resistance for AC and DC stress. If this trend is confirmed in further comparative tests, the use of DC tests will be an important supplement to AC tests.

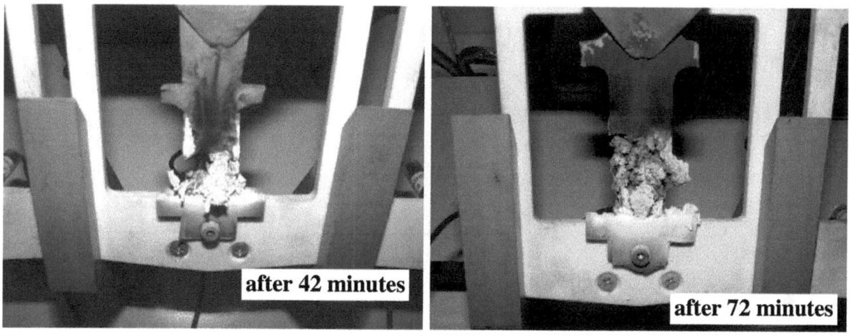

Fig. 11.20 Results of a brief Inclined Plane Test at 4.5 kV on housing specimens

Fig. 11.21 Ignited EPDM shed segment immediately after activating the Inclined Plane Test at 4.5 kV

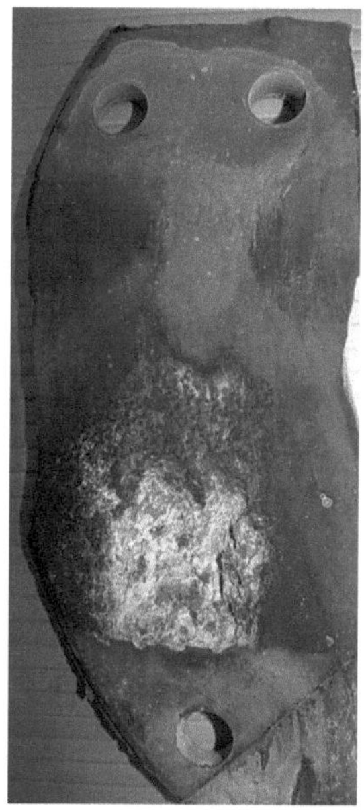

The Chinese standard DL/T810-2002 [36] describes both a salt fog test and an Inclined Plane Test with DC stressing. The salt fog test is based on the specifications contained in the first edition of IEC (6)1109 [7] and defines a DC voltage stress of 0.29 kV/cm (positive or negative polarity), a fog conductivity of 10 kg/m^3 and a spray rate of 0.4 l/(hm^3).

The Inclined Plane Test is conducted using only a positive DC voltage. The criterion for successfully passing this test is as follows: The maximum erosion level of 2.5 mm must not be exceeded after 6 h at +4.5 kV. As described in IEC 60587, the electrolyte conductivity is 3.95 Ωm. For the DC test, however, the flow rate was reduced from 0.6 to 0.2 ml/min.

The following can be derived from these current analyses [35, 37, 38]:

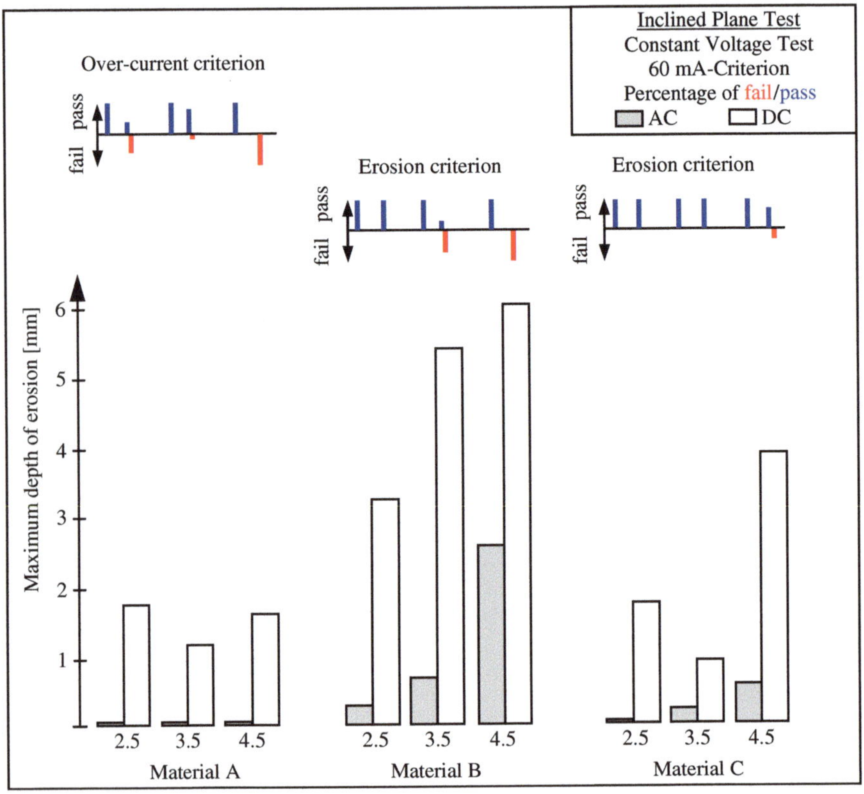

Fig. 11.22 Results of AC and DC Inclined Plane Tests performed on three different material families [35]

- The erosion rate or failure rate for the same testing parameters is higher under DC stressing than under AC stressing.
- In the case of tested insulating materials commercially used for overhead transmission line insulators, a trend whereby the positive polarity results in a higher stress is evident.
- Secondary effects due to the absence of a change in polarity were identified for certain materials, which resulted in the formation of conductive tracks and the premature triggering of the over-current criterion or heavy erosion.

These differences show that conducting the Inclined Plane Test with DC stressing as a supplement to the AC test appears necessary until sufficiently validated results are available in relation to reactions that are partly material dependent.

11.5.2 Inclined Plane Test Principle for Evaluating the Resistance of Hydrophobicity

As part of an international round-robin test by CIGRE Working Group D1.14, the ability to reproduce the results of the Dynamic Drop Test was evaluated [39]. This test is based on the principle of rolling drops on an inclined test specimen surface known as the "inclined plane" (see also Fig. 7.38). The result [40] found that the time until a loss of hydrophobicity was experienced showed good correlation in 4 of the 5 laboratories and only one laboratory measured higher values that differed on average (Fig. 11.23).

The testing parameters defined for the round-robin test were originated in the systematic fundamental analyses of various influencing factors [39], the results of which can be summarised as follows:

- The sliding volume of the droplet essentially depends more on the inclination angle and the level of adhesion at the discharge electrode and less on the material to be tested. Small inclination angles mean a greater volume.
- The greater the droplet volume (sliding volume), the shorter the failure time.
- The surface roughness of a material greatly influences the failure time; smoother surfaces have a longer failure time.
- The greater the level of corrosion associated with the high voltage electrode at the electrolyte discharge point, the greater the droplet volume.
- The higher the electrolyte conductivity and/or stressing, the shorter the failure time.

The results presented in "Effect on the Resistance of Hydrophobicity" [41], which relate to various materials under slightly different testing parameters, are in tune with the results of the round-robin test. Material X1 is a comparable RTV silicone rubber whose mean failure time is lower than the mean failure time in the round-robin test. This is due to higher stressing and a smaller inclination angle. The value for an HTV with silanised ATH and the high value for the base formulation (no filler) are classified accordingly.

These tests have shown that the acceleration associated with evaluating hydrophobicity is higher in comparison to a salt fog test or wheel test. In other words, less time is involved. This finding and the simplicity of the test specimens make it possible to also take measurements in a statistically evaluable volume (e.g. 10–20 test specimens) for the purpose of quality assurance within production. Storing test specimens under defined conditions before the test commences contributes towards good reproducibility of the measurement results.

The results of follow-up studies in [40] have shown that the recovery of hydrophobicity can also be quantified using this method.

Now that a method for evaluating the hydrophobicity transfer with statically applied droplets [40] has been presented, and given the good experiences associated with the (reproducible) temporary suppression of hydrophobicity for

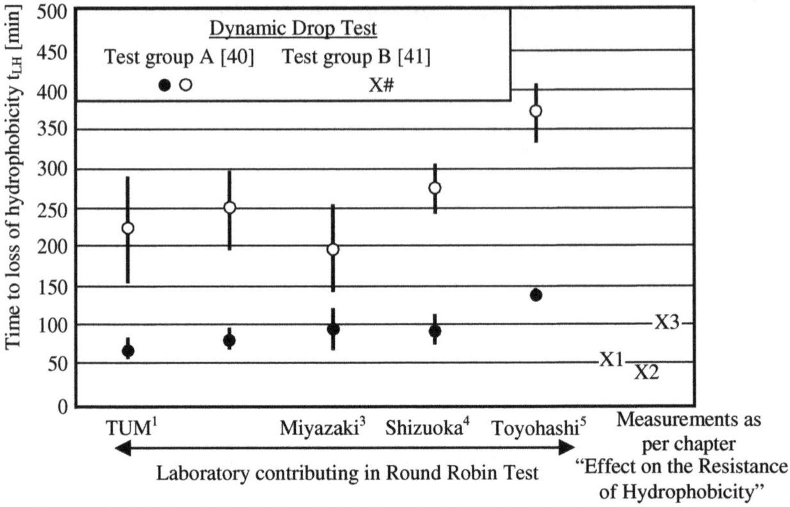

Test parameter	Test group A	Test group B
Specimen inclination relative to horizontal line [°]	60	45
Creepage distance [mm]	50	50
Electrolyte conductivity [mS/cm]	1.5 ± 0.2	1.5 ± 0.2
Flow rate [ml/min]	1.0 ± 0.2	1.0 ± 0.2
Voltage [kV]	o = 4.0 ● = 5.0	x = 5.5
Material	RTV-Silicone Rubber	x1 = RTV-Silicone Rubber
		x2 = HTV- Silicone Rubber with silanised ATH
		x3 = Base polymer (gum)
Criterion for failure I [mA]	2 ± 0.5 for 4 ± 0.5 s	2 ± 0.5 for 4 ± 0.5 s

1 = TU Munich (Germany), 2 = HTWS Zittau (Germany), 3 = University of Miyazaki (Japan), 4 = Shizuoka University (Japan), 5 = Toyohashi University of Technology (Japan), 6 = PFISTERER SEFAG (Switzerland)

Fig. 11.23 Dynamic drop test results from different laboratories [40, 41]

pollution layer tests [17, 42], the next section will describe the option of using the dynamic drop test to evaluate the hydrophobicity transfer.

11.5.3 The Inclined Plane Test Principle for Evaluating the Hydrophobicity Transfer

The simple plate-shaped geometry used in the Dynamic Drop Test was also used here. Furthermore, the test set-up and testing parameters were virtually identical (Fig. 11.24) to those in Fig. 11.23 (Group B) or Section "Effect on the Resistance of

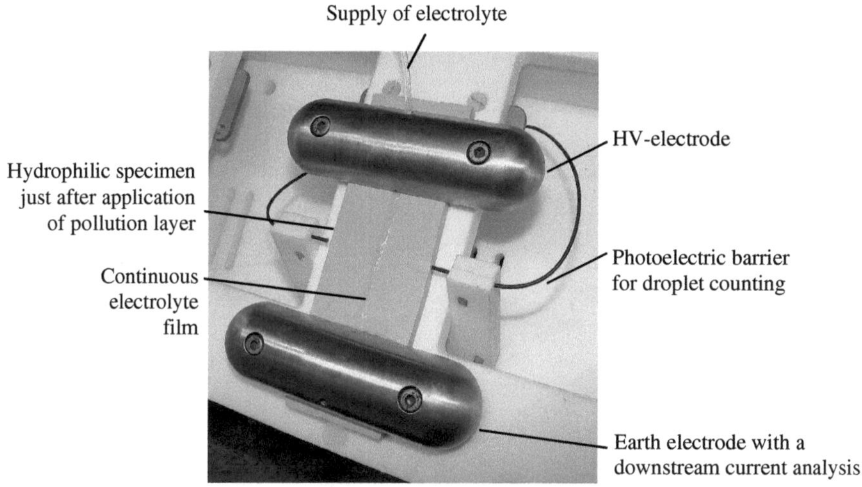

Fig. 11.24 A loss of hydrophobicity after applying a dry pollution layer in a dynamic drop test

Hydrophobicity". However, the surface of the test specimens, which is made from HTV silicone rubber with silanised ATH, has a dry pollution layer. This pollution layer is applied using a paint-brush or a soft sponge; compressed air is then used to lightly blow off the excess pollution layer. There is a complete loss of hydrophobicity immediately after the pollution layer has been applied (Fig. 11.24). In this phase of the method evaluation, the transfer involving each test specimen is evaluated once only. In other words, only one test specimen was used for each measurement. The silicone rubber plates yet to be tested were stored in a desiccator in order to control, in a defined manner, the effect of the ambient humidity on the transfer process [40].

If the length of time in storage is increased, a recovery of hydrophobicity is experienced after the hydrophobicity transfer (Fig. 11.25). Initial quantitative measures have shown that, as expected, the type of dry pollution layer has a major effect on the transfer time (Fig. 11.26). The Sicron® used is a quartz powder that is ground to a grain size of 9 μm (iron-free). The shape of the curve progression, which correlates the time to a loss of hydrophobicity measurement directly with the transfer intensity, corresponds to the increase in contact angles in other measurement campaigns [40]. In contrast to the measurement concerning the static contact angle, please note that, as a result of using rolling droplets in the evaluation, the dynamic receding contact angle is reproduced (see also Fig. 11.6) and there is better correlation with the efficacy of the hydrophobicity transfer when the electrolytic pollution layer starts to flow.

The test was repeated using commercially available Kaolin (60 % of the particle size $\leq$2 μm). Kaolin is a hydrated aluminium silicate with the chemical formula Al_2O_3 $2SiO_2$ $2H_2O$. Depending on the particle size, its specific surface is between 10 and 40 m^2/g. In contrast to Sicron®, an increase during the analysis period (200 h) was barely detectable when Kaolin was used.

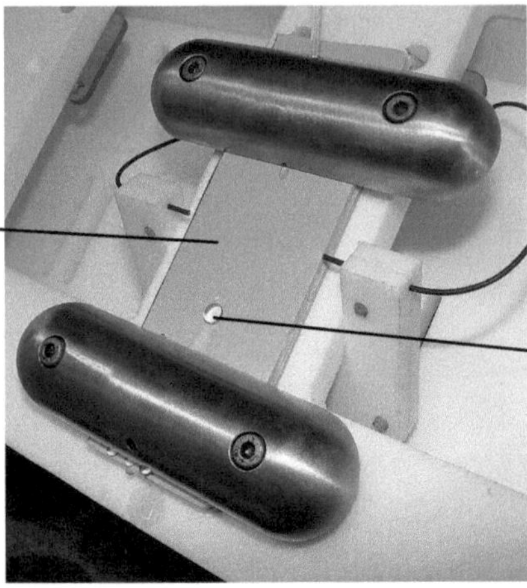

Fig. 11.25 Return of hydrophobicity by transfer in a dynamic drop test after applying a dry pollution layer and subsequent storage

Initial results show that the Inclined Plane Test principle also has the diagnostic potential to evaluate the hydrophobicity transfer. Other systematic analyses are required, in particular, to establish and apply the pollution layer in a reproducible manner.

11.6 Finger-Print Analysis of Polymeric Housing Materials

11.6.1 Background

For many years now, the CIGRE Study Committee D1 has been intensively concerned with materials and applicable test methods for verifying the suitability of such materials for use in power supplies. The topic currently being examined by Working Group D1.27 (Material Properties for New and Nonceramic Insulation) complements this work, namely the properties of new materials and materials currently used in composite insulation systems. One such topic is the material description for finger-printing, an overview of which is provided in the section below. The purpose of this work is to evaluate analytical methods that uniquely identify polymeric materials and to select suitable processes from the many established methods available. Is this topic academic or essential? This topic was predominantly initiated by network users of composite insulators. Furthermore, it helps manufacturers to verify the quality of their products. Its need can be derived from the following considerations:

- To date, there are just a few binding criteria in relation to whether materials and designs released after permissibility tests or design tests are used or whether, as

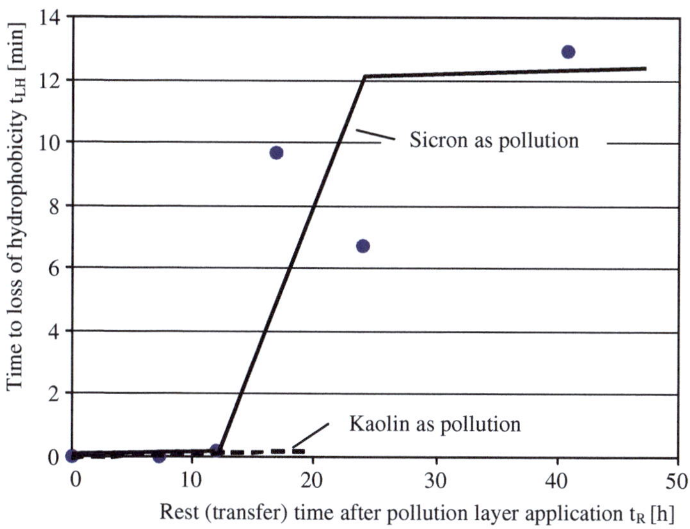

Fig. 11.26 A longer time to failure as a result of the hydrophobicity transfer

a result of cost pressures or the discontinuation of some raw materials, alternative materials of a lower quality or with no long-term experience are processed under the operating conditions of the user.

- Furthermore, "misunderstandings" may arise if general, non-specific material descriptions are provided in specifications/tenders. One such example is where low viscosity liquid silicone was "sold" as HTV silicone rubber because of its typical vulcanisation temperature of 100–120 °C.
- In the case of damage resulting from irreversible ageing phenomena, critical material changes (in comparison to storage samples of the same material batch) can be examined analytically.
- Interestingly, these analytical methods have the potential to be deployed in in-process quality assurance procedures.

11.6.2 Overview of Processes

Even in the German-speaking world, the short names of methods are being used both colloquially and in test reports. Table 11.2 provides a brief overview of the processes currently deployed.

Table 11.2 Methods investigated for material finger-printing

Method	Analysis	Main property, evaluated
Density measurement	Physical	
Hardness measurement	Physical	
DSC	Thermal	Temperatures: Glass transition temperature Crystallisation temperature Melting temperature
TGA	Thermal	Weight change
FTIR	Spectroscopic	Type and quantity of certain chemical bonds
Material testing device	Mechanical	Coefficient of expansion Modulus of elasticity Tensile strength Ultimate elongation Tear resistance

11.6.3 Processes and Examples: A Closer Look

11.6.3.1 Density and Hardness Measurement

Density measurement is an established process for various materials. Density, when measured in accordance with ISO 2781 [43], provides initial information about the filler content relative to the base polymer. Castable SR systems such as RTV und LSR usually have a density of <1.2 g/cm^3, while HTV with ATH values have a density of >1.5 g/cm^3. In the case of hydrophobic materials, it may be necessary to add a wetting agent to water in order to measure the density of the material. Otherwise, air bubbles on the interface between the material surface and water may falsify the result.

The Shore A hardness measurement usually taken for elastomers is a non-destructive test in accordance with DIN 53505 [44]. Materials with a higher filler content tend to have increased hardness.

11.6.3.2 Differential Scanning Calorimetry

Differential scanning calorimetry (DSC) is a thermal process for measuring the exothermic/endothermic heat quantities of a material to be tested. For this purpose, the behaviour of a material sample with a weight of a few mg in a certain temperature range is compared with the behaviour of a reference sample, which is generally an encapsulated air-filled container, also known as a “pan” in-process terminology. Depending on the application and the expected material changes, a temperature range from e.g. −80 to +600 °C may be of interest. A range from −100 to 200 °C is often used for silicone rubber. Figure 11.27 shows a typical graph for HTV silicone rubber. The transition from elastic mode to brittle mode occurs at

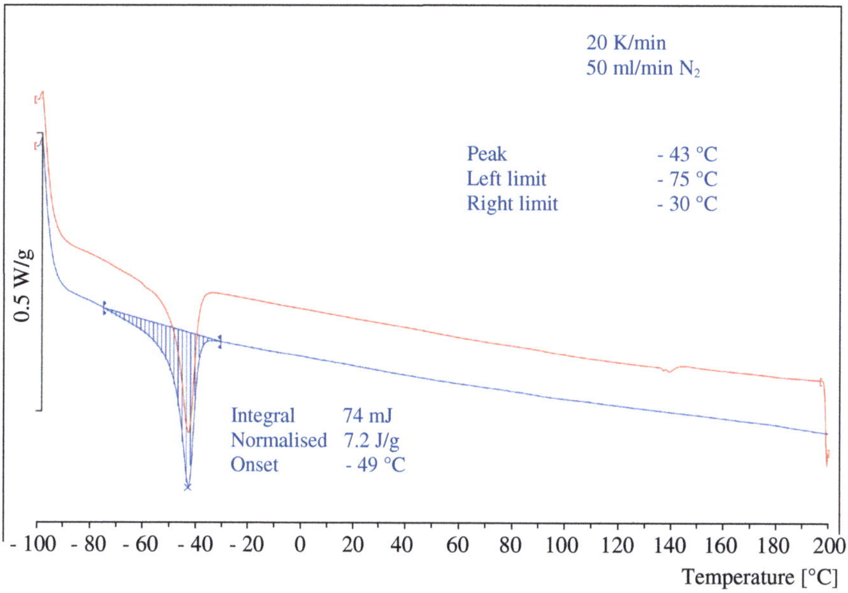

Fig. 11.27 DSC progression for HTV silicone rubber

approx. −42 °C. This "change in state" (known as reversible crystallisation) can easily be diagnosed with this process and corresponds to the results of low temperature tests on composite insulators that have this housing material. Irrespective of the material to be evaluated, two test runs of the same sample is standard practice in order to determine the effect of post-vulcanisation.

When a DSC measurement of GFK rods is taken, this produces a different progression. In this case, the glass transition temperature (softening of the material) is more or less pronounced and can typically lie between 110 °C and 160 °C.

11.6.3.3 Thermo-Gravimetric Analysis

Thermo-gravimetric analysis (TGA), also known as thermal gravimetric analysis, measures the change in mass of a sample depending on the applied temperature per time unit. This test also examines a material sample with a weight of just a few mg. The sample holder is permanently coupled with a microbalance, while a thermocouple is used to measure the temperature in the sample chamber. Typical temperature ranges lie between 25 °C and 1,000 °C.

A comparison of the information content relating to the TGA measurement of various types of silicone rubber is shown in Fig. 11.28: Firstly, the HTV SR filled with ATH shows a reduction in mass that is proportionate to the temperature; the ATH filler provides protection (see also Fig. 7.31) against stress without there being any direct damage to the polymer. As expected, the mass loss of RTV and

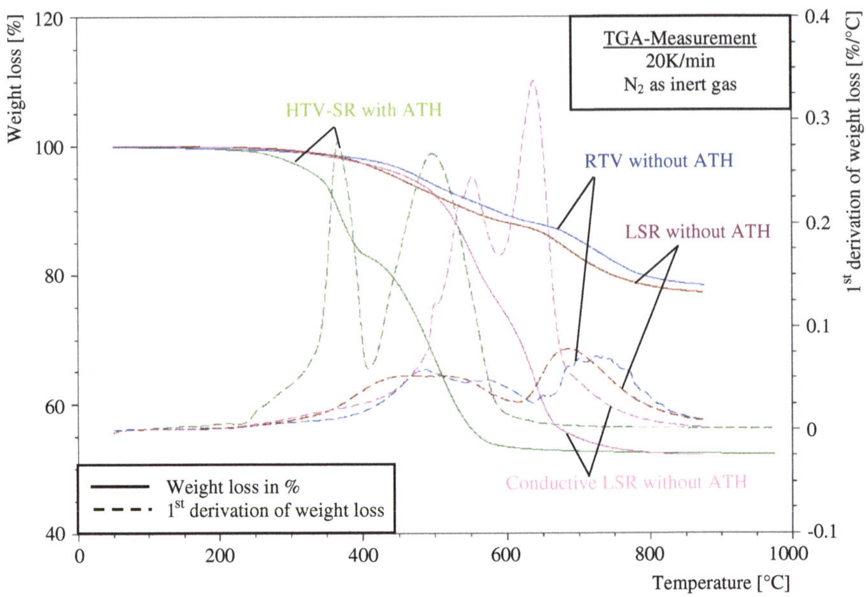

Fig. 11.28 TGA progressions of various types of silicone rubber

LSR (without ATH) starts at higher temperatures, indicated by the first decomposition reactions of the polymer. In the case of a conductive LSR, an intensive thermal decomposition of the special filler (carbon black) starts as of approx. 500 °C. The first derivatives shown in Fig. 11.28 show the respective (maximum) temperature value, which is characterised by a substantial change in mass.

The TGA comparison between the EPDM material (Fig. 11.29) that caught fire in the Inclined Plane Test after a brief period (Fig. 11.21) and a commercially available HTV silicone rubber with ATH that did not ignite in the Inclined Plane Test (Fig. 11.30) is also of interest. In both cases, the TGA test is performed using the inert gas N_2, thus preventing an occurrence of the fire observed under atmospheric conditions. Both insulating materials are filled with ATH and the first water release occurs at above 210 °C. The flammability risk for EPDM is not apparent from the TGA measurement. In other words, the use of only one test method is not sufficient to obtain a representative finger-print. If, however, the TGA measurement is taken under oxygen, there are strong deflagrations as of approx. 350 °C for EPDM, which correspond to the point of ignition in the Inclined Plane Pest.

11.6.3.4 Fourier Transform Infrared Spectroscopy

Fourier transform infrared spectroscopy (FTIR) can be used to determine material bonds that form a dipole moment after they are excited by means of IR radiation or

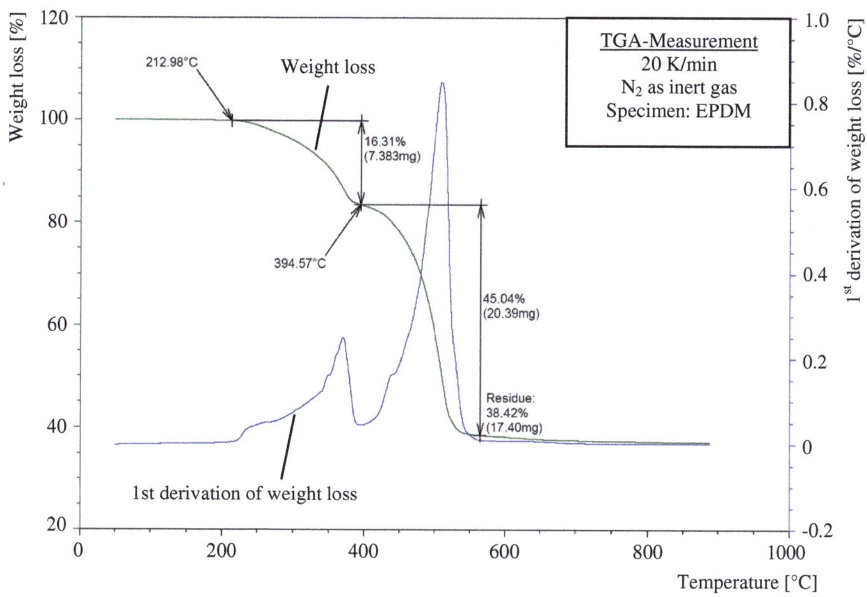

Fig. 11.29 TGA progression of the EPDM that caught fire in the Inclined Plane Test

change an existing dipole moment. The material sample is exposed to IR radiation for this purpose. The radiation spectrum is not recorded on the basis of a gradual change in wavelength. Instead, it is calculated using the Fourier transformation of a measured interferogram. Figure 11.31 shows the typical spectrum for an HTV silicone rubber. The visible peaks in a characteristic wavelength represent certain molecules, e.g. 1,261 cm^{-1} Si-CH_3 (bands). The ordinate can be used to determine the concentration; a large deviation from the 100 % value corresponds to a high concentration. The wavelengths characterised for certain bonds are listed systematically in reference tables.

11.6.3.5 Mechanical Materials Testing

Optimised mechanical materials testing devices (Fig. 11.32) exist for the most varied materials and test specimens. When measuring elastomers, it must be remembered that high elongations can occur until the elastomer fails. By using the special punching gauges and holding fixtures described in DIN 53504 [45] for example, it is possible to make test specimens and to mount them onto a testing device. This simple test make it possible to easily identify material deficits (e.g. under-vulcanisation), the effect of the particle size, and particle treatment in relation to a filler.

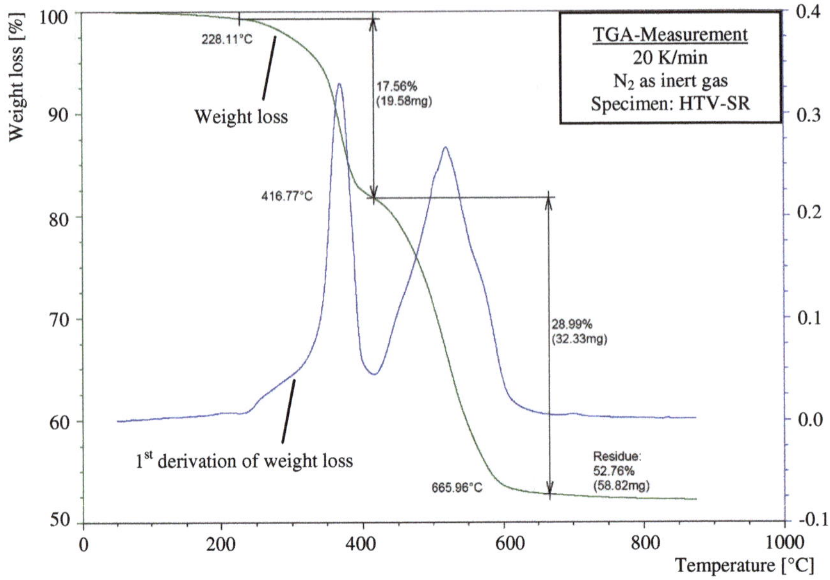

Fig. 11.30 TGA progression for HTV silicone rubber

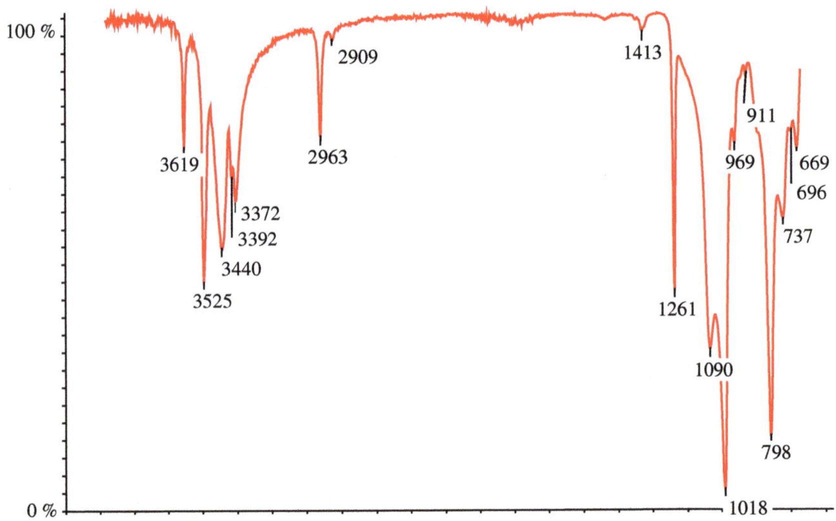

Fig. 11.31 FTIR for HTV silicone rubber

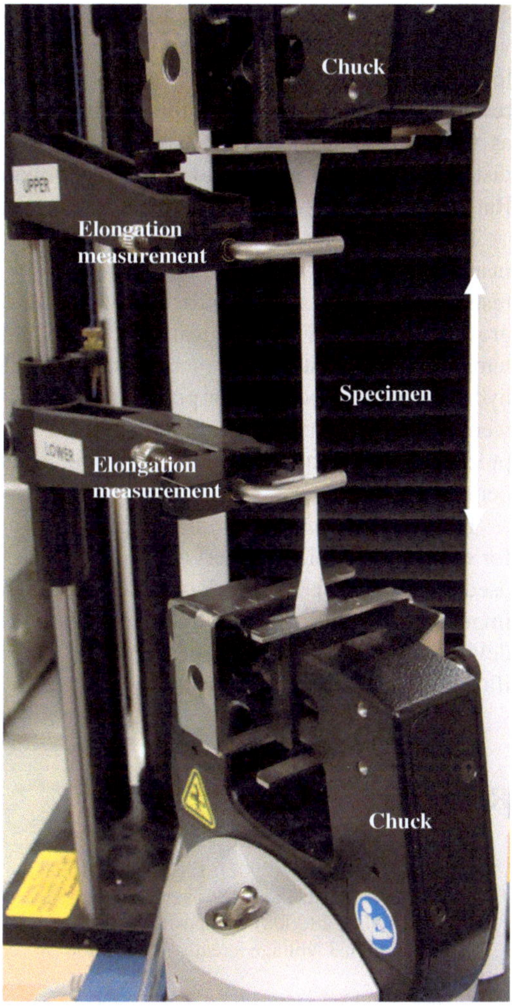

Fig. 11.32 Mechanical multi-functional measurement device

11.7 Summary

This chapter discusses the current standards for composite and/or polymeric insulators in comparison to conventional insulators. It shows that, from a design perspective, some detailed tests apply to one insulator type only. In the case of composite insulators, for example, interface areas are frequently evaluated.

Nowadays, the standards for composite insulators have reached a level that provides both the manufacturer and user with great support when designing reliable, long-lasting components. This is also possible because the standards are subject to revision at specified intervals, thus enabling them to adapt to the latest technology as well as any new service experiences. In the case of insulator strings/sets, a distinction is no longer made between polymeric and conventional insulators. Rather, the entire system is evaluated.

The range of tests that can be conducted on housing materials and composite insulators has been expanded considerably. The evaluation of erosion and tracking resistance continues to be extremely important, especially when one considers the growing use of HVDC applications. In this chapter, new test methods were systematically evaluated in order to make it possible to quantify the dynamic hydrophobicity processes for polymeric insulating materials. The results presented were mainly associated with the inclined plane principle because this simple procedure is of equal interest from both a development and quality assurance perspective.

The description of material properties as finger-prints is an important resource for ensuring reliability in service and continuity among the processing parameters used in production. The work of CIGRE Working Group D1.27 provides important impetus for the test methods to be applied and the selection of suitable easy-to-use devices. The current status of the work undertaken by this group to date was illustrated using examples from the group's own contributions to this topic.

References

1. IEC 60587 Ed. 3 (2007) Electrical insulating materials used under severe ambient conditions—test methods for evaluating resistance to tracking and erosion
2. IEC 61109 Ed. 2 (2008) Composite suspension and tension insulators for a.c. overhead lines with a nominal voltage greater than 1,000 V—definitions, test methods and acceptance criteria
3. CIGRE Working Group B2.03 (2007) Guidance for the establishment of naturally polluted insulator testing stations. Technical Brochure 333
4. Sklenicka V, Zeman I (2001) Utilization and service experience with composite insulators in Czech power system. Proceedings of world congress and exhibition on insulators, arresters and bushings—Shanghai, November 2001
5. Schmuck F, Aitken S, Papailiou KO (2010) A proposal for intensified inspection and acceptance tests of composite insulators as an addition to the guidelines of IEC 61109 Ed. 2: 2008 and IEC 61952 Ed. 2: 2008. IEEE Trans Dielectr Electr Insul 17(2) April
6. CIGRE WG 22.10 (1988) Technical basis for minimal requirement for composite insulators. ELT_088_3
7. P-IEC 61109 Ed1 (1992) Composite insulators for a.c. overhead lines with a nominal voltage greater than 1,000 V—definitions, test methods and acceptance criteria
8. CIGRE Working Group D1.14 (2004) Material properties for non-ceramic outdoor insulation: state of the art. Technical brochure 255

9. IEC 61302 Ed. 1 (1995) Electrical insulating materials—method to evaluate the resistance to tracking and erosion—rotating wheel dip test
10. IEC 61621 Ed. 1 (1997) Dry, solid insulating materials—resistance test to high-voltage, low-current arc discharges
11. IEC 60112 Ed. 4 (2004) Method for the determination of the proof and the comparative tracking indices of solid insulating materials
12. Bärsch R, Kamenz A, Schafferczyk A, Riedl U (1997) Bewertung elektrischer Eigenschaften von glasfaserverstärkten UP-Formstoffen für den Einsatz unter erschwerten klimatischen Umgebungsbedingungen. Tagung Verbundwerkstoffe und Werkstoffverbunde, Kaiserslautern Sept 1997
13. IEC 60343 Ed. 2 (1991) Recommended test methods for determining the relative resistance of insulating materials to breakdown by surface discharges
14. Ma B, Andersson J, Gubanski SM (2010) Evaluating resistance of polymeric materials for outdoor applications to corona and ozone. IEEE Trans Dielectr Electr Insul 17(2) April
15. IEC 60507 Ed 2 (1991) Artificial pollution tests on high-voltage insulators to be used on a.c. systems
16. IEC/TS 61245 Ed. 1 (1993) Artificial pollution tests on high-voltage insulators to be used on d.c. systems
17. CIGRE WG C4.03.03 (2009) Proposal for the round robin pollution test for polymeric insulators. Oct 2009
18. CEN/TS 45545-2 (2009) Railway applications—fire protection on railway vehicles—part 2: requirements for fire behaviour of materials and components
19. EN ISO 4589-2 (1996) Amend. 1 2005: plastics—determination of burning behaviour by oxygen index—part 2: ambient-temperature test
20. EN ISO 5659-2 (2006) Plastics—smoke generation—part 2: determination of optical density by a single-chamber test
21. NFX 70 100-1 (2006) Fire tests—analysis of gaseous effluents—part 1: methods for analysing gases stemming from thermal degradation
22. NFX 70 100-2 (2006) Fire tests—analysis of gaseous effluents—part 2: tubular furnace thermal degradation method
23. Schmuck F, Bärsch R, Pilling J (1996) About an improved evaluation of results by the standardized salt-fog procedure for composite insulators. Eur Trans Electr Power 6:237–243
24. Schmuck F Zur zeitraffenden Alterungsprüfung von Silikongummi-Oberflächen unter Fremdschichtbelastung und simultaner 50-Hz-Spannungsbeanspruchung. Dissertation HTWS Zittau, Germany (1992)
25. Jahn H Zur Bewertung stofflicher und herstellungsbedingter Einflussgrößen auf das Hydrophobie- und Erosionsverhalten von Silikonelastomeroberflächen. Dissertation TU Dresden, Shaker Verlag (2003) ISBN 3-8322-1963-3
26. Lambrecht J Über Verfahren zur Bewertung der Hydrophobieeigenschaften von Silikonelastomer-Formstoffen. Dissertation TU Dresden, Shaker Verlag (2001) ISBN 3-8265-9129-1
27. Ansorge S, Schmuck F, Papailiou KO (2011) Improved silicone rubbers for the use as housing material in composite insulators. Submitted to IEEE
28. Schmidt LE, Kornmann X, Krivda A, Hillborg H (2010) Tracking and erosion resistance of high temperature vulcanizing ATH-free silicone rubber. IEEE Trans Dielectr Electr Insul 17(2) April
29. Krivda A, Schmidt LE, Kornmann X, Ghorbani H, Ghorbandaeipour A, Eriksson M, Hillborg H (2009) Inclined-plane tracking and erosion test according to the IEC 60587 standard. IEEE Electr Insul Mag 25(6) November/December
30. Hochspannungsisoliertechnik: Elektrisches Verhalten und Klassifizierung der Isolierungen 1985

31. Alghamdi ASG, Auckland DW, Risino AJ, Varlow BR (1996) Reducing variability in inclined-plane tracking test results. IEEE Trans Dielectr Electr Insul 3(3):439–443
32. CIGRE Working Group D1.14 (2012) Important material properties of RTV silicone rubber insulator coatings. TB 478 October 2011
33. IEC TR 62039 Ed. 1 (2007) Selection guide for polymeric materials for outdoor use under HV stress
34. Papailiou KO, Schmuck F (2008) Intensified sample testing of composite insulators. MedPower
35. CIGRE Working Group D1.27 (2011) Exploratory RRT—material ranking AC-DC for tracking and erosion. IWD D1-27_082, Contribution of University of Zittau
36. DL/T 810-2002: Electric power industry standard of the People's Republic of China: technical specification for ±500 kV DC long rod composite insulators
37. Heger G, Vermeulen HJ, Holtzhausen JP, Vosloo WL (2010) A comparative study of insulator materials exposed to high voltage AC and DC surface discharges. IEEE Trans Dielectr Electr Insul 17(2):513–520
38. Rowland SM, Bruce GP, Liu Y, Krivda A, Schmidt LE (2011) Use of image analysis in DC inclined plane tracking tests of nano and micro composites. IEEE Trans Dielectr Electr Insul 18(2):365–374
39. Cervinka R, Bärsch R, Exl F, Kindersberger J, Winter HJ (2008) Untersuchungen zur Beständigkeit der Hydrophobie von polymeren Isolierstoffoberflächen und ihrer Wiederkehr mit dem Dynamischen Tropfen-Prüfverfahren. ETG-Fachtagung, Grenzflächen in elektrischen Isoliersystemen, September in Würzburg
40. CIGRE WG D1.14 (2010) Evaluation of dynamic hydrophobicity properties of polymeric materials for non-ceramic outdoor insulation—retention and transfer of hydrophobicity. Technical Brochure 442, December
41. Ansorge S, Schmuck F, Aitken S, Papailiou KO (2010) Improved performance of silicone rubbers for the use in composite insulators. CIGRE Paris, D1-105_2010
42. Gutman I, Dernfalk A (2010) Pollution tests for polymeric insulators made of hydrophobicity transfer materials. IEEE Trans. Dielectr. Electr. Insul 17(2):384–393
43. ISO 2781:2008: Rubber, vulcanized or thermoplastic—Determination of density
44. DIN 53505 2000: Prüfung von Kautschuk und Elastomeren—Härteprüfung nach Shore A und Shore D
45. DIN 53504 2009: Prüfung von Kautschuk und Elastomeren—Bestimmung von Reißfestigkeit, Zugfestigkeit, Reißdehnung und Spannungswerten im Zugversuch
46. IEC 62217 Ed 1 (2005) Polymeric insulators for indoor and outdoor use with a nominal voltage >1,000 V—General definitions, test methods and acceptance criteria
47. IEC/TR 62730 Ed. 1 (voraussichtlich 2012): HV polymeric insulators for indoor and outdoor use—Tracking and erosion testing by wheel test and 5000 h test
48. IEC 61952 Ed. 2.0 (2008) Insulators for overhead lines—composite line post insulators for A.C. systems with a nominal voltage greater than 1,000 V—Definitions, test methods and acceptance criteria
49. IEC 62609-1 Ed. 1: (voraussichtlich 2011): Insulators for overhead lines—composite line post insulators: part 1—definitions, end fittings and designations
50. IEC 61462 Ed. 1.0 (2007) Composite hollow insulators—pressurized and unpressurized insulators for use in electrical equipment with rated voltage greater than 1,000 V—definitions, test methods, acceptance criteria and design recommendations
51. IEC 62231 Ed. 1 (2006) Composite station post insulators for substations with a.c. voltages greater than 1,000 V up to 245 kV—definitions, test methods and acceptance criteria
52. IEC 62231-1 Ed. 1.0 (expected in 2012) Composite station post insulators for substations with ac voltages greater than 1,000 V up to 245 kV—dimensional, mechanical and electrical characteristics

53. IEC 60660 Ed. 2 (1999) Insulators—tests on indoor post insulators of organic material for systems with nominal voltages greater than 1,000 V up to but not including 300 kV
54. IEC 62621 Ed. 1 (2011) Railway applications—fixed installations—electric traction—specific requirements for composite insulators used for overhead contact line systems
55. IEC 61466-1 Ed. 1 (1997) Composite string insulator units for overhead lines with a nominal voltage greater than 1,000 V—part 1: standard strength classes and end fittings
56. IEC 61466-2 Ed. 1.1 Consol. with Amendment 1 (2002) Composite string insulator units for overhead lines with a nominal voltage greater than 1,000 V—part 2: dimensional and electrical characteristics
57. IEC/TS 60815-1 Ed. 1 (2008) Selection and dimensioning of high-voltage insulators intended for use in polluted conditions—part 1: definitions, information and general principles
58. IEC/TS 60815-3 Ed. 1 (2008) Selection and dimensioning of high-voltage insulators intended for use in polluted conditions—part 3: polymer insulators for a.c. systems
59. IEC 60815-4 Ed. 1.0 (expected in 2012) Selection and dimensioning of high-voltage insulators intended for use in polluted conditions—part 4: insulators for d.c. systems
60. IEC 60071-1 Ed 8.1 Consol. with Amendment 1 (2011) Insulation co-ordination—part 1: definitions, principles and rules
61. IEC 60071-2 Ed 3 (1996) Insulation co-ordination—part 2: Application guide
62. IEC 60383-2 Ed.1 (1993) Insulators for overhead lines with a nominal voltage above 1,000 V—part 2: insulator strings and insulator sets for a.c. systems—definitions, test methods and acceptance criteria
63. IEC 61284 Ed. 2 (1997) Overhead lines—requirements and tests for fittings
64. IEC 60437 Ed. 2 (1997) Radio interference test on high-voltage insulators
65. IEC 61467 Ed.1 (2008) Insulators for overhead lines—insulator strings and sets for lines with a nominal voltage greater than 1,000 V—AC power arc tests
66. IEC 60672-1 Ed. 2 (1995) Ceramic and glass insulating materials—part 1: definitions and classification
67. IEC 60672-2 Ed. 2 (1999) Ceramic and glass insulating materials—part 2: methods of test
68. IEC 60672-3 Ed. 2 (1997) Ceramic and glass-insulating materials—part 3: specifications for individual materials
69. IEC 60433 Ed. 3 (1998) Insulators for overhead lines with a nominal voltage above 1,000 V—ceramic insulators for a.c. systems—characteristics of insulator units of the long rod type
70. IEC 60305 Ed. 4 (1995) Insulators for overhead lines with a nominal voltage above 1,000 V—ceramic or glass insulator units for a.c. systems—Characteristics of insulator units of the cap and pin type
71. IEC TR 60797 Ed. 1 (1984) Residual strength of string insulator units of glass or ceramic material for overhead lines after mechanical damage of the dielectric
72. IEC 60720 Ed. 1 (1981) Characteristics of line post insulators
73. IEC/TR 60575 Ed 1 (1977) Thermal-mechanical performance test and mechanical performance test on string insulator units
74. IEC 60383-1 Ed 4 (1993) Insulators for overhead lines with a nominal voltage above 1,000 V—part 1: ceramic or glass insulator units for a.c. systems—definitions, test methods and acceptance criteria
75. IEC/TS 62371 Ed. 1 Characteristics of hollow pressurised and unpressurised ceramic and glass insulators for use in electrical equipment with rated voltages greater than 1,000 V
76. IEC 62155 Ed. 1 (2003) Hollow pressurized and unpressurized ceramic and glass insulators for use in electrical equipment with rated voltages greater than 1,000 V
77. IEC 60273 Ed. 3 (1990) Characteristic of indoor and outdoor post insulators for systems with nominal voltages greater than 1,000 V
78. IEC 60168 Ed. 4.2 (2001) Tests on indoor and outdoor post insulators of ceramic material or glass for systems with nominal voltages greater than 1,000 V

79. IEC 61325 Ed. 1 (1995) Insulators for overhead lines with a nominal voltage above 1,000 V—ceramic or glass insulator units for d.c. systems—definitions, test methods and acceptance criteria
80. IEC 60120 Ed. 3 (1984) Dimensions of ball and socket couplings of string insulator units
81. IEC 60471 Ed. 2 (1977) Dimensions of clevis and tongue couplings of string insulator units
82. IEC/TS 60815-2 Ed. 1 (2008) Selection and dimensioning of high-voltage insulators intended for use in polluted conditions—part 2: ceramic and glass insulators for a.c. systems
83. Bärsch R (2003) Bewertung der Hydrophobie sowie des Kriechstromverhaltens von Silikonelastomeren für Hochspannungs-Freiluftisolatoren. ETG-Fachbericht 93 (2003) S:97–108

Correction to: Composite Post Insulators

Correction to: Chapter 3 in: K. O. Papailiou and F. Schmuck, *Silicone Composite Insulators*, Power Systems, https://doi.org/10.1007/978-3-642-15320-4_3

The original version of this chapter was published with incorrect function in the equation as "tan", which is now corrected to "tanh". The original chapter has been corrected.

The updated version of the chapter can be found at
https://doi.org/10.1007/978-3-642-15320-4_3

K. O. Papailiou and F. Schmuck, *Silicone Composite Insulators*,
Power Systems, https://doi.org/10.1007/978-3-642-15320-4_12

Index

A
400 kV-transmission line Switzerland, 5
Acid, 206, 310
  attack, 213, 386
  hydrofluoric acid, 223
  nitric acid, 210, 245, 310
  resistance, 208, 245
  weight loss, 245
ACIM (Automated Continuous Injection Moulding), 262, 268
Acoustic emission, 248–249, 210
Adhesion test, 424
Advancing contact angle, 244, 460
AE-Signal, 251, 252
Air pressure, 300
Aluminium
  corona ring, 323, 357
  end fitting, 189, 191, 358
  hollow core fitting assembly, 254
  melting loss, 356
  power arc, 355
Aluminium outer layer wire, 354
Analysis, thermo-gravimetric (TGA), 230, 332, 406, 407, 479
ANSI 29.11, 399
ASDD (Apparent Salt Deposit Density), 409
Astrazon, 218

B
Band, 333, 481
Batch, 203, 220, 452
Bending deformation energy, 143
Bending length, 64
Bending moment, 75
Bending test
  examination of cracks, 65, 67
Bird Streamer, 341, 389
Boehmite, 229, 230
Bolt force, 188
Breaking strain
  coordination, 214
Brittle fracture, 206, 210, 212, 293, 386
  resistance, 214
  test, 212, 213
Buckling beam
  bending stress, 152
Buckling test, 109
Boomerang fitting, 108
Bushing, 7, 167, 185, 222

C
Cable termination, 6, 171, 255
Capillary, 208, 218
Casting, 268, 271, 274, 349
Casting, low pressure cavity, 226, 234, 255, 268
CEN/TS 45545, 456
CEPS (Czech Power Utility), 383
CFL (Cantilever Failing Load), 66, 68
CIGRE
  Study Committee B2, 3
  Study Committee D1, 476
  Technical Brochure
    TB 255, 200, 454
    TB 284, 317, 324
    TB 306, 381
    TB 442, 243
    TB 481, 381
  Working Group
  22.10, 454
  B2.03, 317, 371, 381, 386
  B2.21, 218, 381, 399

K. O. Papailiou and F. Schmuck, *Silicone Composite Insulators*, Power Systems,
DOI: 10.1007/978-3-642-15320-4,

C (*cont.*)
C4.03.03, 456
D1.14, 200, 360, 408, 454, 467, 473
D1.27, 245, 476
Coating, 237, 241, 467
Compact line
EMF, 121
medium voltage, 157
Right of Way (ROW), 99
Comparison of stresses
theory vs. experiment
composite post, 88
Composite insulator, 199, 315, 380, 399, 453
first generation, 258
advantage, 2
components, 2
problem, 3
Composite longrod
average failing load, 23
construction, 11
failing load, 20
Compression strength, 221
Conductor, kinetic energy, 146
Construction
composite post, 55
Contact angle, 239, 304, 475
Corona, 287, 293, 314, 316
discharge, 169, 245, 286, 297, 304, 457
effect, 287
extinction voltage, 298
inception voltage, 297, 295
inception, 303
protection, 287, 340, 347, 359
ring, 317, 369
shed, 272, 274, 322, 346
threshold value, 324
treatment, 240
Corrosion, 297, 324, 398
protection, 435
resistance, 205, 206, 438
Covering layer, 272, 423
Crack, 263
acid, 246
capillary force, 218
curing, 209
fibre, 206
formation, 293, 312
surface, 442
visibility, 294
Creep coefficient, 62
Creepage distance, reduction, 409
Crimped fitting
assembly technology, 32
compression, 34
crimp length, 29
crimping, correct, 50
critical areas, 31
design, 29
failing load, 30
parameter, 50
relaxation, 36, 41
roughness, 32
tolerance, 31
von Mises-stress, 50
wall thickness, 30
Crimping, 247, 252
comparison of stresses
analytical vs. FEM, 40
Current density, 357, 359, 361, 370, 457

D
Damage limit
composite post, 62
cyclic load, 78
Damage limit load
composite longrod, 17, 399
composite post, 61
hollow core insulator, 176
DC Voltage, 470, 471
De Tourreil, 15, 17, 22
Deflection
allowable, 81
analytical, 81
in rod centre, 151
permanent, 177
Density
glass, 207
measurement, 478
optical, 456
quality control, 218
resin, 209
steel, aluminium, 356
Design errors, 288, 292, 315
Dew point, 433
Diameter, minimum, 82
Dielectric strength, 208
Differences of manufacture
solid core insulator, 268
hollow core insulator, 269
DIN 53504, 201, 481
DIN 53505, 478
DIN EN ISO 6506, 204
DLL (Damage Limit Load), 22, 66, 69
Drop test, 238, 473
Droplet layer, 456, 457
Dynamic Differential
Calorimetry, 406, 407, 478

E
Emission method, acoustic, 93
EN ISO 4589, 456
EN ISO 5659, 456
End burning point, 342, 346
End fitting
  conical, 15
  crimped fitting, 15
  post, 79
  wedge type, 15
End fitting design
  composite long rod, 28
EPDM (Ethylene Propylene Dien Monomer), 362, 399, 470
EPRI (Electric Power Research Institute), 293, 295, 384, 389
Erosion, 342, 460, 468
  bulk, 227, 457
  erosion resistance, 229, 231, 257, 458
  erosion track, 228
  rate, 232, 472
  surface, 458
ESCA (Electron Spectroscopy for Chemical Analysis), 242
ESDD (Equivalent Salt Deposit Density), 408, 409, 431
ESKOM (South African Utility), 383
Euler angle, 151
Euler limit load, 107, 153
EVA (Ethylene Vinyl Acetate), 362, 399
Exothermal crack, 219
Extrusion, 226, 263

F
Fail safe base, 105
Failing load, 15
  standard deviation, 24
  temperature dependency, 15
Failure rate
  brittle fracture, 213
  composite insulator, 384, 387
  empirical, 446
  glass cap and pin insulator, 387
Fibre diameter, 205, 206
Field
  influenced by cross-arm, 372
  influenced by protective fittings, 373
Field stress electrical, 273
Filament, 205, 206, 208
Filler, 209
  active, 225, 229
  ATH, 225, 229, 232, 310
  CaCO3, 310
  treatment, 231, 235
Film layer, 310, 457, 458
Finger-Print, 407, 476
First ply failure, 176
Flame resistance, 360, 456
Flashover voltage, 354, 403, 456, 458, 462
Floating dead-end, 13
Fog test
  clean fog, 229
  clean fog, modified, 405, 431
  quick salt fog test, 403, 442
  rapid clean fog test, 404
  salt fog, 232, 274, 458, 473
FRP-rod, 15
  breakage, 252, 253, 397
  compression, 150
  curvature, 150
  danger factor, 46
  elasticity constants, 14
  load transfer, 22
  strength, 15, 48
FRP-tube
  material properties, 184
  von Mises stress, 192
FTIR (Fourier Transform Infrared Spectroscopy), 478, 480

G
Galloping, 136
  amplitude, 142
  amplitude reduction, 140
  compression load, 142
  conductor amplitude, 136
  IREQ tests, 25
Glass cap and pin insulator, 288, 348, 387
Glass fibre, 205, 220, 382
  E-CR-Glass, 206, 212
  E-Glass, 206, 210, 212, 386
  glass melt, 205, 206
Glass transition temperature, 254, 370, 478
Gotthard-Basis-Tunnel, 446

H
Hardness measurement, 478
HB40 Classification, 360
Hollow core composite insulator, 112
  components, 166
  damage limit load, 176
  differences of manufacture, 269
  load–deflection curve, 194
  market penetration, 172
  service experience, 174

H (*cont.*)
SIP (Specified Internal Pressure), 178
SML (Specified Mechanical Load), 177
HTV- injection moulding, 260
HTV-Stepping, 260, 414
Hydraulic press, 33
Hydrophobicity, 228, 238, 403, 457
evaluation, 422, 440
loss, 237, 240, 264, 294
recovery, 457
resistance, 236, 237, 473
transfer, 243, 244, 409, 457, 475

I
IEC 60068-2-17, 254
IEC 60071, 455
IEC 60093, 201
IEC 60112, 454
IEC 60120, 203, 351, 370, 455
IEC 60168, 455
IEC 60234-3, 201
IEC 60243-1, 201
IEC 60250, 201
IEC 60273, 455
IEC 60305, 455
IEC 60343, 201, 455
IEC 60383, 203, 403, 425, 455
IEC 60433, 455
IEC 60437, 296, 455
IEC 60471, 203, 455
IEC 60507, 403, 405, 455
IEC 60575, 455
IEC 60587, 201, 363, 452
IEC 60660, 455
IEC 60672, 455
IEC 60695-11-10, 202, 360, 363
IEC 60707, 360
IEC 60720, 455
IEC 60797, 455
IEC 60815, 408, 409, 455
IEC 61006, 202
IEC 61109, 20, 24, 403, 405, 416, 453, 455
IEC 61245, 455
IEC 61284, 296, 455
IEC 61302, 454
IEC 61325, 455
IEC 61462, 176, 179, 254, 455
IEC 61466, 28, 34, 203, 455
IEC 61467, 345, 348, 361, 455
IEC 61621, 202, 308, 405, 454
IEC 61952, 66, 70, 108, 455
IEC 62039, 199, 201, 454, 455
IEC 62073, 399
IEC 62155, 455
IEC 62217, 201, 218, 406, 455
IEC 62231, 455
IEC 62371, 455
IEC 62609, 455
IEC 62621, 455
IEC 62662, 214, 386, 409
IEC 62730, 454, 455
IEEE 1829, 296
IEEE 539, 296
Ignition wire, 349
Inclined Plane Test, 235, 417
Inertia moment, 82
Injection moulding, 226, 259, 268, 414
high pressure, 225
Installation fault, 288, 319, 389, 445
Insulated cross-arm, 57
application diagram, 110
critical wind speed, 119
dynamic tests, 114
FEM-simulation, 108
geometry, 104
loading diagram, 105
loads, 107
pivoted, 107
rigid, 107, 110
snap-through, 115
stability, 117
test, 111
testing arrangement, 106
Integral complete elliptic
of the 2nd kind, 152
Integral elliptic of the 1st kind, 151
Internal pressure, 178
IPS (Interphase Spacer)
advantage, 131
bending moment, 143
bending stress, 144
conductor attachment, flexible, 133
conductor attachment, rigid, 133
creepage distance, 156
ice shedding
compression load, 145
length, variable, 134
outdoor substations, 148
placement, 137
post-buckling behaviour, 153
protection fittings, 155
questionnaire, 129
short-circuit behaviour, 147
triple conductor bundle, 135
ISO 2178, 204
ISO 2781, 478
ISO 34, 201

ISO 3452, 413
ISO 4287, 406
ISO 4892-2, 201
ISO 4892-3, 201

K
Kaiser-Effect, 250
Kaolin, 475

L
Leakage current, 237, 431, 458
Lightning withstand voltage, 453
Load diagram, 73
Load, equivalent tensile, 191
Load–deflection curve, 68
  hollow core insulator, 192
Load-time curve, 16, 18, 19, 23
Load-time test, 402
Longrod composite insulator
  axial stress, 39
  circumferential (hoop) stress, 37, 44
  crimp pressure distribution, 83
  failure, 47
  fatigue test, 28
  pull-out, 42
  pulsed tension, 25
  radial stress, 43
  RML (Routine Mechanical Load), 20
  routine test, 20
  shear stress, 37, 43
  shear stress, maximum, 43
  SML (Specified Mechanical Load), 20
  tensile load, maximum, 41, 49
Lötschberg-Scheitel-Tunnel, 227, 435, 445

M
Mass loss
  TGA (thermo-gravimetric analysis), 333
Max. tensile load
  comparison of stresses
    analytical vs. FEM, 42
  composite longrod, 41
MDCL (Maximum Design Cantilever Load), 66, 68, 213
Metal-composite joint, 32
  hollow core insulator, 185
MML (Maximum Mechanical Load), 177
Modeling (FEM), 42
Modular process, 226, 264, 268
Morning star, 308, 457
Mould cavity, 264
Mould line, 236, 258, 264, 320
MSP (Maximum Specified Pressure), 178
Müller-Rochow Synthesis, 224

N
NF-X 70-100, 456
Noise emission
  acoustic, 210
  analysis, 249, 250
NSDD (Non-Soluble Deposit Density), 408, 431

O
Overcasting, 271
Overmoulding, 268, 271, 352, 414

P
Partial discharge, 225, 228, 286, 417
Pollution layer, 342, 457
  accumulation, 228
  adsorption, 227, 408
  artificial, 244, 408
  behaviour, 264, 316, 431
  class, 331, 410
  composition, 435, 440
  evaluation, 403
  flashover, 341, 354, 456, 462
  hydrophobing, 242
  natural, 404, 409
  test, 474
Porcelain, 315, 341, 348, 370, 387
Post composite insulator
  axial stress, 84
  bending stress, 63
  CFL (Cantilever Failing Load), 66, 68
  comparison of stresses theory vs. experiment, 88
  construction, 55
  cyclic load, 78
  damage limit load, 61
  damage limit, 66
  danger factor, 93
  damage zones, 90
  deflection measurement, 65
  end fitting, 79
  failure, 89
  FRP rod dimensions, 50
  load cycle, 78
  radial stress, 84
  shear stress, 85
  ultimate bending load, 86

P (*cont.*)
Power arc, 341, 398
 horn, 297, 370
 protective fitting, 288, 340, 342, 346, 355, 371
 test, 361
Press-bonded joint, 199, 223, 254
Pressure test, 184
Pultrusion, 205, 209, 215
 speed, 217
Pure resin layer, 212, 213

R
Racket, 288, 367
Railway insulator, 59
Receding contact angle, 243, 244, 303, 310, 475
Recovery, 229, 240, 242
Residual strain, 176, 183
Residual strength, 348
Resin, 206, 214, 221
 bath, 216, 217, 220
 epoxy resin, 209, 220, 362, 399
 polyester resin, 209, 210
 vinyl ester resin, 209, 211
Roughness
 contact angle measurement, 240
 electrode, 460
 influence, 240
 pollution layer, 245
 rod, 270
 surface, 240, 294, 473
Round-Robin-Test, 218, 408, 473
Roving, 205, 206, 209, 216
RTE (French Utility), 383
RTV-Stepping, 255

S
Safe burning zone, 355
Safe failure mode, 69
Sand blasting, 214, 461
Scanning electron microscope (SEM) exposure, 208, 414
SCL (Specified Cantilever Load), 66, 69
Severity classification, 400
Shed
 cracking, 394
 tear, 258
Short circuit
 bending moment, 147
 current, 145
 duration, 146
 load per unit length, 145
 test, 361
Sicron, 334, 475
Silicone gel, 273, 423
Silicone rubber, 224-226, 237, 244, 259, 265, 458
 HTV, 226, 232, 246, 262, 263, 273
 LSR, 226, 229, 246, 362, 401
 RTV, 226, 228, 246, 362, 473
SIP (Specified Internal Pressure), 178
Size, 203, 206, 382
SML (Specified Mechanical Load), 20, 177
Stabilizing cross-arm, 121
Station post, 58
Strain gauge measurement
 hollow core, 193
 longrod, 46
 post, 88
Strength, 432
 electrical, 208, 341
 mechanical, 199, 228, 364, 392
 tensile, 214, 356, 370, 478
Surface tension
 liquid, 300
 resin, 210
 system between gas and liquid, 239
Surge arrester, 168, 182
Survey, 3

T
Test equipment (rig),
 bending test, 88
 hollow core insulator, 193
 tensile test, 48
tex-number, 205, 209, 216
TGA (Thermogravimetric Analysis)
 mass loss, 479
Thickness
 zinc layer, 436
Three-point-bending, 82
Torsional load, 82
Tower
 tennis racket, 161
 Trident, 57
Traceability, 271, 383
Tracking, 231, 258, 342, 395, 457, 460
 resistance, 231, 232, 257, 458
Triple leg construction, 58
Tunnel, 429
 inside, 436
 operation, 203, 243
 service, 243

U
Underrib, 264
UV-Radiation, 257, 304, 309, 314, 457

V
V0-Classification, 360
Vulcanisation
    addition, 306
    condensation, 226
    peroxide, 226

W
Water droplet corona, 273, 300, 317, 324
Wettability class, 399, 408, 425, 431
Wheel test, 236, 237, 454, 458, 473
Winding angle, 199, 221, 254

Z
Zinc layer, 272, 359, 433
    thickness, 203

MIX
Papier aus verantwortungsvollen Quellen
Paper from responsible sources
FSC® C105338
FSC
www.fsc.org

If you have any concerns about our products,
you can contact us on
ProductSafety@springernature.com

In case Publisher is established outside the EU,
the EU authorized representative is:
Springer Nature Customer Service Center GmbH
Europaplatz 3, 69115 Heidelberg, Germany

Printed by Libri Plureos GmbH
in Hamburg, Germany